Integrated Processing Technologies for Food and Agricultural By-Products

Integrated Processing Technologies for Food and Agricultural By-Products

Edited by

Zhongli Pan

Ruihong Zhang

Steven Zicari

ELSEVIER

ACADEMIC PRESS

An imprint of Elsevier

Academic Press is an imprint of Elsevier
125 London Wall, London EC2Y 5AS, United Kingdom
525 B Street, Suite 1650, San Diego, CA 92101, United States
50 Hampshire Street, 5th Floor, Cambridge, MA 02139, United States
The Boulevard, Langford Lane, Kidlington, Oxford OX5 1GB, United Kingdom

Library of Congress Cataloging-in-Publication Data

A catalog record for this book is available from the Library of Congress

British Library Cataloguing-in-Publication Data

A catalogue record for this book is available from the British Library

ISBN: 978-0-12-814138-0

For information on all Academic Press publications
visit our website at https://www.elsevier.com/books-and-journals

Publisher: Charlotte Cockle
Acquisition Editor: Nancy Maragioglio
Editorial Project Manager: Kelsey Connors
Production Project Manager: Vignesh Tamil
Cover Designer: Miles Hitchen

Typeset by SPi Global, India

Contents

Contents

Contributors

Natthiporn Aramrueang Department of Biotechnology, Faculty of Science, Mahidol University, Bangkok, Thailand

Suvaluk Asavasanti Food Technology & Engineering Laboratory, Pilot Plant Development & Training Institute; Food Security and Process Innovation Research Group; Food Engineering Department, Faculty of Engineering, KMUTT, Bangkok, Thailand

Tyler J. Barzee Department of Biological and Agricultural Engineering, University of California - Davis, Davis, CA, United States

Giovana Bonat Celli Department of Food Science, Cornell University, Ithaca, NY, United States

Yike Chen Department of Biological and Agricultural Engineering, University of California - Davis, Davis, CA, United States

Hamed M. El-Mashad Department of Biological and Agricultural Engineering, University of California - Davis, Davis, CA, United States; Department of Agricultural Engineering, Faculty of Agriculture, Mansoura University, El-Mansoura, Egypt

Adriana S. Franca Universidade Federal de Minas Gerais, Belo Horizonte, Brazil

Varsha Gaonkar Iowa State University, Ames, IA, United States

Guangwei Huang Almond Board of California, Modesto, CA, United States

Stephen Kaffka California Biomass Collaborative, University of California - Davis, Davis, CA, United States

Aphinya Khanunthong Office of Agricultural Economics, Ministry of Agriculture and Cooperatives, Bangkok, Thailand

Ragab Khir Department of Biological and Agricultural Engineering, University of California - Davis, Davis, CA, United States; Department of Agricultural Engineering, Faculty of Agriculture, Suez Canal University, Ismailia, Egypt

Alisson Pacheco Kovalesk School of Integrative Plant Science, Section of Horticulture, Cornell University, Geneva, NY, United States

Karen Lapsley Almond Board of California, Modesto, CA, United States

Douglas Larson University of California - Davis, Davis, Department of Agricultural and Resource Economics, Davis, CA, United States

Yao Olive Li California State Polytechnic University, Human Nutrition and Food Science Department, Pomona, CA, United States

Wei Liao Biosystems and Agricultural Engineering, Michigan State University, East Lansing, MI, United States

Yan Liu Biosystems and Agricultural Engineering, Michigan State University, East Lansing, MI, United States

Rebecca Milczarek United States Department of Agriculture—Agricultural Research Service, Healthy Processed Foods Research Unit, Albany, CA, United States

Leandro S. Oliveira Universidade Federal de Minas Gerais, Belo Horizonte, Brazil

Zhongli Pan Department of Biological and Agricultural Engineering, University of California - Davis, Davis, CA, United States

Kurt A. Rosentrater Iowa State University, Ames, IA, United States

Zhenhua Ruan LanzaTech, Skokie, IL, United States

Ivana Sedej United States Department of Agriculture—Agricultural Research Service, Healthy Processed Foods Research Unit, Albany, CA, United States

Xiuzhi S. Sun Department of Grain Science and Industry/Department of Biological and Agricultural Engineering, Kansas State University, Manhattan, KS, United States

Chandrasekar Venkitasamy Department of Biological and Agricultural Engineering, University of California - Davis, Davis, CA, United States

Donghai Wang Department of Biological and Agricultural Engineering, Kansas State University, Manhattan, KS, United States

Xiaoqing Wang Tate & Lyle, Hoffman Estates, IL, United States

Selina Wang University of California - Davis, Olive Center, Davis, CA, United States

Youjie Xu Department of Biological and Agricultural Engineering, Kansas State University, Manhattan, KS, United States

Ruihong Zhang Department of Biological and Agricultural Engineering; Agricultural and Biological Engineering, University of California - Davis, Davis, CA, United States

Liming Zhao Key Laboratory of Meat Processing of Sichuan, Chengdu University, Chengdu; State Key Laboratory of Bioreactor Engineering, R&D Center of Separation and Extraction Technology in Fermentation Industry, East China University of Science and Technology, Shanghai, China

Steven Zicari California Safe Soil, LLC, McClellan, CA, United States

Preface

Improving food and agricultural production efficiencies to feed our globally expanding population is one of the most critical challenges of our time. Approximately one-third of the food produced for humans is wasted and for every pound of food produced, roughly an equal amount of nonfood by-product is also generated, which has a significant environmental and economic impact. While some by-products have found use as animal feed or are combusted for energy, new technologies that integrate conversion of production and processing by-products into higher value food or nonfood products, chemicals, and energy resources will be key in enabling a transition to a more sustainable food system.

This book assembles information on innovations in production, processing, and waste management across food supply chains and serves as a reference which summarizes current technologies that improve value, health, and resource efficiency from specific agricultural and food processing by-products. Current knowledge and technologies for integrating by-product valorization in important food crop systems are highlighted in each chapter. This book is structured differently than many others in this field; each chapter focuses on a specific agricultural product, from production through processing, for crops such as wheat, rice, corn, soybean, tomato, grape, berries, citrus, olives, nuts, coffee, and more. Current production and processing methods, locations, yields, economics, policy drivers, energy sources, or other pertinent information on each specific industry are presented. Descriptions of major by-product sources, including physical and chemical biomass characterizations and associated variability, are discussed in detail. Relevant technologies for value-added processing of by-products which can readily be integrated into current food production systems are presented for each important agricultural industry. Extra attention is paid to technologies which create or extract compounds with nutritional or pharmaceutical value, or those which result in the creation of nonfood products such as polymers, chemicals, or energy. Conversions may be accomplished through physical, biological, chemical, or thermal methods and demonstrated examples of such technologies are highlighted. In order to develop sustainable food systems, it is imperative to maximize the value of by-products from food production and processing systems.

This book serves to inform researchers, engineers, and designers involved in agricultural and food production and processing industries regarding opportunities for integrating by-product recovery and use for improved resource-recovery and value. This book will also interest those not involved with technical aspects of food or agricultural production, but who wish to quickly educate themselves on current food production or processing practices and identify what opportunities exist for improving food and agricultural by-product utilization. Investors, policy makers, educators, marketers, and the general public wishing to gain a greater knowledge of food production systems will greatly appreciate this work.

The editors would like to express their appreciation and gratitude to all the contributing authors who have provided their expert knowledge to this book. We would also like to thank Elsevier publishing personnel who provided us with much support and help in the successful completion of this book.

Zhongli Pan
Ruihong Zhang
Steven Zicari

SECTION 1
Cereals and Soybeans

CHAPTER 1
Wheat

Youjie Xu*, Xiuzhi S. Sun†, Donghai Wang*
*Department of Biological and Agricultural Engineering, Kansas State University, Manhattan, KS, United States, †Department of Grain Science and Industry/ Department of Biological and Agricultural Engineering, Kansas State University, Manhattan, KS, United States

Chapter Outline

1 INTRODUCTION

Wheat (*Triticum spp.*) has a long history of crop domestication, which revolutionized human cultural evolution and led to the emergence of human civilization (Carver, 2009). Modern wheat cultivars mainly consist of two polyploid species such as hexaploid bread wheat (*Triticum aestivum*) and tetraploid hard or durum wheat (*Triticum turgidum*) used for macaroni and low-rising bread. Another kind of cultivated diploid species einkorn wheat (*Triticum momococcum*) is a relic and only exists in some mountainous Mediterranean regions (Shewry, 2009). Approximately 95% of the wheat currently grown worldwide is hexaploid bread wheat and the remaining 5% are mainly tetraploid durum wheat (Shewry, 2009). Wheat is considered as the world's largest and most important cereal crop for human staple food, with an annual production of >700 million tonnes produced globally over the past few years (http://www.fao.org).

Integrated Processing Technologies for Food and Agricultural By-Products
https://doi.org/10.1016/B978-0-12-814138-0.00001-0

The most common characteristics used to classify wheat cultivars is mostly based on kernel color and kernel harness, often described as red or white, and hard or soft. Planting and growing cycles are often used to identify wheat, such as winter wheat and spring wheat (Carver, 2009). Wheat kernel consists of 2%–3% germ, 13%–17% bran, and 80%–85% endosperm (Šramková et al., 2009). Wheat germ is rich in protein (25%) and lipid (8%–13%) and also an important source of Vitamin E. Wheat bran provides a protection layer to the kernel and occupies over 8% of the total weight of the kernel. The endosperm makes up the major part (80%–85% by weight) of the kernel and consists of a protein and starch matrix. Wheat protein content usually ranges from 10% to 18% of the total dry matter.

In addition to human food and livestock feed, wheat and wheat by-product from wheat processing as well as wheat straws also gained interest as renewable resources for biofuels and bio-based products. Wheat is currently the dominant feedstock for the production of bioethanol in the Europe (Brancoli et al., 2018), whereas wheat straw as one of the most abundant agricultural wastes has great potential for the production of liquid or gaseous biofuels (Lopez-Hidalgo et al., 2017). Besides, wheat straw also shows great industrial application for straw particleboard fabrication, arabinoxylans extraction, and bioenergy production. In addition, milling industry by-products, wheat bran and germ, are important sources of health-enhancing bioactive components, meanwhile wheat starch and protein are substantial biopolymers for producing platform chemicals.

In this chapter, we review the industrial and nonfood applications of wheat including bio-based products from wheat straw such as particleboard production, xylans extraction, and bioenergy manufacturing; phytochemicals from wheat bran and germ such as phenolics, carotenoids, β-glucan, vitamins, and dietary fibers; and biochemicals from wheat protein and starch such as biofilms and bioadhesives, as demonstrated in the proposed wheat biorefinery concept for multiple products generation (Fig. 1).

2 BIO-BASED PRODUCTS FROM WHEAT STRAW

Agricultural crop residues such as cereal straw are produced in billions of tonnes annually around the world, representing an abundant, inexpensive, and sustainable resource of lignocellulosic biomass (Sain and Panthapulakkal, 2006). Wheat straw is the main by-product of wheat production and a small portion is used as animal husbandry or household fuel with a major quantity being burnt in the field causing environmental pollution. The utilization of these inexpensive materials as renewable resources for the industrial application, such as straw particleboard fabrication (Cheng et al., 2004; Halvarsson et al., 2008), cellulose nanofiber extraction (Alemdar and Sain, 2008; Reddy and Yang, 2005; Reddy and Yang, 2007), arabinoxylans extraction (Ruzene et al., 2008), and lignocellulosic biofuel production (Ballesteros et al., 2006; Carpenter et al., 2010), will pave a new road by turning agricultural wastes into value-added products and certainly bring additional avenues to the rural economy.

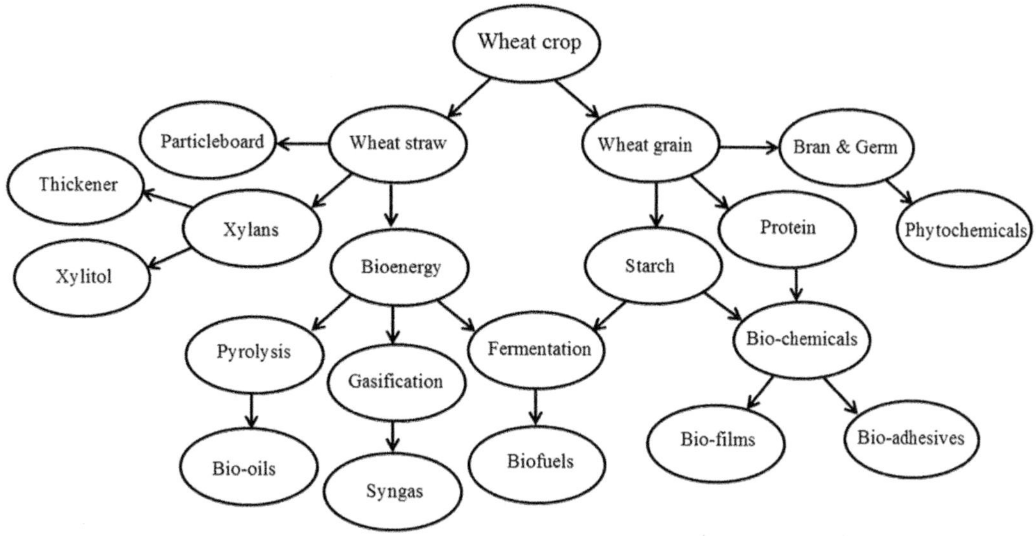

FIG. 1
A wheat biorefinery concept.

2.1 Particleboard

Particleboard has been made with forest products for years. With the decreasing availability of these raw materials and increasing demand for particleboard products, research efforts have been shifted to agriculture-based materials, with the advantages of the built-in insulation, sound suppression characteristics, and low cost (Cheng et al., 2004). Among various agricultural waste materials, wheat straw shows promising potential for manufacturing particleboard. Wheat straw is a natural composite with cellulose, hemicellulose, and lignin as the main composition, and shows great potential to produce particleboard of various density, including low-density (Wang and Sun, 2002), medium-density (Mo et al., 2003), and high-density particleboards (Panthapulakkal and Sain, 2007), with the assistance of different binders such as urea formaldehyde (UF), phenol formaldehyde (PF), methylene diphenyl diisocyanate (MDI) (Halvarsson et al., 2008; Halvarsson et al., 2009; Mo et al., 2003; Tabarsa et al., 2011). Low-density particleboard, typically in the range of 0.2–0.3 g/cm^3, has application potentials in insulation, packaging, filter, and lightweight core materials (Mo et al., 2001). Medium-density particleboard is within a density range of 0.59–0.8 g/cm^3 and has received more research effort to improve its strength. The strength of particleboard depends on the fibers of wheat straw and also on the adhesive bonds between them. UF has excellent compatibility with wood fibers and dominates the major adhesive for wood-based particleboard. However, its application to wheat straw was minimal because the wax and silica components on straw surface interfere with adhesions (Wang and Sun, 2002). Various techniques have been studied to improve the performance of wheat straw particleboard, including different pretreatments of raw materials such as enzyme treatment

(Zhang et al., 2003; Jiang et al., 2009) and chemical treatment (Mo et al., 2001; Zheng et al., 2007), and the addition of different additives during manufacturing (Han et al., 2001; Ye et al., 2007). The surface wax of wheat straw is reduced by treating with enzyme lipases; thus, better properties of manufactured particleboard can be obtained (Zhang et al., 2003; Jiang et al., 2009). Compared with wheat straw made with UF adhesive, particleboard prepared with MDI adhesive had 3–10 times superior mechanical strength. Wheat straw particleboards made with 4% MDI resulted in improved mechanical properties similar to wood-based particleboards and have better water resistance with the increasing density from 0.71 to 0.75 g/cm^3 (Zheng et al., 2007). Due to environmental concern and human health, bio-based adhesives have gained more attention over recent years (Tabarsa et al., 2011). For protein-based adhesives, initial straw moisture content significantly affected the mechanical properties because water functioned as a plasticizer and assisted the protein to unfold to a greater extent and entangle upon heating, resulting in stronger bonding strength when water has evaporated. In contrast, beyond a threshold moisture content, internal cracks will develop within the particleboard due to water vapor trapped inside the composite (Mo et al., 2003).

2.2 Xylans

Polysaccharides make up most of the cell wall to support and to maintain the shape of the plant. Different kinds of polysaccharides are present in the cell walls, and the composition of polysaccharides varies in different species and tissues (Ebringerová, 2005). Traditionally, cell wall polysaccharides have been categorized into cellulose, hemicellulose, and pectin. Hemicelluloses, isolated from wheat straw, usually consist of 20%–30% of the biomass plants, which are viewed as important renewable resources of biopolymers. Hemicelluloses are majorly complex heteropolysaccharides with the structure varying in the nature and degree of branching of the β-1,4-linked xylopyranosyl main chain (Gatenholm and Tenkanen, 2003). Hemicelluloses are heteropolysaccharides with homopolymeric backbone chains of 1,4-linked β-D-xylopyranose units, which contain xylose, arabinose, glucuronic acid or its 4-*O*-ether, and acetic, ferulic, and *p*-coumaric acids. The distribution pattern of side chains in arabinoxylans which reflects the structure of the polymer chains has major influence on their solubility, interactions with other cell wall polymers, digestibility of enzymes, rheological properties, and other functional properties (Ebringerova and Heinze, 2000). This polysaccharide has potential to substitute and integrate into a variety of industrial applications, such as biofilms, thickeners, adhesives, emulsifiers, stabilizers and binders in food, pharmaceutical, and cosmetics industries (Jacquemin et al., 2012).

The most critical biological role of hemicelluloses is their strengthening ability to the cell wall by interaction with cellulose and lignin. The extraction of the arabinoxylan components from cell wall is restricted by the present lignin network as well as ester and ether linked lignin-carbohydrate complex. Also the internal hydrogen bonding between cell wall components may inhibit isolation of the arabinoxylan components (Sun et al., 1996). The extraction methods to isolate hemicellulose could also affect the functional properties (Sun and Tomkinson, 2002). There are four main isolation methods to extract arabinoxylans: physical, chemical, biological, and physicochemical methods. Heat treatments such as steam explosion, hot water extraction, and microwave irradiation are considered as physicochemical treatments. Chemical methods are typically carried out with the assistance of acids, bases, or organic solvents (Ruzene et al., 2008).

Xylitol, a hydrogenation product of xylose, has attracted much attention due to its potential applications such as food sweetener, a dental caries reducer, and sugar substitutes. This five-carbon sugar alcohol, xylitol, is currently produced by catalytic reduction of the xylose derived mainly from wood hydrolysate, which yielded 50%–60% of the xylan fraction or 8%–15% of the raw material. Alternatively, production of xylitol by fermentation is gaining more attention because of concerns associated with its chemical production (Saha, 2003). Xylitol is mainly used in various food products such as chewing gum, candy, and ice cream due to its strong cooling effect and fresh sensation (Saha, 2003).

2.3 Bioenergy

Biofuel production from agricultural wastes is attracting considerable attention globally as a strategy to diversify energy resources, spur rural economic development, and mitigate greenhouse gas emissions (Demirbas, 2004). Among various agricultural residues, wheat straw is the most abundant feedstock in Europe and the second largest in the world after rice straw, and represents cheap resources of biomass fuels for electricity and heat production (Talebnia et al., 2010). Feedstock supply is a major challenge to the biofuel industry as transportation of low-density biomass involves significant costs; therefore, biomass has been pelletized or briquetted to increase the energy density (Giuntoli et al., 2013). Torrefaction is an effective treatment to improve the combustion properties of biomass and has received much interest in the past two decades (Bridgeman et al., 2008). In addition, biomass can be used to produce diverse biofuels via biochemical conversion processes such as fermentation to produce ethanol (Fonseca et al., 2011), butanol (Qureshi et al., 2013), oil (Yu et al., 2011), and anaerobic digestion to generate methane (Chandra et al., 2012), or via thermochemical pathways such as pyrolysis to produce bio-oil (Li et al., 2016) and gasification to generate syngas (Carpenter et al., 2010).

Ethanol is a suitable alternative transportation fuel to petroleum oil either as a sole fuel for vehicles with modified engines or as an additive in fuel blends to improve the engine performance. Pretreatment to disrupt the rigid cell wall structure, enzymatic hydrolysis to chop down the polysaccharides, and subsequent fermentation are major processes to produce ethanol from lignocellulosic biomass (Zhang et al., 2015). Commercialization of second-generation bioethanol from lignocellulosic biomass is still under development due to low fermentation efficiency and ethanol titers, high enzyme cost, and high water consumption. Advanced technology to achieve high ethanol titer and ethanol yield with low enzyme loadings is imperative. Various pretreatment techniques have been applied to improve the enzymatic digestibility of wheat straw, including dilute acid pretreatment (Saha et al., 2005), steam explosion pretreatment (Ballesteros et al., 2006), alkaline pretreatment (Saha and Cotta, 2006), wet oxidation pretreatment (Schmidt and Thomsen, 1998), and ionic liquid pretreatment (Li et al., 2009). Currently, pilot-scale conversion of wheat straw to bioethanol achieved an ethanol yield of 0.29 g/g of raw material, corresponding to 86% of the theoretical ethanol yield (Saha et al., 2015). High gravity bioconversion (>15% biomass loading) is superior to low-solid loadings as enhanced fermentable sugars and less water consumption are preferred from the economic and environmental standpoints. High ethanol concentration has the advantage of reducing capital and energy costs as a minimum ethanol concentration of 40 g/L is generally required for economical ethanol distillation (Xu and Wang, 2017). There are still research gaps to effectively increase cellulosic ethanol concentration and yield simultaneously and accelerate the commercialization of bioethanol production from lignocellulosic materials.

Pyrolysis is defined as the thermal destruction (400–700°C) of organic materials without the presence of oxygen to produce hydrocarbon-rich gas mixtures, crude bio-oils, and solid residue charcoals (Demirbas, 2004; Lehto et al., 2014). Bio-oil has the potential to replace heavy fuel oil for industrial-scale combustion and district heating (Lehto et al., 2014). Pyrolysis can be applied to produce bio-oil through flash pyrolysis processes and is currently at pilot-scale stages as certain problems such poor thermal stability and corrosivity of the oil remain to be addressed (Demirbas, 2004; Lehto et al., 2014). Gasification is considered as a form of pyrolysis with partial oxidation or partial combustion at high temperatures (700–1000°C) to optimize the gas production. The resultant gas mixtures include carbon monoxide, hydrogen, methane, carbon dioxide, and nitrogen (Demirbas, 2004). Steam reforming of gasified biomass can be used to produce a clean burning syngas with hydrogen and carbon monoxide as main constituents.

New concepts in biomass gasification has been developed recently to integrate gasification, gas cleaning, and upgrading in single reactor unit, and other unique concepts combine pyrolysis and gasification or gasification and combustion in single controlled stages to improve the economic viability and sustainability of biomass utilization (Heidenreich and Foscolo, 2015).

3 PHYTOCHEMICALS FROM WHEAT BRAN AND GERM

Phytochemicals are bioactive and naturally occurring compounds present in edible plants. Cereals have been recognized as important sources of health-enhancing bioactive components such as phenolics, carotenoids, β-glucan, vitamins, and dietary fibers. Phytochemical content and antioxidant activity vary significantly in wheat varieties; therefore, breeding genotypes of cereal grains with high levels of such components is of great interest to cereal consumers. Wheat barn particle size was also found to influence phytochemical extractability and antioxidant properties (Brewer et al., 2014). Wheat bran and germ are conventional milling by-products of wheat grains to extract the endosperm and often combined to form the bran fraction. Physiological effects of wheat bran and germ can be classified into nutritional effects, mechanical effects contributed by fibers on the gastrointestinal tract, and antioxidant effects arising from phytonutrients (Stevenson et al., 2012).

3.1 Phenolics

Phenolics are aromatic compounds with one or more hydroxyl groups and the most common phenolic compounds in cereal grains are phenolic acids and flavonoids (Carver, 2009). Phenolic acids have received much interest due to their antioxidative, anti-inflammatory, antimutagenic, and anticarcinogenic properties as well as their ability to manipulate certain key enzymatic functions in cells (Liyana-Pathirana and Shahidi, 2006). Phenolic acids form a diverse group that can be subdivided into two major groups: hydroxybenzoic acids and hydroxycinnamic acids. The former include p-hydroxybenzoic, protocatechuic, vanillic, syringic, and gallic acids, whereas p-coumaric, caffeic, ferulic, and sinapic acids belong to the latter group (Velioglu et al., 1998). Ferulic acid exists in the free, soluble-conjugated, and bound forms in wheat grains and accounts for approximately 90% of total phenolic acids (Adom and Liu, 2002). Wheat bran is a good source of ferulic acid and majority is esterified to the hemicellulose component of the cell wall. This bound phenolic acid could be released during food processing (Wang et al., 2014) or by enzyme esterase in human small intestine and colon (Price et al., 2008). Ferulic acid content may differ among wheat cultivars. Significant genetic and agronomic variability in ferulic acid content was detected in durum wheat and common wheat (Lempereur et al., 1997). Solvent extraction with the assistance of ultrasound was used to isolate phenolics-rich xylans from wheat bran (Hromádková et al., 2008). Various processing technologies have been developed to improve bioaccessibility and bioavailability of phenolic compounds, including mechanical treatment, thermal treatment, extrusion cooking, and bioprocessing. For details of each technique the readers are referred to a review paper (Wang et al., 2014).

3.2 Carotenoids

Carotenoids are natural pigments with yellow, orange, and red colors and >700 carotenoids have been identified in plant (Burkhardt and Böhm, 2007). Carotenoids are divided into carotenes and oxygenated xanthophylls with only a few found in humans, including α-carotenes, β-carotenes, lycopene, lutein, zeaxanthin, and β-cryptoxanthin (Abdel-Aal et al., 2007). Fruits and vegetables are the main dietary sources of carotenoids, but certain cereals such as wheat contain high amount of carotenoids. Wheat variety and growing environment can significantly alter the antioxidant profiles such as carotenoids in wheat brans (Atienza et al., 2007). Lutein and zeaxanthin are the main carotenoids found in durum wheat and einkorn wheat. In seven tested wheat bran samples, lutein and cryptoxanthin were reported in the range 0.50–1.80 and 0.18–0.64 μg/g, respectively, while the greatest zeaxanthin concentration of 2.19 μg/g was detected. Einkorn wheat is considered as a potential candidate for developing genotypes of high lutein level (Abdel-Aal et al., 2002). These results indicate that large genotypic variance of carotenoid contents may differ the potentials of provitamin and antioxidant roles (Zhou et al., 2004) and opportunities to enhance the carotenoid content through plant breeding (Leenhardt et al., 2006).

In plants, carotenoids play a critical function of harvesting light for photosynthesis and in protecting chlorophyll against oxidative damage, which are related to the reduced risks of cancers, cardiovascular diseases, and aging macular degeneration when consumed by human. Several factors including extraction time, extraction temperature, particle size, agitation, and pre-soaking in water were studied for their influence on extractability of carotenoids from durum wheat and it was found that soaking in water for 5 min prior to organic solvent extraction had the greatest impact on carotenoid yields (Burkhardt and Böhm, 2007).

3.3 β-Glucan

β-Glucan is one of the main non-starch polysaccharides in wheat brans and only occupies approximately 1% of whole wheat grain (Henry, 1987). Compared to cellulose with only β-1,4 linkages, the β-1,3 linkages disrupt the linear β-1,4 linkages to make β-glucan more soluble and viscous. Structural differences in cereal β-glucans are generally characterized by the ratios of trisaccharide-to-tetrasaccharide and wheat, usually has the ratios of 4.2–4.5 (Li et al., 2006). This ratio is higher in wheat compared to other cereal β-glucans, and this unique structural difference often leads to poor solubility and fast gelation. At low concentration, cereal β-glucans show shear thinning behavior, and with increasing concentration, they tend to form gels. Few studies have been conducted to characterize wheat β-glucan due to its relative amount in the whole wheat compared to barley and oat. Alkaline extraction with ammonium sulfate precipitation was applied to isolate wheat β-glucan with high purity (91.6%) from white wheat bran compared to previously reported purity of only 70% (Li et al., 2006).

Cereal β-glucans have some biological effects such as lowering blood cholesterol level, controlling blood sugar, and enhancing the immune system (Wursch and Pi-Sunyer, 1997). The effects of β-glucans on blood cholesterol and sugar lever are highly related to its viscous property functioned as a soluble fiber to bind cholesterol and facilitate their elimination from human body. β-Glucans are also recognized by the immune system and play important roles in host defense and anticancer treatment (Goodridge et al., 2009).

3.4 Vitamin E

Vitamin E is a generic term that refers to eight naturally occurring forms of tocotrienols (α-tocotrienol, β-tocotrienol, γ-tocotrienol, δ-tocotrienol) and tocopherols (α-tocopherol, β-tocopherol, γ-tocopherol, δ-tocopherol). Their basic structures contain a polar chromanol ring linked to an isoprenoid-derived hydrocarbon chain. These molecules display a diversity of biological and physiological properties, for instance, they are potent lipid-soluble antioxidants and can protect plant cells against oxidative pressure such as those arising the breakdown of polyunsaturated fatty acids in seed oils (Cahoon et al., 2003).

A by-product of the flour milling industry, wheat germ is considered as a potential excellent and low-cost source of vitamins, minerals, and dietary fibers (Carver, 2009). Vitamin E compounds are present mostly in wheat germ. As lipid-soluble antioxidants, they disrupt the propagation of reactive oxygen compounds that spread through membranes. These compounds have other physiological effects such as the suppression of free radicals, the prevention of cancers, resistance of aging, and enhancing the response of the immune systems. Supercritical fluid extraction of Vitamin E from wheat germ is superior to conventional liquid extraction because of high extraction efficiency, short extraction time, and zero chemical contamination (Ge et al., 2002).

3.5 Dietary Fiber

Dietary fiber is referred to the components of plant cell that resists digestion by human enzymes, including water soluble and insoluble fibers. The fiber content of wheat grain ranges from 11.6% to 12.7% dry weight, while wheat bran is one of the richest sources of fiber with the level of 36.5%–52.4% (Stevenson et al., 2012). In wheat bran, roughly 1% is soluble fiber and the majority is insoluble fiber, including 46% non-starch polysaccharide such as arabinoxylan, cellulose, and beta-glucan.

Inulin and resistant starch also belong to dietary fiber. Inulin is a polyfructan consisting of a linear β-2,1-linked polyfructose chains with a degree of polymerization of 2–60. Due to its structural conformation, it resists the hydrolysis by human enzymes. Resistant starch can resist intestinal digestion and pass into the lower intestine to be fermented by microflora in the colon, including physically trapped starch, resistant starch granules, and retrograded starch. Resistant starch used in products provides bulk but reduced caloric content of foods, and its consumption promotes lipid oxidation and metabolism.

The recommended dietary fiber intake in Europe is 25g/day based on the AOAC (Association of Official Analytical Chemists') method. The physiological effects of dietary fiber are mainly on gastric emptying and small intestinal transit time, which results in an improved glucose tolerance and a decreased digestion of starch (Roberfroid, 1993). Wheat bran shows beneficial effects on the prevention of diseases such as colorectal cancer, cardiovascular disease, obesity, and some gastrointestinal diseases, including diverticular disease, constipation, and irritable bowel syndrome (Stevenson et al., 2012). Wheat bran can also function as a prebiotic. Prebiotics are referred to nondigestible food ingredients that beneficially affect host health by selectively stimulating the growth or activity of friendly and healthy intestinal bacteria (probiotics) in the colon. Inulin is a preferred food for probiotics including lactobacilli and bifidobacteria to stimulate their growth and improve their balance in the colon (Roberfroid, 1993). Bifidobacteria digest inulin to produce short-chain fatty acids, including acetic, propionic, and butyric acids, to exert systemic effects on lipid metabolism.

4 BIOCHEMICALS FROM WHEAT STARCH AND PROTEIN

Wheat starch can be used for adhesives, coatings, textiles, polymers, and paper additives. Wheat protein, when extracted from other flour components, can be used for adhesives, films, and coating. Peptides from wheat protein can be applied in cosmetics, lotions, skin care products, and biodegradable resins (Day et al., 2006). Low-grade wheat is also used to produce alcohol via fermentation.

4.1 Biofilms

Recently, starch film has been evaluated for use in food package area as the trend of petroleum-based plastics are being substituted by natural and biodegradable polymers due to environmental concern. Native starch granules are insoluble in cold water. However, when heated in water, starch granules absorb water and swell and eventually decompose into small fragments, which is called starch gelatinization. Water absorbed by the amorphous regions of starch granules acts as a plasticizer to destabilize the rigid crystalline structure and upon continuous heating, granules continue to swell and finally the crystalline structure is melted to obtain a homogeneous solution and further to form starch films.

Starch films can be produced from native starch or its main components, amylose and amylopectin, through wet method of solution casting and subsequent drying or dry method of thermoplastic processing (Paes et al., 2008). Starch films obtained by wet methods are difficult to manufacture using industrial processes and also the drying times are too long to allow large-scale production. Alternatively, by dry process starch films could be produced if treated properly. Thermoplastic starch (TPS) could be produced by processing a starch-plasticizer mixture in an extruder at a temperature range of 140–160°C under condition of high shear and high pressure (Carvalho, 2008). The TPS could be repeatedly melted and molded through shear forces and heat, thus the techniques commonly used in the plastics industry can be applied to process TSP. Starch-based films have the advantage of low cost, low permeability to oxygen, and the ability to form a continuous matrix with plasticizer and other additives (Dole et al., 2004). However, as compared to plastic polymers, starch-based films present some issues such as the hydrophilic character and poor mechanical properties (Jiménez et al., 2012). The addition of plasticizer such as glycerol makes starch films more flexible, reduce tensile strength (TS), and increase elongation at break (E%). However, plasticizer addition may bring some drawbacks such increased water vapor and gas permeability.

Wheat protein can be also used for the fabrication of bioplastics alternative to synthetic oxygen-barrier polymers in packaging applications (Micard et al., 2000; Gällstedt et al., 2004). The high contents of hydrogen bonds make wheat protein films brittle and a plasticizer is needed to increase the toughness. From commercial perspectives, compression-mold films are superior to solution-cast process. In extrusion processes, shear force, pressure, temperature, plasticizer, and time are critical parameters that determine the properties of resultant films (Gällstedt et al., 2004). The use of organic or inorganic fillers are an effective way to reinforce the films by forming biocomposites. Flexible strength and crack resistance can be improved by using reinforcements such as flax-fiber-weaves (Wu et al., 2017), hemp fibers (Wretfors et al., 2009), and clay nanocomposites (Olabarrieta et al., 2006).

4.2 Bioadhesives

Formaldehyde-based resins and latex-based systems are dominant adhesives in the wood industry, but environmental friendly adhesives derived from plant proteins are gaining more interest. The properties of thermoplasticity and good film forming allow wheat protein to produce natural adhesives. With the controlled hydrolysis to break the sulfide bonds and the use of plasticizers, the properties of the adhesives can be improved and applied to pressure-sensitive medical bandages and adhesive tapes (Day et al., 2006).

Wheat storage protein, called wheat gluten, is a by-product of wheat starch industry with the annual global production of approximately 400,000 metric tons at the price of less than $0.5/lb. (Nordqvist et al., 2012a). Wheat gluten consists of high molecular weight glutenins and low molecular weight gliadins; both protein fractions are rich in hydrophobic amino acids (Nordqvist et al., 2012a). The glutenins are polypeptide chains linked together by interchain disulfide bonds, whereas most gliadins contain cysteine residues forming intrachain disulfide bonds (Woerdeman et al., 2004). The adhesive properties of gliadins were found to have water resistance inferior to that of glutenins due to the over-penetration of the protein into wood materials (Nordqvist et al., 2012a). Various methods have been tested to improve the adhesive properties of wheat protein, including enzymatic hydrolysis or heat treatment (Nordqvist et al., 2012b), urea and sodium hydroxide modification (El-Wakil et al., 2007), and the addition of cross-linking agents (Khosravi et al., 2014).

Wheat starch is used as an adhesive for postage stamps and can also be used to hold the bottom of paper grocery sacks together. In particular, high-amylose wheat starches can be used in adhesive products and in the production of corrugated board and paper (Jobling, 2004). Starch modification treatments are often applied to improve its adhesive properties, especially for remoistenable adhesives (Eden et al., 1999). Wheat flour containing mainly starch and protein can be used as bio-based adhesives. The pressing temperature of 105°C was found to be sufficient for bonding of spruce wood with wheat four (D'Amico et al., 2010).

4.3 Biofuels

Biofuels are derived biologically from renewable and even waste organic substrates by microorganisms, which shows great potentials to replace fossil fuels (Zhang et al., 2016). Renewable and carbon neutral biofuels are essential to environmental and economic sustainability, for instance, utilization of liquid biofuels (ethanol, butanol, and diesel) as transportation fuels, and gaseous biofuels (methane, hydrogen, and hythane) for power generation are important alternatives to fossil fuel resources. Among all biofuels, bioethanol is the most utilized liquid biofuel either as a fuel or as a gasoline additive, and can be produced using starch-based feedstocks, such as corn in North America and wheat in Europe (Vohra et al., 2014). Conventional industrial bioprocesses utilizing starch as raw material for bioethanol production include well-established wet and dry milling techniques. The starting material for ethanol production in dry milling process is grain flour, whereas in wet milling process, the starch remaining after component separation is the starting source. Approximately 90% of the grain ethanol production is currently contributed by the dry milling process, with the remaining 10% coming from wet mills. The whole conversion process is divided into three stages: liquefaction (90–100°C), saccharification (50–70°C) and fermentation (30–35°C) (Naguleswaran et al., 2012). Enzymes such as α-amylase and glucoamylase are incorporated to facilitate the starch hydrolysis for glucose production in current bioethanol production with the energy-intensive process of gelatinizing starch. Alternatively, recently developed granular starch enzymes are capable of hydrolyzing starch at sub-gelatinization temperature, which could reduce capital and operational costs of industrial plants by approximately 41% and 51%, respectively (Cinelli et al., 2015). Bioethanol production via granular starch hydrolysis is a very promising technology that may be potentially adopted in the conventional starch-to-ethanol processes at large scales in the near future.

Anaerobic digestion (AD) is a well-established technology to produce biogases such as methane and applied as a viable means for providing continuous power generation (Mao et al., 2015). The biodegradation process mainly includes four steps: hydrolysis, acidogenesis, acetogenesis, and methanogenesis (Zhang et al., 2014). High molecular substrates such as carbohydrates, lipids, and protein are hydrolyzed by fermentative bacteria into small molecular units (e.g., glucose, fatty acids, and amino acids). In general, this hydrolysis process is considered as the rate-limiting step in the AD of solid organic materials. Subsequently, these small molecules are degraded into volatile fatty acids (e.g., acetate, propionate, and butyrate) as well as the by-products (e.g., NH_3, CO_2, and H_2S). Finally, these intermediates are digested into acetate, H_2, and CO_2, which could be used by methanogenic bacteria for methane production (Zhang et al., 2014). The performance of AD is mainly affected by temperature, VFA, pH, C/N ratio, ammonia, and metal elements. Food wastes (e.g., waste breads) and fermentation residues (e.g., brewer's spent grain, and bioethanol distillation by-products) are excellent sources for effective methane production via anaerobic microbial growth (Mao et al., 2015).

5 SUMMARY

Massive amounts of agricultural biomass such as wheat straw are burnt in open environmental resulting in the release of harmful gases. In contrast, utilizing wheat by-products for value-added products is a critical step toward sustainable bio-based economics. Wheat straw is the main by-product obtained after wheat harvesting and utilization of these abundant but inexpensive materials as renewable resources for the industrial application such as straw particleboard fabrication, arabinoxylans extraction, and lignocellulosic biofuel production will pave a new road by turning agricultural wastes into value-added products and certainly bring additional avenues to the rural economy. Milling industry by-products wheat bran and germ contain important health-enhancing bioactive components such as phenolics, carotenoids, β-glucan, vitamins, and dietary fibers. Wheat starch and protein as important renewable polymers can be used for adhesives, coatings, textiles, polymers, and paper additives.

The complete characterization of wheat by-products is required and useful for the generation of new products with sound properties. Feedstock utilization could be optimized via the development of biorefinery concepts (Fig. 1), in which high-value phytochemicals should be initially extracted followed by the thermochemical or biochemical conversion to produce biofuels or biochemicals. Integrated bioprocessing strategies are promising ways to reduce production costs and improve process efficiencies; however, it still needs extensive research and improvement to develop productivity and economics.

With the increasing production of wheat over recent years, wheat emerges as a promising substitute of corn for bioethanol production. Advanced processing technology must be sought to extract valuable products from wheat fermentation residues in order to improve the economic viability of exiting bioethanol industry. In addition, integration of cellulosic ethanol production and existing starch-based ethanol production via co-fermentation of wheat straw and wheat grain could boost ethanol concentration and yield, and consequently accelerate the commercialization of ethanol production from lignocellulosic materials such as wheat straw and wheat bran. Overall, upgrading the traditional industry into viable biorefineries for the production of biofuels, biochemicals, and biopolymers is essential to a sustainable and healthy bio-based society in order to substitute petrochemicals.

References

Abdel-Aal, E.M., Young, J., Wood, P., Rabalski, I., 2002. Einkorn: a potential candidate for developing high lutein wheat. Cereal Chem. 79, 455.

Abdel-Aal, E.M., Young, J.C., Rabalski, I., Hucl, P., Fregeau-Reid, J., 2007. Identification and quantification of seed carotenoids in selected wheat species. J. Agric. Food Chem. 55, 787–794.

Adom, K.K., Liu, R.H., 2002. Antioxidant activity of grains. J. Agric. Food Chem. 50, 6182–6187.

Alemdar, A., Sain, M., 2008. Isolation and characterization of nanofibers from agricultural residues–wheat straw and soy hulls. Bioresour. Technol. 99, 1664–1671.

Atienza, S.G., Ballesteros, J., Martín, A., Hornero-Méndez, D., 2007. Genetic variability of carotenoid concentration and degree of esterification among tritordeum (× Tritordeum Ascherson et Graebner) and durum wheat accessions. J. Agric. Food Chem. 55, 4244–4251.

Ballesteros, I., Negro, M.J., Oliva, J.M., Cabañas, A., Manzanares, P., Ballesteros, M., 2006. Ethanol production from steam-explosion pretreated wheat straw. Appl. Biochem. Biotechnol. 130, 496–508.

Brancoli, P., Ferreira, J.A., Bolton, K., Taherzadeh, M.J., 2018. Changes in carbon footprint when integrating production of filamentous fungi in 1st generation ethanol plants. Bioresour. Technol. 249, 1069–1073.

Brewer, L.R., Kubola, J., Siriamornpun, S., Herald, T.J., Shi, Y., 2014. Wheat bran particle size influence on phytochemical extractability and antioxidant properties. Food Chem. 152, 483–490.

Bridgeman, T., Jones, J., Shield, I., Williams, P., 2008. Torrefaction of reed canary grass, wheat straw and willow to enhance solid fuel qualities and combustion properties. Fuel 87, 844–856.

Burkhardt, S., Böhm, V., 2007. Development of a new method for the complete extraction of carotenoids from cereals with special reference to durum wheat (*Triticum durum* Desf.). J. Agric. Food Chem. 55, 8295–8301.

Cahoon, E.B., Hall, S.E., Ripp, K.G., Ganzke, T.S., Hitz, W.D., Coughlan, S.J., 2003. Metabolic redesign of vitamin E biosynthesis in plants for tocotrienol production and increased antioxidant content. Nat. Biotechnol. 21, 1082.

Carpenter, D.L., Bain, R.L., Davis, R.E., Dutta, A., Feik, C.J., Gaston, K.R., Jablonski, W., Phillips, S.D., Nimlos, M.R., 2010. Pilot-scale gasification of corn stover, switchgrass, wheat straw, and wood: 1. Parametric study and comparison with literature. Ind. Eng. Chem. Res. 49, 1859–1871.

Carvalho, A.J., 2008. Starch: Major Sources, Properties and Applications as Thermoplastic Materials. Elsevier, Amsterdam.

Carver, B.F., 2009. Wheat: Science and Trade. John Wiley & Sons.

Chandra, R., Takeuchi, H., Hasegawa, T., 2012. Methane production from lignocellulosic agricultural crop wastes: a review in context to second generation of biofuel production. Renew. Sust. Energ. Rev. 16, 1462–1476.

Cheng, E., Sun, X., Karr, G.S., 2004. Adhesive properties of modified soybean flour in wheat straw particleboard. Compos. A: Appl. Sci. Manuf. 35, 297–302.

Cinelli, B.A., Castilho, L.R., Freire, D.M., Castro, A.M., 2015. A brief review on the emerging technology of ethanol production by cold hydrolysis of raw starch. Fuel 150, 721–729.

D'Amico, S., Hrabalova, M., Müller, U., Berghofer, E., 2010. Bonding of spruce wood with wheat flour glue—effect of press temperature on the adhesive bond strength. Ind. Crop. Prod. 31, 255–260.

Day, L., Augustin, M., Batey, I., Wrigley, C., 2006. Wheat-gluten uses and industry needs. Trends Food Sci. Technol. 17, 82–90.

Demirbas, A., 2004. Combustion characteristics of different biomass fuels. Prog. Energy Combust. Sci. 30, 219–230.

Dole, P., Joly, C., Espuche, E., Alric, I., Gontard, N., 2004. Gas transport properties of starch based films. Carbohydr. Polym. 58, 335–343.

Ebringerová, A., 2005. Structural diversity and application potential of hemicelluloses. Macromol. Symp. 232, 1–12.

Ebringerova, A., Heinze, T., 2000. Xylan and xylan derivatives–biopolymers with valuable properties, 1. naturally occurring xylans structures, isolation procedures and properties. Macromol. Rapid Commun. 21, 542–556.

Eden, J.L., Shi, Y., Nesiewicz, R.J., Wieczorek Jr, J., 1999. Maltodextrin-Based Adhesives.

El-Wakil, N.A., Abou-Zeid, R.E., Fahmy, Y., Mohamed, A., 2007. Modified wheat gluten as a binder in particleboard made from reed. J. Appl. Polym. Sci. 106, 3592–3599.

Fonseca, C., Olofsson, K., Ferreira, C., Runquist, D., Fonseca, L.L., Hahn-Hägerdal, B., Lidén, G., 2011. The glucose/xylose facilitator Gxf1 from *Candida intermedia* expressed in a xylose-fermenting industrial strain of *Saccharomyces cerevisiae* increases xylose uptake in SSCF of wheat straw. Enzym. Microb. Technol. 48, 518–525.

Gällstedt, M., Mattozzi, A., Johansson, E., Hedenqvist, M.S., 2004. Transport and tensile properties of compression-molded wheat gluten films. Biomacromolecules 5, 2020–2028.

Gatenholm, P., Tenkanen, M., 2003. Hemicelluloses: Science and Technology. ACS Publications.

Ge, Y., Yan, H., Hui, B., Ni, Y., Wang, S., Cai, T., 2002. Extraction of natural vitamin E from wheat germ by supercritical carbon dioxide. J. Agric. Food Chem. 50, 685–689.

Giuntoli, J., Boulamanti, A.K., Corrado, S., Motegh, M., Agostini, A., Baxter, D., 2013. Environmental impacts of future bioenergy pathways: the case of electricity from wheat straw bales and pellets. GCB Bioenergy 5, 497–512.

Goodridge, H.S., Wolf, A.J., Underhill, D.M., 2009. β-glucan recognition by the innate immune system. Immunol. Rev. 230, 38–50.

Halvarsson, S., Edlund, H., Norgren, M., 2008. Properties of medium-density fibreboard (MDF) based on wheat straw and melamine modified urea formaldehyde (UMF) resin. Ind. Crop. Prod. 28, 37–46.

Halvarsson, S., Edlund, H., Norgren, M., 2009. Manufacture of non-resin wheat straw fibreboards. Ind. Crop. Prod. 29, 437–445.

Han, G., Kawai, S., Umemura, K., Zhang, M., Honda, T., 2001. Development of high-performance UF-bonded reed and wheat straw medium-density fiberboard. J. Wood Sci. 47, 350–355.

Heidenreich, S., Foscolo, P.U., 2015. New concepts in biomass gasification. Prog. Energy Combust. Sci. 46, 72–95.

Henry, R., 1987. Pentosan and (1→3),(1→4)-β-glucan concentrations in endosperm and wholegrain of wheat, barley, oats and rye. J. Cereal Sci. 6, 253–258.

Hromádková, Z., Košťálová, Z., Ebringerová, A., 2008. Comparison of conventional and ultrasound-assisted extraction of phenolics-rich heteroxylans from wheat bran. Ultrason. Sonochem. 15, 1062–1068.

Jacquemin, L., Zeitoun, R., Sablayrolles, C., Pontalier, P., Rigal, L., 2012. Evaluation of the technical and environmental performances of extraction and purification processes of arabinoxylans from wheat straw and bran. Process Biochem. 47, 373–380.

Jiang, H., Zhang, Y., Wang, X., 2009. Effect of lipases on the surface properties of wheat straw. Ind. Crop. Prod. 30, 304–310.

Jiménez, A., Fabra, M.J., Talens, P., Chiralt, A., 2012. Edible and biodegradable starch films: a review. Food Bioprocess Technol. 5, 2058–2076.

Jobling, S., 2004. Improving starch for food and industrial applications. Curr. Opin. Plant Biol. 7, 210–218.

Khosravi, S., Khabbaz, F., Nordqvist, P., Johansson, M., 2014. Wheat-gluten-based adhesives for particle boards: effect of crosslinking agents. Macromol. Mater. Eng. 299, 116–124.

Leenhardt, F., Lyan, B., Rock, E., Boussard, A., Potus, J., Chanliaud, E., Remesy, C., 2006. Genetic variability of carotenoid concentration, and lipoxygenase and peroxidase activities among cultivated wheat species and bread wheat varieties. Eur. J. Agron. 25, 170–176.

Lehto, J., Oasmaa, A., Solantausta, Y., Kytö, M., Chiaramonti, D., 2014. Review of fuel oil quality and combustion of fast pyrolysis bio-oils from lignocellulosic biomass. Appl. Energy 116, 178–190.

Lempereur, I., Rouau, X., Abecassis, J., 1997. Genetic and agronomic variation in arabinoxylan and ferulic acid contents of durum wheat (*Triticum duruml.*) grain and its milling fractions. J. Cereal Sci. 25, 103–110.

Li, W., Cui, S.W., Kakuda, Y., 2006. Extraction, fractionation, structural and physical characterization of wheat β-D-glucans. Carbohydr. Polym. 63, 408–416.

Li, Q., He, Y., Xian, M., Jun, G., Xu, X., Yang, J., Li, L., 2009. Improving enzymatic hydrolysis of wheat straw using ionic liquid 1-ethyl-3-methyl imidazolium diethyl phosphate pretreatment. Bioresour. Technol. 100, 3570–3575.

Li, C., Aston, J.E., Lacey, J.A., Thompson, V.S., Thompson, D.N., 2016. Impact of feedstock quality and variation on biochemical and thermochemical conversion. Renew. Sust. Energ. Rev. 65, 525–536.

Liyana-Pathirana, C.M., Shahidi, F., 2006. Importance of insoluble-bound phenolics to antioxidant properties of wheat. J. Agric. Food Chem. 54, 1256–1264.

Lopez-Hidalgo, A.M., Sánchez, A., De León-Rodríguez, A., 2017. Simultaneous production of bioethanol and biohydrogen by *Escherichia coli* WDHL using wheat straw hydrolysate as substrate. Fuel 188, 19–27.

Mao, C., Feng, Y., Wang, X., Ren, G., 2015. Review on research achievements of biogas from anaerobic digestion. Renew. Sust. Energ. Rev. 45, 540–555.

Micard, V., Belamri, R., Morel, M., Guilbert, S., 2000. Properties of chemically and physically treated wheat gluten films. J. Agric. Food Chem. 48, 2948–2953.

Mo, X., Hu, J., Sun, X.S., Ratto, J.A., 2001. Compression and tensile strength of low-density straw-protein particleboard. Ind. Crop. Prod. 14, 1–9.

Mo, X., Cheng, E., Wang, D., Sun, X.S., 2003. Physical properties of medium-density wheat straw particleboard using different adhesives. Ind. Crop. Prod. 18, 47–53.

Naguleswaran, S., Li, J., Vasanthan, T., Bressler, D., Hoover, R., 2012. Amylolysis of large and small granules of native triticale, wheat and corn starches using a mixture of α-amylase and glucoamylase. Carbohydr. Polym. 88, 864–874.

Nordqvist, P., Thedjil, D., Khosravi, S., Lawther, M., Malmström, E., Khabbaz, F., 2012a. Wheat gluten fractions as wood adhesives—glutenins versus gliadins. J. Appl. Polym. Sci. 123, 1530–1538.

Nordqvist, P., Lawther, M., Malmström, E., Khabbaz, F., 2012b. Adhesive properties of wheat gluten after enzymatic hydrolysis or heat treatment—a comparative study. Ind. Crop. Prod. 38, 139–145.

Olabarrieta, I., Gällstedt, M., Ispizua, I., Sarasua, J., Hedenqvist, M.S., 2006. Properties of aged montmorillonite–wheat gluten composite films. J. Agric. Food Chem. 54, 1283–1288.

Paes, S.S., Yakimets, I., Mitchell, J.R., 2008. Influence of gelatinization process on functional properties of cassava starch films. Food Hydrocoll. 22, 788–797.

Panthapulakkal, S., Sain, M., 2007. Agro-residue reinforced high-density polyethylene composites: fiber characterization and analysis of composite properties. Compos. A: Appl. Sci. Manuf. 38, 1445–1454.

Price, R.K., Welch, R.W., Lee-Manion, A.M., Bradbury, I., Strain, J., 2008. Total phenolics and antioxidant potential in plasma and urine of humans after consumption of wheat bran. Cereal Chem. 85, 152–157.

Qureshi, N., Saha, B., Cotta, M., Singh, V., 2013. An economic evaluation of biological conversion of wheat straw to butanol: a biofuel. Energy Convers. Manag. 65, 456–462.

Reddy, N., Yang, Y., 2005. Biofibers from agricultural byproducts for industrial applications. Trends Biotechnol. 23, 22–27.

Reddy, N., Yang, Y., 2007. Preparation and characterization of long natural cellulose fibers from wheat straw. J. Agric. Food Chem. 55, 8570–8575.

Roberfroid, M., 1993. Dietary fiber, inulin, and oligofructose: a review comparing their physiological effects. Crit. Rev. Food Sci. Nutr. 33, 103–148.

Ruzene, D.S., Silva, D.P., Vicente, A.A., Gonçalves, A.R., Teixeira, J.A., 2008. An alternative application to the Portuguese agro-industrial residue: wheat straw. Appl. Biochem. Biotechnol. 147, 85–96.

Saha, B.C., 2003. Hemicellulose bioconversion. J. Ind. Microbiol. Biotechnol. 30, 279–291.

Saha, B.C., Cotta, M.A., 2006. Ethanol production from alkaline peroxide pretreated enzymatically saccharified wheat straw. Biotechnol. Prog. 22, 449–453.

Saha, B.C., Iten, L.B., Cotta, M.A., Wu, Y.V., 2005. Dilute acid pretreatment, enzymatic saccharification and fermentation of wheat straw to ethanol. Process Biochem. 40, 3693–3700.

Saha, B.C., Nichols, N.N., Qureshi, N., Kennedy, G.J., Iten, L.B., Cotta, M.A., 2015. Pilot scale conversion of wheat straw to ethanol via simultaneous saccharification and fermentation. Bioresour. Technol. 175, 17–22.

Sain, M., Panthapulakkal, S., 2006. Bioprocess preparation of wheat straw fibers and their characterization. Ind. Crop. Prod. 23, 1–8.

Schmidt, A.S., Thomsen, A.B., 1998. Optimization of wet oxidation pretreatment of wheat straw. Bioresour. Technol. 64, 139–151.

Shewry, P.R., 2009. Wheat. J. Exp. Bot. 60, 1537–1553.

Šramková, Z., Gregová, E., Šturdík, E., 2009. Chemical composition and nutritional quality of wheat grain. Acta Chim. Slov. 2, 115–138.

Stevenson, L., Phillips, F., O'sullivan, K., Walton, J., 2012. Wheat bran: its composition and benefits to health, a European perspective. Int. J. Food Sci. Nutr. 63, 1001–1013.

Sun, R.C., Tomkinson, J., 2002. Characterization of hemicelluloses obtained by classical and ultrasonically assisted extractions from wheat straw. Carbohydr. Polym. 50, 263–271.

Sun, R., Lawther, J.M., Banks, W., 1996. Fractional and structural characterization of wheat straw hemicelluloses. Carbohydr. Polym. 29, 325–331.

Tabarsa, T., Jahanshahi, S., Ashori, A., 2011. Mechanical and physical properties of wheat straw boards bonded with a tannin modified phenol–formaldehyde adhesive. Compos. B Eng. 42, 176–180.

Talebnia, F., Karakashev, D., Angelidaki, I., 2010. Production of bioethanol from wheat straw: an overview on pretreatment, hydrolysis and fermentation. Bioresour. Technol. 101, 4744–4753.

Velioglu, Y., Mazza, G., Gao, L., Oomah, B., 1998. Antioxidant activity and total phenolics in selected fruits, vegetables, and grain products. J. Agric. Food Chem. 46, 4113–4117.

Vohra, M., Manwar, J., Manmode, R., Padgilwar, S., Patil, S., 2014. Bioethanol production: feedstock and current technologies. J. Environ. Chem. Eng. 2, 573–584.

Wang, D., Sun, X.S., 2002. Low density particleboard from wheat straw and corn pith. Ind. Crop. Prod. 15, 43–50.

Wang, T., He, F., Chen, G., 2014. Improving bioaccessibility and bioavailability of phenolic compounds in cereal grains through processing technologies: a concise review. J. Funct. Foods 7, 101–111.

Woerdeman, D.L., Veraverbeke, W.S., Parnas, R.S., Johnson, D., Delcour, J.A., Verpoest, I., Plummer, C.J., 2004. Designing new materials from wheat protein. Biomacromolecules 5, 1262–1269.

Wretfors, C., Cho, S., Hedenqvist, M.S., Marttila, S., Nimmermark, S., Johansson, E., 2009. Use of industrial hemp fibers to reinforce wheat gluten plastics. J. Polym. Environ. 17, 259.

Wu, Q., Rabu, J., Goulin, K., Sainlaud, C., Chen, F., Johansson, E., Olsson, R.T., Hedenqvist, M.S., 2017. Flexible strength-improved and crack-resistant biocomposites based on plasticised wheat gluten reinforced with a flax-fibre-weave. Compos. A: Appl. Sci. Manuf. 94, 61–69.

Wursch, P., Pi-Sunyer, F.X., 1997. The role of viscous soluble fiber in the metabolic control of diabetes. a review with special emphasis on cereals rich in beta-glucan. Diabetes Care 20, 1774–1780.

Xu, Y., Wang, D., 2017. Integrating starchy substrate into cellulosic ethanol production to boost ethanol titers and yields. Appl. Energy 195, 196–203.

Ye, X.P., Julson, J., Kuo, M., Womac, A., Myers, D., 2007. Properties of medium density fiberboards made from renewable biomass. Bioresour. Technol. 98, 1077–1084.

Yu, X., Zheng, Y., Dorgan, K.M., Chen, S., 2011. Oil production by oleaginous yeasts using the hydrolysate from pretreatment of wheat straw with dilute sulfuric acid. Bioresour. Technol. 102, 6134–6140.

Zhang, Y., Lu, X., Pizzi, A., Delmotte, L., 2003. Wheat straw particleboard bonding improvements by enzyme pretreatment. Eur. J. Wood Wood Prod. 61, 49–54.

Zhang, C., Su, H., Baeyens, J., Tan, T., 2014. Reviewing the anaerobic digestion of food waste for biogas production. Renew. Sust. Energ. Rev. 38, 383–392.

Zhang, K., Johnson, L., Prasad, P.V., Pei, Z., Wang, D., 2015. Big bluestem as a bioenergy crop: a review. Renew. Sust. Energ. Rev. 52, 740–756.

Zhang, Z., O'Hara, I.M., Mundree, S., Gao, B., Ball, A.S., Zhu, N., Bai, Z., Jin, B., 2016. Biofuels from food processing wastes. Curr. Opin. Biotechnol. 38, 97–105.

Zheng, Y., Pan, Z., Zhang, R., Jenkins, B.M., Blunk, S., 2007. Particleboard quality characteristics of saline jose tall wheatgrass and chemical treatment effect. Bioresour. Technol. 98, 1304–1310.

Zhou, K., Su, L., Yu, L., 2004. Phytochemicals and antioxidant properties in wheat bran. J. Agric. Food Chem. 52, 6108–6114.

CHAPTER 2
Rice

Ragab Khir*†, Zhongli Pan*
*Department of Biological and Agricultural Engineering, University of California - Davis, Davis, CA, United States, †Department of Agricultural Engineering, Faculty of Agriculture, Suez Canal University, Ismailia, Egypt

Chapter Outline

1 INTRODUCTION

Rice is one of the main cereal crops and a staple food for over half of the world's population. It provides up to 50% of the world's dietary caloric supply and a substantial part of the protein intake for a large portion of people in various regions of the world. Consequently, rice is closely associated with food security and political stability (Khir et al., 2017, 2018). World rice production reached 759.6 million tons (paddy basis) in 2017 (FAO, 2018). In the last 10 years, significant rice production growth (about 10%) has been achieved through the transfer of research and technology initiatives that have produced high-yielding varieties and rapid harvesting and transport capabilities (FAO, 2018).

Rice undergoes several processing operations before it can be consumed by humans. Rice processing covers the operations from harvesting to milling, ultimately producing polished white rice. Due to population and economic growth,

Integrated Processing Technologies for Food and Agricultural By-Products
https://doi.org/10.1016/B978-0-12-814138-0.00002-2

the global demand for rice is expected to remain high for a long time. Therefore, the rice industry should remain sustainable meeting the expected demand, and production of rice by-products will remain high. Rice processing operations yield large quantities of by-products, including straw, husk, bran, and brewers. These by-products are often underutilized and therefore their value is low or lost. The management and reprocessing of such large quantities of by-products are a critical sustainability challenge for the rice industry. The disposal of these by-products could result in negative impacts on the environment, affecting public health while wasting a valuable resource which could be converted to useful value-added products. Consequently, there is a pressing need to find efficient technologies that can be used to process these by-products. Doing so would help advance rural economies, promote diversity in the rice economy, create employment opportunities, reduce environmental pollution, and improve sustainability.

This chapter is divided into three main parts. The first part addresses paddy rice compositions and the rice processing industry. The second part addresses the by-products produced during the rice processing operations and their characteristics. The third part addresses processing technologies used to convert the rice by-products to value-added products with their potentials for food and nonfood applications. The potential of rice by-products as source of energy is included as well. The vision put forth by this chapter is that there is a great need to understand the value of by-products produced during rice processing and state-of-the-art processing technologies in order to expand and advance applications of these wastes at large scale with high efficiency and reduced cost.

2 COMPOSITIONS OF RICE

There are three forms of rice produced during the processing operations. The harvested rice grain is named paddy rice or rough rice. The paddy rice grain is enveloped by the husk composed of two modified leaves: lemma and palea (Fig. 1). The husk is removed during husking (shelling process), and brown rice, composed of the seed coat, nucellus, endosperm, and embryo, is produced. When the outer layers of brown rice are removed during the milling process, the milled (white) rice is produced. The main parts of the rice grain (paddy) are endosperm (70%), rice husk (20%), bran (8%), and rice germ (2%) (Van Hoed et al., 2006).

The majority of rice nutrients, including vitamins, minerals (such as iron), phenolics tocopherols, and γ-oryzanol, are mainly concentrated in the rice bran. Therefore, these rice components are higher in rice bran and brown rice than milled rice (Bao, 2012). Rice has thiamin (B1), riboflavin (B2), niacin, pantothenic acid, folate, and vitamin E, but usually does not have vitamins A, C, and D. Rice grain also contains a number of minerals with the contents varying with different analyses (Heinemann et al., 2005). However, the mineral contents in rice are too low to meet the micronutrient demands for humans that consume rice as staple food. The rice bran has a total free phenolic content ranging from 3.1 to 45.4 mg GAE/g bran, much higher than in brown rice (Goffman and Bergman, 2004). The most common phenolic compounds found in whole rice grains

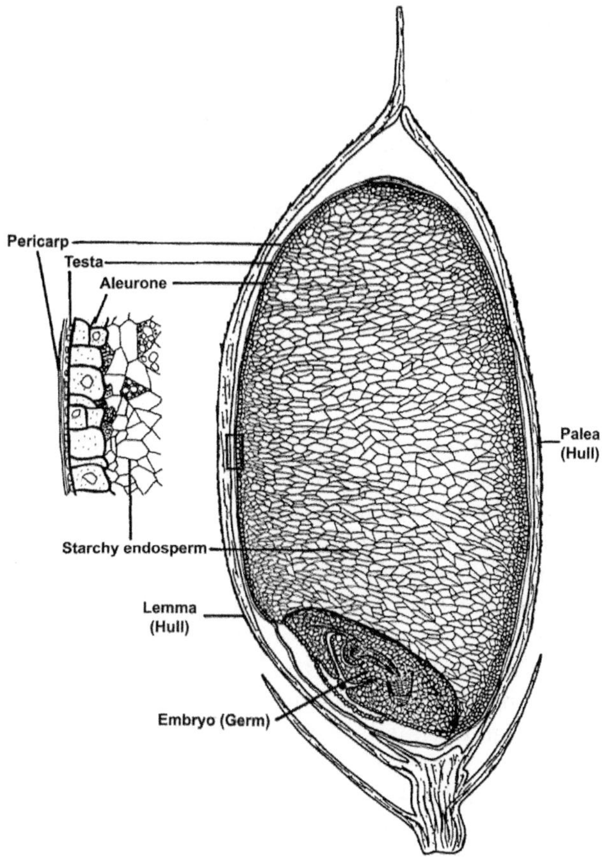

FIG. 1
Structure of rice grain (Saunders, 1986).

are phenolic acids and flavonoids, which are mainly present in the bound form, linked to cell-wall structural components, such as cellulose, lignin, and proteins through ester bonds.

Carbohydrates are the major component of the rice grain. Total starch content of brown rice ranged between 72% and 82% on a dry matter basis (Frei et al., 2003), and around 90% in the milled rice. Dietary fiber includes cellulose, hemicellulose, galactooligosaccharides, pectins, resistant starch, lignin, and substances associated with non-starch polysaccharides and lignin complexes—waxes, phytate, cutin, saponins, suberin, and tannins (AACC, 2001). It constitutes 2.9%–4.0% of brown rice, 0.7%–2.3% of milled rice, or 17%–29% of rice bran, and is mostly insoluble (Juliano and Bechtel, 1985; Abdul-Hamid and Luan, 2000).

The protein content of milled rice ranges from 4.5% to 10.5% or from 5.1% to 11.3% (Champagne et al., 1999) depending on different genotype and environmental effects. The lipid content of brown rice ranges from 2.76% to 3.84% on a dry weight basis depending on different varieties and is affected by the growing environment. The major fatty acids are oleic (18:1) and linoleic (18:2) acids, followed by palmitic acid (16:0). These three fatty acids account for >90% of the total fatty acid content (Kitta et al., 2005).

3 RICE PROCESSING OPERATIONS

Rice processing can be defined as a set of operations, including harvest, drying, and milling. It is conducted to obtain rough (paddy), and then converted to useable forms of food, feed, and/or by-products.

3.1 Harvest

Harvesting operations include cutting, collecting, and threshing the rice crop, and separating and cleaning the grains. Based on the level of mechanization of the operations, the harvesting systems are characterized into traditional and advanced (combine) systems. Traditional harvesting systems consist of manual cutting followed by threshing using simple threshing frames, animals, or tractors running over the crop spread on the ground (Gummert et al., 2018). Postharvest losses in such systems range from 5% to 15% (Chandler Jr., 1979). Combine system gets its name from the fact that it combines all five harvesting operations: cutting, collecting, and threshing the rice crop, and separating and cleaning the grains. Combine harvesters, widely used nowadays in high production countries, potentially reduce harvesting losses to 1%–2% (Gummert et al., 2018). Harvesting operation produces rough rice as a main product and straw as a by-product.

3.2 Drying

Rough rice is normally harvested at moisture content (MC) higher than the required level of 12%–14% (wet basis) for safe storage (Pan et al., 2008). Therefore, rough rice needs to be dried as soon as possible for improving storability and reducing handling costs and losses (Khir et al., 2017). Rough rice drying is a critical postharvest handling process and has a direct effect on rice quality, subsequent handling processes, and commercial value of rice crop. The drying is an important operation in prolonging the storage life of rice by slowing down respiration and preventing deterioration due to molds and insect attack (Khir et al., 2011). Proper drying and storage practices could likely result in 10%–20% increase in rice availability in some developing countries and the increased rice supply could be crucial in nourishing hungry people in these countries (Carl, 1980).

Typical rice drying methods include natural (sun) drying, ambient air drying, and heated air drying. Natural drying is the traditional method widely used by farmers in developing countries (Khir et al., 2017). Considerable losses (10%–25%) occur during sun drying in the field for various reasons, such as rodents, birds, spoilage, and contamination (Helmy et al., 1995). Drying by blowing ambient air at local site conditions is considered as a simple method. It is performed by blowing ambient air with fans under different flow rates through the rice storage location. This method is dependent on weather conditions and slow. Heated air drying is the simplest, most common, and economical industrial process for the drying of grains (Pan et al., 2008, 2011; Khir et al., 2014). This drying method, however, is still a slow process because heated air at a relatively low temperature must be used to preserve the rice milling quality. Due to the relatively low air temperature, the convective drying process is not able to kill the insects in infested rough rice and inactivate the lipase enzyme responsible for lipid degradation during the rice storage period (Pan et al., 2008).

As a new approach, IR heating provides a high heating rate, rapid moisture removal, effective disinfestation and disinfection, good rice milling quality, and storage stability (Pan et al., 2008, 2011). The research also showed that IR drying provided a potential to store brown rice instead of rough rice (Pan et al., 2008, 2011, Khir et al., 2011, 2014; Ding et al., 2015; Wang et al., 2014).

3.3 Milling

Rice is mainly consumed as white (milled) rice. To produce white rice, paddy rice must undergo the milling process. Normally, milling process primarily aims at the separation of the outer tissue to make the endosperm easily accessible for human consumption as food (Khir et al., 2018). The milling process includes several steps, including dehulling or dehusking, whitening, polishing, and separation, to produce partially or well-milled rice. Dehulling is conducted to remove husk from the paddy grains, the resultant finished product of this step is brown rice and husk as a by-product. Whitening is carried out to exclude the bran layer from the brown rice; the end product of this step is white rice and rice bran as by-product. Polishing is performed to remove the loose bran that keeps adhering to the surface of white rice after whitening to obtain refined rice as the end product and bran as by-products. Separation is conducted to separate the broken kernels from whole kernels for the milled grains. The resulting products from separation step are polished rice as a main product, and bran and brewer that is a milled rice kernel that is one quarter to half the size of a full kernel as by-products. Edible rice makes up to 65%–72% with 28%–35% as by-products and waste (Singh et al., 2014).

The most important milling quality indices are total rice yield, head rice yield, and whiteness index (Pan et al., 2008, 2011; Ding et al., 2016). Total rice is the amount of whole and broken kernels of the milled sample expressed as percentage of paddy based on weight. Head rice yield is the amount of whole kernels in the milled sample. The average total yield is between 65% and 75%, whereas the average head yield is between 55% and 65% (Yadav et al., 2018). The extent of removal of bran layers from the rice kernel is characterized in terms of the degree of milling, which determines the whiteness of the rice.

4 BY-PRODUCTS OF RICE PROCESSING

Rice straw, rice husk, rice bran, germ, and brewers are the main rice by-products produced during the processing operations. These by-products have the potential to be used as important sources of raw material that could be utilized as ingredients of value-added products for food and nonfood applications (Esa et al., 2013). The rice straw may range from 0.41 to 3.96 kg for every kilogram of harvested paddy (Koopmans and Koppejan, 1997). On average, the milling of paddy rice has nearly a 69% yield of rice (endosperm) as its major product, and the rest as by-products, including rice husk (20%), rice bran (8%), rice germ (2%), and brewers (1%) (Wells, 1993; Van Hoed et al., 2006; Esa et al., 2013).

Global rice production was estimated at 952 million tons in 2016 (FAO, 2016). It is estimated that it will generate 22.8, 190.4, 76.16, 19.04, and 9.52 million tons of straw, husk, bran, germ, and brewer, respectively (Table 1). However, these by-products are underutilized, and even cause serious pollution problems. Therefore, in recent years, rice by-products have received increased attention as functional ingredients for foods due to their phenolic compounds which can help to lower cholesterol and enact antiatherogenic activity (Wilson et al., 2002) in addition to having high amounts of vitamins, minerals, and fibers. The distinctive characteristics of the rice by-products and their potential utilizations are discussed in the following section.

5 CHARACTERIZATION OF BY-PRODUCTS

5.1 Straw

Rice straw has the characteristics that make it suitable for producing value-added products. Straw can be utilized as raw materials for fuels, chemicals, paper industry, building board, and food industries because of its low cost, wide availability, and low content in nonfibrous materials (Ebeling et al., 1999; Puglia et al., 2003; Gousse et al., 2004; Mohanty et al., 2004; Samir et al., 2004). The chemical and elemental compositions of rice straw are shown in Table 2. It has high cellulose (32%–47%), hemicelluloses (19%–27%), and lignin (5%–24%) contents (Maiorella, 1983; Zamora and Crispin, 1995; Garrote et al., 2002; Saha,

Table 1 Estimated Quantitates of By-Products Produced During Rice Processing Operations

Type of by-products	Origin of by-products	Percentage (%)	Quantity (million tons)
Straw	Harvest	2.4	22.8
Husk	Shelling	20	190.4
Bran	Milling	8	76.16
Germ	Milling	2	19.04
Brewer	Milling	1	9.52

Table 2 Compositions of Rice Straw

Chemical composition	wt% (dry basis)
Cellulose	32–47
Hemicellulose	19–27
Lignin	5–24
Protein	3–5
Starch	1–21
Ash	8–24
Elemental composition	wt% (dry basis)
C	33–58
N	1–3
H	4–7
O	37–54
S	0.1–0.7
Cl (NaCl or KCl)	0.3–3.2
Elemental composition of ash (%)	wt% (dry basis)
SiO_2	74.67
CaO	3.01
MgO	1.75
Na_2O	0.96
K_2O	12.3
K_2O	12.3

Jenkins and Ebeling (1985); Jenkins et al. (1998); Zhang and Zhang (1999); Park et al. (2004); Thy et al. (2004, 2006); Liu et al. (2011); Sheikh et al. (2013); Gu et al. (2014); Duan et al. (2015).

2003). The cellulose and hemicelluloses can be readily hydrolyzed into fermentable sugars (Binod et al., 2010). Rice straw also contains nonstructural carbohydrates, protein, ash, and elemental compositions that are listed in Table 2.

Additionally, physical characteristics of rice straw, such as MC, particle size, bulk density, water holding capacity, heating value, surface morphology, porosity, and thermal conductivity, are important when straw is used for different applications (Zheng et al., 2018). For instance, when rice straw is used as a fuel, high MC decreases biomass heating value which in turn reduces the energy efficiency (Mansaray and Ghaly, 1997). Particle size is an important parameter to affect flowability, heating, diffusion, and reaction rate (Guo et al., 2003; Hernández et al., 2010). Low bulk density of rice residues makes handling processes, such as harvesting, packing, transportation, and storage, more challenging. Moreover, low bulk density is issues for submerged bioconversion processes, such as anaerobic digestion and fermentation (Natarajan et al., 1998). The water holding capacity influences the application of rice residues for soil amendment and as dietary fiber resource (Sangnark and Noomhorm, 2004).

One of the main limiting factors for utilization of rice straw is its natural cell structure. The strong crystalline structure of cellulose in rice straw and the presence of a complex structure of lignin and hemicellulose with cellulose together limit the accessibility of rice straw (Chen et al., 2011). In the cell walls, the cellulose chains bind together using hydrogen bonds to form microfibrils of several nanometers in diameter and millimeters in length. These high tensile strength crystalline microfibrils are the fundamental structural unit in straw cell walls. They are the major contributors to the mechanical strength of the straw cell walls. The microfibrils are bonded by a gel matrix composed of hemicellulose, lignin, and other carbohydrate polymers to form biocomposites (Atanu et al., 2006; Orts et al., 2005; Xu et al., 2007). Various pretreatments have been developed to overcome the complicated structure of the rice straw (Chen et al., 2011). The pretreatment approach is discussed in detail in Section 6.2.6.1.

5.2 Husk

The chemical and physical characteristics of rice husk determine the end uses and vary depending on the type of paddy, crop year, climatic, and geographical conditions (Kumar et al., 2010). Rice husk predominantly contains cellulose, hemicellulose, lignin, protein, fat, carbohydrate, lignin, protein, ash, and dietary fibers. It also contains minerals, such as calcium, copper, iron, potassium, magnesium, manganese, zinc, sodium, phosphorus, and silica (Kumar et al., 2012, 2013; Kuan and Yuen, 2012). The typical organic and chemical compositions, ash constituents and fiber fractions of rice husk are presented in Table 3.

Physical properties of rice husk, such as MC, particle size, bulk density, water holding capacity, heating value, surface morphology, porosity, and thermal conductivity, should be characterized for its efficient use. The bulk density and true density of rice husk depend on the MC. There is a linear relationship between the MC and true density. The true density of unground husk increases from 1021 to 1054 kg/m^3 as MC increases from 11.4% to 21.3% (Mishra et al., 1986).

There is also a relationship between the specific heat and MC of rice husk. Due to the higher specific heat of water, the specific heat of rice husk increases linearly with the MC within the moisture range of 10%–21%. Moreover, the thermal conductivity of rice husk has a linear relationship with MC.

5.3 Bran

Rice bran contains valuable components, such as protein, vitamins, edible oil, fiber, and amino acids with a high nutritional value and functionality (Luh, 1980; Wang et al., 2017). The quality of rice bran is closely related to its physicochemical properties. These physicochemical properties have been studied by many researchers. Proximate compositions of the rice bran are shown in Table 4 (Juliano, 1985; Wang et al., 1999; Hernandez et al., 2000; Tang et al., 2003; Rosniyana et al., 2007; Sereewatthanawut et al., 2008; Garcia et al., 2012). Rice bran has a high inherent nutritional value with a protein content of 12.25%. Protein digestibility of rice bran is >90% (Wang et al., 1999). Rice bran also has a high oil content. Based on its oil content, rice bran is classified into three groups: full-fatted raw bran obtained from milling of raw paddy, full-fatted parboiled bran obtained from milling of parboiled paddy, and De-fatted/

Table 3 Compositions of Rice Husk

Composition	Weight (%, dwb)
Cellulose	31.12
Hemicellulose	22.48
Lignin	22.34
Mineral ash	13.87
Chemical composition	Weight (%, dwb)
Protein	5.49
Fat	0.92
Carbohydrate	15.45
Ash	71.62
Moisture	6.52
Analysis of rice husk ash	Weight % of ash
Si as SiO_2	92.45
Ca as CaO	0.86
Mg as MgO	0.77
Al Trace	1.86
Fe Trace	1.04
Mn Trace	0.41
K as K_2O	1.1
Na as Na_2O	0.78
P as P_2O_5	0.53
S as SO_4	1.13
Dietary fiber fractions	Weight (%, dwb)
Total dietary fiber	59.01
Soluble dietary fiber	1.00
Insoluble dietary fiber	58.01

Kumar et al. (2010); Kumar et al. (2013); Kuan and Yuen (2012).

De-oiled bran obtained after extraction of oil from either raw or parboiled bran (Yadav and Joyner, 2018).

The physical characteristics of rice bran, such as bulk density, are important parameters that determine the behavior of bran during packaging and mixing process. The bulk density is determined by particle geometric forms, porosity, starch, and protein contents, MC, particle distribution, and cell wall structure (Chastril, 1990). Despite of its high nutrition value rice bran is underutilized, except when it is used for producing oil in some countries (Wang et al., 2017).

Utilization of rice bran is severely restricted by the activity of endogenous enzymes, such as lipase, which can hydrolyze the triglyceride into glycerol and free fatty acids (FFAs) (Wang et al., 2017). As the result, the product forms off flavor with a rancid smell and bitter taste and deteriorate the oil quality, thereby rendering the bran and oil unpalatable and unsuitable for human consumption (Luh, 1980). It takes only <1 week to make the FFAs in the oil exceed 10%, the maximum value that is suitable for human consumption (Desikachar,

Table 4 Compositions of Rice Bran

Composition	wt% (dry basis)
Moisture content	9.7
Fat oil	16.75
Protein	12.25
Fiber	6.64
Ash	7.6
Carbohydrate	48.3
Bulk density	0.39 (g/mL)
Water activity	0.53
Amino acids	g amino acid/g protein
Lysine	5.1
Histidine	1.6
Isoleucine	3.75
Leucine	7.7
Threonine	4.05
Tryptophan	0.95

Juliano (1985); Wang et al. (1999); Hernandez et al. (2000); Tang et al. (2003); Rosniyana et al. (2007); Sereewatthanawut et al. (2008); Garcia et al. (2012).

1977; Qian et al., 2014). Therefore, a normal practice is for the bran to undergo a stabilization process to inactivate the potent lipase enzyme immediately after milling in order to extend storage time by controlling the FFAs concentration at a low level (<10%) before oil extraction (Randall et al., 1985). The rice bran stabilization and processing are discussed in details in Section 7.

5.4 Broken Rice and Brewers

Broken rice (brokens) is rice fragments that are <75% of the length of whole kernels with half of the economic value of whole kernels (USDA, 2009). The rice kernel breakage can accrue because of many factors such as rapid moisture absorption of rice in the field, immature kernels, fast drying, milling procedures and degree, and chalkiness (Mhalaskar et al., 2017). Additionally, environmental factors such as relative humidity (RH) and temperature changes and insect infestation during storage can also cause rice breakage (Siebenmorgen et al., 1998).

Differences in the protein content between different sizes of brokens can be attributed to the fact that rice kernels are nonuniform in their composition across the kernel dimension. Smaller brokens have higher protein contents than larger brokens and whole kernels. The chemical properties of brokens are informed by the degree of size fractionation (Mukhopadhyay and Siebenmorgen, 2017).

Brewers mostly consist of broken rice, with only small percentage of rice bran and rice germ. They contain phenolics, vitamin E, and γ-oryzanol, but the contents are significantly lower than those in rice bran (Esa et al., 2013). The health benefit of the brewers has yet to be fully studied.

6 TECHNOLOGIES FOR PRODUCING VALUE-ADDED PRODUCTS FROM THE BY-PRODUCTS

6.1 Food Applications

6.1.1 *PROTEIN SOURCE*

Rice protein contains lysine that is considered hypoallergenic and is therefore favorable for human consumption (Helm and Burks, 1996). Different extraction methods have been studied to extract proteins from rice by-products. Alkali hydrolysis under different conditions has been explored to improve the yield of extracted protein (Lew et al., 1975; Bera and Mukherjee, 1989; Gnanasambandam and Hettiarachchy, 1995). Alkali hydrolysis followed by acid precipitation has been studied by Ansharullah and Chesterman (1997). In this method, deoiled rice bran was suspended in distilled water (1 g in 5 mL of water). Then the suspension was adjusted to pH 11 by sodium hydroxide (NaOH) solution. The reaction was carried out at 30°C and alkali hydrolysis for 45 min. After the reaction, the pH of the suspension was adjusted to pH 7 with hydrochloric acid (HCl) solution to stop the reaction. The residue bran was separated from the soluble product with a vacuum filter using a filter paper, then the soluble portion was assayed for extracted protein. This is a simple method, but the high pH conditions could reduce the protein yield because of degradation phenomenon (Jiamyangyuen et al., 2005). Moreover, high pH conditions could also lead to undesirable effects like molecular cross-linking and rearrangements, resulting in decrease in nutritive value and formation of toxic compounds, such as lysinoalanine (Cheftel et al., 1985; Otterburn, 1989).

A subcritical water hydrolysis method was studied by Sereewatthanawut et al. (2008) as an alternative method for rice protein extraction. In this method, a deoiled rice bran suspended in distilled water (1 g in 5 mL of water) was charged into a batch reactor and heated to 100–220°C by an electric furnace heater for up to 30 min. The pressure in the reactor was estimated to be between 101.35 kPa and 3.97 MPa. Then the reactor was cooled to room temperature by immersion in a cool water bath. The soluble portion was separated from the residue bran with a vacuum filter and assayed for extracted protein. It was reported that the amount of produced protein increased with an increase in temperature. The highest yield of protein was 219 ± 26 mg/g of dry bran at extraction conditions of 200°C for 30 min.

Enzymatic methods have also been used to improve protein extractability at neutral pH. Ansharullah and Chesterman (1997) studied the enzymatic pretreatment using different concentrations of carbohydrases (Viscozyme, 0–120 FBG unit) and (Celluclast 1.5 L, 0–360 NC unit), time of incubation (1–5 h), pH (3.8–5.4), and temperatures of 40°C and 50°C. They found that the maximum protein extraction was obtained when the Viscozyme-Celluclast mixture used at pH 3.8 with incubation time of 5 h and temperature of 40°C. While, at the process temperature of 50°C, the use of Viscozyme alone, under similar conditions, had a significant effect in increasing the efficiency of the protein extraction.

6.1.2 ANIMAL FEED

Rice straw has a high fiber content and, consequently, has been used as animal feed in many countries worldwide (Ranjhan, 1990; Devendra and Thomas, 2002). However, rice straw has a low crude protein and digestibility (Drake et al., 2002). Therefore, pretreatments, including physical, chemical, physio-chemical, and biological treatments, are needed to improve its nutritional value and digestibility. To improve nutrient content, rice straw is normally treated by urea which is safe compared to aqueous ammonia (Halim, 1985). Another approach to improve the digestibility of cellulose is to treat the rice straw with sodium hydroxide or ammonia (Dommershuijzen and Lozare, 1990; Ahuja and Uma, 2018). Additionally, the rice bran can be used in chicken and duckling diets. Dietary inclusion levels of up to 200 g/kg can be tolerated by chickens without depressing performances. For ducklings the maximum inclusion level appears to be below 600 g/kg (Farrell, 1994).

6.1.3 MEDICAL APPLICATIONS

It has been reported that the consumption of rice by-products such as brown rice has some medical benefits. The high concentration of the B-group of vitamins in rice bran has played a major part in advances in nutrition, medicines, and chemistry (Houston, 1972). Rice bran powder offers health benefits like lowering of blood cholesterol (Hegsted et al., 1990; Kahlon et al., 1992), decreasing the incidence of arthrosclerosis disease, and having laxative effect (Saunders, 1990). Rice bran helped in accelerating human growth, facilitating blood circulation, and stimulating hormonal secretion on automatic nervous system and provided beneficial effects on the skin (Yokochi, 1977). Inositol extracted from defatted bran is effective against liver cirrhosis, vascular sclerosis, and high cholesterol (Farrell, 1994; Saunders, 1986).

Additionally, rice bran oil is unique among edible oils due to its healthy antioxidants (lecithin, oryzanol, tocopherol, and tocotrienols). These healthy antioxidants are helpful in protecting the body's cells against the effects of free radicals. The oryzanol impedes melanin pigmentation by restraining the erythema activity of tyrosinase as it stops the ultraviolet rays' transmission at the skin's surface (Patel and Naik, 2004). Lipid oxidative constancy is important in the prevention of pathologic processes, such as cardiovascular diseases, arthritis and inflammatory processes. It has also been reported that rice bran oil was effective in reducing plasma cholesterol and cholesterol absorption thus decreasing early atherosclerosis (Ahuja and Uma, 2018).

6.2 Nonfood Applications

6.2.1 CONSTRUCTION MATERIALS

Rice straw is widely used in making particle boards, fiber, cement bonded boards, and composites for use in construction of buildings. In many parts of the world, straw is used to bind clay and concrete. A mixture of clay and straw, known as cob, is used as a building material. Soil ash bricks that were made from soil and straw ash using emulsified asphalt as a soil stabilizer giving high degree of insulation are light and can absorb atomic radiation (Anonymous, 1955). Pressed rice straw panels offer lightness, strength, good insulating properties, water resistance, and low toxicity. They are degradable and recyclable and have been proposed as ideal construction materials for low cost durable housing at earthquake zones, particularly in developing countries. Particle board performance is mostly related to the properties of adhesives and their compatibility with particles (Zhang and Hu, 2014). Studies on the possibilities of use of rice straw as filler in thermoplastic poly matrix revealed that rice straw fiber system possessed compatible mechanical properties like those of wood composites (Yao et al., 2011). In Japan, Strawnit is a hardboard made from rice straw and used as a building material for thermal insulation (Sakurai, 1977). There has been a significant interest in the production of fiberboard panels and building materials from rice straw in Australia (Quayle, 2016).

Additionally, straw bales have been used to build houses since the invention of the baler in the late 1800s in the United States. Straw-bale construction has recently gained more confidence to be used as structural elements. Several laboratory tests worldwide have shown straw-bale construction can support substantial service loads with proper baling, stacking, and plastering. When baled, straw has an excellent insulation characteristics and more rot resistant than other kinds of straw. It can be used alone or in a post-and-beam construction to build straw-bale houses (Garas et al., 2009). Straw-bale construction holds the promise of sustainable, affordable, and warm housing for cold and straw-growing areas across the globe. The use of straw bales in building is increasing rapidly across the United States, Korea, and China. In People's Republic of China, straw-bale construction is currently in vogue to build houses and other public buildings (Ahuja and Uma, 2018).

6.2.2 *AGRICULTURAL APPLICATIONS*

Rice straw is available in significant quantities to most rice farmers. It contains a large amount of nutrients, that is, about 40% of the nitrogen (N), 30%–35% of the phosphorus (P), 80%–85% of the potassium (K), and 40%–50% of the sulfur (S) taken up by rice and remained in straw at crop maturity (Dobermann and Fairhurst, 2002). Therefore, incorporation of straw into the soil can return most of the nutrients, conserve soil fertility, reduce CO_2 and N_2O emission, and improve plant yield (Pathak et al., 2006; Wang et al., 2014).

Rice straw is an effective substrate for growing mushrooms which are a high-value crop and can be grown year-around (Nader and Robinson, 2010). Although wheat straw has been the primary substrate for mushroom production, rice straw was reported to yield 10% more oyster mushroom under the same cultivation conditions (Zhang et al., 2002; Yang et al., 2013).

Rice straw has also been used in erosion control applications. The use of rice straw in revegetation has advantages and relevant reasons (Vagg, 2013). Cal Recycle California Department of Resources Recycling and Recovery (2014) mentioned that the US Forest Service, along with the United States Department of Agriculture (USDA), contracts helicopter operators to perform the task over devastated areas, usually at a rate of 2.47 tons per hectare creating a blanket of straw approximately 2.47 cm thick. Its high silica content means that it takes significantly longer to break down compared to standard straw options such as wheat or barley. In Australia, councils developing new roads and passes often use straw bales to control sediment runoff and storm water management. Companies in the United States have modernized this process with rice straw by creating a sausage like object up to 7.5 m long and in varying widths up to half a meter in diameter.

6.2.3 *EFFECTIVE ADSORBENTS*

Use of rice by-products, such as husk and husk ash as absorbents, adds value and helps reduce the cost of their disposal. Rice husk has been used as an absorbent because of its specific physical and chemical characteristics. These include its morphology, granular structure, high mechanical strength, carbon and silica content, insolubility in water, chemical stability, and its local availability at almost no cost. Rice husk has also been used as an adsorbent for the removal of various pollutants from water and wastewaters, including dyes, phenols, organic compounds, pesticides, inorganic anions, and heavy metals (Ahmaruzzaman and Gupta, 2011).

Rice husk has been intensively investigated as an adsorbent for the removal of heavy metals such as Cd, Pb, Zn, Cu, Co, Ni, and Au from wastewaters (Chuah et al., 2005). The studies on heavy-metal adsorption using rice husk suggested the applicability of Langmuir adsorption isotherm, which assumes monolayer coverage and constant adsorption energy on the surface of the adsorbent (Mishra et al., 1997). The rice husk ash has >95 wt% of silica with high porosity and large surface area because it retains a cellular structure skeleton. The ash has been reported to have good adsorptive properties and was utilized for the adsorptive removal of metal ions from wastewater. It was used as an adsorbent for the removal of Hg, Cd, Pb, Ni, and Zn from aqueous solutions (Ahmaruzzaman and Gupta, 2011). Rice husk has been investigated as a good adsorbent material for dye removal. The dye removal mechanism was reported to be complex, consisting of both surface adsorption and pore diffusion. Adsorption of dyes on rice husk or rice husk ash was influenced by temperature and pH of solution (Lakshmi et al., 2009; Ponnusami et al., 2006).

Additionally, acid or alkali chemical treatment or thermal treatment of rice husk can help improve their sorption performance, particularly for metal ions, including shortening equilibrium time and increasing pore volume. The modified rice husk exhibited higher adsorption capacities than unmodified rice husk (El-Shafey, 2007; Sahu et al., 2009). Modified rice husk proved potentially useful material for the removal of Cu and Pb from aqueous solutions (Wong et al., 2003).

6.2.4 SILICA PRODUCTION

Rice husk contains nearly 20% silica which is present in hydrated amorphous form. Silica is predominantly in inorganic linkages, but some of the silica is also bonded covalently to the organic compounds which is undissolved in alkali and can withstand very high temperatures (Chandrasekar et al., 2003). On thermal treatment, the silica converts to cristobalite, which is a crystalline form of silica. Conversion of rice husk to ash for pozzolan production can be done below 1000°C (Fabiyi, 2013). Under controlled burning conditions, amorphous silica with high reactivity, ultrafine size and large surface area is produced. This micro silica can be a source for preparing advanced materials in many applications (Krishnarao et al., 2001; Chandrasekar et al., 2003).

There is a growing demand for fine amorphous silica in the production of high-performance cement and concrete used in bridges, marine environments, and nuclear power plants. Soltani et al. (2015) and Kartini (2011) mentioned that rice husk silica is used in high-strength concrete as a substitute for silica fume. They also stated that since ordinary portland cement (OPC) is typically the most expensive constituent of concrete, the replacement of a portion of the constituent with rice husk ash offers improved concrete affordability, particularly for low cost. A mixture of rice husk ash as a pozzolanic component with lime can replace portland cement to a large range for masonry applications (Dass, 1984). The ultrafine size of rice husk ash makes an excellent filler for the microvoids giving high strength to the concrete (Chandrasekhar et al., 2006). The use of rice husk ash to prepare pozzolan prevents the cracking of the concrete and protect it from corrosion by not allowing any leaching agents to break down the material. Saraswathy and Song (2007) found that the incorporation of rice husk ash up to 30% replacement level reduced the chloride penetration, decreased permeability, and improved strength and corrosion resistance properties. Cordeiro et al. (2013) prepared high performance concretes with 0%, 10%, 15%, and 20% of the cement (by mass) replaced by ultrafine rice husk ash. They reported that for all levels of cement replacement, the ultrafine rice husk ash concretes achieved superior performance in the mechanical and durability tests.

Due to amorphous silica having high purity, small particle silica has been widely used in vegetable oil refining, pharmaceutical products, detergents, adhesives, chromatograph column packing, and ceramics (Soltani et al., 2015). It is also used in the production of nanomaterials. The expression "nanodispersed silica" covers a variety of silica forms including sols, gels, suspensions, and pastes. Silica sol or colloidal silica refers to a stable dispersion of colloidal silica particles in water medium. It is used in many applications, such as in refractory materials, binder for inorganic paint, stiffener for hard coating reagents, abrasive particles, adsorbents, and catalyst (Tsai et al., 2006). In vitreous form, silica is used in the manufacture of optical elements and glass wires and a basic material for glass, ceramic, and refractory industries (Turmanova et al., 2012). Silica finds use in reinforcing filter in natural and synthetic rubber and is used in adhesives to enhance bond strength. Other uses include control of porosity in battery separators for controlling oil spillage and used in brake linings. It is also used in cosmetics as an absorbent and thickening agent, in inks as a thickening and matting agent, in tooth paste as an abrasive additive powder, in paper as filler for opacity and brightness, in polishes as a mild abrasive, and in the food industry as an anticaking agent (Ghosh and Bhattacherjee, 2013; Chandrasekar et al., 2003; Sun and Gong, 2001). Silicates have also been used in edible oil refining and frying oil treatment as an alternative to caustic addition. Silicate treatment has the advantage of reduced oil losses and soap contamination in comparison to alkali refining method involving caustic (NaOH) addition to remove FFAs from oil (Kalapathy et al., 2000).

6.2.5 *PULP AND PAPER PRODUCTION*

Recently, production of pulp and paper from rice by-products such rice straw has gained an increased interest. Rodriguez et al. (2010) mentioned that using rice straw for pulp and paper production can achieve the following advantages: (1) minimize the importation of wood and pulp in countries which lack wood resources; (2) increase the profitability of rice farmers through value-added application of byproducts; (3) increase of public interest in paper produced from environmentally friendly and sustainable raw materials. Rice straw is relatively easier to bleach than wood pulp, being less power intensive in pulp preparation, and requires smaller amount of chemical for digestion and is commonly used in pulping industry (Jones, 1975). Consequently, producing paper from rice by-products has become popular in many countries, including Egypt, Indonesia, India, Australia, Europe, Japan, Philippines, and the USA.

The production of paper from rice straw totaled 23,000 tons per year in Indonesia (Dommershuijzen and Lozare, 1990). Forest Products Research and Development Institute (FPRDI) in Los Banos developed the art of paper making from rice straw. Bleached straw pulp and bleached cotton linter pulp can be blended in 1:1 ratio to produce a pulp that can be used in the production of popular varieties of writing paper (Jakata, 1981). Straw pulps on their own are unsuitable for soft papers and are appropriate mainly for writing and printing papers as well as corrugating medium (Zerrudo, 1984). The first industrial rice straw paper based on the new technology was launched in 2011 in Egypt (SciDev Net, 2011). Egyptian researchers developed a new chemical pulping technology to turn rice straw into paper and insecticide for mosquito control.

6.2.6 *SOURCE OF BIOENERGY*

Rice straw and rice husk can be used as a source of renewable energy through biochemical and thermochemical conversion processes. The technology of using of rice straw and husk for producing bioenergy is well established. In biochemical conversion processes, the production of bioenergy from rice by-products generally involves three main steps, pretreatment, enzymatic hydrolysis, and fermentation to produce ethanol (Drapcho et al., 2008). Additionally, thermochemical conversion, including pyrolysis, gasification, and combustion, is used to produce different types of energy, fuel, and chemical products from the rice by-products.

6.2.6.1 Biochemical Conversion Processes

- *Pretreatment*

Pretreatment is an essential step to destruct the strong structures of rice straw and husk and increase efficiency in energy conversion. The pretreatment can be conducted through physical, chemical, biological methods, or a combination of these methods (Sun and Cheng, 2002). It is necessary to develop a low cost, efficient, and environmentally friendly pretreatment techniques for rice straw to be used for energy production.

- *Physical pretreatment*

Physical pretreatment is conducted to increase the accessible surface area and size of pores and decrease the crystallinity and degrees of polymerization of cellulose. This can improve enzymatic hydrolysis and fermentation yields. Densification, grinding and milling, irradiation, and high temperature are commonly methods used for physical treatments. Densification improves the initial bulk density, minimizes the cost of handling, storage and transportation, and provides better combustion efficiency (Ryu et al., 2006). Converting rice straw into fine particles increased sugar and ethanol yields by 103% and 95% on enzymatic hydrolysis and fermentation, compared to untreated rice straw (Chuetor et al., 2015). Irradiation pretreatments, such as proton beam and electron beam irradiations, were found to improve glucose yield of rice straw upon enzymatic hydrolysis from approximately 20 (untreated) to 52%–70% (treated) when the proton and electron irradiation doses were 15 and 80 kGy, respectively (Bak et al., 2009; Kim et al., 2011). To further improve the grindability of biomass, a new technology called torrefaction (high-temperature drying at 200–300°C in an inert atmosphere) has been recently developed for biomass particle size reduction with less energy demand compared to ordinary size reduction methods (Sheikh et al., 2013)

- *Chemical pretreatment*

Compared to physical pretreatment, chemical pretreatment is more often used because it is effective upon enzymatic hydrolysis and final biofuel product yields. The most promising chemicals for pretreatment are acid, alkali, and ammonia. Acid pretreatment can solubilize most hemicellulose and enhance cellulose degradability. Acids being used include hydrochloric acid (Singh et al., 2010), nitric acid (Ravoof et al., 2012), and sulfuric acid (Hsu et al., 2010; Kshirsagar et al., 2015; Zhang et al., 2015). Alkali pretreatment involves the application of alkaline solutions like NaOH or KOH to remove lignin and a part of the hemicelluloses, and efficiently increase the accessibility of enzyme to the cellulose. Alkali treatment appears to be the most effective method in breaking the ester bonds between lignin, hemicellulose, and cellulose, and avoiding fragmentation of the hemicellulose polymers. NaOH pretreatment under optimum conditions was found to increase reducing sugar yield by >100% (Singh et al., 2010). Kim and Han (2012) also optimized NaOH pretreatment of rice straw and achieved about fourfold more sugar yield upon enzymatic hydrolysis compared with raw straw with 2.96% NaOH loading at 81.79°C for 57 min. As a pretreatment reagent ammonia has number of desirable characteristics, it is a nonpolluting and noncorrosive chemical and an effective swelling reagent for lignocellulosic materials. It has high selectivity for reactions with lignin over those with carbohydrates. Its high volatility makes it easy to recover and reuse. Drapcho et al. (2008) conducted a study on pumping ammonia through a bed of biomass maintained at 170°C. They reported that by this process up to 85% delignification and almost theoretical yield of glucose in enzyme hydrolysis can be achieved. Ammonia pretreatment was also effective to improve enzymatic sugar yield of rice straw from 22% to 71% (Ko et al., 2009).

- *Biological pretreatment*

Wood-degrading fungi are used to modify chemical compositions and structures of lignocellulosic biomass, rendering biomass less resistant to biodegradation. Compared to chemical pretreatment, biological pretreatment has the following advantages: (1) low chemical and energy use; (2) low cost; (3) safe and environmentally-friendly method for lignin removal from lignocellulose. Taniguchi et al. (2005) mentioned that the most promising microorganisms for biological pretreatment are white-rot fungi such as *Phanerochaete chrysosporium* and *Pleurotus ostreatus* that belong to class Basidiomycetes. They reported that *P. ostreatus* selectively degraded 41% lignin, and 52% hemicellulose, and 44% cellulose of pretreated rice straw through solubilization with cellulase. Balasubramaniam and Rajarathinam (2013) used *P. ostreatus* to pretreat rice straw to remove 27.85% lignin and 24% cellulose and achieved 2.3% (v/w) ethanol concentration during fermentation. It is important to notice that biological pretreatment usually takes much longer time and needs larger storage space than physical and chemical pretreatments.

- *Enzymatic hydrolysis*

Enzymatic hydrolysis is the second step in the production of ethanol from lignocellulosic materials. It involves cleaving the polymers of cellulose and hemicellulose by using enzymes to convert them into sugars. This step is conducted under mild conditions of low pressure and long retention time. There are several factors affecting the yields of sugars. These factors include particle size, liquid to solid ratio, type and concentration of acid used, temperature, and reaction time, size of macromolecules, degree of polymerization of cellulose, configuration of the cellulose chain, and association of cellulose with other protective polymeric structures within the plant cell wall such as lignin, pectin, hemicellulose, proteins, and mineral elements (Taniguchi et al., 2005).

Many studies have been reported on enzymatic hydrolysis. Valkanas et al. (1998) conducted hydrolysis of rice straw using different acids with varying concentrations (0.5%–1% H_2SO_4, 2%–3% HCl and 0.5%–1% H_3PO_4). They reported that after 3 h retention time, rice straw pentosans converted to a solution of monosaccharides, suitable for fermentation. Moreover, hydrolysis of rice straw using 5%–10% H_2SO_4 at 80–100°C was reported by Valdes and Planes (1983). They found that the best sugar yield was achieved at 100°C with 10% H_2SO_4 for

240 min. Also, Yin et al. (1982) studied the hydrolysis of hemicellulose fraction of rice straw with 2% H_2SO_4 at 110–120°C and reported that >70% of pentoses was successfully hydrolyzed.

- *Fermentation*

Fermentation is the third step in the production of ethanol from the cellulose and hemicellulose fraction of rice straw. Studies on fermentation have been centered on the strain development and selection (e.g., genetic modification making baking yeast *Saccharomyces cerevisiae* be able to ferment both glucose and xylose) and process/reactor design and optimization [e.g., simultaneous saccharification and fermentation (SSF) vs. consolidated bioprocessing] (Binod et al., 2010; Zheng et al., 2018). In general, the fermentation step can be conducted thorough two approaches. The first approach includes SSF while the second approach involves a separate enzymatic hydrolysis and fermentation (SHF) processes. Many microorganisms that can metabolize sugars (glucose and xylose) to ethanol have been identified and studied. SSF results in higher yield of ethanol compared to SHF by minimizing product inhibition and is favored due to its low production costs (Wyman, 1994). However, the difference in optimum temperatures of the hydrolyzing enzymes and fermenting microorganisms is considered one of the disadvantages of this process. It is important to note that the optimum temperature for enzymatic hydrolysis is at 40–50°C, while the microorganisms with good ethanol productivity and yield do not usually tolerate this high temperature. This problem can be avoided by applying thermo-tolerant microorganisms such as *Kluyveromyces marxianus*, *Candida lusitaniae*, and *Zymomonas mobilis* or mixed culture of some microorganisms like *Brettanomyces claussenii* and *S. cerevisiae* (Golias et al., 2002; Spindler et al., 1988).

There are many reports stating that the SSF is superior to the traditional saccharification and subsequent fermentation in the production of ethanol from rice straw because the SSF process can improve ethanol yields by removing end-product inhibition of saccharification process and eliminate the need for separate reactors for saccharification and fermentation (Chadha et al., 1995). In SSF, ethanol yield of rice straw via dilute sulfuric acid pretreatment was 77% by using baking yeast *S. cerevisiae* (Wang et al., 2015). Development of new ethanol fermenting yeasts to enhance yield has been receiving many research interests. Karimi et al. (2006) achieved the highest ethanol yield of 74% from dilute-sulfuric acid pretreated rice straw using *Rhizopus oryzae*.

Additionally, previous studies on strain cocultures showed that the fermentation of glucose in the sugar mixture proceeded efficiently with a traditional glucose-fermenting strain, while the fermentation of xylose was often slow and had low efficiency due to the conflicting oxygen requirements between the two strains and/or the catabolite repression of xylose assimilation caused by the glucose (Grootjen et al., 1991; Kordowska-wiater and Targonski, 2002). Consequently, hundreds of microorganisms, which can metabolize xylose to ethanol, have been identified, including bacteria (*Z. mobilis*), filamentous fungi (*Neurospora crassa*), yeasts (*Pachysolen tannophilus*, *Pichia stipitis*, *Candida shehatae*, *Candida tropicalis*,

Hansenula polymorpha) (Saha and Cotta, 2011; Konishi et al., 2015). Different approaches, including immobilization of one of the strains, coimmobilization of two strains, continuous culture, two stage fermentation in one bioreactor, and separate fermentation in two bioreactors, have been carried out to circumvent these difficulties related to the fermentation of xylose and to improve its fermentation efficiency (Grootjen et al., 1991; deBari et al., 2004; Fu and Peiris, 2008; Laplace et al., 1993; Delgenes et al., 1996).

It has been concluded from the previous studies that rice straw has several characteristics that make it a promising potential for ethanol production. Theoretically, ethanol yield of rice straw can be reach to 110 gal/dry ton biomass which is about 90% of corn grain (DOE, 2015). However, technical and financial challenges result in an actual ethanol yield of rice straw much lower than the theoretical value. To increase the bioethanol yield from rice straw research should focus on improving the efficiency of pretreatment, hydrolysis, and fermentation steps. Pretreatment and hydrolysis are critical for fermentable sugar yield which directly determines the final ethanol yield. Fermentation studies should focus on the strain development and selection. The efforts should be made to overcome the financial barriers.

6.2.6.2 Thermochemical conversion processes

The energy content of rice straw is about 14 MJ/kg (Ebeling and Jenkins, 1985). Thermochemical conversion processes, including combustion, gasification, and pyrolysis, are used to produce different types of energy from rice by-products (Fig. 2).

- *Combustion*

In the direct combustion process, rice straw can either be used alone or mixed with other biomass materials. Currently, combustion is the dominant technology for large-scale energy production from rice straw (Gadde et al., 2008; Park et al., 2014). This process is characterized by two main advantages, (a) the theoretical energy conversion efficiency of direct combustion is higher than that of indirect thermochemical processes such as gasification and pyrolysis (Said et al., 2013); (b) by-products such as fly ash and bottom ash are produced during this

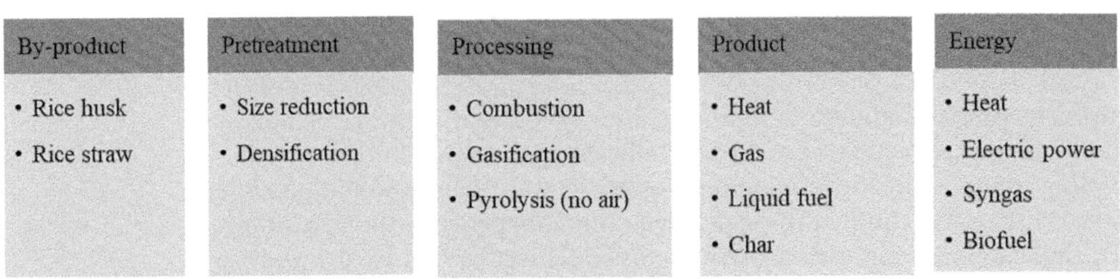

FIG. 2
Energy conversion processes of rice by-products. http://www.knowledgebank.irri.org/step-by-step-production/postharvest/rice-by-products/rice-husk/using-rice-husk-for-energy-production.

process which have an economic value (use in cement and/or brick manufacturing, construction of roads and embankments, etc.). However, rice straw is problematic for burning in most existing combustion systems because of the high contents of alkali, chloride, silica, and sulfur which cause boiler operational problems, such as slagging, fouling, agglomeration, and corrosion. This can substantially increase maintenance costs and reduce efficiency and operating revenues (Calvo et al., 2012). There are two approaches, including leaching and co-combustion, can be used to improve combustion properties of rice straw.

Leaching of rice straw with water leads to significant changes in inorganic composition and substantial improvements in the combustion behavior (Bakker et al., 2002). Bakker and Jenkins (2003) reported that leaching rice straw with water in both natural and industrial leaching processes can reduce the concentrations of undesirable inorganic constituents which cause fouling and slagging. Said et al. (2013) used tap water to wash rice straw and remove 87% chlorine, 50% potassium, and 20% total ash, resulting in 8% increase of high heating value. It is important to emphasis that the economic feasibility of leaching is a big concern.

- *Gasification*

The advantage of rice straw as a feedstock for gasification lies in its broad availability, abundance, and low cost. Gasification process converts rice straw into a mixture of gases such as carbon monoxide, hydrogen and light hydrocarbons, carbon dioxide, and nitrogen. It is considered as one of the most promising technologies because of its ability to rapidly convert large amounts and various kinds of biomass into easily storable and transportable gas or liquid fuel (Gómez-Barea et al., 2009; Matsumoto et al., 2009). Gasification could be an alternative to combustion and pyrolysis to generate a gaseous fuel with improved handling and reduced emission of harmful pollutant, such as NOx (Calvo et al., 2012).

The optimized gasifier design and operation and feed stock treatment could mitigate challenges related to slagging and agglomeration due to the presence of high ash content in rice straw (Shie et al., 2011). Su and Luo (2009) developed a new two-stage gasifier for rice straw gasification and achieved 91% carbon conversion rate and 85% overall gasification efficiency with low tar in produced gas. The gas yield was $1.84\,m^3/kg$ dry straw and the gas heating value was high as $7248\,kJ/m^3$. No agglomeration was found during gasification in the fluidized-bed gasifier in which alumina silicate was partially replaced with MgO in bed because MgO could act as catalyst for tar elimination by repression of the bed agglomeration (Calvo et al., 2012).

- *Pyrolysis*

Pyrolysis has been widely used for converting biomass into fuel gases, liquids, and solids alongside other thermochemical conversion processes. While this process is often used as an independent process to produce biofuels, it can also be used for just the initial stage of combustion and gasification of biomasses. Pyrolysis can be carried out through two approaches, slow pyrolysis and fast pyrolysis. Slow pyrolysis, often referred to as conventional or traditional pyrolysis, aims to maximize the yield of fuel gas through preferred conditions of high temperature, low heating rate (heating rate = ~10°C/m) and long gas resistance time, or to enhance the char production at the low temperature and low heating rate. Fast pyrolysis, also known as flash pyrolysis, aims to maximize the yield of liquid product through the processing conditions of (1) very high heating rate (heating rate = ~1000°C/s) and heat transfer rate, (2) finely ground biomass feed (<1 mm), (3) carefully controlled temperature, and (4) rapid cooling of the pyrolysis vapors to give the bio-crude products. Fast pyrolysis of biomass solid waste is at present considered as an emerging energy technology for liquid-tar and solid-char production (Zheng et al., 2018).

7 RICE BRAN PROCESSING

Rice bran, one of the most important by-products of rice processing, constitutes valuable components as previously mentioned in Section 5.3. However, its utilization is severely restricted by the activity of endogenous enzymes which can hydrolyze the triglyceride into glycerol and FFAs (Wang et al., 2017). Among the enzymes, lipase has merited most attention because it affects the quality and extent of industrial utilization of rice bran. Once rice is milled, bran is separated, and lipase is exposed to the oxygen in air, then the oxidation starts, resulting in hydrolyzing the oil present in bran into glycerol and FFAs. The FFA formation gives the product a rancid smell and bitter taste (Luh, 1980).

To process bran into a food-grade product of a good quality and high industrial value, the activity of the components causing deterioration must be arrested. FFAs concentration in rice bran should be kept at a low level (<10%) before oil extraction. It is important that enzyme inactivation must be complete and irreversible (Barber and Carmen, 1980). Therefore, stabilization of bran is carried out to inactivate the potent lipase enzyme and prevent the oxidative changes that affect the oil quality and ensure suitability for human consumption (Wang et al., 2017).

7.1 Stabilization

Many approaches, including heat, chemical, and extrusion treatments, for bran stabilization have been investigated and reported. The use of heat to stabilize rice bran has been investigated extensively. Rice bran stabilization depends on temperature, duration of heat treatment, MC, pH, and other parameters (Luh, 1980; Tao et al., 1993). Recommended conditions for dry and moist heating treatments for effective stabilization treatments differ dramatically and results are very often inconsistent. In addition to heat treatment, chemical methods (Prabhakar and Venkatesh, 1986), extrusion (Randall et al., 1985), and microwave treatments (Tao et al., 1993) have been studied. The most common method used in the rice industry is extrusion even though it is an energy intensive process. Additionally, air heating could be used, but it may not be able to heat the bran quickly to the needed temperature with desired uniformity (Kim et al., 1987; Pacheco de

Delahaye et al., 2005). The disadvantages in the mentioned methods are: (1) nonuniform and severe processing conditions that can damage valuable components of rice bran and result in high energy consumption, (2) complete and irreversible inactivation of enzymes may be not achieved, and (3) these stabilization methods are typically carried out right after the milling process, which creates operational difficulties for certain rice mills.

Infrared (IR) radiation heating was found to be distinctly different from the other conventional heating methods and has many advantages: (1) high energy efficiency and improved product quality; (2) infrared radiation penetrates directly into the product without heating the surroundings; (3) uniform heating of the product; (4) leveling of the moisture profiles in the product and low product deterioration; (5) ease of control; (6) inexpensive equipment, a long service life and low maintenance cost; (7) easily adapted to conventional dryers (Das et al., 2004; Sharma et al., 2005; Amaratunga et al., 2005; Pan and Atungulu, 2010; Khir and Pan, 2013). Consequently, IR heating has a promising potential to be used as an effective method to achieve an effective stabilization for rice bran.

Khir and Pan (2013) have reported that IR heating has potential to be used as an efficient, alternative method to conventional heating methods for achieving a high heating rate and effective stabilization of rice bran. They found that IR heating of rice bran to a temperature of 100°C followed by tempering treatment for 4 h has shown promising potential to completely inactivate lipase enzyme and reduce the FFA concentration to under 4% for a storage time of >4 weeks compared to 7 days for the control. Wang et al. (2017) found that the IR drying of rough rice to 60°C followed by tempering treatment for 4 or 5 h resulted in significant reduction in lipase activity, particularly for rice with high initial MC. They also reported that the storage time of stabilized rice bran before reaching FFA concentration of 10% could be extended to 38 days under two passes of rice drying compared to 7 days for the control. The results showed that IR heating can achieve simultaneous drying and effective stabilization for rice bran and provide an effective way for the utilization of rice bran without concerning the quality of rice bran oil. IR drying of rough rice by heating to a temperature of 60°C followed by tempering treatment and natural cooling can achieve a high moisture diffusivity and drying rate. Simultaneously, good milling and quality, effective disinfestation and disinfection, extended shelf life of both rough and brown, and effective stabilization of rice bran were obtained (Khir and Pan, 2013; Ding et al., 2015, 2016; Khir et al., 2011, 2014; Pan et al., 2008, 2011; Wang et al., 2014).

7.2 Oil Extraction

High-grade edible oil with high nutrition value can be extracted from rice bran. The most popular oil extraction methods are described in Tables 5 and 6. After extraction, the refining process is conducted to purify the crude oil for improving its taste, brightness, and color. The detailed description to the refining process is presented in Table 7.

Table 5 Extraction Methods for Rice Bran Oil

Extraction methods	Description	Advantages	Disadvantages	Reference
Enzymatic method	This method uses enzymes, such as cellulose and pectinase, to disrupt the cell walls of rice bran granules to increase the oil yield. Then solvents, such as hexane, are used to extract the oil. The oil yield is significantly affected by enzyme concentration	Oil yield can be increased by increase enzyme concentration	Cost is high	Johnson and Lusas (1983); Christensen (1991); Rosenthal et al. (1996)
Aqueous extraction	In this method, rice bran and water are mixed in a container to a required bran to water ratio. The mixture is continuously stirred for a specific time at a constant temperature and pH. Then the cream, supernatant, and solid phase (or defatted bran) are separated by using a centrifuging process. The separated cream is stored in a refrigerator. During processing time, the cream is boiled gently to break emulsion until oil appears. The aqueous phase (clear and oil free) is discarded, and the oil is decanted. Then the oil is kept in a hot air oven for 2 h at 105°C until constant weight is obtained	It is an environmental-friendly process	Because of the low solubility of oil in water, the efficiency of extraction is low	Amarasinghe et al. (2009)
Hexane extraction	Stabilized rice bran is mixed with n-hexane (food grade) directly at 20°C at 1:2 (w/w rice bran to solvent) ratio or pre-heated at 60°C at 1:3 (w/w rice bran to solvent) ratio in capped flasks, which are immersed in a constant-temperature water bath at 40°C or 60°C for predetermined time. Then vacuum evaporation using a solvent is conducted to produce crude rice bran oil	This method provides high oil extractability (98%) and it is easy to be operated	Oil extracted with this method has a high content of FFA, wax, and unsaponifiable matter and also shows a dark greenish brown to light yellow coloration. Additionally, it has health and environment concerns	Minsh et al. (1990); Krishna (1992); Yoon and Kim (1994); Amarasinghe and Gangodavilage (2004); Yadav and Joyner (2018)

Table 6 Extraction Methods for Rice Bran Oil

Extraction method	Description	Advantages	Disadvantages	Reference
Supercritical fluid extraction	This method uses fluids that possess good solubility properties of liquids as well as diffusion properties of gases. This results in improved lipid solubility and mass transfer due to high diffusion and low viscosities. The oil yield increased by increasing temperature at isobaric conditions. By using this method, the oil yield is approximately four times higher than the highest yield obtained by extraction with liquid organic solvents. Additionally, the odor and the flavor of extracted oil are superior to that obtained by n-hexane extraction	It has been recognized as a very environmental-friendly process.	It has limitation in term of high cost of equipment for extraction	Ramsay et al. (1991); Kim et al. (1999); Xu and Godber (2000); Dunford et al. (2003); Giuseppe et al. (2003); Bhosle and Subramanian (2005); Danielski et al. (2005)
Subcritical water extraction	This method is used to simultaneously inactivate lipase enzyme existing in rice bran and extract of its oil. It is carried out in the temperature range between 120°C and 240°C for residence time ranging between 10 and 20 min in a batch reactor	This method provides a green, simple, and nonflammable medium to inactivate the lipase enzyme and simultaneously extract oil for rice bran	Scaling-up of this method has not yet been described	Lerma-García et al. (2009); Pourali et al. (2009)
Microwave-assisted extraction	Microwave-assisted extraction method is used to rapidly extract oil from rice bran by using n-hexane, ethanol, or isopropanol as solvents. This system achieves a maximum extraction efficiency at 73°C and 21 min. It extracted >90% of recoverable oil from rice bran	It can be scaled up for industrial applications	It needs solvents	Zigoneanu et al. (2008); Terigar et al. (2011)
Ultrasonic-assisted extraction	The extraction of bioactive compounds under ultrasound irradiation (20–100 kHz) can achieve extraction yields ranged from 11% to 20.2%. Optimal ultrasonic-assisted extraction (UAE) conditions are identified as solvent percentage of 65%–67% ethanol, temperature of 51–54°C, and time of 40–45 min	It can offer high reproducibility in shorter times, simplified manipulation, reduced solvent consumption, temperature, and lower energy input	Limited scaling up feasibility	Cravotto et al. (2004); Chemat et al. (2008); Tabaraki and Nateghi (2011)

Table 7 Refining Steps for Rice Bran Oil

Refining step	Purpose	Description	Reference
Dewaxing	Because wax existed in extracted oil has high melting point and negatively affects the oil quality. Wax must be removed from the oil	Dewaxing can be conducted by filtering or centrifuging crude oil at low temperatures. Also, wax can be removed for bran oil by freezing out refined oil, followed by filtration	Chang et al. (1980)
Deacidification	As high FFA content is responsible for greater oil refining loss and the darker color of processed oil. Therefore, deacidification is conducted to neutralize the free fatty acids and transform them into soap	Definite amounts of dilute, aqueous, caustic solution and oil are combined in a mixing unit. Then the mixture is passed into a heating unit to be heated to temperature of 55–75°C. After heating, centrifugal process is applied to separate the soap stock from the neutral oil	Yoon and Kim (1994); De and Bhattacharyya (1998); Chang et al. (1980); Gopala Krishna et al. (2001)
Bleaching	This step is conducted to remove the greenish cast caused by the presence of chlorophyll. This step results in a product acceptable for consumption	Crude oil is treated with 5% saturated aqueous solution of organic acids such as oxalic or citric acids at room temperature for 30 min. Then the mixture is heated and allowed to settle down. The oily layer is washed with hot water and bleached with mixture of activated carbon at 110°C for 20 min. After cooling and filtering, a light color oil is obtained	Chang et al. (1980)
Deodorizing	This step is to remove the objectionable odors caused by peroxides, aldehydes, and ketones. Additionally, it provides an additional advantage of destroying carotenoid and some other pigments and reducing the free fatty acids	The deodorization is achieved by blowing live a steam through the heated oil at 220–250°C under vacuum (3–5 mmHg). Then the oil is filtered and cooled down under vacuum to about 50°C. The step produces oil which is extremely tasteless and odorless	Chang et al. (1980); Rajan and Krishna (2014)
Winterizing	This step is carried out to remove saturated glycerides which become soluble in the oil at medium and higher temperatures, but precipitate in cold conditions resulting in a turbid oil	This step involves holding the oil at temperature of 30–35°C for a determined period. Then the temperature is lowered slowly at a uniform rate to 15°C within about 12 h with agitation and then to 4–5°C to enable the higher melting components to form crystals. Then the oil is filtrated in a cool room	Chang et al. (1980); Strieder et al. (2017)

The main technical challenge is rapid deterioration of the oil in the bran after separation from the brown rice as mentioned earlier and the subsequent deterioration in oil quality after extraction. Oil extracted from stored rice bran contains a high percentage of FFAs which are difficult to be removed in refining, resulting in high refining losses.

It important to note that rice bran oil has been extensively studied for a long period of time. The advantages of rice bran oil, such as antioxidants, low cholesterol, and hot flash relief in menopause, are fueling the global rice bran oil market. Other factors that also drive the global rice bran oil market include its diversified applications, numerous cooking benefits compared to other edible oils, growing health consciousness, aggressive promotions by manufacturers, and increasing penetration levels in both developed and emerging markets (Tables 5–7).

8 CONCLUSIONS

The production and processing of rice are accompanied with the generation of large quantities of by-products, such as straw, husk, bran, and germ. The management and reprocessing of such by-products is a critical challenge to the rice industry. The disposal of such wastes can cause severe environmental issues. Understanding the value of these by-products and finding appropriate processing methods and technologies to convert them into value-added products are important for both eliminating environmental concerns and bring benefits to the sustainability of the rice industry. The current state of rice processing methods, value of rice by-products, and processing technologies converting these wastes to value-added products have been addressed in this chapter. Although many studies have been conducted on processing technologies dealing with the by-products, most of the technologies had difficulties for scaling up to industrial scale. The main reason is economic feasibility. Therefore, further research is needed to address the issues that limit the value-added utilizations of the by-products.

References

AACC (American Association for Clinical Chemistry), 2001. The definition of dietary fiber. Cereal Foods World 46, 112.

Abdul-Hamid, A., Luan, Y.S., 2000. Functional properties of dietary fibre prepared from defatted rice bran. Food Chem. 68, 15–19.

Ahmaruzzaman, M., Gupta, V.K., 2011. Rice husk and its ash as low-cost adsorbents in water and wastewater treatment. Ind. Eng. Chem. Res. 50, 13589–13613.

Ahuja, S.C., Uma, A., 2018. Non-food applications of multiple components in rice. In: Pan, Z., Khir, R. (Eds.), Advances in Science & Engineering of Rice. DEStech Publications, Inc, Pennsylvania, USA, pp. 491–586.

Amarasinghe, B.M.W.P.K., Gangodavilage, N.C., 2004. Rice bran oil extraction in Sri Lanka: data for process equipment design. Food Bioprod. Process. 82 (C1), 54–59.

Amarasinghe, B.M.W.P.K., Kumarasiri, M.P.M., Gangodavilage, N.C., 2009. Effect of method of stabilization on aqueous extraction of rice bran oil. Food Bioprod. Process. 87, 108–114.

Amaratunga, K., Pan, Z., Zheng, X., Thompson, J.F., 2005. Comparison of drying characteristics and quality of rough rice dried with infrared and heated air. In: American Society of Agricultural Engineers Meetings Papers, St. Joseph, MI, USA, pp. 1–10. ASAE Paper No. 056005.

Anonymous, 1955. Bricks from rice hull ash. Ind. Eng. Chem. 47 (10), 11A–12A.

Ansharullah, J.A.H., Chesterman, C.F., 1997. Application of carbohydrases in extracting protein from rice bran. J. Sci. Food Agric. 74, 141–146.

Atanu, B., Badal, C.S., John, W.L., Shogren, R.L., Willett, J.L., 2006. Process for obtaining cellulose acetate from agricultural by-products. Carbohydr. Polym. 64, 134–137.

Bak, J.S., Ko, J.K., Han, Y.H., Lee, B.C., Choi, I.-G., Kim, K.H., 2009. Improved enzymatic hydrolysis yield of rice straw using electron beam irradiation pretreatment. Bioresour. Technol. 100, 1285–1290.

Bakker, R.R., Jenkins, B.M., 2003. Feasibility of collecting naturally leached rice straw for thermal conversion. Biomass Bioenergy 25, 597–614.

Bakker, R.R., Jenkins, B.M., Williams, R.B., 2002. Fluidized bed combustion of leached rice straw. Energy Fuel 16, 356–365.

Balasubramaniam, M.K., Rajarathinam, R., 2013. Implementation of white rot fungal pretreated rice straw for sustainable biotechnology production by Saccharomyces cerevisiae. Int. J. Eng. Res. Technol. 2, 4047–4053.

Bao, J., 2012. Nutraceutical properties and health benefits of rice. In: Liangli, Y., Tsao, R. (Eds.), Cereals and Pulses: Nutraceutical Properties and Health Benefits. John Wiley and Sons, Ltd., Publication, Iowa, USA, pp. 37–57.

Barber, S., Carmen, B.D.B., 1980. Rice bran: chemistry and technology. In: Luh, B.S. (Ed.), Rice: Production & Utilization. AVI Publishing Company, Inc, Westport, Connecticut, pp. 790–870.

Bera, M.B., Mukherjee, R.K., 1989. Solubility, emulsifying and foaming properties of rice bran protein concentrates. J. Food Sci. 54, 142–143.

Bhosle, B.M., Subramanian, R., 2005. New approaches in deacidification of edible oils—a review. J. Food Eng. 69, 481–494.

Binod, P., Raveendran, S., Reeta, R.S., Surender, V., Lalitha, D., Satya, N., Noble, K., Rajeev, K.S., Ashok, P., 2010. Bioethanol production from rice straw: an overview. Bioresour. Technol. 101, 4767–4774.

California Department of Resources Recycling and Recovery, 2014. Public Affairs Office, Sacramento, CA. www.calrecycle.ca.gov/Publications/1-800.

Calvo, L.F., Gil, M.V., Otero, M., Morán, A., García, A.I., 2012. Gasification of rice straw in a fluidized-bed gasifier for syngas application in close-coupled boiler-gasifier systems. Bioresour. Technol. 109, 206–214.

Carl, W.H., 1980. Drying and Storage of Agricultural Crops. AVI Publishing Company, JNC, Westport, CT, USA.

Chadha, B.S., Kanwar, S.S., Garcha, H.S., 1995. Simultaneous saccharification and fermentation of rice straw into ethanol. Acta Microbiol. Immunol. Hung. 42, 71–75.

Champagne, E.T., Bett-Garber, K.L., Vinyard, B.T., McClung, A.M., Barton, I.I.F.E., Moldenhauer, K., Linscombe, S., McKenzie, K., 1999. Correlation between cooked rice texture and rapid visco analyzer measurements. Cereal Chem. 76, 764–771.

Chandler Jr., R.F., 1979. Rice in the Tropics: A Guide to the Development of National Programs. IADS Development Oriented Literature Series, Westview Press, Inc, USA.

Chandrasekar, S., Satyanarayana, K.G., Pramada, P.N., Ragahavan, P., 2003. Processing, properties and applications of reactive silica from rice husk—an overview. J. Mater. Sci. 38, 3159–3168.

Chandrasekhar, S., Pramada, P., Majeed, J., 2006. Effect of calcination temperature and heating rate on the optical properties and reactivity of rice husk ash. J. Mater. Sci. 41, 7926–7933.

Chang, S.C., Saunders, R.M., Luh, B.S., 1980. Rice oil: chemistry and technology. In: Luh, B.S. (Ed.), Rice: Production and Utilization. AVI Publising Company, INC, Westport, CT, USA, pp. 764–789.

Chastril, J., 1990. Chemical and physicochemical of rice during storage at different temperatures. J. Cereal Chem., 71–84.

Cheftel, J.C., Cuq, J.L., Lorient, D., 1985. In: Fennema, O.R. (Ed.), Food Chemistry. Marcel Dekker, New York.

Chemat, F., Tomao, V., Virot, M., 2008. Ultrasound-assisted extraction in food analysis. In: Ötles, S. (Ed.), Handbook of Food Analysis Instruments. CRC Press, Boca Raton, FL, pp. 85–103.

Chen, X., Yuc, J., Zhang, Z., Lu, C., 2011. Study on structure and thermal stability properties of cellulose fibers from rice straw. Carbohydr. Polym. 85, 245–250.

Christensen, F.M., 1991. Extraction by aqueous enzymatic processes. Inform 2, 984–987.

Chuah, T.G., Jumasiah, A., Azni, I., Katayon, S., Choong, S.Y.T., 2005. Rice husk as a potentially low-cost biosorbent for heavy metal and due removal: an overview. Desalination 175, 305–316.

Chuetor, S., Luque, R., Barron, C., Solhy, A., Rouau, X., Barakat, A., 2015. Innovative combined dry fractionation technologies for rice straw valorization to biofuels. Green Chem. 17, 926–936.

Cordeiro, G.C., Filho, R.D.T., Fairbairn, E.M.R., 2013. Use of ultrafine rice husk ash with high-carbon content as pozzolan in high performance concrete. Mater. Struct. 42 (7), 983–992.

Cravotto, G., Binello, A., Merizzi, G., Avogadro, M., 2004. Improving solvent-free extraction of policosanol from rice bran by high-intensity ultrasound treatment. Eur. J. Lipid Sci. Technol. 106, 147–151.

Danielski, L., Zetzl, C., Hense, H., Brunner, G., 2005. A process line for the production of raffinated rice oil from rice bran. J. Supercrit. Fluids 34, 133–141.

Das, I., Das, S.K., Bal, S., 2004. Drying performance of a batch type vibration aided infrared dryer. J. Food Eng. 64, 129–133.

Dass, A., 1984. Pozzolanic behaviour of rice husk ash. Batiment Int. Build. Res. Pract. 12 (5), 307–311.

De, B.K., Bhattacharyya, D.K., 1998. Physical refining of rice bran oil in relation to degumming & dewaxing. JAOCS 75, 1683–1686.

deBari, I., Cuna, D., Nanna, F., Braccio, G., 2004. Ethanol production in immobilized cell bioreactors from mixed sugar syrups and enzymatic hydrolysates of steam-exploded biomass. Appl. Biochem. Biotechnol. 114, 539–557.

Delgenes, J.P., Laplace, J.M., Moletta, R., Navarro, J.M., 1996. Comparative study of separated fermentations and cofermentation processes to produce ethanol from hardwood derived hydrolysates. Biomass Bioenergy 11, 353–360.

Desikachar, H.S.R., 1977. Preservation of Byproducts of Rice Milling. In: Proceeding of the Rice by-Products Utilization, Valencia, Spain.

Devendra, C., Thomas, D., 2002. Crop-animal interactions in mixed farming systems in Asia. Agric. Syst. 71, 27–40.

Ding, D., Khir, R., Pan, Z., Zhao, L., Tu, K., El-Mashad, H., McHugh, T.H., 2015. Improvement in shelf life of rough and brown rice using infrared radiation heating. Food Bioprocess Technol. 8, 1149–1159.

Ding, D., Khir, R., Pan, Z., Wood, D.F., Tu, K., El-Mashad, H., Berrios, J., 2016. Improvement in storage stability of infrared-dried rough rice. Food Bioprocess Technol. https://doi.org/10.1007/s11947-016-1690-5.

Dobermann, A., Fairhurst, T., 2002. Rice straw management. Better Crops Int. 16 (Special Supplement), 7–11.

DOE, 2015. Alternative Fuels Data Center. http://www.afdc.energy.gov/fuels/ethanol_feedstocks.html.

Dommershuijzen, J.J., Lozare, J.V., 1990. Rice by-Product Utilization in Selected Countries in Asia. Regional Network for Intercountry Cooperation on Postharvest Technology and Quality Control of Food Grains, p. 120.

Drake, D.J., Nader, G., Forero, L., 2002. Feeding Rice Straw to Cattle. http://anrcatalog.ucdavis.edu/pdf/8079.pdf.

Drapcho, C.M., Nghim, N.P., Walker, T., 2008. Biofuels Engineering Process Technology. McGraw-Hill.

Duan, F., Chyang, C.S., Zhang, L.H., Yin, S.F., 2015. Bed agglomeration characteristics of rice straw combustion in a vortexing fluidized-bed combustor. Bioresour. Technol. 183, 195–202.

Dunford, N.T., Teel, J.A., King, J.W., 2003. A continuous countercurrent supercritical fluid deacidification process for phytosterol ester fortification in rice bran oil. Food Res. Int. 36, 175–181.

Ebeling, J., Jenkins, B., 1985. Physical and chemical properties of biomass fuels. Trans. ASAE 28, 898–902.

Ebeling, T., Paillet, M., Borsali, R., Diat, O., Dufresne, A., Cavaille, J.Y., et al., 1999. Shear-induced orientation phenomena in suspensions of cellulose microcrystals, revealed by small angle X-ray scattering. Langmuir 15, 6123–6126.

El-Shafey, E.I., 2007. Sorption of cd (II) and se (IV) from aqueous solution using modified rice husk. J. Hazard. Mater. 147, 546–555.

Esa, N.M., Ling, T.B., Peng, L.S., 2013. By-products of rice processing: an overview of health benefits and applications. J. Rice Res. 1, 107. https://doi.org/10.4172/jrr.1000107.

Fabiyi, J.S., 2013. Suitability of Portland cement and rice husk ash pozzolan systems for cement bonded composites production. J. Mater. Environ. Sci. 4 (6), 848–854.

Farrell, D.J., 1994. Utilization of rice bran in diets for domestic fowl and ducklings. World's Poult. Sci. J. 50 (2), 115–131.

Food and Agriculture Organization (FAO). 2016. FAOSTAT Data. http://www.fao.org./faostat/en/#data/QC.

Food and Agriculture Organization of United Nation (FAO), Rice Market Monitor, April 2018. Vol. XXI, Issue No. 1. http://ww.fao.org/economic/RMM.

Frei, M., Siddhuraju, P., Becker, K., 2003. Studies on the in vitro starch digestibility and the glycemic index of six different indigenous rice cultivars from the Philippines. Food Chem. 83, 395–402.

Fu, N., Peiris, P., 2008. Co-fermentation of a mixture of glucose and xylose to ethanol by Zymomonas mobilis and Pachysolen tannophilus. World J. Microbiol. Biotechnol. 24, 1091–1097.

Gadde, B., Menke, C., Siemers, W., Pipatmanomai, S., 2008. Technologies for energy use of rice straw: a review. Int. Rice Res. Notes 33.

Garas, G.L., Allam Hala, M.E., Kady, M.E.L., El Alfy, A.H., 2009. Compressibility of single un-rendered rice straw bales: characteristics of bales used for building. ARPN J. Eng. Appl. Sci. 4 (5), 1819–1830.

Garcia, M.C., Benassi, M.D.T., Soares Júnior, M.S., 2012. Physicochemical and sensory profile of rice bran roasted in microwave. Food Sci. Technol. (Campinas) 32 (4), 754–761.

Garrote, G., Dominguez, H., Parajo, J.C., 2002. Autohydrolysis of corncob: study of non-isothermal operation for xylooligosaccharide production. J. Food Eng. 52, 211–218.

Ghosh, R., Bhattacherjee, S., 2013. A review study on precipitated silica and activated carbon from rice husk. J. Chem. Eng. Process Technol. 4, 4–7.

Giuseppe, P., Miniati, E., Montanari, L., Fantozzi, P., 2003. Improving the value of rice by-products by SEF. J. Supercrit. Fluids 26, 63–71.

Gnanasambandam, R., Hettiarachchy, N.S., 1995. Protein concentrates from unstabilized and stabilized rice bran: preparation and properties. J. Food Sci. 60, 1066–1069. 1074.

Goffman, F.D., Bergman, C.J., 2004. Rice kernel phenolic content and its relationship with antiradical efficiency. J. Sci. Food Agric. 84, 1235–1240.

Golias, H., Dumsday, G.J., Stanley, G.A., Pamment, N.B., 2002. Evaluation of a recombinant Klebsiella oxytoca strain for ethanol production from cellulose by simultaneous accharification and fermentation: comparison with native cellobiose-utilising yeast strains and performance in co-culture with thermotolerant yeast and Z. mobilis. J. Biotechnol. 96, 155–168.

Gómez-Barea, A., Vilches, L.F., Leiva, C., Campoy, M., Fernández-Pereira, C.P., 2009. Optimization and ash recycling in fluidised bed waste gasification. Chem. Eng. J. 146, 227–236.

Gopala Krishna, A.G., Khatoon, S., Sheila, P.M., Sarmandal, C.V., Indira, T.N., et al., 2001. Effect of refining crude rice bran oil on the retention of oryzanol in the refined oil. JAOCS 78, 127–131.

Gousse, C., Chanzy, H., Cerrada, M.L., Fleury, E., 2004. Surface silylation of cellulose microfibrils: preparation and rheological properties. Polymer 45, 1569–1575.

Grootjen, D.R.J., Meijlink, L.H.H.M., Vleesenbeek, R., Lans, R.G.J.M.V.D., Luyben, K.C.A.M., 1991. Cofermentaion of glucose and xylose with immobilized Pichia stipitis in combination with Saccharomyces cerevisiae. Enzym. Microb. Technol. 13, 530–536.

Gu, Y., Chen, X., Liu, Z., Zhou, X., Zhang, Y., 2014. Effect of inoculum sources on the anaerobic digestion of rice straw. Bioresour. Technol. 158, 149–155.

Gummert, M., Quilty, J., Van Hung, N., Vial, L., 2018. Engineering and management of rice harvesting. In: Pan, Z., Khir, R. (Eds.), Advances in Science & Engineering of Rice. DEStech Publications, Inc, Pennsylvania, USA, pp. 67–102.

Guo, Y., Yu, K., Wang, Z., Xu, H., 2003. Effects of activation conditions on preparation of porous carbon from rice husk. Carbon 41, 1645–1648.

Halim, A., 1985. Rice and politics in Bangladesh. J. East Asian Stud. 24, 57–65.

Hegsted, M., Windhauser, M., Lester, S., 1990. Stabilized rice bran and oat bran lower cholesterol in humans. FASEB J. 4, A368.

Heinemann, R.J.B., Fagundes, P.L., Pinto, E.A., Penteado, M.V.C., Lanfer-Marquez, U.M., 2005. Comparative study of nutrient composition of commercial brown, parboiled and milled rice from Brazil. J. Food Compos. Anal. 18, 287–296.

Helm, R.M., Burks, A.W., 1996. Hypoallergenicity of rice protein. Cereal Foods World 41, 839–843.

Helmy, M.A., Radwan, S.M., El-Kholy, M., 1995. Effect of different paddy drying methods of rice milling quality. MISR J. Agric. Eng. 12 (2), 496–509.

Hernandez, N., Rodriguez-Alegría, M.E., Gonzalez, F., Lopez-Munguia, A., 2000. Enzymatic treatment of rice bran to improve processing. JAOCS 77, 177–180.

Hernández, J.J., Aranda-Almansa, G., Bula, A., 2010. Gasification of biomass wastes in an entrained flow gasifier: effect of the particle size and the residence time. Fuel Process. Technol. 91, 681–692.

Houston, D.F., 1972. Rice hulls. 301-352. In: Houston, D.F. (Ed.), Rice Chemistry and Technology, first ed. American Association of Cereal Chemists, St. Paul, Minnesota, USA.

Hsu, T., Guo, G., Chen, W., Hwang, W., 2010. Effect of dilute acid pretreatment of rice straw on structural properties and enzymatic hydrolysis. Bioresour. Technol. 101, 4907–4913.

Jakata, D.N., 1981. Rice straw pulping and bleaching in small paper mills in India. In: UNIDO International Experts Group Meeting Pulp and Paper Technology, Manila, Philippines, 1980. United Nations Industrial Development Organization, Vienna, p. 10.

Jenkins, B.M., Ebeling, J.M., 1985. Thermochemical properties of biomass fuels. Calif. Agric. 39, 14–16.

Jenkins, B.M., Baxter, L.L., Miles Jr., T.R., Miles, T.R., 1998. Combustion properties of biomass. Fuel Process. Technol. 54, 17–46.

Jiamyangyuen, S., Srijesdaruk, V., Harper, W.J., 2005. Extraction of rice bran protein concentrate and its application in bread. Songklanakarin J. Sci. Technol. 27 (1), 55–64.

Johnson, L.A., Lusas, E.W., 1983. Comparison of alternative solvents for oil extraction. J. Am. Oil Chem. Soc. 60, 229–242.

Jones, F.T., 1975. Exploring rice straw by scanning electron microscope. Microscope 23, 37–46.

Juliano, B.O., 1985. Grain structure, composition and consumers. In: Rice in Human Nutrition. International Rice Research Institute and Food and Agriculture Organization of the United Nations, Rome, Italy.

Juliano, B.O., Bechtel, D.B., 1985. The rice grain and its gross composition. In: Juliano, B.O. (Ed.), Rice, Chemistry and Technology, second ed. AACC, St. Paul, MN, USA, pp. 17–57.

Kahlon, T.S., Saunders, R.M., Sayre, R.M., Chow, F., Chiu, M., Betschart, A., 1992. Cholesterol lowering effects of rice bran and rice bran oil fractions in hypercholesrolemic hamsters. Cereal Chem. 69, 485–489.

Kalapathy, U., Proctor, A., Shultz, J., 2000. A simple method for production of pure silica from rice hull ash. Bioresour. Technol. 73 (3), 257–262.

Karimi, K., Emtiazi, G., Taherzadeh, M.J., 2006. Ethanol production from dilute acid pretreated rice straw by simultaneous saccharification and fermentation with Mucor indicus, Rhizopus oryzae, and Saccharomyces cerevisiae. Enzyme Microbial Technol. 40, 138–144.

Kartini, K., 2011. Rice husk ash—pozzolanic material for sustainability. Int. J. Appl. Sci. Technol. 1 (6), 169–178.

Khir, R., Pan, Z., 2013. Development of New Techniques for Improved Shelf Life of Rough and Brown Rice and Stabilization of Rice Bran. Research Progress Report. California Rice Research Board, Yuba City, CA. Available at: http://carrb.com/AnnualRpts.htm.

Khir, R., Pan, Z., Salim, A.d., Hartsough, B.R., Mohamed, S., 2011. Moisture diffusivity of rough rice under infrared radiation drying. LWT-Food Sci. Technol. 44 (4), 1126–1132.

Khir, R., Pan, Z., Thompson, J.F., El-Sayed, A.S., Hartsough, B.R., El-Amir, M.S., 2014. Moisture removal characteristics of thin layer rough rice under sequenced infrared radiation heating and cooling. J. Food Process. Preserv. 38 (1), 430–440.

Khir, R., Venkitasamy, C., Pan, Z., 2017. Infrared heating for improved drying efficiency, food safety and quality of rice. In: Shi, J., Liu, D. (Eds.), Food Physical Processing Science Advances and Prospects. Zhejiang University Press, Hangzhou, China, pp. 167–192.

Khir, R., Venkitasamy, C., Pan, Z., 2018. Rice fotifications. In: Pan, Z., Khir, R. (Eds.), Advances in Science & Engineering of Rice. DEStech Publications, Inc, Pennsylvania, USA, pp. 391–413.

Kim, I., Han, J., 2012. Optimization of alkaline pretreatment conditions for enhancing glucose yield of rice straw by response surface methodology. Biomass Bioenergy 46, 210–217.

Kim, C.J., Byun, S.M., Cheigh, H.S., Kwon, T.W., 1987. Comparison of solvent extraction characteristics of rice bran pretreated by hot air drying, steam cooking and extrusion. J. Am. Oil Chem. Soc. 64 (4), 514–516.

Kim, H.J., Lee, S.B., Park, K.A., Hong, I.K., 1999. Characterization of extraction and separation of rice bran oil rich in EFA using SFE process. Sep. Purif. Technol. 15, 1–8.

Kim, D.J., Oh, S.K., Chun, A., Yoon, M.R., Hong, H.C., 2011. Evaluation of biological activities of rice husk extracts. J. Food Sci. Nutr. 16, 179–183.

Kitta, K., Ebihara, M., Iizuka, T., Yoshikawa, R., Isshiki, K., Kawamoto, S., 2005. Variations in lipid content and fatty acid composition of major non-glutinous rice cultivars in Japan. J. Food Compos. Anal. 18, 269–278.

Ko, J.K., Bak, J.S., Jung, M.W., Lee, H.J., Choi, I., Kim, T.H., Kim, K.H., 2009. Ethanol production from rice straw using optimized aqueous-ammonia soaking pretreatment and simultaneous saccharification and fermentation processes. Bioresour. Technol. 100, 4374–4380.

Konishi, J., Fukuda, A., Mutaguchi, K., 2015. Xylose fermentation by Saccharomyces cerevisiae using endogenous xylose-assimilating genes. Biotechnol. Lett. 37 (8), 1623–1630.

Koopmans, A., Koppejan, J., 1997. Agricultural and forest residues-generation, utilization and availability. In: Regional Consultation on Modern Applications of Biomass Energy.

Kordowska-wiater, M., Targonski, Z., 2002. Ethanol fermentation on glucose/xylose mixture by co-cultivation of restricted glucose catabolite repressed mutants of Pichia stipitis with respiratory deficient mutants of Saccharomyces cerevisiae. Acta Microbiol. Pol. 51, 345–352.

Krishna, A.G.G., 1992. A method of bleaching rice bran oil with silica gel. JACOS 69, 1257–1259.

Krishnarao, R.V., Godkhindi, M.M., Mukunda, P.G., 2001. Formation of SiCwhiskers from compacts of raw husks. J. Mater. Sci. 29, 2741–2744.

Kshirsagar, S.D., Waghmare, P.R., Loni, P.C., Patil, S., Govindwar, S., 2015. Dilute acid pretreatment of rice straw, structural characterization and optimization of enzymatic hydrolysis conditions by response surface methodology. RSC Adv. https://doi.org/10.1039/C5RA04430H.

Kuan, C.Y., Yuen, K.H., 2012. Physical, chemical and physicochemical characterization of rice husk. Br. Food J. 114 (6), 853–867.

Kumar, P.S., Ramakrishnan, K., Kirupha, S.D., Sivanesan, S., 2010. Thermodynamic and kinetic studies of cadmium adsorption from aqueous solution onto rice husk. Braz. J. Chem. Eng. 27 (2), 347–355.

Kumar, A., Mohanta, K., Kumar, D., Parkash, O., 2012. Properties and industrial applications of rice husk: a review. Int. J. Emerg. Technol. Adv. Eng. 2, 86–90.

Kumar, S., Sangwan, P., Dhankhar, R., Mor, V., Bidra, S., 2013. Utilization of rice husk and their ash: a review. Res. J. Chem. Environ. Sci. 1 (5), 126–129.

Lakshmi, U.R., Srivastava, V.C., Mall, I.D., Lataye, D.H., 2009. Rice husk ash as an effective adsorbent: evaluation of adsorptive characteristics for indigo carmine dye. J. Environ. Manag. 90 (2), 710–720.

Laplace, J.M., Delgenes, J.P., Moletta, R., Navarro, J.M., 1993. Ethanol production from glucose and xylose by separated and co-culture processes using high cell density systems. Process Biochem. 28, 519–525.

Lerma-García, M.J., Herrero-Martínez, J.M., Simó-Alfonso, E.F., Carla, R.B., Ramis-Ramos, G., 2009. Composition, industrial processing and applications of rice bran γ-oryzanol. Food Chem. 115, 389–404.

Lew, E.J.L., Houston, D.F., Fellers, D.A., 1975. A note on protein concentrate from full fat rice bran. Cereal Chem. 52, 748.

Liu, Z., Xu, A., Zhao, T., 2011. Energy from combustion of rice straw: status and challenges to China. Energy Power Eng. 3, 325–331.

Luh, B.S., 1980. Rice: Production and Utilization. AVI Publishing Company, INC, Westport, CT.

Maiorella, B.I., 1983. Ethanol industrial chemicals. Biochem. Fuels, 861–914.

Mansaray, K.G., Ghaly, A.E., 1997. Physical and thermochemical properties of rice husk. Energy Sources 19 (9), 989–1004.

Matsumoto, K., Keiji, T., Toshimitsu, L., Tomoko, O., Masakazu, N., 2009. Gasification reaction kinetics on biomass char obtained as a by-product of gasification in an entrained-flow gasifier with steam and oxygen at 900–1000°C. Fuel 88 (3), 519–527.

Mhalaskar, S.R., Thorat, S.S., Deshmukh, Y.R., 2017. Broken rice—a novel substrate for the production of food bio-colours through solid state fermentation. Int. J. Pure Appl. Biosci. 5 (2), 467–478.

Minsh, S.K., Kakar, A., Surnua, P.S., 1990. Effect of different degumming agents on the physico chemical characteristic of rice bran oil. J. Food Sci. Technol. 32, 280–283.

Mishra, P., Chakraverty, A., Banerjee, H.D., 1986. Studies on physical and thermal properties of rice husk related to its industrial application. J. Mater. Sci. 21 (6), 2129–2132.

Mishra, S.P., Tiwari, D., Dubey, R.S., 1997. The uptake behaviour of rice (Jaya) husk in the removal of Zn (II) ions—a radiotracer study. Appl. Radiat. Isot. 48, 877–882.

Mohanty, A.K., Drzal, L.T., Askeland, P., Misra, M., 2004. Effects of alkali treatment on the structure, morphology and thermal properties of native grass fibers as reinforcements for polymer matrix composites. J. Mater. Sci. 39, 1051–1054.

Mukhopadhyay, S., Siebenmorgen, T.J., 2017. Physical and functional 388 characteristics of broken rice kernels created by rapid moisture adsorption. Cereal Chem. 94 (3), 539–545.

Nader, G.A., Robinson, P.H., 2010. Rice Producer's Guide to Marketing Rice Straw. http://anrcatalog.ucdavis.edu/pdf/8425.pdf.

Natarajan, E., Nordin, A., Rao, A.N., 1998. Overview of combustion and gasification of rice husk in fluidized bed reactors. Biomass Bioenergy 14, 533–546.

Orts, W.J., Shey, J., Imam, S.H., Glenn, G.M., Guttman, M.E., Revol, J.F., 2005. Application of cellulose microfibrils in polymer nanocomposites. J. Polym. Environ. 13, 301–306.

Otterburn, M.S., 1989. In: Phillips, R.D., Finley, J.W. (Eds.), Protein Quality and the Effects of Processing. New York, Marcel Dekker.

Pacheco de Delahaye, E., Jimenez, P., Perez, E., 2005. Effect of enrichment with high content dietary fiber stabilized rice bran flour on chemical and functional properties of storage frozen pizzas. J. Food Eng. 8 (1), 1–7.

Pan, Z., Atungulu, G.G., 2010. Infrared dry blanching. In: Pan, Z., Atungulu, G.G. (Eds.), Infrared Heating for Food and Agricultural Processing. CRC Press, Florida, USA, pp. 169–201.

Pan, Z., Khir, R., Godfrey, L.D., Lewis, R., Thompson, J.F., Salim, A., 2008. Feasibility of simultaneous rough rice drying and disinfestations by infrared radiation heating and rice milling quality. J. Food Eng. 84 (3), 469–479.

Pan, Z., Khir, R., Bett-Garber, K.L., Champagne, E.T., Thompson, J.F., Salim, A., 2011. Drying characteristics and quality of rough rice under infrared radiation heating. Trans. ASABE 54 (1), 203–210.

Park, Y.-K., Jeon, J.-K., Kim, S., Kim, J.-S., 2004. Bio-oil from rice straw by pyrolysis using fluidized bed and char removal system. Prepr. Pap. Am. Chem. Soc. Div. Fuel Chem. 49, 800–801.

Park, J., Lee, Y., Ryu, C., Park, Y.-K., 2014. Slow pyrolysis of rice straw: analysis of products properties, carbon and energy yields. Bioresour. Technol. 155, 63–70.

Patel, M., Naik, S.N., 2004. Gamma-orzanol from rice brain oil—a review. J. Sci. Ind. Res. 63, 569–578.

Pathak, H., Singh, R., Bhatia, A., Jain, N., 2006. Recycling of rice straw to improve wheat yield and soil fertility and reduce atmospheric pollution. Paddy Water Environ. 4, 111–117.

Ponnusami, V., Krithika, V., Madhuram, R., Srivastava, S.N., 2006. Biosorption of reactive dye using acid-treated rice husk: factorial design analysis. J. Hazard. Mater. 142 (12), 397–403.

Pourali, O., Salak, A.F., Yoshida, H., 2009. A rapid and ecofriendly treatment technique for rice bran oil stabilization and extraction under sub-critical water condition. In: Proceedings of the World Congress on Engineering and Computer Science (WCECS), San Francisco, USA. vol. 1.

Prabhakar, J.V., Venkatesh, K.V.L., 1986. A simple chemical method for stabilization of rice bran. Chem. Mater. Sci. 63 (5), 644–646.

Puglia, D., Tomassucci, A., Kenny, J.M., 2003. Processing, properties and stability of biodegradable composites based on mater-bi-(R) and cellulose fibres. Polym. Adv. Technol. 14, 749–756.

Qian, J.Y., Gu, Y.P., Jiang, W., Chen, W., 2014. Inactivating effect of pulsed electric field on lipase in brown rice. Innov. Food Sci. Emerg. Technol. 22, 89–94.

Quayle W.C. 2016. Alternative Management of Rice Straw—A Position Paper for the Rice Industry. RIRDC Publication No.16/008 Project No. PRJ-009170.

Rajan, R.G., Krishna, G.A.G., 2014. A simple method for purification of deodorizer distillate from Indian rice (Oryza Sativa) bran oil and preparation of phytosterols. Grasas Aceites 65 (4), 1–7.

Ramsay, M.E., Hsu, J.T., Novak, R.A., Reightler, W.J., 1991. Processing rice bran by supercritical fluid extraction. Food Technol. 30, 98–104.

Randall, J.M., Sayre, R.N., Schultz, W.G., Fong, R.Y., Mossman, A.P., Tribelhorn, R.E., 1985. Rice bran stabilization by extrusion cooking for extraction of edible oil. J. Food Sci. 50, 361–364.

Ranjhan, S.K., 1990. Agro Industrial By-Products and Non-Conventional Feeds for Livestock Feeding. Indian Council of Agricultural Research, New Delhi, India, p. 127.

Ravoof, S.A., Prateepa, K., Supassri, T., Chittibabu, S., 2012. Enhancing enzymatic hydrolysis of rice straw using microwave-assisted nitric acid pretreatment. Int. J. Med. Biosci. 1, 13–17.

Rodriguez, A., Sanchez, R., Requejo, A., Ferrer, A., 2010. Feasibility of rice straw as a raw material for the production of soda cellulose pulp. J. Clean. Prod. 18, 1084–1091.

Rosenthal, A., Pyle, D.L., Niranjan, K., 1996. Aqueous & enzymatic processes for edible oil extraction. Enzym. Microb. Technol. 19, 402–420.

Rosniyana, A., Hashifah, M.A., Norin, S.S., 2007. The physico-chemical properties and nutritional composition of rice bran produced at different milling degrees of rice. J. Tropical Agric. Food Sci. 35 (1), 99.

Ryu, C., Yang, Y.B., Khor, A., Yates, N.E., Sharifi, V.N., Swithenbank, J., 2006. Effect offuel properties on biomass combustion: part I. Experiments—fuel type, equivalence ratio and particle size. Fuel 85, 1039–1046.

Saha, B.C., 2003. Hemicellulose bioconversion. Ind. Microbiol. Biotechnol. 30, 279–291.

Saha, B.C., Cotta, M.A., 2011. Continuous ethanol production from wheat straw hydrolysate by recombinant ethanologenic Escherichia coli strain FBR5. Appl. Microbiol. Biotechnol. 90 (2), 477–487.

Sahu, J.N., Agarwal, S., Meikap, B.C., Biswas, M.N., 2009. Performance of a modified multi-stage bubble column reactor for lead (II) and biological oxygen demand removal from wastewater using activated rice husk. J. Hazard. Mater. 161 (1), 317–324.

Said, N., Bishara, T., Garcia-Maraver, A., Zamorano, M., 2013. Effect of water washing on the thermal behavior of rice straw. Waste Manag. 33, 2250–2256.

Sakurai, J., 1977. Utilization of rice by-product in Japan. In: Barber, S., Tortosa, (Eds.), Rice Bran Utilization: Food and Feed. Rice By-Product Utilization International Conference, Valencia, Spain, 1974. In: vol. 4. Inst. Agroquim. Technol. Aliment., Valencia, Spain, pp. 101–113.

Samir, M.A.S.A., Alloin, F., Sanchez, J.Y., Dufresne, A., 2004. Cross-linked nanocomposite polymer electrolytes reinforced with cellulose whiskers. Macromolecules 37, 4839–4844.

Sangnark, A., Noomhorm, A., 2004. Chemical, physical and baking properties of dietary fiber prepared from rice straw. Food Res. Int. 37, 66–74.

Saraswathy, V., Song, H.W., 2007. Corrosion performance of rice husk ash blended concrete. Constr. Build. Mater. 21, 1779–1784.

Saunders, R.M., 1986. Rice bran: composition and potential food uses. Food Rev. Int. 1 (3), 465–495.

Saunders, R.M., 1990. The properties of rice bran as a food stuff. Cereal Foods World 35, 632–636.

SciDev Net, 2011. http://www.scidev.net/global/climate-change/news/egyptiantech-turns-rice-straw-into-paper-insecticide.html.

Sereewatthanawut, I., Prapintip, S., Watchiraruji, K., Goto, M., Sasaki, M., Shotipruk, A., 2008. Extraction of protein and amino acids from deoiled rice bran by subcritical water hydrolysis. Bioresour. Technol. 99, 55–561.

Sharma, G.P., Verma, R.C., Pathare, P.B., 2005. Thin-layer infrared radiation drying of onion slices. J. Food Eng. 67, 361–366.

Sheikh, M.M.I., Kim, C.-H., Park, H.-J., Kim, S.H., Kim, G.-C., Lee, J.-Y., Sim, S.W., Kim, J.W., 2013. Effect of torrefaction for the pretreatment of rice straw for ethanol production. J. Sci. Food Agric. 93, 3198–3204.

Shie, J.L., Chang, C.-Y., Chen, C.-S., Shaw, D.-G., Chen, Y.-H., Kuan, W.-H., Ma, H.-K., 2011. Energy life cycle assessment of rice straw bio-energy derived from potential gasification technologies. Bioresour. Technol. 102, 6735–6741.

Siebenmorgen, T.J., Nehus, Z.T., Archer, T.R., 1998. Milled rice breakage due to environmental conditions. Cereal Chem. 75, 149–152.

Singh, A., Singh, N., Bishnoi, N.R., 2010. Enzymatic hydrolysis of chemically pretreated rice straw by two indigenous fungal strains: a comparative study. J. Sci. Ind. Res. 69, 232–237.

Singh, A., Das, M., Bal, S., Banerjee, R., 2014. Rice processing. In: Ferreira Guiné, R., Correia, P.P.M. (Eds.), Engineering Aspects of Cereal and Cereal-Based Products. CRC Press Taylor & Francis Group, LLC Folrida, USA, pp. 71–97.

Soltani, N., Bahrami, A., Pech-Canul, M.I., González, L.A., 2015. Review on the physicochemical treatments of rice husk for productionof advanced materials. Chem. Eng. J. 264, 899–935.

Spindler, D.D., Wyman, C.E., Mohagheghi, A., Gorhmann, K., 1988. Thermotolerant yeast for simultaneous saccharification and fermentation of cellulose to ethanol. Appl. Biochem. Biotechnol. 17, 279–293.

Strieder, M.M., Pinheiro, C.P., Borba, V.S., Pohndorf, R.S., Cadaval, T.R.S., Pinto, L.A.A., 2017. Bleaching optimization and winterization step evaluation in the refinement of rice bran oil. Sep. Purif. Technol. 175, 72–78.

Su, Y., Luo, Y., 2009. Experiment on rice straw gasification in a two-stage gasifier. In: Asia-Pacific Power and Energy Engineering Conference, 2009 (APPEEC 2009), pp. 1–4.

Sun, Y., Cheng, J., 2002. Hydrolysis of lignocellulosic materials for ethanol production: a review. Bioresour. Technol. 83, 1–11.

Sun, L., Gong, K., 2001. Review, silicon-based materials from rice husks and their applications. Ind. Eng. Chem. Res. 40, 5861–5877.

Tabaraki, R., Nateghi, A., 2011. Optimization of ultrasonic-assisted extraction of natural antioxidants from rice bran using response surface methodology. Ultrason. Sonochem. 18, 1279–1286.

Tang, S., Hettiarachchy, N.S., Horax, R., Eswaranandam, S., 2003. Physicochemical properties and functionality of rice bran protein hydrolyzate prepared from heat-stabilized defatted rice bran with the aid of enzymes. J. Food Sci. 68 (1), 152–157.

Taniguchi, M., Suzuki, H., Watanabe, D., Sakai, K., Hoshino, K., Tanaka, T., 2005. Evaluation of pretreatment with Pleurotus ostreatus for enzymatic hydrolysis of rice straw. J. Biosci. Bioeng. 100, 637–643.

Tao, J., Rao, R., Liuzzo, J., 1993. Microwave heating for rice bran stabilization. J. Microw. Power Electromagn. Energy. 38 (3), 156–164.

Terigar, B.G., Balasubramanian, S., Sabliov, C.M., Lima, M., Boldor, D., 2011. Soybean and rice bran oil extraction in a continuous microwave system: from laboratory-to pilot-scale. J. Food Eng. 104, 208–217.

Thy, P., Jenkins, B.M., Williams, R.B., Lesher, C.E., 2004. Slag formation and potassium volatilization from rice straw blended wood fuel. Prepr. Pap. -Am. Chem. Soc. Div. Fuel Chem. 49, 89–91.

Thy, P., Jenkins, B.M., Lesher, C.E., Grundvig, S., 2006. Compositional constraints on slag formation and potassium volatilization from rice straw blended wood fuel. Fuel Process. Technol. 87, 383–408.

Tsai, M., Huang, P.Y., Yang, C.H., 2006. Formation mechanisms of colloidal silica sodium silicate. J. Nanoparticle Res. 8 (6), 943–949.

Turmanova, S., Svetlana, G., Lyubomir, V.L., 2012. Obtaining some polymer composites filled with rice husks ash—a review. Int. J. Chem. 4 (4), 62–89.

USDA, 2009. United States Standards for Rice. Online publication, U.S. Department Agriculture, Grain Inspection, Packers, and Stockyards Administration, Federal Grain InspectionService, Washington DC. https://www.gipsa.usda.gov/fgis/standards/ricestandars.pdf.

Vagg A., 2013. Rice Straw Utilization Value Adding and Alternative Uses for the Australian Rice Industry, Rural Industries, Research & Development Corporation, Nuffield Australia No. 1314.

Valdes, G., Planes, L.R., 1983. Study of the hydrolysis of rice straw with sulfuric acid under moderate conditions. Rev. Cienc. Quim. 14, 11–19.

Valkanas, G. N., Valkanas, N.P., Vlyssides, A G., Theodoropoulos, A. G., 1998. Method of Production of Ethl Alcohol. U.S.P.5766895.

Van Hoed, V., Depaemelaere, G., Vila Ayala, J., Santiwattana, P., Verhé, R., De Greyt, W., 2006. Influence of chemical refining on the major and minor components of rice bran oil. J. Am. Oil Chem. Soc. 83, 315–321.

Wang, M., Hettiarachchy, N.S., Qi, M., Burks, W., Seibenmorgen, T., 1999. Preparation and functional properties of rice bran protein isolate. J. Agric. Food Chem. 47, 411–416.

Wang, B., Khir, R., Pan, Z., El-Mashad, H., Atungulu, G.G., Ma, H., 2014. Effective disinfection of rough rice using infrared radiation heating. J. Food Prot. 77 (9), 1538–1545.

Wang, G., Tan, L., Sun, Z.-Y., Gou, Z.-X., Tang, Y.-Q., Kida, K., 2015. Production of bioethanol from rice straw by simultaneous saccharification and fermentation of whole pretreated slurry using Saccharomyces cerevisiae KF-7. Environ. Prog. Sustain. Energy 34, 582–588.

Wang, T., Khir, R., Pan, Z., Yuan, Q., 2017. Simultaneous rough rice drying and rice bran stabilization using infrared radiation heating. LWT Food Sci. Technol. 78, 281–288.

Wells, J.H., 1993. Utilization of rice bran and oil in human diets. Louisiana Agric. 36, 4–8.

Wilson, T.A., Idreis, H.M., Taylor, C.M., Nicolosi, R.J., 2002. Whole fat rice bran reduces the development of early aortic atherosclerosis in hypercholesterolemic hamsters compared with wheat bran. Nutr. Res. 22, 1319–1332.

Wong, K.K., Lee, C.K., Low, K.S., Haron, M.J., 2003. Removal of Cu and Pb by tartaric acid modified rice husk from aqueous solutions. Chemosphere 50 (1), 23–28.

Wyman, C.E., 1994. Ethanol from lignocellulosic biomass: technology, economics, and opportunities. Bio/Technology 50, 3–16.

Xu, Z.M., Godber, J.S., 2000. Comparison of supercritical–fluid and solvent extraction methods in extracting c-oryzanol from rice bran. J. Am. Oil Chem. Soc. 77, 547–551.

Xu, Z., Wang, Q.H., Jiang, Z., Yang, X.X., Ji, Y.Z., 2007. Enzymatic hydrolysis of pretreated soybean straw. Biomass Bioenergy 31, 162–167.

Yadav, B.K., Joyner, J., 2018. Rice bran processing and utilization. In: Pan, Z., Khir, R. (Eds.), Advances in Science & Engineering of Rice. DEStech Publications, Inc, Pennsylvania, USA, pp. 447–480.

Yadav, B.K., Khir, R., Pan, Z., Joyener, J., 2018. Rice milling. In: Pan, Z., Khir, R. (Eds.), Advances in Science & Engineering of Rice. DEStech Publications, Inc, Pennsylvania, USA, pp. 283–335.

Yang, W., Guo, F., Wang, Z., 2013. Yield and size of oyster mushroom grown on rice/wheat straw basal substrate supplemented with cotton seed hull. Saudi J. Biol. Sci. 20, 333–338.

Yao, F., Wu, Q., Liu, H., Lei, Y., Zhou, D., 2011. Rice straw fiber reinforced high density polyethylene composite: effect of coupled compactibilizing and toughening treatment. J. Appl. Polym. Sci. 119 (4), 2214–2222.

Yin, F., Hwang, A., Yu, N., Hao, P., 1982. Hydrolysis of agricultural wastes. In: Proceedings of Resource Recovery Solid Wastes Conference, pp. 447–456.

Yokochi, K., 1977. Rice bran processing for production of rice bran oil and characteristics and uses of the oil and de-oiled bran. In: Proceedings of Rice By-Products Utilization, International Conference, Valencia, Spain. vol. III. Insituto de Agroquimica Y Technologia de Alimentos, Spain, pp. 1–38. 71 p.

Yoon, S.H., Kim, S.K., 1994. Oxidative stability of high FFA rice bran oil at different stages of refining. JAOCS 71, 224–229.

Zamora, R., Crispin, J.A.S., 1995. Production of an acid extract of rice straw. Acta Cient. Venez. 46, 135–139.

Zerrudo, J.V., 1984. Rice straw for bond and printing paper. NSTA Technol. J. Natl. Sci. Dev. Board 9, 25–29.

Zhang, L., Hu, Y., 2014. Novel lignocellulosic hybrid particleboard composites made from rice straws and coir fibers. Mater. Des. 55, 19–26.

Zhang, R.H., Zhang, Z., 1999. Biogasification of rice straw with an anaerobic phased solids digester system. Bioresour. Technol. 68, 235–245.

Zhang, R., Li, X., Fadel, J.G., 2002. Oyster mushroom cultivation with rice and wheat straw. Bioresour. Technol. 82, 277–284.

Zhang, H., Wang, Y., Lu, F., Chai, L., Shao, L., He, P., 2015. Effect of dilute acid pretreatment on physicochemical characteristics and consolidated bioprocessing of rice straw. Waste Biomass Valor 6, 217–223.

Zheng, Y., Venkitasamy, C., Pan, Z., 2018. Rice straw and hulls. In: Pan, Z., Khir, R. (Eds.), Advances in Science & Engineering of Rice. DEStech Publications, Inc, Pennsylvania, USA, pp. 587–648.

Zigoneanu, I.G., Williams, L., Xu, Z., et al., 2008. Determination of antioxidant components in rice bran oil extracted by microwave-assisted method. Bioresour. Technol. 99, 4910–4918.

Further Reading

Sayre, R.N., Nayyar, D.K., Saunders, R.M., 1985. Extraction and refining of edible oil from extrusion-stabilized rice bran. J. Am. Oil Chem. Soc. 62 (6), 1040–1043.

CHAPTER 3
Corn

Zhenhua Ruan*, **Xiaoqing Wang**[†], **Yan Liu**[‡], **Wei Liao**[‡]
*LanzaTech, Skokie, IL, United States, [†]Tate & Lyle, Hoffman Estates, IL, United States, [‡]Biosystems and Agricultural Engineering, Michigan State University, East Lansing, MI, United States

Chapter Outline

1 INTRODUCTION: CORN AND ITS BY-PRODUCTS

Corn as one of the major cereal crops is widely used to produce human food, animal feed, and industrial products (e.g., cornstarch, cereals, adhesives, sweetener, and alcohol) (Davis, 2001). The world corn production was around 1.03 billion tons in 2017. More than 80% of the corn is produced by eight countries/regions leading by the United States with 37% of the total, followed by China (21%), Brazil (8%), European Union (6%), Argentina (4%), Ukraine (3%), India (2%), and Mexico (2%) (Capehart et al., 2017). The majority of the US corn is produced in the heartland region, which generally produces one-third of the US corn (Capehart, 2017). The land used to produce corn in the United States (more than 90 million acres) is around 18% of the world total. The national average of corn yield is 9.6 metric ton per hectare (Nielsen, 2012), which is approximately double the global average (World Agricultural Production, 2017).

Dent corn (*Zea mays* var. *indentata*) is the dominant variety of corn cultivated in the United States. A kernel of dent corn contains approximately 74% carbohydrate (mostly starch), 9% protein, 7% total dietary fiber, 4.7% lipid (oil), and 10% water. It is a main carbohydrate provider as well as a good mineral source

Integrated Processing Technologies for Food and Agricultural By-Products
https://doi.org/10.1016/B978-0-12-814138-0.00003-4

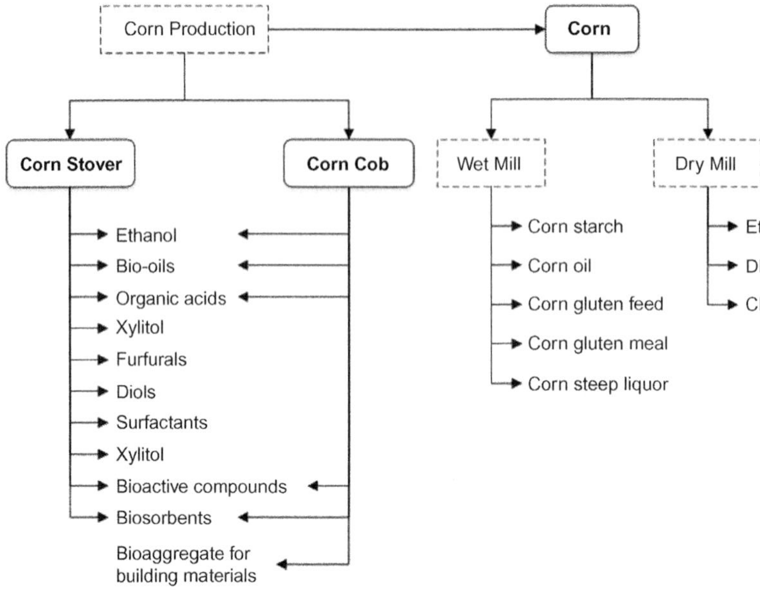

FIG. 1
Flowchart of corn processing and by-products production*. *: Dash frames are for corn processing.

(especially magnesium, phosphorus, potassium), vitamins (mainly vitamin A), and folate for human and animal consumption (A.R.S. United States Department of Agriculture, 2016). More than 50% of corn produced in the United States is used as animal feed, which accounts for more than 95% of the feed grains for livestock; approximately 40% is for food and industrial uses; and the remaining 10% is exported (Capehart and Liefert, 2017).

During corn harvesting and processing, large amounts of by-products such as corn stover, corncob, corn germ, condensed distillers solubles (CDS), corn distillers dried grains with solubles (DDGS), and processing wastewater are generated (Fig. 1). Among these by-products, corn stover and corncob are from corn harvesting. Other by-products are from corn processing of ethanol and starch production. Value-added utilization of these by-products can make a significant contribution to improve the efficiency of corn industry, bring revenue back to rural communities, and benefit the environment. This chapter focuses on using the corn by-products to produce a variety of fuel and chemical products.

2 CORN STOVER

Corn stover is the stalks, leaves, and husks that remain in the field after corn harvest. It is mainly composed of cellulose (~35% w/w), hemicellulose (~20% w/w), and lignin (~12% w/w). In the United States, a small percentage of corn stover is left in the field and integrated into soil with tillage to maintain soil productivity (DeJong-Hughes and Coulter, 2009). The rest of corn stover is harvested and widely used as animal feeds for ruminants and livestock bedding. The global annual corn production is about 1 billion ton currently, with a harvest index ranges from 47% to 56%, about half the weight of the above-ground part of the corn plant is stover. Thus, corn stover is produced at a rate of 1 dry kg per dry kg of corn grain, so the global corn stover production is around 1 billion tons (Li et al., 2014). Due to high carbohydrate content, corn stover is considered as one of the key lignocellulosic feedstocks for biofuels production (e.g., ethanol and bio-oil) (Perlack and Stokes, 2011). It has been reported that corn stover available in North America could potentially produce 38.4 gigaliter year^{-1} of bioethanol (Kim and Dale, 2004). Besides biofuels, corn stover can also be used to produce other value-added chemicals, such as chemical building blocks from glucan and xylan [e.g., 4 diacids (succinic and malic acid), dicarboxylic acid (DCA), propionic acid (PA), lactic acid, levulinic acid, xylitol, furfural, 5-(hydroxymethyl)furfural (HMF), 5-(chloromethyl)furfural, and 1,4-pentanediol, enzymes, and biosufactants], and antioxidants from lignin.

Organic acids as important chemical blocks are widely used in a variety of industries. They can be produced from carbohydrates in corn stover. For instance, succinic acid is widely used in food, chemical, and pharmaceutical industries, which has been considered as one of the most important platform chemicals to produce polymers, acidity regulator, and flavoring agent (Zheng et al., 2010). *Escherichia coli* AFP184 (Nghiem et al., 2016) or *Actinobacillus succinogenes* (Zheng et al., 2010) has been used to biologically convert corn stover hydrolysate rich in glucose and xylose into succinic acid. Malic acid is another important additive serving as acidulant and taste enhancer in food and beverage industry. A recent study reported that a mutant strain of *Thermobifida fusca* can directly ferment milled corn stover to produce malic acid (Deng et al., 2016). Aliphatic DCAs have been recognized as important platform chemicals and intermediates for biodegradable polymers, among which, the α,ω-DCAs with chain length more than 10 carbons could potentially lead to novel polymers with better performance and biodegradability (Ngo et al., 2006). An integrated biological and chemical process has been recently developed to convert corn stover into long-chain DCAs (Fang et al., 2016; Zhao et al., 2017). PA is a potential building block for C3-based chemicals. A native PA-producing strain *Propionibacterium acidipropionici* was used to evaluate PA production on corn stover hydrolysate, which led to a PA titer of 64.7 g/L and a PA productivity of 0.77 g/L/h (Wang et al., 2017a). D-lactic acid is an important precursor for polylactic acid (PLA) synthesis. PLA has a great potential in foods, environmental protection, medicine, and other application fields. Wang and coauthors reported a cost-effective and feasible way of converting corn stover into D-lactate. Up to 18 g/L D-lactic acid was collected from the corn stover hydrolysate (Wang et al., 2017b). Levulinic acid is another important chemical intermediate that can also be produced from corn stover. Levulinic acid is used to synthesize food flavoring agents, plasticizers, herbicides, antifreeze, pharmaceutical, resins, polymers, solvents, and fuel additives or drop-in fuels. Enzymatic hydrolysis of dilute acid pretreated corn stover was integrated with a simultaneous isomerization and reactive extraction process for high-yield levulinic acid production (Alipour and Omidvarborna, 2017; Alipour and Omidvarborna, 2016).

Xylitol is well known as an artificial sweetener with fewer calories than sucrose, and widely used in pharmaceutical, food, and odontological industries. It has been reported that corn stover hydrolysate was mixed with crude glycerol to support growth of *C. tropicals* XK12K to synthesize xylitol (Hong et al., 2016). Corn stover pretreated by vapor diethyl oxalate can also be fermented by mutant *Pichia stipites* to produce xylitol (Rodrigues et al., 2011).

Furfural, 5-(hydroxymethyl)furfural (HMF), and 5-(chlormethyl)furfural (CMF) as precursors for fuel and polymer production can be generated from pretreated corn stover. The direct conversion of both cellulose and hemicellulose in raw corn stover to furfural was developed by employing a unique heterogeneous strong acid catalyst (SC-CaC t-700) in different solvents (Li et al., 2017). The furfural yield of 93% was achieved when the catalytic conversion was carried out in gama-valerolactone (Li et al., 2017). The coupling of metal halides with tetrahydrofuran as a green solvent was also used to coproduce furfural and 5-HMF from corn stover (Cai et al., 2014). High-yield HMF production can also be achieved by the aforementioned process of a simultaneous isomerization and reactive extraction on corn stover hydrolysate (Alipour, 2016). As for CMF production, concentrated hydrochloric acid was used to treat corn stover and generate CMF, and then organic solvent was applied to extract CMF from the reaction solution (Zhang et al., 2017). A reproducible yield of 63% CMF from corn stover at a solid loading of 10% *w/v* was achieved with furfural and levulinic acid as the coproducts (Zhang et al., 2017).

Corn stover can be used to produce biogenic diols, especially 1,4-pentanediol (1,4-PDO) that is the intermediate to produce high-strength biodegradable polyester, pharmaceuticals, fine chemicals, and cosmetics (Geilen et al., 2010). Corn stover was catalytically converted into gamma-valerolactone and further upgraded to 1,4-PDO (Ahn and Han, 2017).

In addition, corn stover is considered as a promising substrate for the production of industrially important enzymes (e.g., cellulases and xylanases) that are widely used in animal feed, food, textile, brewery, wine, laundry, as well as pulp and paper industries. Cellulase/xylanase production from untreated corn stover often adopts solid-state fermentation technology. White rot fungi *Phanerochaete chrysosporium, Sporotrichum pulverulentum, Bjerkandera adusta, Pleurotus ostreatus, T. zonatus,* and *T. trogii* (Tirado-González et al., 2016), or mutant *Aspergillus tereus* AUMC10138 (Isaac and Abu-Tahon, 2015) were cultured on untreated corn stover to excrete enzymes. Corn stover hydrolysates from different chemical/thermal pretreatment methods were also used to produce cellulase/xylanase. Mutant *Trichoderma reesei* EBUV03 was cultured on steam-explosion treated corn stover (Meng et al., 2016). *T. reesei* Rut C-30 produced enzymes on ammonia fiber expansion, dilute acid, and dilute alkali pretreated corn stover (Zhang et al., 2014; Culbertson et al., 2013). *Aspergillus niger* NRRL567 accumulated the enzymes on raw corn stover (Ghori et al., 2011).

Corn stover hydrolysate can also serve as a carbon source to produce surfactants. It has been reported that *Candida bombicola* can produce up to 52.1 g/L sophorolipid from corn stover hydrolysate (Samad et al., 2017). Sophorolilpid as a biosurfactant is widely used in cosmetics, hygiene, and medical industries.

Moreover, corn stover contains bioactive compounds, particularly phenolic compounds from lignin. Extraction of such compounds is important for the value addition to corn stover (Buranov and Mazza, 2009). Ethanol, alkali, and acid are often used as extraction agents. The assessment of bioactivities of the phenolic compounds in the extracts suggest that corn stover could be a potential source of natural antioxidants (Vazquez-Olivo et al., 2017).

3 CORNCOB

Corncob is the hard cylindrical cores that bear the kernels of corn. In a typical corn field, the dry corncobs per acre are around 1500 pounds, which is about 20% of the total corn residues including both corn cob and corn stover (DeJong-Hughes and Coulter, 2009). Similar to corn stover, corncob contains cellulose (~70%), hemicellulose (~22%), and lignin (8%) (Lu and Chen, 2014), though cellulose content in corncob is significantly higher than corn stover, and lignin content is much lower. Corncob has been used as animal feed (Wachirapakorn et al., 2016) as well as a feedstock for energy generation (e.g., fuel ethanol, biodiesel (Kahr et al., 2015), pyrolysis bio-oil (Shariff et al., 2016; Demiral et al., 2012; Oh et al., 2013), and gasification power (Biagini et al., 2014)). As for value-added product production, due to the high cellulose content, corncob can be used as a carbohydrate source to produce aforementioned chemical building blocks, enzymes (Olajuyigbe and Ogunyewo, 2016; Elegbede and Lateef, 2017; Desai and Iyer, 2017), and surfactants. Besides these products, corncob has several other unique applications.

Corncob is considered as a source of cellulose-rich fiber. The cellulose was extracted from corncob using hot sodium hydroxide and hydrogen peroxide bleaching. Microfibrillated cellulose was obtained from the extracted cellulose and cross-linked with poly(lactic acid), which was applied as a reinforcing agent to improve the mechanical properties of poly(lactic acid) (Deejam and Charuchinda, 2015).

Corncob has been used to produce biosorbents for the removal of contaminants in wastewater, such as heavy metals and dyes (Abdelfattah et al., 2016). The corncob-derived silica gel modified 3-aminopropyltriethpxysilane (ATPS) was tested for the adsorption of Cu (II) (Purwanto et al., 2017). The powdered corncob was used to pre-concentrate and recover uranium (VI) from liquid waste with a high removal efficiency of 98.5% (Mahmoud, 2015). The activated carbon prepared from corncob was found to have a good Cr (VI) adsorptive capacity (Tang et al., 2016). A new xanthated biosorbent from corncob was developed and successfully applied for the removal of Cr (III) ions from the wastewater of a chrome plant and methylene blue from river water (Kostić et al., 2017;

Miyah et al., 2016). Corncob and TiO_2 were immobilized onto a thin film to greatly enhance the filtration efficiency of fine particle separation during wastewater treatment (Gan et al., 2017). Corncob and its biochar were also applied for removing inorganic nitrogen Mn (VII) ions from aqueous solution (Wu et al., 2016; Norozi and Haghdoost, 2016).

Corncob is characterized as resource for bioaggregate-based building materials. It was reported that corncob can be used as pore forming agent in the production of porous ceramic brick (Njeumen Nkayem et al., 2016). The porous bricks can greatly enhance the insulation and soundproofing properties of building materials. It was found that the concrete made with 10% corncob ash has the optimum flexural and compressive strength, and could be applied in low-cost housing construction (Ikponmwosa et al., 2015).

Corncob has also been studied as a lignocellulosic feedstock to produce antibiotics. Tetracyclines were produced from solid-state fermentation of corncob by microbes such as *Streptomyces* sp. OXCI, *Streptomyces rimosus* NRRL B2659, *S. rimosus* NRRL B2234, *Streptomyces alboflavus* NRRL B1273, *Streptomyces aureofaciens* NRRL B2183, and *Streptomyces vendagensis* ATCC 25507 (Asagbra et al., 2005). Tetracyclines are broad-spectrum antibiotics used to treat infection and for plant disease control.

In addition, corncob from special corn (i.e., purple corn) is found to be a great source of bioactive compounds such as anthocyanin, flavonoids, and phenolics. The anthocyanin is conventionally extracted by ethanol, while phenolics and flavonoids are mainly extracted using water. Microwave-assisted extraction was performed to obtain anthocyanins from Chinese purple corncob, which greatly improved extraction efficiency and rate (Yang and Zhai, 2010). These bioactive compounds have the potential applications as colorants, natural antioxidants, and food ingredients (Monroy et al., 2016a; Monroy et al., 2016b).

4 COPRODUCTS AND WASTEWATER OF WET-MILLING PROCESS

Wet milling is a widely adopted corn processing technology to obtain the maximum amount of starch from corn kernel (Fig. 2A). During wet milling, corn is generally fractionated into four major compounds including starch, germ, protein, and fiber (Rausch and Belyea, 2006). Firstly, corn is steeped in a dilute sulfur dioxide solution that hydrolyzes and softens its kernel. During steeping, soluble nutrients from corn leach into water. This steep water is evaporated to concentrate the soluble nutrients for corn steep liquor (CSL). Secondly, corn germ and fiber fractions are removed from soaked kernel. Corn oil is recovered from the germ, while the remaining part of the germ is processed to produce corn germ meal, a great source of protein and used as feed. The fiber fraction is removed by the screens and combined with heavy steep water to form corn gluten feed. Thirdly, the remaining slurry is separated into starch and gluten protein fraction according to the difference in density. The gluten protein is concentrated

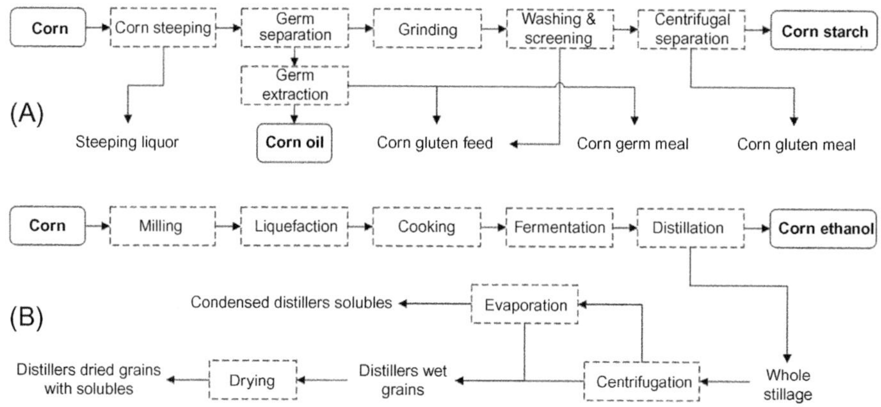

FIG. 2
Corn milling processes*: (A) wet milling process; and (B) dry milling process. *: Dash line frames are the unit operations. Solid line frames are the corn and main corn products.

and dried to obtain corn gluten meal, a protein-rich feed. Lastly, after further purification of protein, pure starch is obtained. Some of the starch is then washed and dried, or modified, and marketed to the food, paper, and textile industries. The remaining starch can be processed into sweetener or fuel ethanol (Rausch and Belyea, 2006).

4.1 Coproducts

Besides cornstarch and corn oil as the main products of wet milling process, the process generates multiple coproducts of corn gluten feed, corn gluten meal, corn germ meal, and CSL. One bushel of corn (56 pounds) processed by a wet mill generates approximately 12.9 pounds of corn gluten feed, 3.1 pounds of corn gluten meal, and 0.5 pounds of corn germ meal (Hoffman, 2011). Together these coproducts represent about 25%–30% of the corn, as the dent corn generally has 70%–75% of starch and 3%–4% of oil (A.R.S. United States Department of Agriculture, 2016). CSL is sometimes combined with the corn gluten feed for cattle or sold separately as a liquid protein source. Corn gluten feed usually consists of corn bran, CSL, and germ meal (Hoffman, 2011). It typically has a dry matter of 88%, which contains about 21% crude protein, 2% lipid, and 10% crude fiber, minerals, and amino acids (Batal and Dale, 2016). Corn gluten feed is widely used in feeds for ruminant, poultry, swine, and pet foods (Li et al., 2012). Corn germ meal is rich in protein. More than 20% of proteins were contained in the germ meal. The protein could be extracted by adding alpha amylase during oil extraction. The extracted corn germ protein has a purity of around 62%. The extracted protein contains seven amino acids essential for human body including threonine, valine, leucine, isoleucine, methionine, phenylalanine, and lysine (Wang et al., 2014), which demonstrated potential of being an amino acid supplement for food production.

4.2 Wastewater

During wet milling, a large amount of wastewater is generated. The cornstarch wastewater features high chemical oxygen demand (COD), which commonly includes starch, glucose, protein, vitamins, and inorganic salts, and is considered as a nutrient source for microbial cultivation (Wang et al., 2016). The corn processing wastewater generally includes CSL, corn gluten water (CGW), and glucose wastewater (GW).

CSL was found to be an important source of bioactive compounds (i.e., surfactants and antioxidants) that have potential to be applied in food, cosmetic, and personal care fields (Rodriguez-Lopez et al., 2016). The bioactive compounds can be obtained using liquid–liquid extraction and hydrothermal treatment of CSL. It has also been reported that corn-steeping liquor served as the nutrient and water sources to support probiotics production by microbial fermentation (e.g., bifidobactera, lactic bacteria) (Ryan et al., 2006; Petrova et al., 2010; Liu et al., 2015; Mironescu et al., 2016). Moreover, CSL was tested as a substrate for fungal cultivation of chitin and chitosan (Bueter et al., 2013; Nwe and Stevens, 2004; Berger et al., 2014; Jasti, 2006; Jasti et al., 2008). The chitin and chitosan obtained using fungal fermentation had homogeneous characteristics and consistent quality, compared with that derived from exoskeletons of marine crustaceans.

Corn processing wastewater was also used to support microbial growth and generate lactate and hydrogen. A simultaneous saccharification and fermentation process using a fungus, *Rhizopus arrhizus* 36,017, was investigated to produce lactate from corn processing wastewater (Jin et al., 2005). Lactate yields up to 0.94–0.97 g/g of starch or sugars in the corn wastewater have been achieved. A recent study evaluated the performance of biohydrogen production using a mixture of CSL and CGW as nitrogen source while GW as the carbon source, which suggested that hydrogen production from mixed cornstarch processing wastewaters (CSL + CGW + GW) was a feasible way for both wastewater treatment and bioenergy production (Wang et al., 2016).

5 CORN ETHANOL AND COPRODUCT GENERATION FROM DRY MILLING PROCESS

Approximately 5289 million bushels of corn were used for fuel ethanol production in the United States in 2016, which was about 37% of the total annual corn production (N.A.S.S. United States Department of Agriculture, 2017). About 90% of the corn for fuel ethanol production is dry milled (Fig. 2B). During dry milling process, the ground corn is mixed with fresh and recycled water to form starch slurry, and then adjusted to a proper pH and temperature to facilitate the enzymatic liquefaction of dextrin release. After liquefaction, the mash is cooled to around 30 °C and sent to a fermenter where glucoamylase is added to convert the dextrin into fermentable mono-sugar, glucose. *Saccharomyces cerevisiae*, a classic brewing yeast, is used to produce ethanol. The fermenting mash (beer) is then sent to distillation section to collect ethanol. The corn protein and other nutrients added during yeast ethanol fermentation is then released into the wastewater (stillage) (Weigel et al., 1997). The water and solids (protein, lipid, and fiber) are collected from the distillation column bottom as whole stillage, which is then centrifuged to separate the solids from the liquid (thin stillage). Some of the thin stillage with a high moisture content can be recycled back to the beginning of the process to reduce the usage of fresh water. The remaining solids can be sold directly or concentrated to become CDS. The CDS can be used as a feed for cattle. It can also be combined with the residual solids and dried to become DDGS, which is a prominent source of protein, lipid, and fiber, and also an excellent source of minerals (mainly phosphorous and potassium) and vitamins as well as amino acids. Corn distillers' DDGS are the major coproduct in dry milling, each bushel (56 pounds) of corn used in dry-mill ethanol production generates about 17.5 pounds of DDGS containing 10% moisture (Mathews Jr and McConnell, 2009), Due to its high nutrient value, DDGS are incorporated into rations for many livestocks such as cattle, swine, poultry, especially those requiring the feed with high nutrient density (Weigel et al., 1997).

Sometimes, higher levels of oil in DDGS are undesirable and would have a negative effect on the feed quality, such as the interference with milk production in milk cattle and bacon texture in DDGS-fed swine (Wang et al., 2009). There were reports about the corn oil extraction from dry milling process to make the obtained DDGS suitable for the feed markets including milk cattle and swine. Improved oil recovery yield was obtained by the physical process including grinding and flaking (Lamsal and Johnson, 2012), or physical/chemical process (heating, pH adjustment) prior to corn dry milling (Majoni et al., 2011). During aqueous cellulase oil extraction, a modified process with the addition of alpha amylase and gluco-amylase prior to dry-fractionated germ was found to be able to produce 20%–30% more oil compared with that without amylases. The study indicated that aqueous cellulase corn germ oil extraction with additional amylases would be preferable and profitable if the crude corn oil cost exceeds $1.1/kg, despite the additional cost of the amylases (Dickey et al., 2011). It was found that the addition of non-starch hydrolase, protease, and phytase at their optimized conditions and process stages could enhance the oil recovery for corn dry milling process. The resulting low oil DDGS would be favored for feed application in monogastric animals (Luangthongkam et al., 2015).

6 CONCLUSION

Value-added processing of corn beyond starch, oil, and ethanol production is critical to improve the efficiency of corn industry and facilitate addressing the food safety and energy security challenges. Corn processing residues of corn stover, corncob, and wastewaters should be systematically studied as the feedstock to generate more valuable products. A future corn processing system should include field residue handling and utilization, corn processing, and processing residue utilization to produce a suite of products ranging from starch, fuels, polymer, food/feed, and pharmaceutical and nutraceutical products, so that an environmentally friendly, technically sound, and economically feasible solution could be established to sustain corn industry.

References

A.R.S. United States Department of Agriculture. 2016. Basic Report 20014, Corn Grain, Yellow, in A.R.S. United States Department of Agriculture (Ed.).

Abdelfattah, I., El Sayed, F., Almedolab, A., 2016. Removal of heavy metals from wastewater using corn cob. Res. J. Pharm., Biol. Chem. Sci. 7 (2), 239–248.

Ahn, Y.C., Han, J., 2017. Catalytic production of 1,4-pentanediol from corn stover. Bioresour. Technol. 245, 442–448.

Alipour, S., 2016. High yield 5-(hydroxymethyl)furfural production from biomass sugars under facile reaction conditions: a hybrid enzyme- and chemo-catalytic technology. Green Chem. 18 (18), 4990–4998.

Alipour, S., Omidvarborna, H., 2016. High concentration levulinic acid production from corn stover. RSC Adv. 6 (112), 111616–111621.

Alipour, S., Omidvarborna, H., 2017. Enzymatic and catalytic hybrid method for levulinic acid synthesis from biomass sugars. J. Clean. Prod. 143 (Suppl. C), 490–496.

Asagbra, A.E., Sanni, A.I., Oyewole, O.B., 2005. Solid-state fermentation production of tetracycline by Streptomyces strains using some agricultural wastes as substrate. World J. Microbiol. Biotechnol. 21 (2), 107–114.

Batal, A. (Sanderson Farms), Dale, N. (University of Georgia, Athens, GA), 2016. Feedstuffs Ingredient Analysis Table.

Berger, L.R.R., Stamford, T.C.M., Stamford-Arnaud, T.M., Franco, L.D., do Nascimento, A.E., Cavalcante, H.M.D., Macedo, R.O., de Campos-Takaki, G.M., 2014. Effect of corn steep liquor (CSL) and cassava wastewater (CW) on chitin and chitosan production by Cunninghamella elegans and their physicochemical characteristics and cytotoxicity. Molecules 19 (3), 2771–2792.

Biagini, E., Barontini, F., Tognotti, L., 2014. Gasification of agricultural residues in a demonstrative plant: corn cobs. Bioresour. Technol. 173 (Suppl. C), 110–116.

Bueter, C.L., Specht, C.A., Levitz, S.M., 2013. Innate sensing of chitin and chitosan. PLoS Pathog. 9 (1).

Buranov, A.U., Mazza, G., 2009. Extraction and purification of ferulic acid from flax shives, wheat and corn bran by alkaline hydrolysis and pressurised solvents. Food Chem. 115 (4), 1542–1548.

Cai, C.M., Nagane, N., Kumar, R., Wyman, C.E., 2014. Coupling metal halides with a co-solvent to produce furfural and 5-HMF at high yields directly from lignocellulosic biomass as an integrated biofuels strategy. Green Chem. 16 (8), 3819–3829.

T. Capehart, 2017. Corn and Other Feed Grains, in E.R.S. United States Department of Agriculture (Ed.). https://www.ers.usda.gov/topics/crops/corn-and-other-feedgrains/feedgrains-sector-at-a-glance/.

T. Capehart, O. Liefert, 2017. Feed Outlook, in E.R.S. United States Department of Agriculture (Ed.). https://www.ers.usda.gov/topics/crops/corn-and-other-feedgrains/feedgrains-sector-at-a-glance/.

T. Capehart, O. Liefert, and D.W. Olson, 2017. Feed Outlook: August 2017, in U.S.D.o. Agriculture (Ed.). https://www.ers.usda.gov/publications/pub-details/?pubid=84705.

Culbertson, A., Jin, M., da Costa Sousa, L., Dale, B.E., Balan, V., 2013. In-house cellulase production from AFEX™ pretreated corn stover using Trichoderma reesei RUT C-30. RSC Adv. 3 (48), 25960–25969.

Davis, K., 2001. Corn milling, processing and generation of co-products. In: 62nd Minnesota Nutrition Conference, Minnesota Corn Growers Association Technical Symposium, Bloomington, MN.

Deejam, P., Charuchinda, S., 2015. Mechanical properties of poly(lactic acid) sheet reinforced with microfibrillated cellulose from corn cobs. In: Adiguzel, O., Zhou, J. (Eds.), EDP Sciences.

DeJong-Hughes, J., and J. Coulter, 2009. Considerations for Corn Residue Harvest in Minnesota. https://www.extension.umn.edu/agriculture/corn/harvest/considerations-for-corn-residue-harvest.

Demiral, İ., Eryazıcı, A., Şensöz, S., 2012. Bio-oil production from pyrolysis of corncob (Zea mays L.). Biomass Bioenergy 36 (Suppl. C), 43–49.

Deng, Y., Mao, Y., Zhang, X., 2016. Metabolic engineering of a laboratory-evolved Thermobifida fusca muC strain for malic acid production on cellulose and minimal treated lignocellulosic biomass. Biotechnol. Prog. 32 (1), 14–20.

Desai, D.I., Iyer, B.D., 2017. Utilization of corn cob waste for cellulase-free xylanase production by Aspergillus niger DX-23: medium optimization and strain improvement. Waste Biomass Valor. 8 (1), 103–113.

Dickey, L.C., Johnston, D.B., Kurantz, M.J., McAloon, A., Moreau, R.A., 2011. Modification of aqueous enzymatic oil extraction to increase the yield of corn oil from dry fractionated corn germ. Ind. Crop. Prod. 34 (1), 845–850.

Elegbede, J.A., Lateef, A., 2017. Valorization of corn-cob by fungal isolates for Production of xylanase in submerged and solid state fermentation media and potential biotechnological applications. Waste Biomass Valor.

Fang, H., Zhao, C., Kong, Q., Zou, Z., Chen, N., 2016. Comprehensive utilization and conversion of lignocellulosic biomass for the production of long chain α,ω-dicarboxylic acids. Energy 116 (Pt 1), 177–189.

Gan, H.Y., Leow, L.E., Ong, S.T., 2017. Utilization of corn cob and TiO2 photocatalyst thin films for dyes removal. Acta Chim. Slov. 64 (1), 144–158.

Geilen, F.M.A., Engendahl, B., Harwardt, A., Marquardt, W., Klankermayer, J., Leitner, W., 2010. Selective and flexible transformation of biomass-derived platform chemicals by a multifunctional catalytic system. Angew. Chem. Int. Ed. 49 (32), 5510–5514.

Ghori, M.I., Ahmed, S., Malana, M.A., Jamil, A., 2011. Corn stover-enhanced cellulase production by Aspergillus niger NRRL 567. Afr. J. Biotechnol. 10 (31), 5878–5886.

Hoffman, L.A., 2011. Market Issues And Prospects for US distillers' Grains Supply, Use, and Price Relationships. Diane Publishing.

Hong, E., Kim, J., Rhie, S., Ha, S.-J., Kim, J., Ryu, Y., 2016. Optimization of dilute sulfuric acid pretreatment of corn stover for enhanced xylose recovery and xylitol production. Biotechnol. Bioprocess Eng. 21 (5), 612–619.

Ikponmwosa, E.E., Salau, M.A., Kaigama, W.B., 2015. In: Rucevskis, S. (Ed.), Evaluation of strength characteristics of laterized concrete with corn cob ash (CCA) blended cement. Institute of Physics Publishing.

Isaac, G.S., Abu-Tahon, M.A., 2015. Enhanced alkaline cellulases production by the thermohalophilic aspergillus terreus AUMC 10138 mutated by physical and chemical mutagens using corn stover as substrate. Braz. J. Microbiol. 46 (4), 1269–1277.

Jasti, N., 2006. Attached Growth Fungal System for Corn Wet milling Wastewater Treatment.

Jasti, N., Khanal, S.K., Pometto, A.L., van Leeuwen, J., 2008. Converting corn wet-milling effluent into high-value fungal biomass in a biofilm reactor. Biotechnol. Bioeng. 101 (6), 1223–1233.

Jin, B., Yin, P., Ma, Y., Zhao, L., 2005. Production of lactic acid and fungal biomass by Rhizopus fungi from food processing waste streams. J. Ind. Microbiol. Biotechnol. 32 (11), 678–686.

Kahr, H., Pointner, M., Krennhuber, K., Wallner, B., Jäger, A., 2015. Lipid production from diverse oleaginous yeasts from steam exploded corn cobs. Agron. Res. 13 (2), 318–327.

Kim, S., Dale, B.E., 2004. Global potential bioethanol production from wasted crops and crop residues. Biomass Bioenergy 26 (4), 361–375.

Kostić, M., Đorđević, M., Mitrović, J., Velinov, N., Bojić, D., Antonijević, M., Bojić, A., 2017. Removal of cationic pollutants from water by xanthated corn cob: optimization, kinetics, thermodynamics, and prediction of purification process. Environ. Sci. Pollut. Res. 24 (21), 17790–17804.

Lamsal, B.P., Johnson, L.A., 2012. Flaking as a corn preparation technique for dry-grind ethanol production using raw starch hydrolysis. J. Cereal Sci. 56 (2), 253–259.

Li, M.H., Robinson, E.H., Bosworth, B.G., Oberle, D.F., Lucas, P.M., 2012. Use of corn gluten feed and cottonseed meal to replace soybean meal and corn in diets for pond-raised channel catfish. North Am. J. Aquac. 74 (2), 153–158.

Li, H., Xu, L., Liu, W., Fang, M., Wang, N., 2014. Assessment of the nutritive value of whole corn stover and its morphological fractions. Asian Australas. J. Anim. Sci. 27 (2), 194.

Li, W., Zhu, Y., Lu, Y., Liu, Q., Guan, S., Chang, H.M., Jameel, H., Ma, L., 2017. Enhanced furfural production from raw corn stover employing a novel heterogeneous acid catalyst. Bioresour. Technol. 245, 258–265.

Liu, S., Ren, F., Zhao, L., Jiang, L., Hao, Y., Jin, J., Zhang, M., Guo, H., Lei, X., Sun, E., Liu, H., 2015. Starch and starch hydrolysates are favorable carbon sources for bifidobacteria in the human gut. BMC Microbiol. 15 (1), 54.

Lu, J.-J., Chen, W.-H., 2014. Product yields and characteristics of corncob waste under various torrefaction atmospheres. Energies 7 (1), 13–27.

Luangthongkam, P., Fang, L., Noomhorm, A., Lamsal, B., 2015. Addition of cellulolytic enzymes and phytase for improving ethanol fermentation performance and oil recovery in corn dry grind process. Ind. Crop. Prod. 77 (Suppl. C), 803–808.

Mahmoud, M.A., 2015. Design of batch process for preconcentration and recovery of U(VI) from liquid waste by powdered corn cobs. J. Environ. Chem. Eng. 3 (3), 2136–2144.

Majoni, S., Wang, T., Johnson, L.A., 2011. Physical and chemical processes to enhance oil recovery from condensed corn distillers solubles. J. Am. Oil Chem. Soc. 88 (3), 425–434.

K.H. Mathews Jr, M.J. McConnell, 2009. Ethanol Co-Product Use in US Cattle Feeding. https://www.biofuelscoproducts.umn.edu/sites/biodieselfeeds.cfans.umn.edu/files/cfans_asset_417549.pdf.

Meng, Z., Xie, Q., Liu, Y., Zhang, X., Xiang, H., 2016. Cellulase production and saccharification of steam-exploded corn stover by mutant strain T. reesei EBUV-3. Int. J. Agric. Biol. 18 (1), 213–217.

Mironescu, M., Mironescu, I.D., Georgescu, C., 2016. Investigations on using wastewater from corn processing as substrate for probiotics. J. Hyg. Eng. Des. 15, 66–71.

Miyah, Y., Lahrichi, A., Idrissi, M., 2016. Removal of cationic dye -methylene bleu- from aqueous solution by adsorption onto corn cob powder calcined. J. Mater. Environ. Sci. 7 (1), 96–104.

Monroy, Y.M., Rodrigues, R.A.F., Sartoratto, A., Cabral, F.A., 2016a. Extraction of bioactive compounds from cob and pericarp of purple corn (Zea mays L.) by sequential extraction in fixed bed extractor using supercritical CO_2, ethanol, and water as solvents. J. Supercrit. Fluids 107 (Suppl. C), 250–259.

Monroy, Y.M., Rodrigues, R.A.F., Sartoratto, A., Cabral, F.A., 2016b. Optimization of the extraction of phenolic compounds from purple corn cob (Zea mays L.) by sequential extraction using supercritical carbon dioxide, ethanol and water as solvents. J. Supercrit. Fluids 116 (Suppl. C), 10–19.

N.A.S.S. United States Department of Agriculture. 2017. Grain Crushings and Co-Products Production 2016 Summary, in N.A.S.S. United States Department of Agriculture (Ed.).

Nghiem, N.P., Senske, G.E., Kim, T.H., 2016. Pretreatment of corn stover by low moisture anhydrous ammonia (LMAA) in a pilot-scale reactor and bioconversion to fuel ethanol and industrial chemicals. Appl. Biochem. Biotechnol. 179 (1), 111–125.

Ngo, H.L., Jones, K., Foglia, T.A., 2006. Metathesis of unsaturated fatty acids: synthesis of long-chain unsaturated-alpha,omega-dicarboxylic acids. J. Am. Oil Chem. Soc. 83 (7), 629–634.

Nielsen, R.B., 2012. Advanced Farming Systems and New Technologies for the Maize Industry. Purdue University.

Njeumen Nkayem, D.E., Mbey, J.A., Kenne Diffo, B.B., Njopwouo, D., 2016. Preliminary study on the use of corn cob as pore forming agent in lightweight clay bricks: physical and mechanical features. J. Build. Eng. 5 (Suppl. C), 254–259.

Norozi, F., Haghdoost, G., 2016. Application of corn cob as a natural adsorbent for the removal of Mn (VII) ions from aqueous solutions. Orient. J. Chem. 32 (4), 2263–2268.

Nwe, N., Stevens, W.F., 2004. Effect of urea on fungal chitosan production in solid substrate fermentation. Process Biochem. 39 (11), 1639–1642.

Oh, S.-J., Jung, S.-H., Kim, J.-S., 2013. Co-production of furfural and acetic acid from corncob using ZnCl2 through fast pyrolysis in a fluidized bed reactor. Bioresour. Technol. 144 (Suppl. C), 172–178.

Olajuyigbe, F.M., Ogunyewo, O.A., 2016. Enhanced production and physicochemical properties of thermostable crude cellulase from Sporothrix carnis grown on corn cob. Biocatal. Agric. Biotechnol. 7 (Suppl. C), 110–117.

Perlack, R.D., Stokes, B.J., 2011. U.S. Billion-Ton Update: Biomass Supply for a Bioenergy and Bioproducts Industry. U.S. Department of Energy, Oak Ridge National Laboratory, Oak Ridge, TN, p. 227.

Petrova, P., Emanuilova, M., Petrov, K., 2010. Amylolytic lactobacillus strains from bulgarian fermented beverage boza. Z. Naturforsch. 65 (3–4), 218–224.

Purwanto, A., Yusmaniar, Ferdiani, F., Damayanti, R., 2017. Synthesis and adsorption of silica gel modified 3-aminopropyltriethoxysilane (APTS) from corn cobs against Cu(II) in water. AIP Conf. Proc. 1823(1).

Rausch, K.D., Belyea, R.L., 2006. The future of coproducts from corn processing. Appl. Biochem. Biotechnol. 128 (1), 47–86.

Rodrigues, R.C.L.B., Kenealy, W.R., Jeffries, T.W., 2011. Xylitol production from DEO hydrolysate of corn stover by Pichia stipitis YS-30. J. Ind. Microbiol. Biotechnol. 38 (10), 1649–1655.

Rodriguez-Lopez, L., Vecino, X., Barbosa-Pereira, L., Moldes, A.B., Cruz, J.M., 2016. A multifunctional extract from corn steep liquor: antioxidant and surfactant activities. Food Funct. 7 (9), 3724–3732.

Ryan, S.M., Fitzgerald, G.F., van Sinderen, D., 2006. Screening for and identification of starch-, amylopectin-, and pullulan-degrading activities in bifidobacterial strains. Appl. Environ. Microbiol. 72 (8), 5289–5296.

Samad, A., Zhang, J., Chen, D., Chen, X., Tucker, M., Liang, Y., 2017. Sweet sorghum bagasse and corn stover serving as substrates for producing sophorolipids. J. Ind. Microbiol. Biotechnol. 44 (3), 353–362.

Shariff, A., Aziz, N.S.M., Ismail, N.I., Abdullah, N., 2016. Corn cob as a potential feedstock for slow pyrolysis of biomass. J. Phys. Sci. 27 (2), 123–137.

Tang, S., Chen, Y., Xie, R., Jiang, W., Jiang, Y., 2016. Preparation of activated carbon from corn cob and its adsorption behavior on Cr(VI) removal. Water Sci. Technol. 73 (11), 2654–2661.

Tirado-González, D.N., Jáuregui-Rincón, J., Tirado-Estrada, G.G., Martínez-Hernández, P.A., Guevara-Lara, F., Miranda-Romero, L.A., 2016. Production of cellulases and xylanases by white-rot fungi cultured in corn stover media for ruminant feed applications. Anim. Feed Sci. Technol. 221 (Pt A), 147–156.

Vazquez-Olivo, G., López-Martínez, L.X., Contreras-Angulo, L., Heredia, J.B., 2017. Antioxidant capacity of lignin and phenolic compounds from corn stover. Waste Biomass Valor.

Wachirapakorn, C., Pilachai, K., Wanapat, M., Pakdee, P., Cherdthong, A., 2016. Effect of ground corn cobs as a fiber source in total mixed ration on feed intake, milk yield and milk composition in tropical lactating crossbred Holstein cows. Anim. Nutr. 2 (4), 334–338.

Wang, H., Wang, T., Pometto, A.L., Johnson, L.A., 2009. A laboratory decanting procedure to simulate whole stillage separation in dry-grind corn ethanol process. J. Am. Oil Chem. Soc. 86 (12), 1241–1250.

Wang, X., Ding, J., Lu, B., Zhang, D., 2014. The extraction of protein from corn germ and components analysis. J. Chin. Cereals Oils Assoc. 29 (2), 62–66.

Wang, S., Zhang, T., Su, H., 2016. Enhanced hydrogen production from corn starch wastewater as nitrogen source by mixed cultures. Renew. Energy 96 (Pt B), 1135–1141.

Wang, X., Salvachúa, D., Sànchez, V.N.I., Michener, W.E., Bratis, A.D., Dorgan, J.R., Beckham, G.T., 2017a. Propionic acid production from corn stover hydrolysate by Propionibacterium acidipropionici. Biotechnol. Biofuels 10(1).

Wang, X., Wang, G., Yu, X., Chen, H., Sun, Y., Chen, G., 2017b. Pretreatment of corn stover by solid acid for D-lactic acid fermentation. Bioresour. Technol. 239, 490–495.

Weigel, J., Loy, D., Kilmer, L., 1997. Feed Co-Products of the Dry Corn Milling Process, Featuring Distillers Dried Grains. Iowa Department of Agriculture and Land Stewardship, Iowa Corn Growers Association, Renewable Fuels Association, National Corn Growers Association.

World Agricultural Production, 2017. Office of Global Analysis, Foreign Agricultural Service/USDA.

Wu, L.J., Wang, C.X., Zhang, F., Cui, J.G., 2016. The adsorption characters of inorganic nitrogen in aqueous solution by maize straw-and corn cob-derived biochars. Zhongguo Huanjing Kexue 36 (1), 74–81.

Yang, Z., Zhai, W., 2010. Optimization of microwave-assisted extraction of anthocyanins from purple corn (Zea mays L.) cob and identification with HPLC–MS. Innov. Food Sci. Emerg. Technol. 11 (3), 470–476.

Zhang, L., Wang, X., Ruan, Z., Liu, Y., Niu, X., Yue, Z., Li, Z., Liao, W., Liu, Y., 2014. Fungal cellulase/xylanase production and corresponding hydrolysis using pretreated corn stover as substrates. Appl. Biochem. Biotechnol. 172 (2), 1045–1054.

Zhang, X., Eren, N.M., Kreke, T., Mosier, N.S., Engelberth, A.S., Kilaz, G., 2017. Concentrated HCl catalyzed 5-(chloromethyl)furfural production from corn stover of varying particle sizes. Bioenergy Res, 1–7.

Zhao, C., Fang, H., Chen, S., 2017. Single cell oil production by Trichosporon cutaneum from steam-exploded corn stover and its upgradation for production of long-chain α,ω-dicarboxylic acids. Biotechnol. Biofuels 10(1).

Zheng, P., Fang, L., Xu, Y., Dong, J.-J., Ni, Y., Sun, Z.-H., 2010. Succinic acid production from corn stover by simultaneous saccharification and fermentation using Actinobacillus succinogenes. Bioresour. Technol. 101 (20), 7889–7894.

CHAPTER 4
Soybean

Varsha Gaonkar, Kurt A. Rosentrater
Iowa State University, Ames, IA, United States

Chapter Outline

1 INTRODUCTION

Oilseeds, especially soybeans, are commonly used for the production of oil and protein components—which are then used for the manufacture of human foods, animal feeds, and industrial products. Soybeans have dominated the oilseed industry after World War II, and now typically represent ~58%–59% of the world's oilseed production (National Oilseed Processors Association, 2018). In recent years, soybean oil usage has increased due to the growth in the production of biodiesel, biopolymers, industrial chemicals, as well as use in human foods and animal feeds (Fig. 1). Annually in the United States, soybeans are produced in >70–75 million acres, resulting in nearly 3 billion bushels (>300 million tons) in total, and soybean trade generates nearly 15 billion US dollars every year (Smith, 2018). The other two nations that produce almost as much

Integrated Processing Technologies for Food and Agricultural By-Products
https://doi.org/10.1016/B978-0-12-814138-0.00004-6

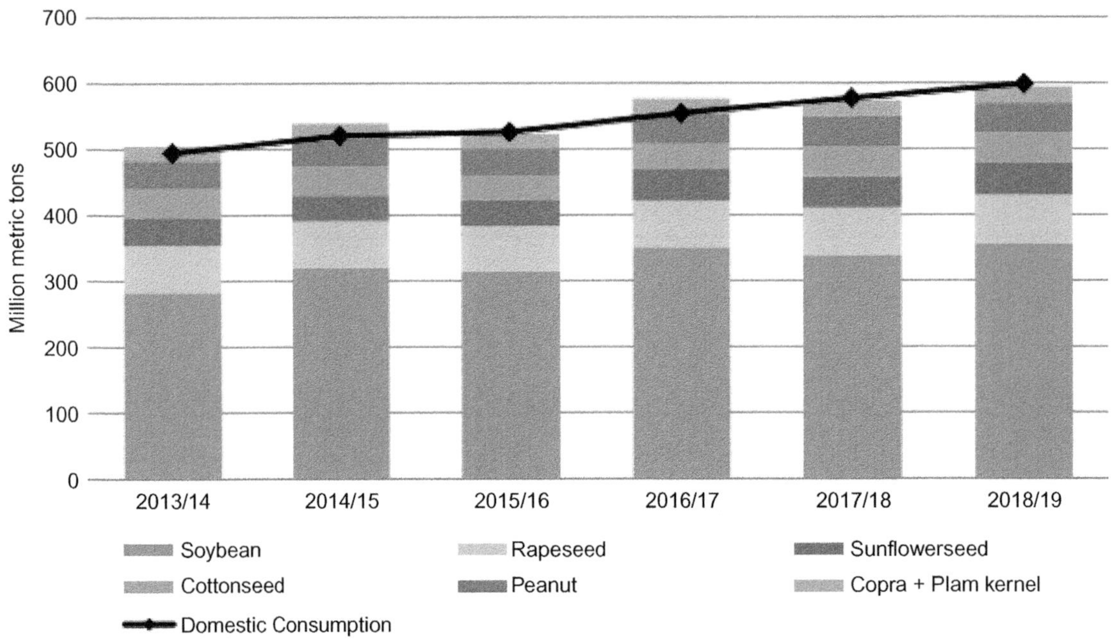

FIG. 1

Changes in the global production and consumption of oilseeds over time. Source: USDA. 2018. World Agricultural Supply and Demand Estimates. USDA, Washington, DC. Available online: http://www.usda.gov/oce/commodity/wasde/.

soybeans as the United States are Brazil and Argentina. According to a report by USDA (2018), global soybean production has increased by at least 100 million metric tons over the last decade.

Soybeans have a variety of end uses (Fig. 2). Major uses include oil for human food preparation (e.g., cooking oil and margarine) and protein-rich meals (e.g., soybean meal, soy protein concentrate, soy protein isolate) which are used as animal feed ingredients. But many other uses exist as well, including various human foods produced directly from the soybeans, industrial applications such as biodiesel, ink, biocomposites and bioplastics, adhesives, waxes, candles, foams, and hydraulic fluids, just to name a few.

The utilization of soybeans for various products depends not only on the composition of the soybeans themselves (Table 1), but also on the chemico-physico characteristics of the soybeans, which impact the ability of processing operations (which will be discussed later in this chapter) to segregate/concentrate oils from proteins and fibers, as well as the products which can be made from them (Table 1).

Because utilization of soybeans most often depends on the separation of oil compounds from proteins and other constituents, this chapter focuses on various

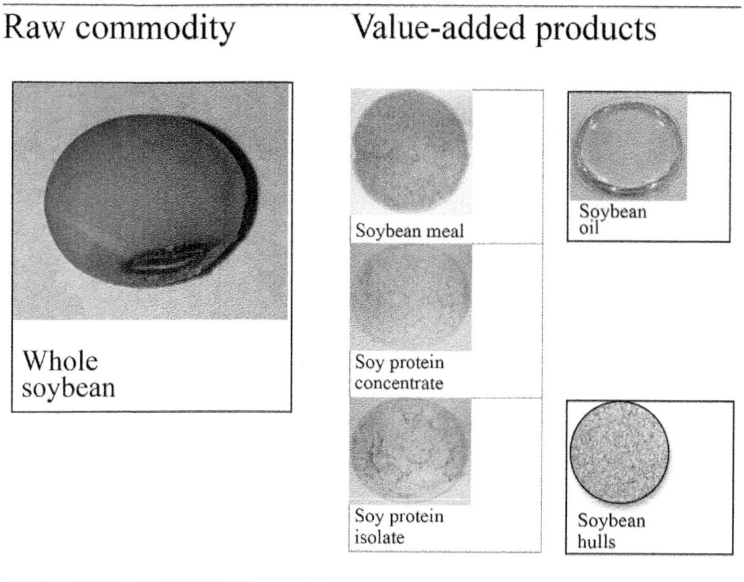

FIG. 2
Some key products manufactured from the soybean include various protein meals (for animal feeds and other industrial applications) as well as soybean oil (for human foods, biodiesel, industrial products, and animal feeds).

Table 1 Chemical Compositions of Soybeans and Some Common Soy Products

	Protein (%)	Oil (%)	Fiber (%)	Total sugars (mg/g)	Ash (%)
Whole Soybeans	27–44	16–24	13–23	22–141	4–6
Soybean meal	46–56	3–9	8–20	14–18	6–9
Soy protein concentrate	65–72	0.1–1	3.5–5		
Soy protein isolate	90–92	0.5–1	0.1–0.2		
Soy hulls	1–18	0.5–6	55–92		4–6

Adapted from Johnson, L. A., White, P. J., Galloway, R., 2008. Soybeans: Chemistry, Production, Processing and Utilization. Urbana, IL: AOCS Press and DairyOne. 2018. Interactive Feed Composition Library. Available online: http://dairyone.com/analytical-services/feed-and-forage/feed-composition-library/interactive-feed-composition-library/.

processing methods for separating soybean oils from proteins and fibers. These approaches include expelling and solvent extraction processes, aqueous and enzyme-assisted aqueous processes, chemical methods, enzymes, and fermentation. While not comprehensive in nature, it should serve as a solid introduction to these topics.

2 CONVENTIONAL OIL EXTRACTION PROCESSES

After extracting the oil from the soybeans, protein constitutes the majority of the residual cake (also known as meal). Due to its high protein and high energy content, soybean meal has become a predominant nitrogen (i.e., protein) ingredient in many animal feeds, especially for monogastric livestock and poultry diets. The nutritional content of soybean meal is often more consistent and is devoid of antinutrient factors compared to other protein and oilseed meals (Waldroup, 2018). High demand for soybean meal and oil is currently driving soybean production around the world, and soybean cultivation is expected to continue to grow in the coming years. And, due to enhanced purchasing power of economically developing countries, the consumption of vegetable oil is predicted to grow as well (both palm oil and soybean oil) as the livestock and poultry sectors grow exponentially.

Conventional oil/protein separation methods used in the oilseed industries include solvent extraction, expeller pressing, and hydraulic pressing. The oldest method, hydraulic pressing, is labor intensive, less efficient, and its use has diminished over the years. Pressing can be combined with solvent extraction in a two-stage extraction process. When the raw material cannot be processed directly, a pre-pressing method helps to prepare the material for separation.

Expeller (extrusion) pressing, on the other hand, uses a continuous or interrupted-flight screw mounted in a barrel to press oil out of the soybeans during the extrusion process, instead of direct pressing via a hydraulic cylinder (i.e., in the hydraulic press method). This process can remove approximately 80%–90% of the oil from the seeds, so the soybean meal tends to be higher in oil content. Since the seeds are under high pressure and undergo friction between the screw and barrel, the process generates a high amount of heat (~140–210°F). Slow and low resistance expelling can result in lower process temperatures (~122°F), but this results in low output and efficiency. The expelling efficiency largely depends on the type of oilseed, temperature of the process, pressing configuration in the extruder, and pretreatment of the raw material. Oilseeds can be pretreated by heating while keeping the temperature below 104°F. While high heat increases the oil extraction efficiency, it decreases its shelf life (Farm Energy, 2015).

Many years ago the following equation was developed to calculate the oil yield after pressing (Koo, 1938):

$$W = CW_o P^{1/2} t^{1/6} v^{-z/2} \tag{1}$$

where

W = Oil yield (wt%)

C = Oilseed constant

W_o = Oil content of the seed (wt%)

P = Pressing pressure (MPa)

t = Time (h)

υ = Kinematic viscosity of oil at given temperature (m^2/s)

z = Exponent of kinematic viscosity varying from 1/6 to ½

For soybeans, $z = 1/2$, $C = 5.4 \times 10^3$ and $W_o = 19.5$.

Solvent extraction is a widely used extraction process due to its high efficiency and low cost (Sawada et al., 2014). Solvent extraction, on other hand, has proven to be excellent for a wide range of oilseeds, oil contents, and processing conditions. The most common solvent used industrially is hexane, which is subsequently separated from the oil by evaporation and distillation (Rosenthal et al., 1996). Solvent extraction is often the preferred method when edible oil needs to be extracted from an oilseed, and the process is extremely efficient, often yielding >95% of the oil from the seed, and the process can recover >95% of the hexane, which can be used again. Despite all the advantages of this process, it is well known that hexane is a highly flammable liquid and can pose health hazards to the workers in the extraction plant—exposure can causes severe damage to the nervous system from either short-term exposure to high levels as well as low exposures over an extended time period. Thus there appears to be a need for a safer and more eco-friendly extraction process and/or extraction medium—thus some of the alternative processes discussed in this chapter.

Once extracted using a solvent (most often hexane), the crude oil undergoes degumming (to remove phosphatides and gums), alkali treatment (to remove metallic pro-oxidants, free fatty acids), bleaching (to remove soap particles and pigments), and deodorization (to remove off-flavors) (Brekke, 1980) as shown in Fig. 3.

Studies have been conducted to test if supercritical fluid, specifically supercritical CO_2 (SC-CO_2), could replace hexane. Oil from full-fat flaked soybeans was extracted using hexane and SC-CO_2 and the yield obtained was 20% and 19.9%, respectively. Along with the yield, the percentage of free fatty acid, peroxide value, and unsaponifiables of SC-CO_2 (0.5%, 0.2%, and 0.5%, respectively) were comparable to hexane too. The phosphorus content of oil decreased by approximately 10-fold and the chromatography refining loss reduced by 3.8% when SC-CO_2 was used (Friedrich and List, 1982). It was also observed that when the pressure at which CO_2 is supplied was increased from 5000 to 8000 psig, the extraction efficiency increased. At higher pressures, the solubility of oil in CO_2 increased with increasing temperature (from 50°C to 60°C at 6000 psig) (Friedrich et al., 1982).

Other solvents like isopropanol were studied individually and in combination with hexane. After 3 h of extraction, oil extraction yield of isopropanol was

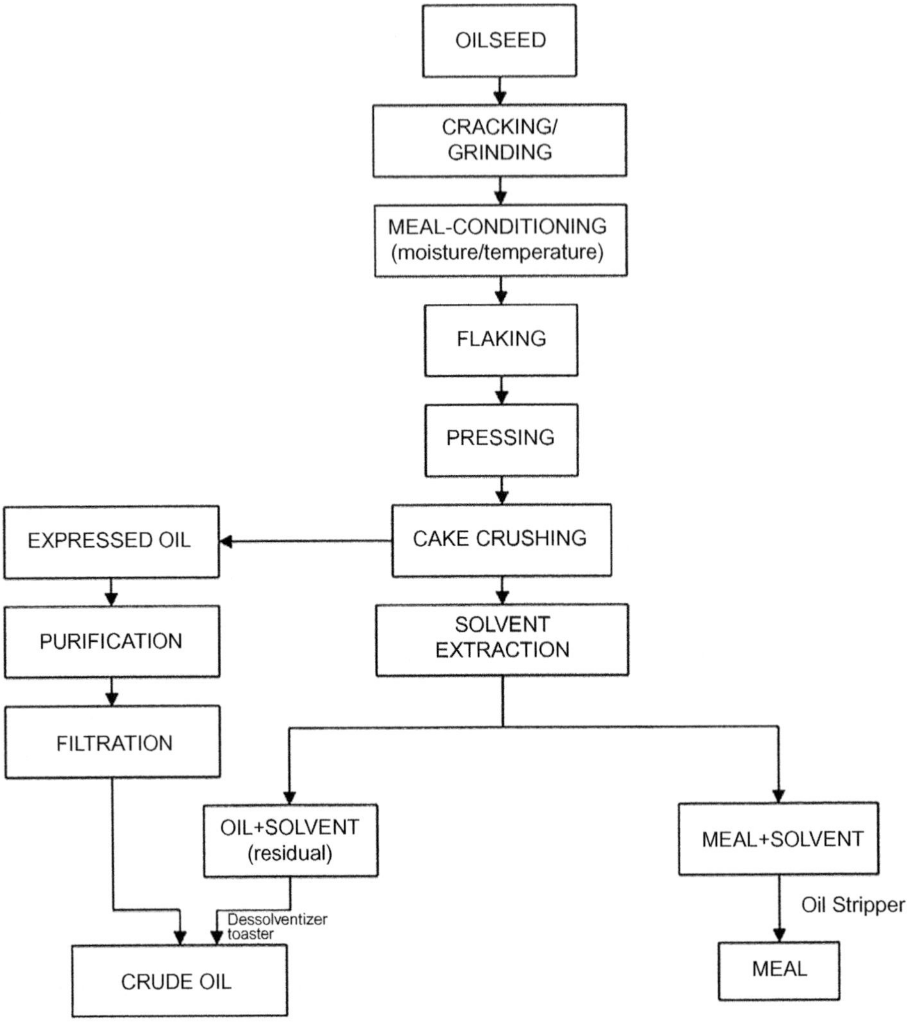

FIG. 3
Solvent extraction and pressing (expelling) are the most common methods used to extract oil from an oilseed. Based on Rosenthal, A., Pyle, D. L., Niranjan, K., 1996. Aqueous and enzymatic processes for edible oil extraction. Enzyme Microb. Technol. 19(6), 402–420.

28%, hexane 34%, and mixed solvent (isopropanol + hexane) ~40%. Various levels of ultrasound waves were tested while using hexane and it was observed that with increasing intensity, the oil yield increased. This is the result of cavities produced by compression and shearing while ultrasound waves are passed through the flaked soybeans. The yield was >45% when ultrasound of intensity 47.6 W/cm^2 was used (Li et al., 2004).

Ethanol was also investigated as a potential solvent. Absolute ethanol could achieve the yield of 21% at 60°C. Ethanol was diluted with deionized water at 6% mass basis and its yield was >20% only at 90°C. Protein extraction decreased with dilution (Sawada et al., 2014).

3 ADDING ENZYMES TO SOLVENT-BASED EXTRACTION

Sherba et al. (1972) first investigated enzymatic extraction when they fractionated soybean using protease as reported by Rosenthal. Fullbrook then worked on aqueous hydrolysis of oilseed followed by the addition of solvent and in simultaneous presence of solvent. His results showed that oil extraction was more efficient when the solvent was present during the aqueous hydrolysis.

The extraction of oil with mixture of enzymes (pectinase, cellulase, and hemicellulase) was also conducted where the solvent used was petroleum ether (Olsen, 1988). Enzymes partially hydrolyze the cell wall and increase the cell permeability. When the canola flakes are autoclaved and moisture adjusted and then subjected to enzyme mixture along with different kinds of carbohydrase, the yield of canola oil and extraction time varies. It is then followed by drying and hexane extraction. The use of the enzymes increases the yield by 45% and the time of extraction decreases comparatively (Sosulski et al., 1988).

To enhance extraction of antioxidant, pectin polysaccharide is processed with enzyme mixture. Enzymes have shown to increase the extraction to 7.4 g/kg from 1.7 g/kg of raw material (dry weight) (Gan and Latiff, 2010). For enhanced extraction of lycopene from tomatoes, cellulase and pectinase are used under optimal condition that increased the yield by 2.5-fold (Choudhari and Ananthanarayan, 2007).

Enzymatic treatment along with hexane extraction has been studied for low moisture content soybean. For soybeans with moisture content between 15% and 20%, cellulose and multi-effect enzymes were added before solvent extraction and they were added simultaneously for soybeans with moisture content lower than 12%. While the total oil extracted from untreated sample (only solvent extraction; no enzymes) was 79.50%, it was 84.85% for 10% moisture soybean that was treated with enzymes and solvent at 0.75 E/S ratio (Dominguez et al., 1995).

4 AQUEOUS EXTRACTION PROCESS

The aqueous extraction process (AEP) may be an environmentally cleaner alternative to hexane extraction, as the only medium used for extraction is water. In addition, in the AEP, protein isolates are extracted along with edible oil. The damage done to protein is often negligible, and the elimination of solvent can make the process more cost effective. Adequate treatment of effluent and relatively low oil yield are amongst the drawbacks of AEP, unfortunately (Rosenthal et al., 1996).

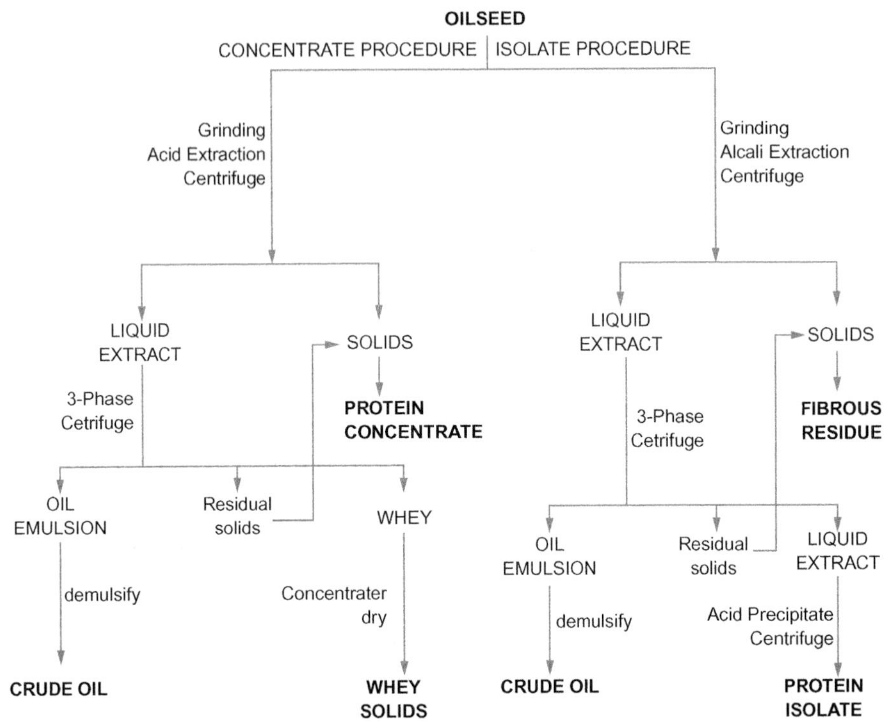

FIG. 4
Steps involved in the aqueous extraction process (AEP). Two alternatives are shown. Based on Rosenthal, A., Pyle, D. L., Niranjan, K. (1996). Aqueous and enzymatic processes for edible oil extraction. Enzyme Microb. Technol. 19(6), 402–420.

In this process, oil is extracted from the oilseeds with the help of water using the principle of dissolution of oil in water. The oilseeds are conditioned, ground, and oil is extracted by boiling water, which floats on the surface. Oil is removed and dried. Instead of boiling water, oil and protein isolates can be separated using centrifuge too. This prevents protein denaturation caused by high temperatures.

Based on the pH of the extraction medium, protein can be collected from solids as concentrate or from liquid phase as isolates (Fig. 4).

For the extraction of oil from soy flour, temperature, solid-liquid ratio, particle size, agitation speed, and pH are the deciding factors. It is shown that at pH of 4.5, the oil and protein yield of AEP of soy flour is very low since at that pH, soy proteins have very low solubility. The oil and protein yield decreases with increasing particle size too (Rosenthal et al., 1998).

Mechanism of crushing the soybean grains majorly affects the particle size that in turn affects the oil yield. Milling, flaking, milling + flaking (flour from flakes), and extrusion were tested and complete cellular disruption was achieved only in extrusion. Light microscopy of residual matter after 2 h of aqueous extraction of flour from flakes showed coalesced oil droplets, dissolved proteins, and very little residual matter in extracellular space. In the case of extruded soybean, oil droplets were found in the solid matrix. The oil yield was 75% for flour from flakes and 68% in the case of extruded material (Campbell and Glatz, 2009). Combining flaking and extrusion would ensure complete cell disruption and increase the oil yield.

The oil yield of AEP is usually low when compared to solvent extraction since 100% recovery of oil from skim is not achieved and some oil stays un-extracted in the fiber-rich fraction. This problem can be addressed by using subcritical water (water with temperature >100°C but <374°C at pressure between 1 and 8 MPa; condensed form), which can be used to extract proteins since the polarity of water decreases at subcritical conditions making the extraction of bioactive compounds easier. Oil yield varied from 38.2% to 83.9% in the case of extruded soybean flakes and between 4.2% and 50.2% for un-extruded soybean flakes when the solids to liquid ratio changed between 1:3.3 and 1:10, temperature varied between 66°C and 234°C, and the extraction time varied between 13 and 47 min. Temperature affects the solubility and dissolution of solute in solvent. When heated, the denatured proteins sequester the oil particles and at temperature >150°C, protein extractability was slightly higher in the case of extruded flakes. Conditions that increased the oil and protein yield in extruded flakes and just the flakes were 150°C and 66°C, respectively, and solid-to-liquid ratio was 1:11.7 for both (Ndlela et al., 2012).

5 ENZYME-ASSISTED AQUEOUS EXTRACTION PROCESS

Inadequate rupturing of the cell walls (and thus difficult oil removal) has been thought to be the reason for low yield during the AEP process. Flaking and extruding can help enhance oil extraction: extruding has been shown to increase the extraction by up to 23% in AEP (Campbell, 2010). Furthermore, it has been shown that the addition of hydrolytic enzymes (up to 3%), extracted from *Aspergillus niger*, to AEP of soybeans could increase the oil extraction yield to 90% (Fullbrook, 1983). The carbohydrase breaks down the cotyledon cell wall structures and cell membranes, while protease hydrolyzes the lipophilic proteins, and the lipids are easily released for extraction. Thus enzyme-assisted aqueous extraction (EAEP) has been shown to be an effective alternative to solvent extraction.

The addition of enzymes like cellulase, pectinase, and protease to the AEP for enzymatic breakdown of the cell components makes the process EAEP process. It has been widely used in the past to extract natural pigments, flavors, medicinal compounds, polysaccharides, and oils. The advantage is accelerated extraction, enhanced recovery, energy efficient process and it is eco-friendly process as solvent usage is reduced (Puri et al., 2012). Phenolic compounds have been extracted from citrus peel using EAEP. The citrus peel was grounded and pretreated with Cellulase MX, Cellulase CL, and Kleerase AFP (food grade enzymes). After centrifugation and filtration, the filtrate is extracted by evaporation of solvent using rotary evaporator (Li et al., 2006).

Enzymes have been incorporated in the rural extraction processes too. The copra meal was finely milled and slurry with water was prepared. The slurry then was pretreated with enzymes like protease and pectinase from *A. niger*, cellulase/hemicellulase from *Trichoderma reesei* and α-amylase from *Aspergillus oryzae* and incubated for 6 h at 37°C. Water floatation technique was used to extract the oil. Extraction yield reportedly increased by 50% when compared to the controls (Tano-Debrah and Ohta, 1997).

Enzyme mixture depends on the meal that is being used. For peanut oil, cellulase, protease, and α-1,4-galacturonide glucanohydrolase has been shown to increase the yield by 6%–10%. Protizyme, containing papain, trypsin, and chymotrypsin, has been shown to increase the peanut oil yield from 44% to 92% when the shaking speed was 80 rpm (Sharma et al., 2002).

In a study Campbell and Glatz (2009) examined AEP and determined that the addition of a protease (Protex 7L) increased the oil yield from flour produced from flakes and from extruded soybean flakes, as tabulated in Table 2.

It was clear with the light microscopy image of protease hydrolyzed residual matter that most of the protein was dissolved and the material is loose and amorphous. Cellulase was shown to have no effect on oil yield in the case of extruded material and this indicates that extrusion process ensures complete cell disruption (Campbell and Glatz, 2009).

In a study conducted by de Moura et al. (2008), soybeans were extruded at around 100°C and 100 rpm screw speed in a twin-screw extruder. The extruded flakes were added to water to achieve 1:10 solid-to-liquid ratio. Proteases (Protex 6L-alkaline protease—at 0.5% and 1.0% and Protex 7L-neutral protease—at 0.5%) were added at optimum concentration, temperature, and pH. The slurry was centrifuged to separate out insolubles after the extraction and the liquid fraction was processed to separate the free oil. A process diagram of this is displayed in Fig. 5.

Table 2 Oil Yield (in %) With and Without Enzyme (Protease-Protex 7L) Use

Oil yield (%)	Soybean flour from flakes	Soybean extruded flakes
AEP (without enzymes)	75	68
AEP (with Protex 7L)	79	88–96

Adapted from Campbell, K.A., Glatz, C.E., 2009. Mechanisms of aqueous extraction of soybean oil. J. Agric. Food Chem. 57, 10904–10912.

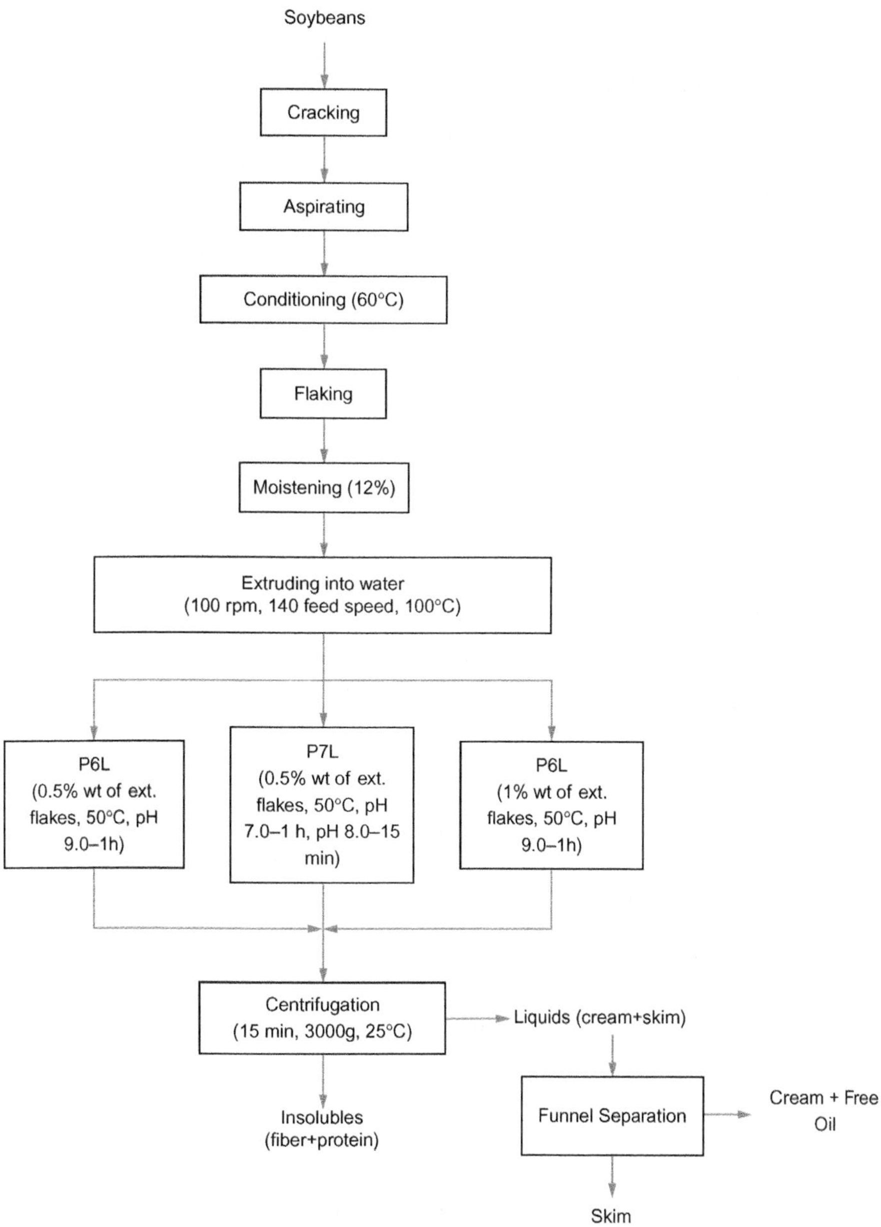

FIG. 5

Flow chart of the enzyme-assisted extraction process of soybean where two different proteases were compared (Protex 6L and Protex 7L). Based on de Moura, J. M. L. N., Campbell, A. K., Mahfuz, A. A., Jung, A. S., Glatz, A. C. E., Johnson, A. L., 2008. Enzyme-assisted aqueous extraction of oil and protein from soybeans and cream de-emulsification. J. Am. Oil Chem. Soc. 85(10): 985–995. doi: 10.1007/s11746-008-1282-2.

Table 3 Oil and Protein Yields Resulting From Different Enzyme Treatments on Extruded Soybean Flakes, Along With Size of Peptides Produced

	Protex 6L 0.5%	Protex 6L 1%	Protex 7L
Oil yield (%)	96	97	93
Protein extraction yield (%)	85	87	73
Molecular weights of peptides extracted	<30 kDa	–	>54.1 kDa
Dry matter extraction yield (%)	77	79	71

All values are rounded-up.
Adapted from de Moura, J. M. L. N., Campbell, A. K., Mahfuz, A. A., Jung, A. S., Glatz, A. C. E., Johnson, A. L., 2008. Enzyme-assisted aqueous extraction of oil and protein from soybeans and cream de-emulsification. J. Am. Oil Chem. Soc. 85(10), 985–995. doi: 10.1007/s11746-008-1282-2.

The high oil yield with Protex 6L (Table 3) might be due to better selection of soybean varieties, extrusion parameters, and/or enzymes. The amount of free oil obtained using Protex 6L (0.5%) was twice the amount obtained using Protex 7L. It was observed that conditions favoring protein and oil extraction were similar because protein networks and oleosin membranes capture oil particles that are released when the network is broken down. High amount of solids were found in skim fraction in the case of 0.5% Protex 6L (58%).

6 FACTORS AFFECTING OIL AND PROTEIN RECOVERY

6.1 Pre-Extraction Steps

Grinding and flaking operations determine the particle size of the soybean during processing. The smaller the particle size, the easier it is for water-soluble particles to separate out. Enzyme dissolution also becomes easier when the size is small. Based on the moisture content of the seeds, they are either ground dry or wet. These operations along with extrusion determine the stability of emulsion. In the absence of enzymes, the oil recovery from dry pellets of soybean was 50%, which increased to 60% on grinding the pellets to about 2–3 mm in diameter and the oil recovery further increased to 75% when the pellets were extruded under water at 100°C (Lamsal et al., 2006a, 2006b).

The EAEP of extruded full-fat soy flakes yielded more oil when compared to AEP of full-fat soy flours and separatory funnel procedure was used to quantify the oil in each fraction (Lamsal and Johnson, 2007). The free oil obtained by EAEP of extruded flakes was 8 times that of AEP of flour (Table 4) indicating that the extrusion and use of enzymes for extraction has significant effects on the oil yield. The size of oil droplets produced by extruded flakes was 2.25 times larger than droplets produced by AEP of soy flour. Increase in size results in higher terminal

Table 4 Distribution of Oil, Protein, and Solids Amongst Various Fractions of AEP of Soy Flour and EAEP Extruded Flakes

	Fractions	Cream	Skim	Insolubles	Free oil
Oil yield (%)	AEP of full-fat soy flour	45	15	35	2
	EAEP of full-fat extruded soy flakes	60	13	13	16
Protein yield (%)	AEP of full-fat soy flour	2	85	19	–
	EAEP of full-fat extruded soy flakes	1	79	23	–
Dry matter (%)	AEP of full-fat soy flour	12	53	34	1
	EAEP of full-fat extruded soy flakes	15	53	30	4

All the values are rounded up.
Adapted from Lamsal, B.P., Johnson, L.A., 2007. Separating oil from aqueous extraction fraction of soybeans. J. Am. Oil Chem. Soc. 85(8), 785–792.

velocity and the oil particles rise up, and coalesce making more and more free oil available (Lamsal and Johnson, 2007).

Temperature while extrusion plays a major role in protein extraction and in turn oil extraction. High-temperature denatures proteins, breaking the protein networks, hence releasing the oil trapped. At temperatures higher than 100°C, frothing was observed which leads to oil sequestering, thereby decreasing the oil yield. Temperature of extruder barrel has been shown to positively affect the oil and protein yield till 100°C (at 12% moisture content and 100 rpm screw speed; oil yield 55%) and negatively affect it at temperatures 120°C (at 12% moisture content and 100 rpm screw speed; oil yield 45%) and higher (Lamsal et al., 2006a, 2006b).

It is clear from Table 5 that increases in temperature have negative effects on both the oil and protein yields. Moisture content has positive effect on the dependent variables since the diffusion of oil and its release becomes easier. The screw speed did not seem to have much effect on the oil and protein yield since at 14% moisture content and 100°C temperature, change in rpm did not matter.

6.2 Enzymes Used for Extraction and De-Emulsification

Since enzymes have been shown to increase the oil and protein yield, their concentration, pH of the slurry when added, and temperature is very important. It is not necessary that every enzyme may give good results. Cellulase, for example, has not shown any effect on the oil yield of extruded soy flakes. Proteases, on other hand, can degrade the peptides surrounding oil molecules called oleosin, making oil extraction easier.

Table 5 Effect of Temperature of Extruder Barrel, Moisture Content of Flakes and Screw Speed on Oil and Protein Yield

Temperature of barrel (°C)	Moisture content of extruded soybean flakes (%)	Extruder screw speed of (rpm)	Oil yield (%)	Protein yield (%)
100	12	100	53	51
100	12	150	50	54
100	**14**	**100**	**55**	**59**
120	12	100	46	33
100	14	150	55	54
120	14	150	42	35

All values are rounded up. Bold formatted numbers have the highest oil and protein yields.
Adapted from Lamsal, B.P., Reitmeier, C., Murphy, P.A., Johnson, L.A., 2006a. Enzymatic hydrolysis of extruded-expelled soy protein and resulting functional properties. J. Am. Oil Chem. Soc. 83(8), 732–737 and Lamsal, B.P., Murphy, P.A., Johnson, L.A., 2006b. Flaking and extrusion as a mechanical treatment for enzyme-assisted aqueous extraction of oil from soybeans. J. Am. Oil Chem. Soc. 83(11), 973–979.

Kapchie et al. (2008) compared the effect of enzyme cocktail (pectinase, cellulase, and Multifect CX3L) at different concentrations and total time of application with the control (no enzymes) on oleosomes. The extraction was performed by blending the hydrated soybean flour with buffers in a Waring blender. As the blending time increased, the oil yield in control decreased since it created an emulsion in the upper layer after centrifugation. In the case of enzyme-assisted extraction, the oil yield increased with increasing blending time. With increase in enzyme concentration, the total oil yield increased and the oil content in residue decreased (Table 6).

When similar studies were carried out at pilot plant scale, after 10 h of centrifugation, 93% of oil was recovered from the oleosome fraction. The enzyme cocktail used here was slightly different from the laboratory scale: pectinase, Multifect CX B, and Multifect CX G. Laboratory-scale experiments were performed again with the new set of enzymes and the oil yield was found to be 77%. Horizontal decanter centrifuge used in pilot plant in comparison to series bench top centrifuge used in laboratory scale was better at mixing and recirculating slurry for maximum oil release. Protein content was highest in supernatant after centrifugation (~26%) and the distribution of glycinin and β-conglycinin fractions were studied in supernatant, precipitate (oleosome), and initial soy flour and there was a slight decrease in percentage of total mass in supernatant of these protein subunits in pilot plant when compared to laboratory scale (Towa et al., 2011a, 2011b).

When 0.5% Multifect Protease enzyme was added to soybean flakes, the oil yield increased from 46% to 71% and for extruded flakes, it increased from 56% to 88%. Extrusion on its own increased the yield by 10% (Lamsal et al., 2006a, 2006b).

Table 6 Changes in Oil Yield With the Enzyme Concentration

Oil yield (%)	Control (0% enzyme)	0.6% enzymes (v/w)	1.5% enzymes (v/w)	3.0% enzymes (v/w)
45-s blending time	36	–	–	36
180-s blending time (1st extraction)	38	52	59	64
180-s blending time (2nd extraction)	7	10	14	14
180-s blending time (3rd extraction)	2	1	8	5
180-s blending time (4th extraction)	1	1	3	3
Summation of Yield (%) at 180-s blending	49	57	82	85

The Enzyme Combination Used here was Comprised of Pectinase, Cellulase, and Multifect CX3L. All values are rounded up.
Adapted from Kapchie, V., Wei, D., Hauck, C., Murphy, P.A., 2008. Enzyme-assisted aqueous extraction of oleosomes from soybeans (Glycine max). J. Agric. Food Chem. 56, 1766–1771.

Protex 6L when added to isolated oleosomes at 0.25% dosage recovered lesser oil (3h of hydrolysis and 30min of destabilization time; oil yield of 65%) when compared to 2.5% of Protex 6L (3h of hydrolysis and 30min of destabilization time; oil yield of 85%). The yield increased with increase in hydrolysis and destabilization time. Maximum yield of 90% was reached at 2.5% of Protex 6L, 18h of hydrolysis, and 3h of destabilization time (Towa et al., 2011a, 2011b).

The presence of substances like free fatty acids or phospholipids decreases the quality of oil and it was observed that while the percentage of free fatty acid of hexane-extracted freeze dried soybean oil was 1.18%, enzyme-assisted aqueous extracted soybean oil had only 0.11%–0.18% of fatty acids. This was because enzymatic hydrolysis occurred at high temperatures and basic pH at which the free fatty acids would neutralize and precipitate out when centrifuged. These fatty acids affect the oxidative stability index (OSI) of the oil too. While the OSI of soybean oil extracted by EAEP was lower (12h) than the crude soybean oil (27h), it was slightly higher when compared to hexane extracted soybean oil (9h) and almost similar to commercial sold soybean oil (11h) (Towa et al., 2011a, 2011b).

When different food grade enzymes [endopeptidase: Multifect Neutral (MN), Bromelain (BR) and exopeptidase: exo. C] were tested for their ability to emulsify soy flour hydrolysates, the emulsification capacity (gram of oil/g of protein) decreased from 1935 (control-no enzyme) to 1702 for MN, 1456 for BR, and 1288 for exo. C (Lamsal et al., 2006a, 2006b).

Table 7 Effects of Extrusion and Enzymes on Oil and Protein Yield

Processing prior to extraction	Extraction process	Oil yield (%)	Protein yield (%)
Full fat flakes	Aqueous extraction without enzymes	60	74
Full fat flakes	Aqueous extraction with enzymes (Protex 7L at 0.5% dosage)	60	76
Extruded flakes	Aqueous extraction without enzymes	68	45
Extruded flakes	Aqueous extraction with enzymes (Protex 7L at 0.5% dosage)	90	75

All values are rounded-up.
Adapted from Jung, S., Maurer, D., Johnson, L.A., 2009. Factors affecting emulsion stability and quality of oil recovered from enzyme-assisted aqueous extraction of soybeans. Bioresour. Technol. 100(21), 5340–5347.

While reviewing the literature of oil extraction from soybean, it was determined that two main factors which significantly affect the oil and protein yield of the process were extrusion and addition of enzymes as mentioned above. Extrusion mainly ruptures the cell wall and makes the proteins available for enzymes (proteases) to breakdown and for water to flow and carry the free oil. Cellulases are ineffective when the extraction is preceded by extrusion.

Table 7 shows that the oil yield of 15% moisture soy flakes increases to 90% from 60% when extruded at 100 rpm, at 100°C, before extraction, and 0.5% proteases are added. The solubility of isolated proteins increased when the flakes were extruded from 71% to 94% (Jung et al., 2009).

7 DOWNSTREAM PROCESSING

Generally, in AEP, oil and protein are demulsified and then centrifuged to separate into layers of aqueous and oil phase. Centrifugation can cause creaming at times but it gives complete separation through phase inversion (Rosenthal et al., 1996). In EAEP of soybean, the free oil + cream fraction undergoes de-emulsification to separate residual skim and to obtain free oil. Centrifugation separates the de-emulsified fraction into three distinct layers—free oil layer, intermediate layer, and skim layer (Figs. 6 and 7) (Campbell and Glatz, 2010; Wu et al., 2009). De-emulsification can be done physically (freeze-thaw, heating), chemically (pH 4.5), or with enzymes (like phospholipase C, Protex 6L, or enzyme cocktail).

When heating at 95°C, freeze-thaw, Lysomax/G-zyme cocktail (enzyme cocktail) and phospholipase C were tested for cream de-emulsification against the control and except for heating, every other treatments were very effective in destabilizing

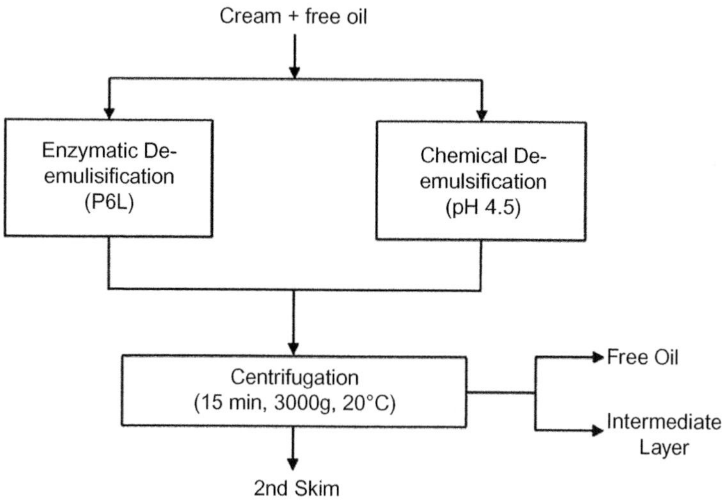

FIG. 6

Flow diagram of cream de-emulsification. Based on de Moura, J. M. L. N., Campbell, A. K., Mahfuz, A. A., Jung, A. S., Glatz, A. C. E., Johnson, A. L., 2008. Enzyme-assisted aqueous extraction of oil and protein from soybeans and cream de-emulsification. J. Am. Oil Chem. Soc. 85(10), 985–995. 10.1007/s11746-008-1282-2.

the emulsion. While freeze-thawing cause denaturation of soy protein, breaking the emulsion and releasing the oil as free oil, Lysomax and phospholipase breaks the ester bond between the fatty acid groups. Freeze-thaw yielded around 86% of free oil and phospholipase recovered 73% of oil as free oil (Lamsal and Johnson, 2007). The effect of pH was tested on the stability of cream and free oil emulsion. Control was held at pH 8, which is the initial cream pH. Yield of free oil increased with decreasing pH and maximum (100% free oil yield) was reached at pH 4–4.5 as shown in Figs. 6 and 7.

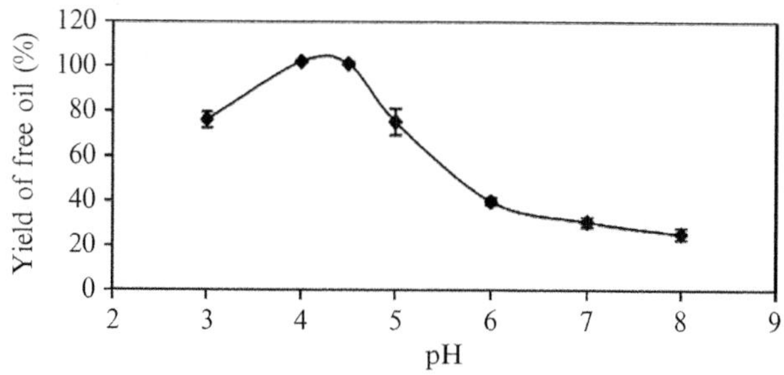

FIG. 7

Free oil yield as the pH of (cream + free oil) fraction changes. Based upon Wu, J., L.A. Johnson, S. Jung., 2009. De-emulsification of oil-rich emulsion from enzyme-assisted aqueous extraction of extruded soybean flakes. Bioresour. Technol. 100(2), 527–533.

LysoMax was compared to Protex 51FP to test its impact on cream destabilization and it was observed that Protex 51FP yielded 88% of oil at 0.2% level, which was more than twice of that LysoMax achieved at the same concentration (Wu et al., 2009).

According to de Moura et al. (2008), the free oil yield in percentage is calculated as

$$\text{Free Oil Yield}(\%) = \frac{[\text{Free oil (g)} + \text{hexane washed free oil (g)}]}{[\text{cream (g)} \times \text{oil content } (\%) \text{in}\{\text{cream} + \text{free oil fraction}\}]} \tag{2}$$

Pasteur pipette was used to collect free oil and it was quantified using hexane following the procedure mentioned by Lamsal and Johnson (2007).

Of note 2.5% of Protex 6L (P6L) was used for de-emulsifying cream produced by Protex 6L and Protex 7L extraction of soybeans. Maximum free oil yield for Protex 7L-cream was 91% and for Protex 6L-cream was 100%. When the pH of cream extracted with Protex 6L was set to 4.5, temperature was maintained at 25°C, and no agitation was provided, 100% free oil yield was achieved (de Moura et al., 2008).

Cream de-emulsification is an important step in EAEP and enzymes like proteases prove to be efficient in destabilizing the cream and separating out skim from the free oil. Research has been conducted to integrate EAEP and the de-emulsification step, and will be discussed in the following section.

8 TWO-STAGE EAEP AND SCALE-UP

According to de Moura et al. (2011a, 2011b), work on scaling up of EAEP has resulted in high oil and protein extraction yield when the solid-to-liquid ratios (1:10) was relatively low and extraction was performed in a single stage. When scaled up, at same ratio, the water required is too large which leads to large skim production. To enhance the protein and oil extraction, the amount of water required has to be reduced without compromising the efficiency of extraction. This is where two-stage countercurrent EAEP comes into play. The extruded flakes are processed through two stages of extraction and liquid fraction of second stage is recycled to first. Slurry from first stage is centrifuged to remove insoluble particle. Liquid phase is further separated into skim and cream as shown in Fig. 8. The insoluble part is fed again into the extractor. Second stage slurry is centrifuged and the liquid phase is recycled to first stage (de Moura et al., 2008).

The oil content in skim and insoluble decreased slightly in two-stage EAEP, which is desirable as it decreases downstream processing efforts. With increase in solid content in two-stage EAEP (solid-to-liquid ratio 1:5), the amount of water used decreased reducing the oil content in skim fraction and the extraction efficiency improved by 2%.

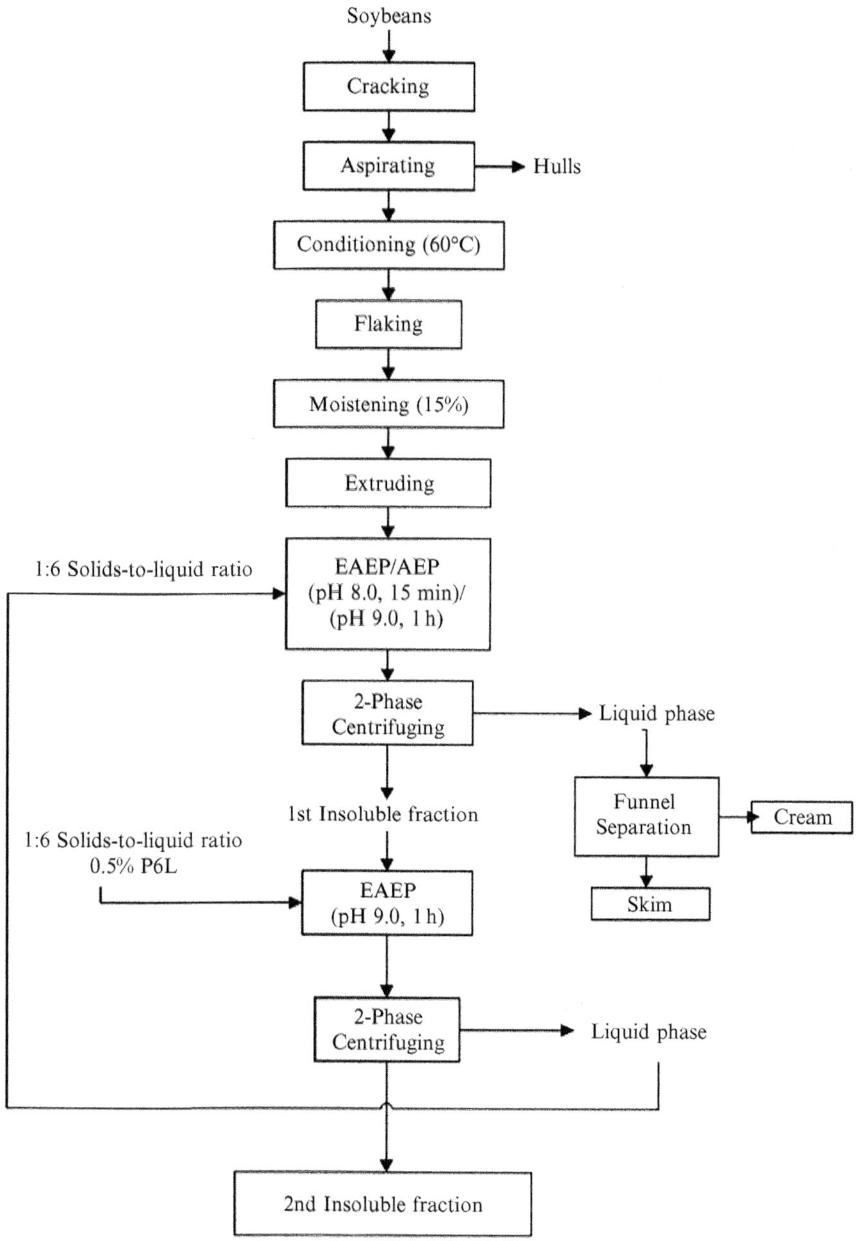

FIG. 8

Two-stage enzyme-assisted aqueous extraction process of soybeans. Based on de Moura, J. M. L. N., Campbell, A. K., Mahfuz, A. A., Jung, A. S., Glatz, A. C. E., Johnson, A. L. 2008. Enzyme-assisted aqueous extraction of oil and protein from soybeans and cream de-emulsification. J. Am. Oil Chem. Soc. 85(10), 985–995. doi: 10.1007/s11746-008-1282-2.

Table 8 Comparison of Single Stage EAEP to Two-Stage EAEP and Scale-Up of Two-Stage EAEP in Terms of Oil Yield, Protein Yield and Solids Yield

	Type of EAEP	Cream	Skim	Insoluble	Total extracted
Oil yield (%)	Standard single-stage EAEP	61	14	4	96
	Two-stage EAEP	85	13	2	98
	Two-stage EAEP-scale up (pH 8.0, 15 min)	76	23	1	**99**
	Two-stage EAEP-scale up (pH 9.0, 60 min)	86	12	1	**99**
Protein yield (%)	Standard single-stage EAEP	1	87	13	87
	Two-stage EAEP	0	91	8	92
	Two-stage EAEP-scale up (pH 8.0, 15 min)	7	87	6	94
	Two-stage EAEP-scale up (pH 9.0, 60 min)	9	87	4	**96**
Solids yield (%)	Standard single-stage EAEP	17	56	23	77
	Two-stage EAEP	23	57	20	80
	Two-stage EAEP-scale up (pH 8.0, 15 min)	23	59	17	83
	Two-stage EAEP-scale up (pH 9.0, 60 min)	28	56	16	**84**

All values are in percentage. Bold numbers are higher amongst respective yields.

Adapted from de Moura, J. M. L. N., Campbell, A. K., Mahfuz, A. A., Jung, A. S., Glatz, A. C. E., Johnson, A. L., 2008. Enzyme-assisted aqueous extraction of oil and protein from soybeans and cream de-emulsification. J. Am. Oil Chem. Soc. 85(10), 985–995. doi: 10.1007/s11746-008-1282-2.

When scaled up from 0.08 kg extruded flakes to 1 kg extruded flakes, the oil, protein, and solid extraction yield of single stage EAEP remained the same. The amount of oil in the skim fraction increased from 14% at laboratory scale to 20% at pilot-plant scale-up. When the two-stage EAEP was scaled-up, there was a slight increase in all the three extraction yields and the oil content in the skim fraction was higher (Table 8). The amount of oil in skim was reduced by changing the extraction conditions (pH 8.0 for 60 min) (de Moura et al., 2008).

Concurrent two-stage EAEP was integrated with cream de-emulsification with three stages in total—two extraction stages and one de-emulsification stage. The enzyme added in de-emulsification step was reused in the second stage of EAEP and the enzyme recovered after the second stage was recycled in the first stage of EAEP (Fig. 9) (de Moura et al., 2011a, 2011b).

There was a slight decrease in the oil and protein yield when compared to two-stage EAEP, as illustrated in Table 9. Since the first liquid fraction was settled overnight, the weight and viscosity increased making it difficult to separate

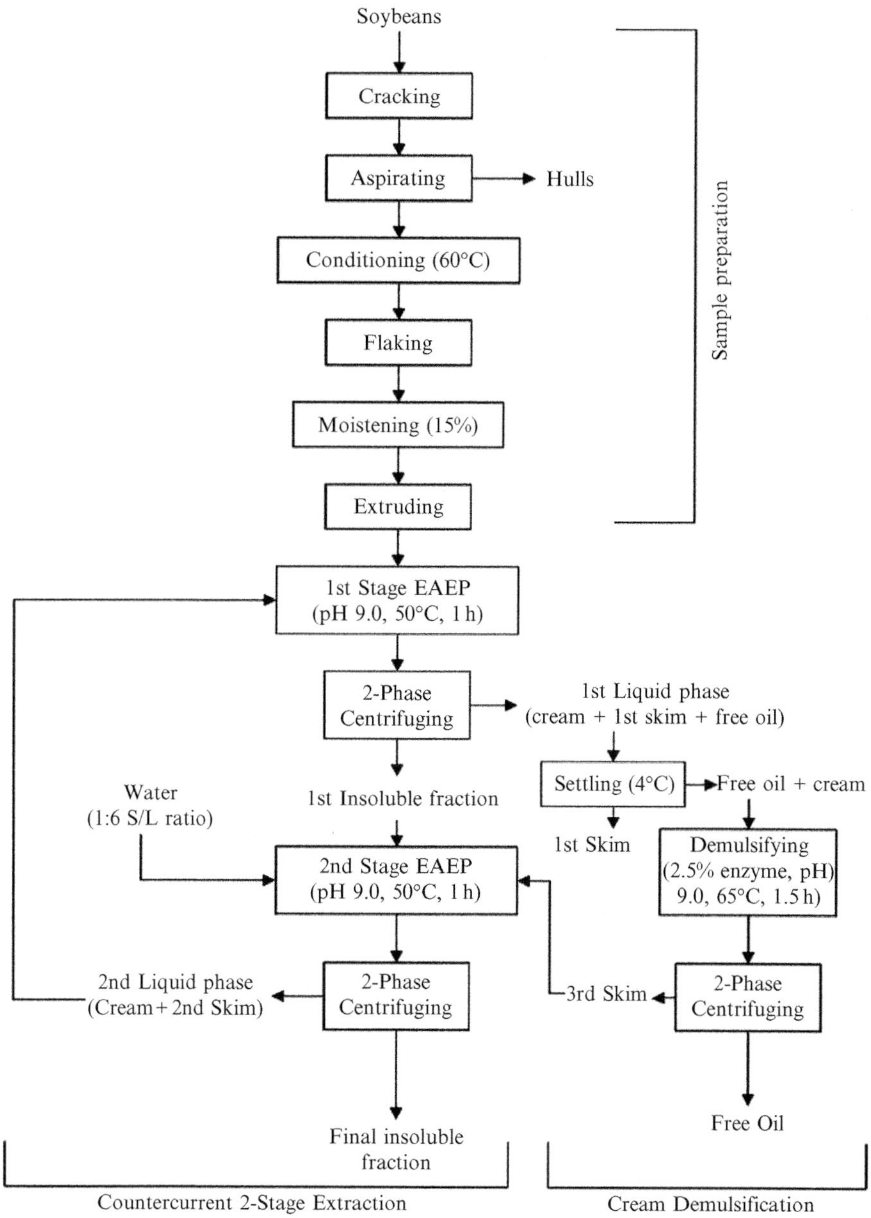

FIG. 9

Integrated concurrent two-stage enzyme-assisted aqueous extraction process of soybeans and cream de-emulsification (based on de Moura et al., 2011a, 2011b).

Table 9 Yields of Integrated Two-Stage EAEP and Cream De-Emulsification

	Cream	Skim	Insoluble	Total extracted
Oil yield (%)	64	32	4	96
Protein yield (%)	8	81	11	89
Solids yield (%)	21	59	18	81

Adapted from de Moura, J.M.L.N., Maurer, D., Jung, S., Johnson, L.A., 2011a. Integrated countercurrent two-stage extraction and cream de-emulsification in enzyme-assisted aqueous extraction of soybeans. J. Am. Oil Chem. Soc. 88(7), 1045–1051 and de Moura, J.M.L.N., Maurer, D., Jung, S., Johnson, L.A., 2011b. Pilot-plant proof-of-concept for countercurrent two-stage enzyme-assisted aqueous extraction processing of soybeans. J. Am. Oil Chem. Soc. 88(10), 1649–1658.

the cream. The separated cream needed enzymes 1.5 times more than usual. When the amount of protease increases, it results in extensive hydrolysis of proteins and there are high chances that new interactions between peptides and oil may have stabilized the emulsion instead of destabilizing it (de Moura et al., 2011a, 2011b).

The amount of enzyme required in this process is very high and the yield obtained is comparatively lower than two-stage EAEP. Although this process was developed to reduce the overall enzyme use, the amount of enzymes required in de-emulsifying third skim is greater than the amount required in two-stage EAEP.

9 SOLVENT EXTRACTION VERSUS AEP AND EAEP

9.1 Environmental Aspects

Cheng et al. (2016) conducted environmental impact (EI) assessment of soybean oil extraction by hexane-based oil extraction method and EAEP. Hexane extraction had an input component EI value of approximately 39 with hexane contributing 92% to it. EAEP, on the other hand, had an input component impact value of ~10 in a multiplying system. The general environment impact (GEI) value of hexane-based extraction and EAEP were ~3.5 and 1.0, respectively.

The output component EI of EAEP was higher when compared to hexane-based extraction (~42 and 36) with soy skim contributing 76% to the EI of EAEP. Soy skim, coproduct of EAEP, is produced in large quantity and as mentioned before, contains around 15%–20% of extracted oil. Since the downstream processing of soy skim is difficult, finding a commercial use of it can reduce its impact. Hexane-based extraction still had the highest GEI of output components (~3.4).

At the rate of 1 kg of soybean oil produced, the greenhouse gas (GHG) emissions of EAEP were higher. Pretreatment of soybean is an essential step in EAEP for higher oil and protein yield and these steps consume electricity 3 times more than the hexane-based extraction method, which drives up the process's GHG emission (Cheng et al., 2016).

9.2 Economic Aspects

For the period 2010–2014, the total plant direct cost (TPDC) of hexane-based oil extraction process accounted for ~44%–48% of total capital investment. While start-up cost was just ~4%, the working capital accounted for 14.95% of the total investment. In the 2015 estimate, TPDC takes up 45% of the total capital cost of EAEP with 13% working capital and 4% start-up capital. In hexane-based extraction, hexane and soybeans are the two raw materials used which accounted for 89.75% of total annual operating cost in 2010–2014. At a large scale of EAEP, materials contribute 83%–88% of the total operating cost: 60.45% soybean, 30.68% enzymes, and remaining water and ammonium hydroxide. While hexane-based extraction is capital intensive, EAEP has higher operating costs. Hexane recovery rate is as high as 95%, which reduces annual consumption rate and saves the operating cost. Recycling enzymes can largely improve the economic feasibility of EAEP.

While soybean oil contributed ~33% toward the total revenue of hexane-based oil extraction, it accounts for just 23.84% in EAEP. The economic aspects of EAEP can be improved if the coproducts of the process can be sold for integrated corn-soy fermentation.

10 CHEMICAL TREATMENTS OF SOY PROTEINS

The digestibility of soy proteins can be improved through various chemical treatments. The major form in which protein is stored in soybeans is glycinin, which contains various disulfide bonds. A trait seen in most plant inhibitor proteins is the presence of intramolecular disulfides (Garicia-Olmedo et al., 1987). Inhibitors, which reduce or prevent protein digestion, most commonly seen in soybeans are the Kunitz trypsin inhibitor (i.e., trypsin inhibitor) and Bowman-Birk inhibitor (i.e., chymotrypsin inhibitor) (Jiao et al., 1992). The number of disulfide bonds involved in the Bowman-Birk inhibitor and the Kunitz trypsin inhibitor are 7 and 2, respectively (Birk, 1976; Wilson, 1988). Treating soy proteins with high temperatures causes denaturation of a portion of the anti-nutritional factors, but is often not enough to completely inactivate them. For example, the trypsin inhibitor was inactivated only 80% by heating soy protein for half an hour at a temperature of 120°C (Friedman et al., 1991). Thus, additional treatments are necessary. During chemical treatments, there may be a change in the three-dimensional structure of the protein due to reduction that makes the proteins more available for digestion (Herkelman et al., 1991).

For example, Wang et al. (2009) studied the use of mild temperatures along with sulfite and metasulfite chemical treatments in order to increase the digestibility of soy protein. Soy white flour produced through hexane extraction was used for the study. Results showed that the samples, which had not been subjected to heat, had lower sulfhydryl content after treatment with the reducing agents, which was attributed to the reoxidation of sulfhydryl compounds. Due to the inactivation of trypsin inhibitors by sodium metabisulfite and sodium sulfite, the in vitro digestibility was higher for the samples that were treated when compared to the untreated samples; this was true even at a mild temperature of 55°C. An increase in either of the two chemicals resulted in higher digestibility values, but the digestibility when sodium metabisulfite was used was higher. Maximum digestibility was recorded at 100°C. Digestibility at 100°C was 3 times greater than the digestibility at 80°C (Wang et al., 2009).

The reduction in disulfide bonds has also been achieved by treating soy protein with NADP-thioredoxin (NTS) (Jiao et al., 1992). They found that treating the proteins with NTS at low temperature resulted in the inactivation of the inhibitors through reduction (i.e., NADPH-reduced). In another study, Faris et al. (2008) used thioredoxin to reduce the disulfide bonds in soy white flour. They found that sulfhydryl content was much higher in the samples that were treated with NTS (35.44 μmol/g) when compared to the untreated soy white flour (7.69 μmol/g) or the control (8.66 μmol/g). It was determined that the soy white flour that was treated with NTS showed an increase in digestibility (10.06%) when compared to the untreated white soy (7.78%) or the control samples (6.38%) (Faris et al., 2008).

11 FERMENTATION AND ENZYME TREATMENTS OF SOYBEANS

Fermentation is a method widely used for processing various substrates to produce a variety of biochemicals. Enzymatic treatments, both before and during fermentation, are often used to break down the substrate into a simpler form, which can increase protein availability and digestibility. There have been a few studies that have attempted to improve soy functionality using enzymes and fermentation treatments.

Fermentation can be accomplished by several methods, including submerged fermentation and solid-state fermentation. Submerged fermentation involves the dissolution of the nutrients in purified water (Chahal, 1991). Solid-state fermentation is a type of fermentation conducted without the presence of any free water in the substrate, into which the microorganisms have been placed (Lonsane et al., 1985). Solid-state fermentation has a few advantages over submerged fermentation, including potentially lower cost, better product recovery, no foaming problems, and reduction of the amount of waste water generated through the process. The disadvantages of solid-state fermentation, however, are contamination by other bacteria, control of moisture levels, and controlling the increase in heat buildup during fermentation (Lonsane et al., 1985; Hasseltine, 1972).

Teng et al. (2012) studied solid-state fermentation of soybean meal using *Bacillus subtilis* and *A. oryzae*. The soybean meal was combined with wheat bran in a ratio of 3:1, with a moisture content of 67%. After 72 h of fermentation they found that the soluble protein had increased by 63.11% and 19.4% for the meal with *Bacillus subtilis* and *A. oryzae*, respectively. The amino acid content, namely arginine, serine, threonine, aspartic acid, alanine, and glycine contents, increased by 50.67%, 45.6%, 34.55%, 22.25%, 21.23%, and 18.12%, respectively. Clearly, *Bacillus subtilis* was a better choice for the fermentation process, as it resulted in a much higher soluble protein and amino acid content post-fermentation.

Solid-state fermentation was used by Lio and Wang (2012) to break down the fiber in soybean products, in order to increase digestibility and improve its use for non-ruminant animal feeds. They observed that the highest activity of xylanase was 757.4 IU/g when *Trichoderma reesei* and *Phanerochaete chrysosporium* were inoculated and incubated for a period of 36 h, followed by *A. oryzae* for an additional 108 h (Lio and Wang, 2012).

Kiers et al. (2000) also studied the effect of fermentation on digestibility of soybeans. Fermentation resulted in the degradation of soybean macromolecules to a large extent, which resulted in an increase in the amount of water-soluble components. It also lowered the molecular weight of the compounds due to the biochemical changes. They found that the in vitro digestibility increased from 29% to 33%–43% after 48 h of fermentation (Kiers et al., 2000).

Enzyme modification of soy is typically accomplished by the addition of enzymes to the substrate directly, or by inoculating the medium with microorganisms, which in turn ferment the medium and thus produces enzymes (Cowieson et al., 2006). Enzymes such as β-glucosidase, α-galactosidase, phytases, and proteases can be used to reduce the effects of inhibitors, antinutritional factors, and oligosaccharides (Bedford, 2000).

Soy isoflavones are found in the form of isoflavone glycosides (Izumi et al., 2000). Isoflavones are known for their health benefits, and they are being studied for their role in health and wellness with respect to cardiovascular disease, osteoporosis, and other negative health symptoms (Delmonte et al., 2006). Yang et al. (2009) studied the extent to which β-glucosidase (*P. thermophile* β-glucosidases and commercial almond-based β-glucosidase) could be used for deglycosylation of isoflavone glycosides, in order to be able to utilize the benefits of the isoflavones. Five types of soybean (daidzin, genistin, daidzein, genistein, and glycitein) were used for the study. They also studied the thermostability of the β-glucosidase. In their results, they reported that the isoflavone glycoside conversions in soybean flour by *P. thermophile* β-glycosidase to aglycones were 98%, 95.8%, and 99.3% of hydrolysis of diadzin, glycitin, and genistin, respectively. It was also found that the deglycosylation of isoflavone was higher in *P. thermophile* β-glucosidase than in the commercial almond β-glucosidase. Additionally, the thermal stability of the enzyme at 50°C was high, and retained 95% of its initial activity, even after 8 h (Yang et al., 2009). da Silva et al. (2011) studied the effect of fermentation using *A. oryzae* on the conversion of isoflavones from glycosides to aglycones and found that fermented, autoclaved whole soy flour contained 75.51% isoflavone aglycones after 48 h of fermentation.

12 ADDITIONAL PRODUCTS AND BY-PRODUCTS

Soybeans are one of the most versatile biological materials. In fact, soybeans can be processed into a myriad of products beyond just concentrated protein and lipids (the primary focus of this chapter has mostly been on protein and oil separation approaches). For example, soybeans can be processed into a variety of human foods, industrial products, biofuels, and bio-based chemicals. It is worth briefly noting some of these here.

Soybeans have a long history of use as a human food ingredient, and have been used in China (where soybeans were first domesticated) for thousands of years. One of the most popular non-fermented soy-based foods is tofu, which is made from soymilk. Fig. 10 illustrates the major processing steps—modern factories replicate traditional methods of production.

The cleaned, wet soybeans are subjected to a wet grinding step using either a stone grinder or hammer mill. Cooking then occurs at approximately 100°C for 10–20 min. The cooking step can substantially deactivate lipoxygenase, which is responsible for the "beany" flavor in the finished products. Coagulation generally occurs at 70–80°C for up to 30 min. Calcium sulfate is the most common coagulant used. Okara is the primary by-product of soymilk and tofu manufacturing, and it is composed of insoluble materials (especially fiber), which can be filtered out of the soymilk stream. It is primarily used as an animal feed. Table 10 provides some typical composition values for these products. Tofu is often used as a replacement or substitute for meat or cheese in various prepared

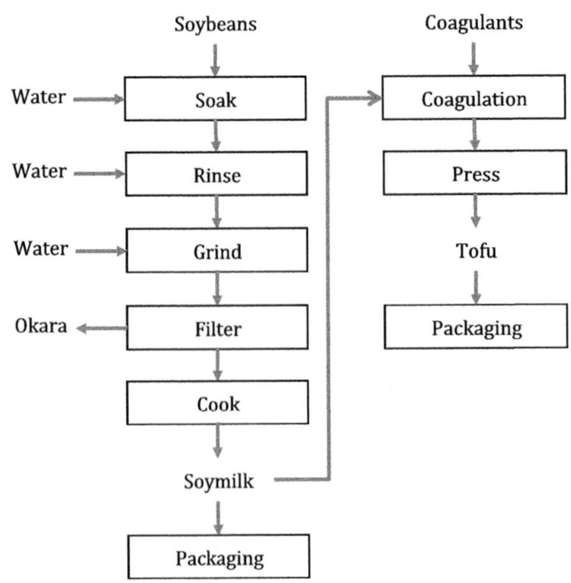

FIG. 10
Production of soymilk and tofu.

Table 10 Composition of Products from Soymilk and Tofu Manufacturing

	Moisture (%)	Protein (%)	Lipid (%)	Other (carbohydrates, fiber, minerals, etc.)
Soymilk (wet basis)	90	3.5	3	3
Tofu (wet basis)	85	8	4	3
(dry basis)	0	50	27	23
Okara (wet basis)	80	4	16	
(dry basis)	0	28	10	62

Based upon Johnson, L. A., White, P. J., Galloway, R., 2008. Soybeans: Chemistry, Production, Processing and Utilization. Urbana, IL: AOCS Press.

dishes. Soymilk can also be used as a substitute for cow or human milk. Additional information about soy-based foods can be found in Liu (2008).

There are a variety of industrial products that can be produced from soy proteins as well as soy lipids. For example, proteins can be used in the production of wood adhesives, plastics, foams, paper coatings, inks, and cosmetics, to name a few. Soybean lipids are often used to manufacture lubricants, soaps, surfactants, inks, plastics, plasticizers, and a variety of bio-based chemicals (especially oleochemicals), as well as for many other applications. Additional information about various soy-based industrial products can be found in Schmitz et al. (2008).

In recent years, the use of soybean oil to produce biofuels (specifically biodiesel) has become quite popular. It is a relatively simple process, whereby triacylglycerol is transesterified using methanol and an alkaline catalyst. For each 10 kg of soybean oil, which will react with 1 kg of methanol in a reactor vessel, about 10 kg of biodiesel (methyl ester) is produced, and 1 kg of crude glycerine ($C_3H_8O_3$) is produced as a by-product. Glycerine is readily separated from the biodiesel via centrifugation due to substantial differences in densities; afterwards, residual-free fatty acids and methanol are removed from the glycerine. The methyl ester stream is then subjected to a series of cleanup operations, whereby residual methanol, acid, and water are removed. Fig. 11 illustrates some of these production steps. Glycerin has a multitude of industrial applications, including foams, lubricants, cosmetics, as well as human food uses. Moreover, in recent years various bio-based plastics applications and animal feed ingredients have been developed as well. Further information about soy-based biodiesel can be found in Van Gerpen and Knothe (2008).

This section briefly introduced some of the additional products that soybeans can be processed into. Exhaustive discussions about food, biofuel, and other industrial uses for soy proteins and lipids can be found in Erickson (1995) and Johnson et al. (2008), to which the reader is referred for more information.

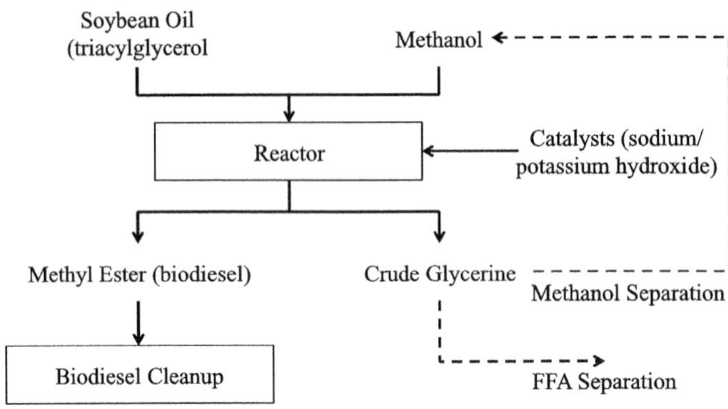

FIG. 11
Production of biodiesel and glycerine.

13 CONCLUSIONS

This chapter provides a basic review of the major processes that can be used for soybean processing—specifically to disrupt the protein-oil matrix in soybeans and subsequently separate them—including solvent extraction, mechanical expelling, aqueous and EAEP, fermentation, and chemical extraction. Mechanical and enzymatic pretreatments of soybeans improve total oil yield, and the process of de-emulsification of cream has improved with the evolution of the aqueous oil extraction process. Further, the oil in the cream fraction can be demulsified either enzymatically or with pH adjustment, a process that is less time and energy consuming when compared to the degumming process used in solvent extraction. While the highest reported total oil yield of EAEP has equaled and even exceeded the oil yield of the solvent-based extraction process (~99%), the free and demulsified oil amount to only 85%. The oil present in the skim fraction is regarded as lost due to the difficulty in separating oil from the fraction, and a commercial use of the skim fraction needs to be identified.

References

Bedford, M.R., 2000. Exogenous enzymes in monogastric nutrition—their current value and future benefits. Anim. Feed Sci. Technol. 86 (1–2), 1–13.

Birk, Y., 1976. Proteinase inhibitors from plant sources. Methods Enzymol. 45, 695–707.

Brekke, O.L., 1980. Edible oil processing. An introduction. In: Erickson, D.R. (Ed.), Handbook of Soy Oil Processing and Utilization. American Soybean Association and American Oil Chemists Society, St. Louis.

Campbell, K.A., 2010. Protein and Oil Recoveries From Enzyme-Assisted Aqueous Extraction of Soybeans and Sunflower Seed. Graduate Theses and Dissertations. Paper 11172.

Campbell, K.A., Glatz, C.E., 2009. Mechanisms of aqueous extraction of soybean oil. J. Agric. Food Chem. 57, 10904–10912.

Campbell, K.A., Glatz, C.E., 2010. Protein recovery from enzyme-assisted aqueous extraction of soybean. Biotechnol. Prog. 26 (2), 488–495.

Chahal, D.S., 1991. Chapter 9: Production of Trichoderma reesei cellulase system with high hydrolytic potential by solid-state fermentation. In: Enzymes in Biomass Conversion. American Chemical Society, Washington DC, pp. 111–122.

Cheng, M., Zhang, W., Rosentrater, K., Sekhon, J., Wang, T., Jung, S., Johnson, L., 2016. Environmental impact analysis of soybean oil production from expelling, hexane extraction and enzyme assisted aqueous extraction. In: An ASABE Meeting Presentation. https://doi.org/10.13031/aim/20162459781.

Choudhari, S.M., Ananthanarayan, L., 2007. Enzyme aided extraction of lycopene from tomato tissues. Food Chem. 102, 77–81.

Cowieson, A.J., Hurby, M., Pierson, E.E., 2006. Evolving enzyme technology: impact on commercial poultry nutrition. Nutr. Res. Rev. 19 (1), 90–103.

da Silva, L.H., Celeghini, R.M.S., Chang, Y.K., 2011. Effect of the fermentation of whole soybean flour on the conversion of isoflavones from glycosides to aglycones. Food Chem. 123 (3), 640–644.

de Moura, J.M.L.N., Campbell, A.K., Mahfuz, A.A., Jung, A.S., Glatz, A.C.E., Johnson, A.L., 2008. Enzyme-assisted aqueous extraction of oil and protein from soybeans and cream de-emulsification. J. Am. Oil Chem. Soc. 85 (10), 985–995. https://doi.org/10.1007/s11746-008-1282-2.

de Moura, J.M.L.N., Maurer, D., Jung, S., Johnson, L.A., 2011a. Integrated countercurrent two-stage extraction and cream de-emulsification in enzyme-assisted aqueous extraction of soybeans. J. Am. Oil Chem. Soc. 88 (7), 1045–1051.

de Moura, J.M.L.N., Maurer, D., Jung, S., Johnson, L.A., 2011b. Pilot-plant proof-of-concept for countercurrent two-stage enzyme-assisted aqueous extraction processing of soybeans. J. Am. Oil Chem. Soc. 88 (10), 1649–1658.

Delmonte, P., Perry, J., Rader, J.I., 2006. Determination of isoflavones in dietary supplements containing soy, red clover and kudzu: extraction followed by basic or acid hydrolysis. J. Chromatogr. A 1107 (1–2), 59–69.

Dominguez, H., Nunez, M.J., Lema, J.M., 1995. Enzyme-assisted hexane extraction of soya bean oil. Food Chem. 54, 223–231.

Erickson, D.R., 1995. Practical Handbook of Soybean Processing and Utilization. AOCS Press, Champaign, IL.

Faris, R.J., Wang, H., Wang, T., 2008. Improving digestibility of soy flour by reducing disulfide bonds with thioredoxin. J. Agric. Food Chem. 56 (16), 7146–7150.

Farm Energy. 2015. Mechanical Extraction Processing Technology for Biodiesel. Retrieved from http://articles.extension.org:80/pages/26911/mechanical-extraction-processing-technology-for-biodiesel.

Friedman, M., Brandon, D.L., Bates, A.H., Hymowitz, T., 1991. Comparison of a commercial soybean cultivar and isoline lacking the Kunitz trypsin inhibitor: composition, nutritional value, and effects of heating. J. Agric. Food Chem. 39 (2), 327–335.

Friedrich, J.P., List, G.R., 1982. Characterization of soybean oil extracted by supercritical carbon dioxide and hexane. J. Agric. Food Chem. 30, 192–193.

Friedrich, J.P., List, G.R., Heakin, A.J., 1982. Petroleum-free extraction of oil from soybean with supercritical CO_2. J. Am. Oil Chem. Soc. 9 (7), 288–292.

Fullbrook, P.D., 1983. The use of enzymes in the processing of oilseeds. J. Am. Oil Chem. Soc. 60, 428A–430A.

Gan, C.Y., Latiff, A.A., 2010. Extraction of antioxidant pecticpolysaccharide from mangosteen (Garcinia mangostana) rind: optimization using response surface methodology. Carbohydr. Polym. 83, 600–607.

Garicia-Olmedo, F., Salcedo, G., Sanchez-Monge, R., Gomez, L., Royo, J., Carbonero, P., 1987. Plant proteinaceous inhibitors of proteinases and α-amylases. In: Miflin, B. (Ed.), Oxford Surveys Plant Molecular and Cell Biology. In: vol. 4. Oxford University Press-ISPMB, New York, pp. 275–334.

Hasseltine, C.W., 1972. Biotechnology report: solid state fermentations. Biotechnol. Bioeng. 14 (4), 517–532.

Herkelman, K.L., Cromwell, G.L., Stahly, T.S., 1991. Effects of heating time and sodium metabisulfite on the nutritional value of full-fat soybeans for chicks. J. Anim. Sci. 69 (11), 4477–4486.

Izumi, T., Piskula, M.K., Osawa, S., Obata, A., Tobe, K., Saito, M., Kataoka, S., Kubota, Y., Kikuchi, M., 2000. Soy isoflavone aglycones are absorbed faster and in higher amounts than their glucosides in humans. J. Nutr. 130 (7), 1695–1699.

Jiao, J.A., Yee, B.C., Kobrehel, K., Buchanan, B.B., 1992. Effect of thioredoxin-linked reduction on the activity and stability of the Kunitz and Bowman-Birk trypsin inhibitor proteins. J. Agric. Food Chem. 40 (12), 2333–2336.

Johnson, L.A., White, P.J., Galloway, R., 2008. Soybeans: Chemistry, Production, Processing and Utilization. AOCS Press, Urbana, IL.

Jung, S., Maurer, D., Johnson, L.A., 2009. Factors affecting emulsion stability and quality of oil recovered from enzyme-assisted aqueous extraction of soybeans. Bioresour. Technol. 100 (21), 5340–5347.

Kapchie, V., Wei, D., Hauck, C., Murphy, P.A., 2008. Enzyme-assisted aqueous extraction of oleosomes from soybeans (Glycine max). J. Agric. Food Chem. 56, 1766–1771.

Kiers, J.L., Van Laeken, A.E., Rombouts, F.M., Nout, M.J., 2000. In vitro digestibility of bacillus fermented soya bean. Int. J. Food Microbiol. 60 (2–3), 163–169.

Koo, E.C., 1938. Studies on expression of vegetable oils IV. Expression of rapeseed oil. J. Chem. Eng. 5, 69–73.

Lamsal, B.P., Johnson, L.A., 2007. Separating oil from aqueous extraction fraction of soybeans. J. Am. Oil Chem. Soc. 85 (8), 785–792.

Lamsal, B.P., Reitmeier, C., Murphy, P.A., Johnson, L.A., 2006a. Enzymatic hydrolysis of extruded-expelled soy protein and resulting functional properties. J. Am. Oil Chem. Soc. 83 (8), 732–737.

Lamsal, B.P., Murphy, P.A., Johnson, L.A., 2006b. Flaking and extrusion as a mechanical treatment for enzyme-assisted aqueous extraction of oil from soybeans. J. Am. Oil Chem. Soc. 83 (11), 973–979.

Li, H., Pordesimo, L., Weiss, J., 2004. High intensity ultrasound-assisted extraction of oil from soybeans. Food Res. Int. 37 (7), 731–738.

Li, B.B., Smith, B., Hossain, M., 2006. Extraction of phenolics from citrus peels II. Enzyme-assisted extraction method. Sep. Purif. Technol. 48, 189–196. https://doi.org/10.1016/j.seppur.2005.07.019.

Lio, J.Y., Wang, T., 2012. Solid-state fermentation of soybean and corn processing coproducts for potential feed improvement. J. Agric. Food Chem. 60 (31), 7702–7709.

Liu, K., 2008. Food use of whole soybeans. In: Johnson, L.A., White, P.J., Galloway, R. (Eds.), Soybeans: Chemistry, Production, Processing and Utilization. AOCS Press, Urbana, IL.

Lonsane, B.K., Ghildyal, N.P., Butiatman, S., Ramakrishma, S.V., 1985. Engineering aspects of solid state fermentation. Enzyme Microb. Technol. 7 (6), 258–265.

National Oilseed Processors Association. 2018. Available online: http://www.nopa.org.

Ndlela, S.C., de Moura, J.M.L.N., Olson, N.K., Johnson, L.A., 2012. Aqueous extraction of oil and protein from soybeans with subcritical water. J. Am. Oil Chem. Soc. 89 (6), 1145–1153.

Olsen, H.S., 1988. Aqueous enzymatic extraction of oil from seeds. vol. 1. Food Science and Technology in Industrial Development, Bangkok, pp. 30–37.

Puri, M., Sharma, D., Barrow, C.J., 2012. Enzyme-assisted extraction of bioactives from plants. Trends Biotechnol. https://doi.org/10.1016/j.tibtech.2011.06.014.

Rosenthal, A., Pyle, D.L., Niranjan, K., 1996. Aqueous and enzymatic processes for edible oil extraction. Enzyme Microb. Technol. 19 (6), 402–420.

Rosenthal, A., Pyle, D.L., Niranjan, K., 1998. Simultaneous aqueous extraction of oil and protein from soybean: mechanism for process design. Food Bioprod. Process. 76, 224–230.

Sawada, M.M., Venâncio, L.L., Toda, T.A., Rodrigues, E.C., 2014. Effects of different alcoholic extraction conditions on soybean oil yield, fatty acid composition and protein solubility of defatted meal. Food Res. Int. 62, 662–670.

Schmitz, J.F., Erhan, S.Z., Sharma, B.K., Johnson, L.A., Myers, D.J., 2008. Biobased products from soybeans. In: Johnson, L.A., White, P.J., Galloway, R. (Eds.), Soybeans: Chemistry, Production, Processing and Utilization. AOCS Press, Urbana, IL.

Sharma, A., Khare, S.K., Gupta, M.N., 2002. Enzyme-assisted aqueous extraction of peanut oil. J. Am. Oil Chem. Soc. 79, 215–218.

Sherba, S. E., Steigerwalt, R. B., Faith, W. T., Smythe Jr., C. V., 1972. Soybean Fractionation Employing a Protease. U.S. patent 3640725 A.

Smith, K., 2018. Soybean History and Future. Soybean Meal Info Center, United Soybean Board. Retrieved from, https://www.soymeal.org/wp-content/uploads/2018/04/soybeans_history_and_future.pdf.

Sosulski, K., Sosulski, F.W., Coxworth, E., 1988. Carbohydrase hydrolysis of canola to enhance oil extraction with hexane. J. Am. Oil Chem. Soc. 65 (3), 357–361.

Tano-Debrah, K., Ohta, Y., 1997. Aqueous extraction of coconut oil by an enzyme-assisted process. J. Sci. Food Agric. 74 (4), 497–502.

Teng, D., Gao, M., Yang, Y., Liu, B., Tian, Z., Wang, J., 2012. Bio-modification of soybean meal with *Bacillus subtilis* or *Aspergillus oryzae*. Biocatal. Agric. Biotechnol. 1 (1), 32–38.

Towa, L.T., Kapchie, V.N., Hauck, C.C., Wang, H., Murphy, P.A., 2011a. Pilot plant recovery of soybean oleosome fractions by an enzyme-assisted aqueous process. J. Am. Oil Chem. Soc. 88, 733–741.

Towa, L.T., Kapchie, V.N., Wang, G., Hauck, C., Murphy, P.A., 2011b. Quantity and quality of free oil recovered from enzymatically disrupted soybean oleosomes. J. Am. Oil Chem. Soc. 88, 1581–1591.

USDA, 2018. World Agricultural Supply and Demand Estimates. USDA, Washington, DC. Available online, http://www.usda.gov/oce/commodity/wasde/.

Van Gerpen, J., Knothe, G., 2008. Bioenergy and biofuels from soybeans. In: Johnson, L.A., White, P.J., Galloway, R. (Eds.), Soybeans: Chemistry, Production, Processing and Utilization. AOCS Press, Urbana, IL.

Waldroup, P. W. 2018. Soybean Meal—Demand. Poultry Science Department, University of Arkansas with Soybean Meal Info Center, United Soybean Board. Retrieved from https://www.soymeal.org/wp-content/uploads/2018/04/soybean_meal_demand.pdf.

Wang, H., Faris, R.J., Wang, T., Spurlock, M.E., Gabler, N., 2009. Increased in vitro and in vivo digestibility of soy proteins by chemical modification of disulfide bonds. J. Am. Oil Chem. Soc. 86 (11), 1093–1099.

Wilson, K.A., 1988. The proteolysis of trypsin inhibitors in legume seeds. Crit. Rev. Biotechnol. 8, 197–216.

Wu, J., Johnson, L.A., Jung, S., 2009. De-emulsification of oil-rich emulsion from enzyme-assisted aqueous extraction of extruded soybean flakes. Bioresour. Technol. 100 (2), 527–533.

Yang, S., Wang, L., Yan, Q., Jiang, Z., Li, L., 2009. Hydrolysis of soybean isoflavone glycosides by a thermostable β-glucosidase from *Paecilomyces thermophila*. Food Chem. 115 (2009), 1247–1252.

Further Reading

Campbell, K.A., Glatz, C.E., Johnson, L.A., Jung, S., de Moura, J.M.N., Kapchie, V., Murphy, P., 2011. Advances in aqueous extraction processing of soybeans. J. Am. Oil Chem. Soc. 88, 449–465.

DairyOne. 2018. Interactive Feed Composition Library. Available online: http://dairyone.com/analytical-services/feed-and-forage/feed-composition-library/interactive-feed-composition-library/.

Jung, S., Mahfuz, A., 2009. Low-temperature dry extrusion and high-pressure processing prior to enzyme-assisted aqueous extraction of full-fat soybean flakes. Food Chem. 114, 947–954.

Jung, S., Lamsal, B.P., Stepien, V., Johnson, L.A., Murphy, P.A., 2006. Functionality of soy proteins produced by enzyme-assisted extraction. J. Am. Oil Chem. Soc. 83 (1), 71–78.

Karki, B., Maurer, D., Jung, S., 2011a. Efficiency of pretreatments for optimal enzymatic saccharification of soybean insoluble fractions. Bioresour. Technol. 102, 6522–6528.

Karki, B., Maurer, D., Kim, T.H., Jung, S., 2011b. Comparison and optimization of enzymatic saccharification of soybean fibers recovered from aqueous extractions. Bioresour. Technol. 102, 1228–1233.

Karki, B., Maurer, D., Kim, H., Jung, S., 2012. Ethanol production from soybean fiber: a co-product of soybean oil extraction. J. Am. Oil Chem. Soc. 89 (1), 1–9.

Leite, J.M., De Moura, N., De Almeida, A.N.M., Johnson, A.L.A., 2009. Scale-up of enzyme-assisted aqueous extraction processing of soybeans. J. Am. Oil Chem. Soc. 86, 809–815. https://doi.org/10.1007/s11746-009-1406-3.

Nobrega de Moura, J.M.L., Johnson, L.A., 2008. Two-stage countercurrent enzyme-assisted aqueous extraction of oil and protein from soybeans. J. Am. Oil Chem. Soc. 86 (3), 283–289.

Nobrega de Moura, J.M.L., Campbell, K., Mahfuz, A., Jung, S., Glatz, C.E., Johnson, L.A., 2008. Enzyme-assisted aqueous extraction of soybeans and cream de-emulsification. J. Am. Oil Chem. Soc. 85 (10), 985–995.

Nobrega de Moura, J.M.L., de Almeida, N.M., Johnson, L.A., 2009. Scale-up of enzyme-assisted aqueous extraction of soybeans. J. Am. Oil Chem. Soc. 86 (8), 809–815.

Sharma, A., Khare, S.K., Gupta, M.N., 2001. Enzyme-assisted aqueous extraction of rice bran oil. J. Am. Oil Chem. Soc. 78, 949–951.

Yao, L., Wang, T., Wang, H., 2011. Effect of soy skim from soybean aqueous processing on the performance of corn ethanol fermentation. Bioresour. Technol. 102 (19), 8727–9334.

Yao, L., Lee, S., Wang, T., de Moura, J.M.L.N., Johnson, L.A., 2012. Effects of fermentation conditions on corn-soy co-fermentation for fuel ethanol production. Bioresour. Technol. 120, 140–148.

SECTION 2
Fruits

CHAPTER 5
Tomato

Hamed M. El Mashad*[†], Liming Zhao[‡], Ruihong Zhang*, Zhongli Pan*
*Department of Biological and Agricultural Engineering, University of California - Davis, Davis, CA, United States, [†]Department of Agricultural Engineering, Faculty of Agriculture, Mansoura University, El-Mansoura, Egypt, [‡]State Key Laboratory of Bioreactor Engineering, R&D Center of Separation and Extraction Technology in Fermentation Industry, East China University of Science and Technology, Shanghai, China

Chapter Outline

1 INTRODUCTION

Tomato (*Lycopersicon esculentum*) is one of the most important crops produced in the world. It is cultivated in warm climates or in greenhouses that provide suitable cultivation conditions. Ideal temperatures for tomato growth should be 21–29.5°C during the day and 18.3–21.1°C at night (Kelley and Boyhan, 2013a). Very high or very low temperatures can negatively affect the yield and quality of tomatoes. The tomato is classified as a vegetable and can be cultivated by transplanting on a variety of soil types. Many varieties with different crop traits (e.g., yield, maturity time, shape, insect resistance, etc.) are being cultivated

Integrated Processing Technologies for Food and Agricultural By-Products
https://doi.org/10.1016/B978-0-12-814138-0.00005-8

under different environmental conditions. The world's total harvested area, total production, and yield of table and processing tomatoes in the period of 1994–2014 are shown in Fig. 1. As can be seen, the world tomato production steadily increased, reaching 170 million tons in 2014. The harvested area increased from 3.1 million tons in 1994 to 5.0 million tons in 2014. The average yield steadily increased, reaching approximately 33 ton/ha in 2008, then slightly decreased afterwards. Varieties of processing tomato are characterized by relatively higher yields than those of table tomato. Varieties such as H5608 and H5508 are top yielding varieties in California with 51 and 47 tons/acre, respectively (UC extension, 2012). During the period of 1999–2009, processing tomatoes ranged from 22% to 28% of the total tomato production worldwide (Tomatoland Information Services, 2011). The estimated amount of processed tomato produced worldwide in 2017 was 38 million tons. The United States, China, Italy, Spain, and Turkey are the major producers for tomato processing (WPTC, 2017). California produces over 30% of the world's tomato production. It is the leading producer of processing tomato with approximately 96% (11.47 million metric tons) of the US production (NASS-USDA, 2017; WPTC, 2017). Indiana, Michigan, and Ohio are other major producers of processing tomato (Kelley and Boyhan, 2013a).

Although tomato is an important cash crop in many countries, it requires high levels of inputs and labor. Most tomatoes are transplanted from greenhouse-grown plants to fields that need to be well prepared (Kelley and Boyhan, 2013b). Most cultivated tomatoes require around 75 days from transplanting

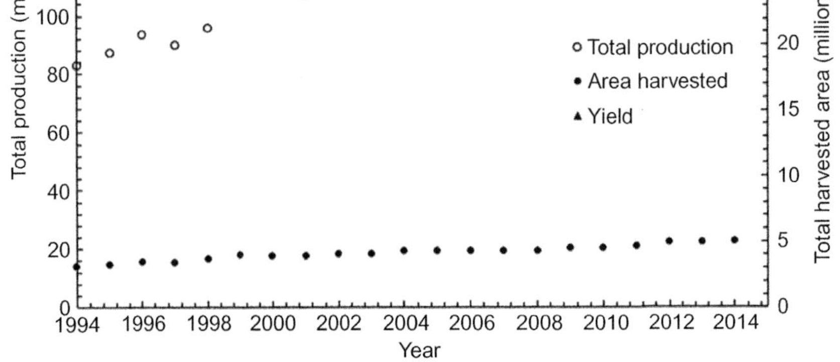

FIG. 1

Total world production and harvested area, and yield of tomato in the period 1994–2014 (FAO, 2017).

to harvest (Kelley and Boyhan, 2013a). The harvesting of table tomato is carried out over several weeks, while processing tomato is harvested all at once. Producers determine the harvesting time of processing tomato based on the maximum percentage of red ripe fruits. Other producers harvest tomato after a certain number of days after full bloom (Gould, 1992). In most cases, the received tomato at processing facilities can contain up to 5% green tomato. In most developed countries, nearly all processing tomatoes are mechanically harvested. However, in many developing countries, hand harvesting of tomatoes is still a common practice. The advantages of mechanical harvesting include efficiency (through saving time from labor and harvesting) and low cost. However, it can cause soil contamination, microbiological contamination, and mechanical damage to tomatoes. Tomato is transported from the fields to the processing plants in either bulk trailers or wagon loads, plastic boxes, or lug boxes. Bulk trailers and wagon loads are used with mechanical harvesting in most developed countries, while other transportation means are used in the developing countries.

The energy requirements and greenhouse gas (GHG) emissions from tomato production and processing depend on the cultivation practices, regions, and product type. Brodt et al. (2013) applied life cycle analysis (LCA) to quantify the energy and water use and GHG emissions in canned paste and canned diced tomatoes that were organically and conventionally grown. The study compared the processed tomato products grown, processed, and consumed within the Great Lakes region of the United States, and tomato products produced in California and then shipped to the Great Lakes region for consumption. Results indicated that total energy use and GHG emissions per kg of tomato paste or diced tomato were very similar regardless of the state of origin or cultivation method. Water use per servings basis was six times higher in California grown products. Bulk packaging and transportation of California grown tomatoes to Michigan added $<10\%$ to total energy consumption and GHG emissions. Compared to Michigan production, the field production of tomatoes in California used approximately 78% of the energy and emitted only 67% of the GHG emissions per kg of final paste produced. This was attributed to regional biophysical variables.

Reducing water and energy consumption and GHG emissions are key indicators of achieving sustainability in tomato production and processing. To increase the sustainability of tomato production, several innovative technologies can be used to optimize harvest timing, and identify disease incidences during production that help in producing more sanitary raw materials and final products. These technologies include precision agriculture, DNA profiling, and satellite imaging (Barrett, 2015). To achieve sustainable processing of tomatoes, technologies such as ohmic, microwave, and pulsed electric fields, as well as infrared processing may reduce the consumption of water and chemicals in tomato processing while maintaining high-quality products in terms of better color, flavor, nutrient content, and product integrity (Barrett, 2015; Pan et al., 2016). Reducing processing wastes and producing value-added products can contribute to achieving sustainable tomato processing.

2 TOMATO PROCESSING AND PRODUCTION OF ITS WASTE

Anatomically, the tomato consists of epicarp, pericarp, locules, and seeds (Weier et al., 1982). The epicarp is a waterproof cuticle that is known as the peel or skin. The pericarp is the fleshy part of the fruit. It is located between the epicarp and locula and consists of endocarp, mesocarp, and exocarp. Locules are the cavity part of the fruit that is surrounded by the pericarp. The seeds are located inside the locular cavities and enclosed in gelatinous membranes. A cross-sectional view of the tomato is shown in Fig. 2. The majority (>80%) of processed tomatoes are consumed in the form of tomato paste, juice, puree, sauce, and catsup (Gould, 1992).

Tomato processing starts with an unloading station where tomatoes are unloaded from transporting trucks. The tomatoes are hydraulically transported to an X-ray and optical sorter to remove defective and undesirable fruits. The tomato is then either processed into a paste or whole peeled and diced. To produce diced and whole peeled tomatoes, the tomatoes are peeled to remove the skin. Steam or lye peeling is commercially applied to loosen the skin from the flesh before peels are mechanically removed. Other peeling methods such as infrared peeling are being developed (Pan et al., 2015). After peeling, defective or unpeeled tomatoes are removed using two sequential manual and optical sorting instruments. The tomato is diced and blended with tomato juice. Diced tomato in juice is sterilized using steam. Finally, the product is filled under aseptic conditions in either macro packs (big bags in drums) for storage or shipping, or in small tins or bags for retail. Intact

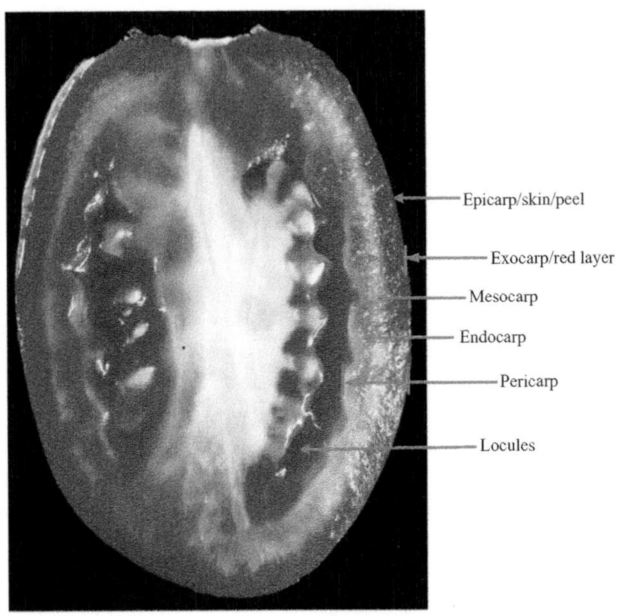

FIG. 2
A cross-sectional view of a rom tomato cultivar (Vidyarthi, 2017).

peeled tomato is used to produce whole peeled tomatoes. To produce tomato paste, received tomatoes are chopped and crushed into small pieces. During the crushing and chopping, tomato pieces simultaneously undergo a cold break (65–70°C) or hot break (>90°C) in a chopper box and tubular heater, to inactivate pectolytic enzymes and increase the efficiency of pectin extraction. The hot, crushed tomatoes pass through a pulper/finisher unit, where seeds and skins are removed. The removed wastes are pressed to increase juice yield. The produced juice from waste is centrifuged to remove impurities before it is added to the refined juice. The refined juice is then concentrated using continuous double or triple effect evaporators. The produced paste is aseptically packed in small cans or big bags. Tomato juice is produced in similar steps as tomato paste except concentration is not applied. The characteristics of ripe and green tomatoes and selected tomato products are shown in Table 1. As can be seen, green tomato has slightly different characteristics than ripe tomato. Different products have different characteristics.

Tomato processing generates tomato pomace that accounts for 3%–5% (wet basis (w.b.)) of fresh tomato (Ruiz Celma et al., 2009). Tomato pomace is the

Table 1 Characteristics of Ripe and Green Tomato, and Commercially Processed Tomato Products (USDA, 2017)

				Canned products				
Nutrient	**Unit/ 100 g**	**Ripe tomato**	**Green tomato**	**Paste**	**Sauce**	**Juice**	**powder**	**Crushed tomato**
Proximate analysis								
Water	g	94.5	93.0	73.5	91.3	94.2	3.1	89.4
Energy	kcal	18.0	23.0	82.0	24.0	17.0	302.0	32.0
Protein	g	0.9	1.2	4.3	1.2	0.9	12.9	1.6
Total lipid (fat)	g	0.2	0.2	0.5	0.3	0.3	0.4	0.3
Carbohydrate	g	3.9	5.1	18.9	5.3	3.5	74.7	7.3
Fiber, total dietary	g	1.2	1.1	4.1	1.5	0.4	16.5	1.9
Sugars, total	g	2.6	4.0	12.2	3.6	2.6	43.9	4.4
Minerals								
Calcium, Ca	mg	10.0	13.0	36.0	14.0	10.0	166.0	34.0
Iron, Fe	mg	0.3	0.5	3.0	1.0	0.4	4.6	1.3
Magnesium, Mg	mg	11.0	10.0	42.0	15.0	11.0	178.0	20.0
Phosphorus, P	mg	24.0	28.0	83.0	27.0	19.0	295.0	32.0
Potassium, K	mg	237.0	204.0	1014.0	297.0	217.0	1927.0	293.0
Sodium, Na	mg	5.0	13.0	59.0	474.0	10.0	134.0	186.0
Zinc, Zn	mg	0.2	0.1	0.6	0.2	0.1	1.7	0.3
Vitamins								
Vitamin C,	mg	13.7	23.4	21.9	7.0	70.1	116.7	9.2
Thiamin	mg	0.04	0.06	0.06	0.02	0.10	0.91	0.08
Riboflavin	mg	0.0	0.0	0.2	0.1	0.1	0.8	0.1

Continued

Table 1 Characteristics of Ripe and Green Tomato, and Commercially Processed Tomato Products (USDA, 2017)—cont'd

				Canned products				
Nutrient	**Unit/100g**	**Ripe tomato**	**Green tomato**	**Paste**	**Sauce**	**Juice**	**powder**	**Crushed tomato**
Niacin	mg	0.6	0.5	3.1	1.0	0.7	9.1	1.2
Vitamin B-6	mg	0.1	0.1	0.2	0.1	0.1	0.5	0.2
Folate, DFE	µg	15.0	9.0	12.0	9.0	20.0	120.0	13.0
Vitamin B-12	µg	0.0	0.0	0.0	0.0	0.0	0.0	0.0
Vitamin A, RAE	µg	42.0	32.0	76.0	22.0	23.0	862.0	11.0
Vitamin A, IU	IU	833.0	642.0	1525.0	435.0	450.0	17,247.0	215.0
Vitamin E (α-tocopherol)	mg	0.5	0.4	4.3	1.4	0.3	12.3	1.3
Vitamin D (D2+D3)	µg	0.0	0.0	0.0	0.0	0.0	0.0	0.0
Vitamin K	µg	7.9	10.1	11.4	2.8	2.3	48.8	5.3
Lipids								
Fatty acids, total saturated	g	0.03	0.03	0.10	0.04	0.02	0.06	0.04
Fatty acids, total monounsaturated	g	0.03	0.03	0.07	0.05	0.01	0.07	0.04
Fatty acids, total polyunsaturated	g	0.08	0.08	0.16	0.12	0.03	0.18	0.11

major constituent of pulper waste. The waste consists mainly of seed, skin (peel), and pulp that represent 33%, 27%, and 40% (wet basis), respectively (Sogi and Bawa, 1998). For three commercial tomato pomace samples collected from plants that process different tomato varieties using cold and hot break processes, Shao et al. (2013a) determined the amount of peel in pomace to range from 44.8% to 74.6% and seed in pomace to range from 25.4% to 45.2%. Typically, tomato processing waste (TW) is used as a livestock feed and/or a soil amendment or dumped in landfills, which can cause negative environmental impacts due to the biochemical degradation of its organic matter. This waste is nutrient rich and can be used to produce valuable products. Shao et al. (2013b) found that tomato pomace, tomato seed oil, and defatted tomato seeds reduced hepatic total cholesterol content in male Golden Syrian hamsters. Defatted tomato seeds decreased plasma total cholesterol and low-density lipoprotein (LDL) cholesterol concentrations. Extracted pigments from tomato by-products could be used for dyeing textile fabrics (Baaka et al., 2017). The production of a certain value-added product depends on the characteristics of the waste.

3 CHARACTERISTICS OF TOMATO WASTE

TW, like other biomass materials, has a low bulk density and volumetric heat capacity. The low bulk density of TW increases the costs of transportation and storage. Brachi (2016) reported bulk and true densities of 125 and 1050 kg/m^3

for air-dried tomato peels, respectively. Therefore, densification by pelletization or briquetting is necessary to increase the economic benefits of thermochemical processes. Celma et al. (2012) pelletized tomato skin and seeds. The moisture content of the biomass affected the physical properties of the produced pellets. The highest bulk density, durability, hardness, and energy content were determined to be 350 kg/m^3, 91.2%, 88 N, and 8 GJ/m^3, respectively, for the pellets with a moisture content of 9.1% (w.b.).

The characteristics of tomato waste depend on the variety and the processes involved in the development of tomato products. Al-Wandawi et al. (1985) found that tomato skins, produced from the Pearson variety, comprise >40% of the total solid of TW. Tomato seeds contained 27% lipids with oleic and palmitic as the predominant fatty acids. After hexane extraction of lipids, the seed flake contained about 40% protein with threonine and lysine as major essential amino acids. Del Valle et al. (2006) analyzed 21 samples of tomato pomace that were collected from 10 tomato processors in Spain. Samples were collected to represent different steps of past production including after pulper, after finisher, and before and after turbo press. Results showed that tomato pomace contained 59.0%, 25.7%, 19.3%, 7.6%, 5.9%, and 3.9% (dry weight basis (d.b.)) of neutral detergent fiber, total sugars, protein, pectin, total fat, and mineral, respectively. There were no significant differences among the collected samples at different steps. Shao et al. (2013a) found that tomato pomace generated from cold and hot break processes contained insoluble dietary fiber, soluble dietary fiber, protein, fat, and ash of 48.5%–64.8%, 8.9%–10.0%, 15.1%–22.7%, 8.4%–16.2%, and 2.9%–4.4%, respectively. They also found that defatted tomato seeds, from hot break processes, contained essential amino acids including histidine, glutamic acid, and glycine with concentrations of 23.4%, 14.3%, and 14.2% (of the total amino acids), respectively. The total unsaturated fatty acids content of the tomato oil was 80% and major fatty acids were linoleic (C18:2), oleic (C18:1), and linolenic (C18:3) with concentrations of 53.7%, 23.8%, and 2.1%, respectively. The total saturated fatty acids content was 20%, and palmitic (C16:0) and stearic (C18:0) represented the major fatty acids with the concentrations of 13.7%, and 5.4%, respectively.

4 TECHNOLOGIES FOR THE PRODUCTION OF VALUABLE PRODUCTS FROM TW

4.1 Technologies for Energy Production

4.1.1 BIOCHEMICAL PROCESSES

Biochemical processes such as anaerobic digestion and fermentation are suitable for the production of biogas and ethanol from TW with high moisture contents. During the anaerobic digestion process, organic matter is converted, in the absence of oxygen, by enzymes and anaerobic microorganisms into biogas that is composed mainly of methane (60%–70% v/v) and carbon dioxide (30%–40%) with trace amounts of other gases (e.g., H_2S) and water vapor.

Biogas can be used after removal of H_2S to produce electricity or compressed renewable gas. The performance and stability of anaerobic digesters depend on system configuration, operational parameters such as temperature and retention time, and the characteristics of the biomass materials such as TW. Detailed descriptions of the operational conditions and applied anaerobic digestion systems have been presented by many researchers (e.g., Zhang and El-Mashad, 2007).

Biogas and methane yields during the batch digestion of tomato skin and seeds at 40°C for 40 days were 424 and 218 L/kg VS, respectively (Dinuccio et al., 2010). During the mesophilic anaerobic digestion (38°C) of TW in a continuous-flow stirred-tank reactor (CSTR), a biogas yield of 398 L/kg VS with a methane content of 52% could be obtained at a hydraulic retention time (HRT) of 10 days (González-González and Cuadros, 2013). Methane production from the anaerobic digestion of TW in semicontinuous digester operated at 33°C ranged from 5 to 310 mL/g VS at HRTs ranging from 8 to 32 days (Sarada and Joseph, 1993). About 50% of cellulose and 65% of hemicellulose were degraded at a HRT of 32 days. While, at an HRT of 8 days, their degradations were 18% and 27%, respectively. The same authors (Sarada and Joseph, 1994) reported a methane yield of 420 m^3/kg VS when they operated the system at an HRT of 24 days, a loading rate of 4.5 kg/m^3/day, and a temperature of 35°C. Although the production of ethanol from food waste is more valuable than biogas production (Huang et al., 2015), little research has been conducted to study the production of ethanol from TW.

Tomato pomace has a recalcitrant nature due to the presence of lignin (Weiss et al., 1997). Therefore, a pretreatment is needed prior to converting it to biogas and ethanol using biochemical processes. Elbashiti et al. (2016) found that microwave assisted 5% H_2SO_4 and 7% HCl for 3 h produced high yields of glucose. The pretreated biomass was used to produce ethanol using immobilized *Saccharomyces cerevisiae*. The maximum ethanol yield of 543.51 mg/g could be obtained from the tomato waste pretreated with microwave-assisted 7% HCl. Allison et al. (2016) studied the effect of ionic liquid 1-ethyl-3-methylimidazolium acetate on the subsequent enzymatic hydrolysis of tomato pomace. The pretreatment at 130°C for 2 h significantly improved the reduction of sugar yield by 21% compared to the unpretreated pomace. The pretreated pomace was anaerobically digested in batch digesters at 55°C. Results showed that there was no significant effect of the pretreatment on methane yield. However, the pretreated biomass at temperatures of 100–130°C resulted in a significantly higher methane content than both the unpretreated and pretreated ones at 160°C. Calabrò et al. (2015) found that alkaline pretreatment did not increase methane yield from TW. An average methane yield of 320 mL/g VS could be determined from the pretreated TW after a digestion time of 30 days.

4.1.2 THERMOCHEMICAL PROCESSES

Thermochemical technologies for bioenergy production use high temperatures either at ambient or high pressures to convert organic matter into heat, synthetic gas, and hydrocarbon fuels or ash depending on the technology and its operational conditions. Gaseous and liquid fuels produced from thermochemical processes can be further processed and upgraded (e.g., using catalytic thermochemical processes) to produce alcohols and light diesel (Cantrell et al., 2007). Thermochemical processes include direct combustion (i.e., incineration), torrefaction, pyrolysis, gasification, and hydrothermal degradation. Compared with biochemical processes, thermochemical processes have the following advantages (Cantrell et al., 2007): (1) thermochemical reactors are more compact; (2) these processes are flexible and can simultaneously process several materials; (3) processing time in the thermochemical processes is short (in the order of minutes); and (4) the applied high temperatures destroy pathogenic microorganisms.

Mangut et al. (2006) characterized and conducted a thermogravimetric analysis of tomato peels and seeds. Results showed that both materials had high volatile contents, high heating values, and low sulfur and ash contents. Therefore, the authors postulated that these waste materials can be used as a source for bioenergy production using thermochemical technologies with less negative impacts on the environment and equipment. The authors also developed a kinetic model for the pyrolysis of hemicellulose, cellulose, lignin, and oil contained in the tomato waste. On the other hand, Rossini et al. (2013) mentioned that high concentrations of chlorine and sulfur cause corrosion in processing equipment if they are not managed. Kraiem et al. (2016) used TW to produce biomass pellets for energy production from boilers. The produced pellets had a lower heating value of 19.5 kJ/kg that was higher than that (16.4 kJ/kg) of wood pellets. Combustion and boiler efficiency of the tomato waste pellets were comparable to those of wood pellets. However, tomato waste pellets produced higher emissions of NOx, SO_2, and particulate particles and less volatile organic compounds than wood pellets did.

Most of the thermochemical processes such as incineration, pyrolysis, and gasification are suitable for biomass materials with low moisture contents (<40%, w.b.). For wet materials such as TW, hydrothermal processing could be applied without the need for drying. In hydrothermal liquefaction processing, biomass materials with high moisture contents are processed under a high temperature (250–374°C) pressurized (4–22 MPa) water environment to produce carbohydrate, liquid hydrocarbons called biocrude, or gaseous products (Elliott et al., 2015). Sabio et al. (2016) studied the effects of temperature, residence time, and biomass/water ratio on the hydrothermal carbonization of tomato peel. Results showed that increasing the temperature and residence time decreased solid yields of hydrochar and increased its high heating value. The biomass/water ratio played a secondary role on the hydrochar. Increasing the biomass/water ratio increased the solid yield of hydrochar and its higher heating value. Toscano et al. (2015) applied torrefaction of tomato peels at temperatures that ranged from 214°C to 316°C and retention times that ranged from 30 to 60 min. Mass yield, energy yield, and energy densification ranges were 94.7%–69.9%, 98.0%–86.0%, and 1.04–1.23, respectively. Increasing the temperature resulted in a more biologically stable and standardized material that could be used as a solid energy. Brachi (2016) investigated the effect of temperature (200°C, 240°C, and 285°C) and holding time (5, 15, and 30 min) on mass yield and the properties of the solid products during the fluidized bed torrefaction of TW. It was observed that higher temperatures and longer holding times increased the calorific value, reduced O/C, and improved hydrophobicity of the torrefied biomass. Increasing the temperature and torrefaction time resulted in a decreased yield. The lowest yield of the torrefied biomass (75% of biomass dry ash free basis) was obtained at the torrefacion temperature of 285°C and time of 30 min, respectively. The higher hydrophobicity means a higher resistance to regaining moisture during storage.

4.2 Technologies for Carotenoids Recovery From TW

4.2.1 *CHARACTERISTICS OF TOMATO CAROTENOIDS*

Lycopene and other carotenoids are the main reasons for the red color of tomatoes. Most of the lycopene and carotenoids are found in the outer pericarp (McCollum, 1955). Using acetone extraction, Al-Wandawi et al. (1985) determined the carotenoid content of different parts of the tomato. Results showed that β-carotene and lycopene were the major carotenoids. Lycopene content was 3.35, 16.79, 11.98, and 0.04 mg/100 g wet sample for whole mature tomato, tomato paste, tomato skins, and tomato seeds, respectively. On the other hand, the corresponding β-carotene content was 0.13, 1.06, 0.30, and 0.19 mg/100g wet sample, respectively. Lycopene, ψ, ψ-carotene, is an acyclic tetraterpenic hydrocarbon with 13 carbon-carbon double bonds, 11 of which are conjugated bonds arranged in a linear array (Fig. 3).

Lycopene content in fresh tomatoes ranged from 0.72 to 20 mg/100 g (w.b.) (Shi and Le Maguer, 2000). The concentration of lycopene in fresh tomatoes varied with the variety, maturity stage, and environmental conditions under which tomato cultivation and maturity of the fruit occurred. Shao et al. (2013a) mentioned that lycopene contents in tomato pomace ranged from 98.1 to 172.2 mg/kg. Lycopene is one of the most effective antioxidants (Shi and Le Maguer, 2000; Zuorro et al., 2011). It along with other carotenoids can be used as pharmaceutical compounds against cancers and lung diseases. Lycopene is commercially produced by extraction and concentration from whole tomato fruits. The available lycopene in the market proves that it is a very expensive antioxidant

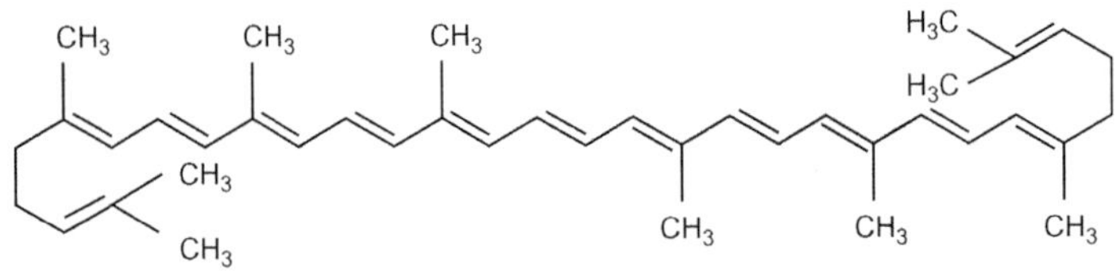

FIG. 3
Chemical structure of lycopene (Zuorro et al., 2011).

(Zuorro et al., 2011). Therefore, alternative sources of lycopene are being investigated; among them is tomato pomace.

Handling and processing of carotenoids should be carefully carried out. This is because they are sensitive to heat, oxygen, light, and acids (Sharma and Le Maguer, 1996b). Shi and Le Maguer (2002) mentioned that lycopene in fresh tomato fruits is found in the all-trans configuration. It is degraded via isomerization to cis-isomers due to thermal and oxidation effects during tomato processing. The number of cis-isomers increased with processing temperature and time. Lycopene degradation affects the sensory quality of final products, such as color and health benefits. The bioavailability of lycopene in processed tomato products is higher than in unprocessed fresh tomato. This is because the body absorption of cis-isomers is higher than that of all-trans isomers. The degradation of lycopene and other carotenoids is due to oxidation, which depends on the presence of oxygen, light, enzymes, metal ions, and co-oxidation with lipid hydroperoxides (Topal et al., 2006). The production of lycopene nanoemulsions could protect the antioxidant activity and improve the bioaccessibility of lycopene-enriched tomato extract (Ha et al., 2015). Higher antiradical efficiency and antioxidant activity were obtained for lycopene nanoemulsions with droplet sizes between 100 and 200 nm than those with sizes smaller than 100 nm. The highest in vitro bioaccessibility was obtained for nanoemulsions with droplets smaller than 100 nm.

Lycopene represents about 84% of the hydrocarbon carotenoids and polyenes in tomato. Other components (e.g., phytoene, phytofluene, α-carotene, and β-carotene) are minor or negligible (Curl, 1961). It is concentrated (72%–92% of total) in the water-insoluble fraction of tomato skins (Sharma and Le Maguer, 1996a). Therefore, separation of skins and seeds is important to maximize the production of valuable products like lycopene. Shao et al. (2013a) applied dry separation using sieves and wet separation with water floatation to separate the seeds and peel of tomato pomace. Results showed that both separation methods were effective for pomace separation. However, a significant loss of micronutrients was determined for the wet separation method due to water leachate. Kaur et al. (2005) applied a flotation-cum-sedimentation system to separate tomato peels and seeds from pomace at different feeding rates. Increasing the feed rate from 6 to 12 kg/min resulted in a reduction of the separation efficiency from 71% to 58% for peel and from 65% to 42% for seeds. Higher quality of seeds and skins was obtained at low feed rates.

Several methods are applied for extracting carotenoids from tomato including conventional solvent extraction, supercritical fluid (CO_2) extraction, and ultrasonic-assisted extraction (UAE). Strati and Oreopoulou (2014) reviewed the operational conditions and performance of different methods commonly applied to extract carotenoids from tomato processing by-products. Efficient and safe separation technologies are being studied to separate lycopene from TW.

4.2.2 SOLVENTS EXTRACTION

Solvent extraction is a common method to extract valuable compounds from fruits and vegetables including lycopene. However, it requires a long extraction time, large amounts of organic solvents, and high-energy inputs for the removal of solvents from extracts. Lycopene is nonpolar; therefore, it is insoluble in water and highly soluble in organic solvents. This leads to difficulties in the separation of lycopene from the extracting organic solvent due to the nonpolar nature of lycopene (Zuknik et al., 2012). Thus, solvent residues may be found in the final extracted product (Saldana et al., 2010). Solvent extraction can be carried out in a single step or multiple steps. Lavecchia and Zuorro (2007) patented a single-step extraction method using a blend of hexane, ethanol, and acetone in a 45:35:20 (vol%) ratio to optimize lycopene extraction at a temperature of 45°C and extraction time of 30 min. Taungbodhitham et al. (1998) mentioned that ethanol and hexane are the most suitable solvents for carotenoid extraction from vegetables and fruits. A good recovery (96%) of lycopene from canned tomato juice could be obtained after two extractions using a mixture (4:3, v:v) of ethanol: hexane. Poojary and Passamonti (2015) studied the kinetic of lycopene extraction from TW. Results showed that the highest lycopene yield of 4.03 mg/100 g was obtained at a temperature of 30°C using an acetone: n-hexane (1:3, v:v) mixture as the extraction solvent. High extraction yield of carotenoids (81 μg/g) from tomato waste was obtained with a mixture of 60% hexane and 40% acetone at 35°C (Baaka et al., 2017). Carotenoid extraction yields from TW were 243, 52, and 46 mg/kg using ethyl lactate, acetone, and ethyl acetate, respectively (Strati and Oreopoulou, 2011). Ethyl lactate is a biodegradable and environmentally friendly solvent that can be produced from the fermentation of carbohydrate feedstock (Strati and Oreopoulou, 2011, 2014).

Phinney et al. (2017) investigated a three-step extraction of lycopene from caustic (pH = 12.5) TW slurry. The slurry was produced from lye peeling after removal of the seed and skin using a decanter. Prior to extraction, membrane filtration was applied to reduce the alkalinity of the caustic slurry. Membrane filtration not only neutralizes the material, but it removes salts and unwanted Millard browning compounds. Acetone, ethanol, and hexane were used as direct or blended extracting solvents. The extraction was performed in three sequential steps: the first step aimed to disrupt the cellular structure, the second step aimed at extracting lycopene from the matrix into a hexane fraction, and the third step aimed to separate the nonpolar from the aqueous layer. The highest extraction efficiency of 94.5% was obtained using an ethanol and acetone mixture in the first step and a hexane and acetone mixture in the second step. Water was used in the third step.

4.2.3 ULTRASONIC AND MICROWAVE-ASSISTED EXTRACTION

Ultrasound and microwave-assisted extraction of lycopene is faster and more efficient than conventional extraction methods. Lianfu and Zelong (2008) optimized the operational parameters of ultrasonic and microwave-assisted extraction (UMAE) and UAE for the extraction of lycopene from tomato paste. The determined optimal conditions for UMAE were 98 W microwave power together with 40 kHz ultrasonic processing, a ratio of solvents to tomato paste of 10.6:1 (*V*/*W*), and an extracting time of 367 s. For UAE, the extracting temperature was 86.4°C, the ratio of the solvents to tomato paste was 8.0:1 (V/W), and the extracting time was 29.1 min. Lycopene yields at the optimal operational conditions were 97.4% for UMAE and 89.4% for UAE.

Anarjan and Jouybanb (2017) applied ultrasonic to extract lycopene from tomato waste, excluding seeds, using different hexane, acetone, and ethyl acetate. The extracted lycopene was converted into nanoparticles using ultrasound-assisted nano-precipitation technique. Using acetone alone as a solvent, the produced nanoparticles had low lycopene content due to the high reactivity of acetone. The best characteristics (particle sizes of 73.5 nm, a polydispersity of 0.11, and lycopene contents of 7615 mg/L) of the nanoparticles were obtained using a mixture of 31% acetone, 38% ethyl acetate, and 31% hexane. However, the produced nanoparticles had a limited stability during storage for four weeks at 4.0 ±2°C. Eh and Teoh (2012) found that sonication enhanced the efficiency and shortened the extraction time of lycopene extraction from tomatoes. Using n-hexane, acetone, and ethanol as solvents, the highest lycopene yield of 99% was obtained at 47.6°C after an extraction time of 46 min using a ratio of solvent to freeze-dried tomato sample (*v*/*w*) of 74.4:1. The purity of total lycopene obtained was 98.3% with b-carotene constituting 1.7% of the extract. Compared with nonsonicated extraction, ultrasound enhances the extraction yield of lycopene by 26%. The positive effect of the ultrasound was attributed to the greater penetration of solvent into material cells. This results in disruption of biological cells and releases cell contents (Dolatowski et al., 2007).

4.2.4 SUPERCRITICAL FLUID EXTRACTION

Supercritical fluid extraction was considered a promising technology for lycopene extraction (Zuknik et al., 2012). In the supercritical fluid, extraction liquid or gases are used as solvents at temperatures and pressures above their critical temperature and pressure (Ciurlia et al., 2009). Carbon dioxide is a common gas utilized in supercritical fluid extraction because of its low critical temperature (31°C) which reduces the potential of lycopene degradation (Saldana et al., 2010). Lianfu and Zelong (2008) mentioned that while conventional extraction methods are slow and have low efficiencies, supercritical CO_2 extraction needs expensive equipment and has a high-energy consumption.

Topal et al. (2006) applied supercritical carbon dioxide (SC-CO_2) extraction at a pressure of 40 MPa, temperature of 373 K, and CO_2 flow rate of 2.5 mL/min for the extraction of pure lycopene from small, dried tomato skin waste. Results showed that a total of 1.18 mg of lycopene/g dried skin, 94% of the total carotenoid content, could be obtained. Protecting the system from light and oxygen minimized degradation of lycopene. The authors mentioned that the most difficult challenges for the commercial application of the supercritical extraction are the increased solubility of lycopene in the supercritical CO_2 and the limited transformation of the trans-lycopene to cis-isomers under the high temperature and pressure. Barros et al. (2017) developed an extraction process for SC-CO_2 to extract avocado oil and use it as a cosolvent with SC-CO_2 for lycopene extraction from tomato pomace. Experiments were carried out using different proportions of the raw materials at 50°C and 400 bar. Lycopene recovery increased with the increase of the proportion of avocado oil. This is due to the effect of oil as an extract and the dilution of lycopene in the extraction broth. Other experiments were carried out using equal parts of avocado pulp and tomato pomace at different temperatures (40°C, 50°C, and 60°C) and pressures (200, 300, and 400 bar). The highest lycopene recovery was obtained at the pressure of 400 bar, independent of temperature.

4.2.5 ENZYMES-ASSISTED EXTRACTION

Enzymatic hydrolysis of tomato peels was studied as a pretreatment for lycopene extraction. Lavecchia and Zuorro (2008) applied enzymatic hydrolysis of tomato peels, for 1 h, prior to lycopene extraction using hexane, ethyl acetate, and a mixture of hexane:acetone:ethanol 50%:25%:25% (*v*/v) as solvents. Enzymes with pectinolytic, cellulolytic, and hemicellulolytic activities were applied. Lycopene recovery from enzymatically treated peels was approximately 77%, 74%, and 98% of the total lycopene content (450 mg/100 g of dry tomato peels) for hexane, ethyl acetate, and the mixture of hexane:acetone:ethanol solvents. The lycopene recovery for untreated samples was 3%, 4.7%, and 29%, respectively. Although the advantages of using enzymatic hydrolysis include increased lycopene yields and small amounts of solvent and energy consumption, enzyme costs might be a key obstacle for large-scale lycopene recovery.

Choudhari and Ananthanarayan (2006) studied the effects cellulase and pectinase enzymes on the hydrolysis of whole tomato, tomato peel, laboratory pulper waste (seeds and peel residues), and TW (peel with some tomato tissues and seeds) prior to lycopene extraction using petroleum ether and acetone (1:1 v/v) as a solvent. Pectinase was more effective than cellulase in increasing the lycopene yield from whole tomato, tomato peels, and pulper waste. However, for industrial wastes, cellulase enzyme was more effective than pectinase in lycopene extraction. The highest lycopene yield was obtained from tomato peels hydrolyzed by the pectinase enzyme. Zuorro et al. (2011) optimized the conditions (temperature, enzyme load, and extraction time) for the enzyme-assisted extraction of lycopene from TW peel. Peclyve PR and Cellulyve 50LC enzymes with pectinolytic and cellulolytic activities were used in the experiments. Results showed that using a mixture composed of equal volumes of Peclyve PR and Cellulyve lycopene recovery increased by 8–18-folds compared with the nonhydrolyzed peels. After enzymatic hydrolysis, hexane was used to extract lycopene. The optimal temperature, enzyme load, and extraction time were 30°C, 0.16 kg/kg skin (84% moisture content), and 3.18 h, respectively.

4.2.6 OTHER METHODS FOR LYCOPENE EXTRACTION

In addition to the aforementioned extraction methods, there are a few emerging technologies for lycopene extraction from tomato and its processing wastes. Naviglio et al. (2008) applied a new process for lycopene extraction from tomato skin residues in a water dispersion using high-pressure extraction (0.7–0.9 MPa). In this process, pressure/depressure cycles were applied on the water dispersion that led to the formation of lycopene aggregates, which could be easily purified by solid-phase extraction using small amounts of methanol. Then, lycopene with a purity of 98% was obtained after recovery with a small quantity of acetone. Lycopene recovery of 2.8 mg/kg tomato waste could be obtained after 4 h of extraction. Although recovery of lycopene was low (10%), the authors mentioned that the produced lycopene could be sold at 10% of the commercial price of lycopene. They also mentioned that the process did not consume large amounts of chemicals and the produced material after lycopene can be safely used as an animal feed. Jun (2006) applied high hydrostatic pressure (HHP) to extract lycopene from tomato paste waste at room temperature. The highest lycopene recovery (92%) was achieved after 1 min extraction at a pressure of 500 MPa, ethanol concentration of 75%, and a solid/liquid ratio (g/mL) of 1:6. Sarkar and Kaul (2014) studied the effect of freeze-drying and hot air drying (at 70°C for 8 h) on lycopene extraction from tomato peels. About 90% and 64% of lycopene content of peels were extracted from the freeze-dried and hot air-dried peels, respectively. The authors attributed the porous structure generated during freeze-drying to improving solvent extraction, while hot air drying resulted in a hard case and firm surface characteristics.

4.3 Production of Tomato Seed Oil

4.3.1 CHARACTERISTICS OF TOMATO SEED OIL

Tomato seed oil can be used as a food additive, biofuel, and as a source of valuable products such as antioxidants. Małecka (2002) studied the effect of the addition of unsaponifiables of tomato seeds, oat grains, and wheat germ oils on the stability of refined rapeseed oil. Results showed that addition of 0.3% of unsaponifiables separated from tomato seed oil achieved the highest antioxidative activity. Unsaponifiables included α- and γ-tocopherols, brasicasterol, campesterol, stigmasterol, β-sitosterol, Δ-5-avenasterol, and 4-methylsterol (citrostadienol). Demirbas (2010) revealed that the fatty acid composition of the seed oil was similar to that of soybean. Tomato seed oil contained high levels (68%–78%) of unsaturated fatty acids, such as oleic acid, linoleic acid, and linolenic acid (Demirbas, 2010; Rossini et al., 2013). It also contained 328.7 mg/kg, 315.9 mg/100 g, 252.3 mg [equivalent gallic acid]/kg, and 9.5 μg β-carotene/g of tocopherols, phytosterol, total phenolic compounds, and totals carotenoids, respectively (da Silva and Neuza Jorge, 2014). Purification and bleaching of tomato oil did not affect the physchochemical properties of the oil, but they reduced the tocopherol contents by about 18% (Lazos et al., 1998).

4.3.2 EXTRACTION OF TOMATO SEED OIL

Several methods are applied to extract oil from tomato seeds. Demirbas (2010) extracted tomato oil using Soxhlet and supercritical acetone extraction. The unsaturated fatty acid contents in the Soxhlet and supercritical acetone extraction were 78.7% and 68.6%, respectively. This was due to the partial thermal degradation of the unsaturated fatty acids under the high temperature (513 K) that was applied in the supercritical acetone extraction. Soxhlet-extracted oil had higher acid, saponification, and iodine values than the supercritical acetone extraction. However, it had lower peroxide value, moisture and volatility (Table 2).

Hexane was used to extract tomato oil at 40–60°C (Giannelos et al., 2005). The authors determined different properties of tomato oil as a renewable energy source (Table 3). Results showed that tomato oil had low sulfur and ash contents. Tomato oil also had poor cold flow properties and high density, high iodine value, and high kinematic viscosity compared to diesel fuels. Cetane number, density, cold flow properties, and energy content of tomato oil were similar to other vegetable oils. Zuorro et al. (2013) produced tomato oil enriched in lycopene. Oil was extracted from seeds using the cold press method. Oil recovery was 153 kg of oil/ton of dry seeds, which represented 53% of the oil contained in seeds. The latter was 30.8% of the dry weight of the seeds. The extracted oil was

Table 2 Properties of Tomato Oils Extracted Using Soxhlet and Supercritical Acetone Extraction (Demirbas, 2010)

Property	Soxhlet acetone extraction	Supercritical acetone extraction
Acid value (mg KOH/g oil)	0.241	0.006
Saponification value (mg KOH/g oil)	190.2	183.6
Iodine value (g/100 g oil)	126.8	109.7
Peroxide value (mmol peroxide/kg oil)	15.0	36.2
Moisture and volatility (%)	0.052	4.18
Specific gravity (g/mL)	0.918	0.919
Refractive index	1.473	1.471
Color	Yellow	Reddish yellow

Table 3 Properties of Tomato Oil

Kinematic viscosity at 40°C, mm^2/s	Cetan number	Energy content, MJ/kg	Cloud point (°C)	Pour point (°C)	Flash point (°C)	Density (kg/m^3)	Sulfur (wt%)	Saponification value (mg KOH/g)	Iodine number	Ash, %
28.0	41.0	35.9	−8.9	−16.1	189	915.1	0.004	195	124.0	0.034

done by incorporation of a tomato oleoresin derived from the peels. Oleoresin was extracted from peels using hexane after pretreatment with polygalacturonase and pectin methylesterase. Oleoresin yield was 25 kg/ton with an average lycopene content of 6.8% (w.b.).

Eller et al. (2010) compared three methods for the extraction of tomato seed oil: SC-CO_2, accelerated solvent extraction using n-hexane (Hexane ASE) and accelerated solvent extraction using ethanol (Ethanol ASE). The highest yield (23.1%) was obtained with Ethanol ASE. The yield of Hexane ASE was 20.0% and SC-CO_2 was 17.3%. The high yield with ethanol may be due to the extraction of more polar compounds than with hexane extraction. The total phytosterol content of oil was 12.3, 10.81, and 10.55 mg/g oil for the SC-CO_2, hexane, and ethanol extraction, respectively. The total tocopherol contents were 1.11, 1.08, and 0.94 mg/g, respectively. Moreover, β-carotene and lycopene concentrations were the highest for the Hexane ASE and Ethanol ASE treatments with a negligible content for the SC-CO_2 extraction. Shen and Xu (2005) applied supercritical CO_2 to extract oil from tomato seeds. The maximum oil yield of 96.3% could be obtained at a pressure of 30 MPa, extraction time of 2 h, and temperature of 50°C. Shao et al. (2015) found a high thermal stability of tomato seed oil after heating for 50 h at 180°C. The high thermal stability was attributed to the presence of the antioxidants such as tocopherol and polyphenol. They also found that storing tomato seed oil at high temperature (25–35°C), exposure to light, and large headspace volume significantly accelerated the deterioration of tomato seed oil during storage, among which the temperature had the most pronounced effect followed by light and headspace volume. It is suggested that tomato seed oil should be packaged with a minimal light exposure and stored at a low temperature. The study provided important information for the quality evaluation and control of tomato seed oil and has a great significance for value-added utilization of a by-product from the tomato processing industry.

4.4 Production of Protein and Pectin From Tomato Waste

Protein content in whole tomato seed meal ranges from 22.9% to 23.7% (d.b.), and it ranges from 26.7% to 27.5% in defatted meal. Ripeness stage does not affect protein content of the seeds. Sogi et al. (2002) revealed that the protein content of whole tomato meal, de-oiled meal, and alkali-, water-, and salt-extracted protein concentrate were 24.99%, 33.07%, 71.32%, 72.45%, and 69.56% (dry basis), respectively. They also found that water and fat absorption of tomato protein concentrates were lower than those of tomato meals. The bulk density of the salt-extracted protein concentrate was the highest. Tomato meal had a neutral pH and its protein concentrates had an acidic pH. The latter was due to the HCl used for the precipitation of isolated proteins. Both tomato meal and protein concentrate had poor foaming capacity and stability, but good emulsion capacity. To improve the foaming capacity and stability, sugar or salt could be added. Tomato seed protein contains 10 amino acids; among them, the eight amino acids that are essential for adult humans are present, making tomato seeds a useful human food supplement. However, there are limited amounts of sulfur bearing amino acids (Brodowski and Geisman, 1980). In addition to the application of tomato seed protein in human food, protein-rich isolates were used for the production of lactic acid (Mechmeche et al., 2017b). A positive correlation was determined between protein concentration and lactic acid production.

Protein isolate was prepared using tomato seed meal (Sarkar and Kaul, 2014). The protein efficiency ratio of the protein isolate was 2.7, making it a good quality protein. The protein isolate contained all essential amino acids, including lysine. It had low, negligible levels of antinutritional factors such as phytate and trypsin inhibitory activity. The antinutritional factors reduce the bioavailability of lycopene. They can be removed and/or destroyed by processing conditions. Mechmeche et al. (2017a) optimized the extraction conditions (water to material ratio, extraction time, and mixing time) of protein isolate from the defatted tomato seed meal. The maximum protein yield was 80% of its content in the seeds, obtained with 82.8 (*v*/*w*) water to raw material ratio, 49.8 h extraction time, and 24.6 min mixing time. The defatted tomato seed protein isolate had most of the essential amino acids and a high antioxidant capacity, which are important properties for the health food/nutraceutical/pharmaceutical industry.

Tomato waste was used to extract pectin using ammonium oxalate/oxalic acid under conventional and ultrasound conditions at 60°C and 80°C in two sequential steps (Grassino et al., 2016a). In the conventional extraction, the first step was conducted for 24 h. Subsequently, the second-step extraction was conducted for 12 h using a fresh volume of solvent. In the ultrasound extraction method, different sonication times (15–90 min) were applied in both steps. Results showed that ultrasound had little effect on pectin yield in both extraction steps. Better quality of pectin was obtained with sonication at shorter extraction times. Grassino et al. (2016b) studied the production of pectin from tomato peel and its application as a corrosion inhibitor. Pectin was extracted in two steps with ammonium oxalate and oxalic acid, under reflux. The highest pectin yield of 32% was obtained. In addition, using it as a food additive and gelling and thickening agent, the produced pectin was an efficient corrosion inhibitor for tin compared to the commercially available apple pectin, even at low concentrations.

4.5 Production of Other Value-Added Products From TW

Tomato pomace was used as a cattle feed and as a cover material in biosolarization for soil pest control in eggplant fields in California. Based on LCA, using tomato pomace in biosolarization could save 19.8 kg CO_2-e and 0.52 GJ per ton of wet pomace (Oldfield et al., 2017). Poli et al. (2007) hydrolyzed TW to produce a polysaccharide that was then converted into bioplastics after mixing with alginate. Sayğılı and Güzel (2016) investigated the production of activated carbon from TW. The effects of impregnation ratios of $ZnCl_2$/TW (1:1, 2:1, 4:1, 6:1, and 8:1), carbonization temperature (400, 500, 600, and 800°C), and carbonization time (0.5, 1.0, 2.0, and 4 h) on the characteristics of the produced activated carbon were studied. The best operating conditions that produced activated carbon with the highest surface area and pore volume were 6:1, 600°C, and 1 h for $ZnCl_2$/TW, carbonization temperature, and carbonization time, respectively. The produced activated carbon optimal activated carbon could effectively be used for removal of cationic and anionic dyes.

4.6 Production of Alternative Food Products From Green Tomato

Green tomato represents 3%–5% of processing tomatoes at harvest. The green tomatoes received at the processing facilities are mixed with red tomato to produce tomato paste. Increased amounts of green tomatoes reduce the quality of tomato paste. Therefore, processors reject loads with >4% green tomato and they employ extra sorting to reduce the amount of green tomato before paste production. Green tomatoes have higher concentrations of the glycoalkaloids α-tomatine and dehydrotomatine than red ones (Friedman and Levin, 1995, 1998). The presence of these compounds increases the importance of green tomato as a healthy food. Friedman et al. (2000) found that feeding hamsters with green and red tomato reduced LDL cholesterol by 59% and 44%, respectively. The corresponding reductions in plasma triglyceride levels were 47% and 31%, respectively. Compared with a control diet devoid of tomato, the LDL/high-density lipoproteins ratio decreased by 59% in the animals fed the green tomato diets. The positive effects of the diet containing tomatoes were attributed to the presence of fiber, protein, free amino acids, lycopene and other antioxidants, polyunsaturated fat, and tomatine. Tomatine contents were 743 and 0.7 mg/kg dry weight of green and red tomato, respectively. β-carotene and lycopene increase with tomato ripening, while α-tomatine and chlorophyll (a and b) decrease with tomato ripening (Yamashoji and Onoda, 2016).

Green tomatoes could be used to produce healthy, crispy snacks using freeze-drying, sequential infrared and hot air drying, or sequential infrared radiation and freeze-drying (Pan et al., 2008). It can also be used to produce green tomato juice, which can be a healthy alternative to sugar-rich juices. To our knowledge, little information is available in the literature on developing healthy products from green tomatoes.

5 SUMMARY, FUTURE RESEARCH NEEDS, AND PERSPECTIVES

Tomato processing produces huge amounts of wastes that are rich in lipids, proteins, micro and macronutrients, and carotenoids. There is a great potential to use this waste to produce several value-added products including bioenergy, edible oil and protein, antioxidants such as lycopene, pectin, and anticorrosion of tin. Several publications are available on the production of each of these value-added products. TW, like other vegetables and fruits, are seasonally produced for about 3 months. Therefore, the economic benefits of the production of value-added products from this waste depend on the price of the produced products, the production scale, the availability of processing facilities that utilize the waste, and storage and transportation methods of the waste. For the suitability of tomato pomace for bioenergy production via biochemical processes, research is still needed to optimize the yield of biomethane, biohydrogen, and ethanol under high solid contents. Moreover, there is a need to study the effect of co-processing of tomato waste with other materials for the production of bioenergy. Research should be conducted to optimize the operational conditions of the thermochemical processes (pyrolysis, gasification, and hydroliquefaction) for converting tomato waste into synthetic gas, bio-oil, and biochar. There is also a research need to study the physical and chemical properties of the products of the thermochemical processes of tomato waste and compare them with those of other biomass materials.

Maximizing the recovery yields of lycopene and other carotenoids depend on the method and its operational conditions. Environment-friendly solvents such as ethyl lactate should be sought and the operational conditions such as extraction temperature and time, and the ratio between solvent and TWs should be optimized. More research is needed to optimize the operational conditions and scale-up ultrasound-assisted extraction of lycopene and other value-added products from TW. LCA studies are needed to compare different value-added products that are created from TWs. The economics of different value-added products need to be appraised. A sensitivity analysis of the factors (e.g., capacity of processing facility, type of value-added product, and processing inputs) affecting the price of each value-added product should be conducted.

Green tomato has high concentrations of the glycoalkaloids α-tomatine and dehydrotomatine, which make it as a healthy food. However, there is a dearth of information on the utilization of green tomatoes to produce juice and crispy snacks. The operational conditions of ultrafiltration and pasteurization to produce green tomato juice need to be optimized. Moreover, the operational conditions of sequential infrared and hot air drying, and infrared and freeze-drying to produce crispy green tomato chips need to be optimized. The quality of the crispy snacks and other types of products needs to be evaluated by measuring the activity of peroxidase, texture, color, total suspended solids, α-tomatidine, lycopene, vitamin, antioxidants, and fiber contents.

References

Allison, B.J., Cádiz, J.C., Karuna, N., Jeoh, T., Simmons, C.W., 2016. The effect of ionic liquid pretreatment on the bioconversion of tomato processing waste to fermentable sugars and biogas. Appl. Biochem. Biotechnol. 179 (7), 1227–1247.

Al-Wandawi, H., Abdul-Rahman, M., Al-Shaikhly, K., 1985. Tomato processing wastes as essential raw materials source. J. Agric. Food Chem. 33 (2), 804–807.

Anarjan, N., Jouybanb, A., 2017. Preparation of lycopene nanodispersions from tomato processing waste: effects of organic phase composition. Food Bioprod. Process. 103, 104–113.

Baaka, N., El Ksibi, I., Mhenni, M.F., 2017. Optimization of the recovery of carotenoids from tomato processing wastes: application on textile dyeing and assessment of its antioxidant activity. Nat. Prod. Res. 31 (2), 196–203.

Barrett, D.M., 2015. Future innovations in tomato processing. In: Battalani, A. et al., (Ed.), Acta Horticulturae 1081. Proceedings of XIIIth 4 IS on Processing Tomato. ISHS.

Barros, H.D.F.Q., Grimaldi, R., Cabral, F.A., 2017. Lycopene-rich avocado oil obtained by simultaneous supercritical extraction from avocado pulp and tomato pomace. J. Supercrit. Fluids 120 (1), 1–6.

Brachi, P., 2016. Fluidized Bed Torrefaction of Agro-Industrial Residues: The Case Study of Residues From Campania Region, Italy (Ph.D. thesis). Universita Degli Studi di Salerno, Italy.

Brodowski, D., Geisman, J.R., 1980. Protein content and amino acid composition of protein of seeds from tomato at various stages of ripeness. J. Food Sci. 45, 228–229, 235.

Brodt, S., Kramer, K.J., Kendall, A., Feenstra, G., 2013. Comparing environmental impacts of regional and national-scale food supply chains: a case study of processed tomato. Food Policy 42, 106–114.

Calabrò, P.S., Greco, R., Evangelou, A., Komilis, D., 2015. Anaerobic digestion of tomato processing waste: effect of alkaline pretreatment. J. Environ. Manag. 163, 49–52.

Cantrell, K., Ro, K., Mahajan, D., Anjom, M., Hunt, P.G., 2007. Role of thermochemical conversion in livestock waste-to-energy treatments: obstacles and opportunities. Ind. Eng. Chem. Res. 46, 8918–8927.

Celma, A.R., Cuadros, F., López-Rodríguez, F., 2012. Characterization of pellets from industrial tomato residues. Food Bioprod. Process. 90, 700–706.

Choudhari, S.M., Ananthanarayan, L., 2006. Enzyme aided extraction of lycopene from tomato tissues. Food Chem. 102, 77–81.

Ciurlia, L., Bleve, M., Rescio, L., 2009. Supercritical carbon dioxide co-extraction of tomato (Lycopersicum esculentum L.) and hazelnuts (Corylus avellana L.): a new procedure in obtaining a source of natural lycopene. J. Supercrit. Fluids 49, 338–344.

Curl, A.L., 1961. The Xanthophylls of tomatoes. Food Sci., 106–111.

da Silva, A.C., Neuza Jorge, N., 2014. Bioactive compounds of the lipid fractions of agro-industrial waste. Food Res. Int. 66, 493–500.

Del Valle, M., Cámara, M., Torija, M.-E., 2006. Chemical characterization of tomato pomace. J. Sci. Food Agric. 86 (2), 1232–1236.

Demirbas, A., 2010. Oil, micronutrient and heavy metal contents of tomato. Food Chem. 118, 504–507.

Dinuccio, E., Balsari, P., Gioelli, F., Menardo, S., 2010. Evaluation of the biogas productivity potential of some Italian agro-industrial biomasses. Bioresour. Technol. 101, 3780–3783.

Dolatowski, Z.J., Stadnik, J., Stasiak, D., 2007. Applications of ultrasound in food technology. Acta Sci. Pol. Technol. Aliment. 6 (3), 89–99.

Eh, A.L.S., Teoh, S.G., 2012. Novel modified ultrasonication technique for the extraction of lycopene from tomato. Ultrason. Sonochem. 19 (1), 151–159.

Elbashiti, T., Alkafarna, A., Elkahlout, K., 2016. Bioethanol production by immobilized saccharomyces cerevisiae using different lignocellulosic materials. IUG J. Nat. Stud. 25 (1), 30–39.

Eller, F.J., Moser, J.K., Kenar, J.A., Taylor, S.L., 2010. Extraction and analysis of tomato seed oil. J. Am. Oil Chem. Soc. 87, 755–762.

Elliott, D.C., Biller, P., Ross, A.B., Schmidt, A.J., Jones, S.B., 2015. Hydrothermal liquefaction of biomass: developments from batch to continuous process. Bioresour. Technol. 178, 147–156.

FAO, 2017. http://www.fao.org/faostat/en/#data/QC. (Accessed 10 July 2017).

Friedman, M., Levin, C.E., 1995. Tomatine content of tomato and tomato products determined by HPLC with pulsed amperometric detection. J. Agric. Food Chem. 43, 1507–1511.

Friedman, M., Levin, C.E., 1998. Dehydromatine content of tomato. J. Agric. Food Chem. 46, 4571–4576.

Friedman, M., Fitch, T.E., Levin, C.E., Yokoyama, W.H., 2000. Feeding tomato to hamsters reduces their plasma low-density lipoprotein cholesterol and triglycerides. J. Food Sci. 65 (2), 897–900.

Giannelos, P.N., Sxizas, S., Lois, E., Zannikos, F., Anastopoulos, G., 2005. Physical, chemical and fuel related properties of tomato seed oil for evaluating its direct use in diesel engines. Ind. Crop. Prod. 22, 193–199.

González-González, A., Cuadros, F., 2013. Continuous biomethanization of agrifood industry waste: a case study in Spain. Process Biochem. 48, 920–925.

Gould, W.A., 1992. Tomato Production, Processing and Technology. Baltimore: CTI Publications, Baltimore, MD., p. 154.

Grassino, A.N., Brnčić, M., Vikić-Topić, D., Roca, S., Dent, M., Brnčić, S.R., 2016a. Ultrasound assisted extraction and characterization of pectin from tomato waste. Food Chem. 198, 93–100.

Grassino, A.N., Halambek, J., Djaković, S., Brnčić, S.R., Dent, M., Grabarić, Z., 2016b. Utilization of tomato peel waste from canning factory as a potential source for pectin production and application as tin corrosion inhibitor. Food Hydrocoll. 52, 265–274.

Ha, T.V.A., Kim, S., Choi, Y., Kwak, H.S., Lee, S.J., Wen, J., Oey, I., Ko, S., 2015. Antioxidant activity and bioaccessibility of size-different nanoemulsions for lycopene-enriched tomato extract. Food Chem. 178, 115–121.

Huang, H., Qureshi, N., Chen, M.H., Liu, W., Singh, V., 2015. Ethanol production from food waste at high solids content with vacuum recovery technology. J. Agric. Food Chem. 63, 2760–2766.

Jun, X., 2006. Application of high hydrostatic pressure processing of food to extracting lycopene from tomato paste waste. High Pressure Res. Int. J. 26 (1), 33–41.

Kaur, D., Sogi, D.S., Garg, S.K., Bawa, A.S., 2005. Flotation cum- sedimentation system for skin and seed separation from tomato pomace. J. Food Eng. 71 (4), 341–344.

Kelley, W.T., Boyhan, G., 2013a. History, significance, classification and growth. In: Commercial Tomato Production Handbook. UGA extension. https://secure.caes.uga.edu/extension/publications/files/pdf/B%201312_6.PDF. (Accessed 1 July 2017).

Kelley, W.T., Boyhan, G., 2013b. Culture and varieties. In: Commercial Tomato Production Handbook. UGA extension. https://secure.caes.uga.edu/extension/publications/files/pdf/B%201312_6.PDF. (Accessed 1 July 2017).

Kraiem, N., Lajili, M., Limousy, L., Said, R., Jeguirim, M., 2016. Energy recovery from Tunisian agri-food wastes: evaluation of combustion performance and emissions characteristics of green pellets prepared from tomato residues and grape marc. Energy 107, 409–418.

Lavecchia, R., Zuorro, A., 2007. Process for the Extraction of Lycopene. US Patent App. 12/513,300, 371(19). Retrieved from, http://www.google.com/patents/US20100055261.

Lavecchia, R., Zuorro, A., 2008. Improved lycopene extraction from tomato peels using cell-wall degrading enzymes. Eur. Food Res. Technol. 228, 153–158.

Lazos, E.S., Tsaknis, J., Lalas, S., 1998. Characteristics and composition of tomato seed oil. Grasas Aceites 49, 440–445.

Lianfu, Z., Zelong, L., 2008. Optimization and comparison of ultrasound/microwave assisted extraction (UMAE) and ultrasonic assisted extraction (UAE) of lycopene from tomato. Ultrason. Sonochem. 15, 731–737.

Małecka, M., 2002. Antioxidant properties of the unsaponifiable matter isolated from tomato seeds, oat grains and wheat germ oil. Food Chem. 79, 327–330.

Mangut, V., Sabio, E., Gañán, J., González, J.F., Ramiro, A., González, C.M., Román, S., Al-Kassir, A., 2006. Thermogravimetric study of the pyrolysis of biomass residues from tomato processing industry. Fuel Process. Technol. 87, 109–115.

McCollum, J.P., 1955. Distribution of carotenoids in the tomato. Food Sci. 20 (1), 55–59.

Mechmeche, M., Kachouri, F., Chouabi, M., Ksontini, H., Setti, K., Hamdi, M., 2017a. Optimization of extraction parameters of protein isolate from tomato seed using response surface methodology. Food Anal. Methods 10 (3), 809–819.

Mechmeche, M., Kachouri, F., Yaghlane, H.B., Ksontini, H., Setti, K., Hamdi, M., 2017b. Kinetic analysis and mathematical modeling of growth parameters of Lactobacillus plantarum in protein-rich isolates from tomato seed. Food Sci. Technol. Int. 23 (2), 128–141.

Naviglio, D., Caruso, T., Iannece, P., Aragòn, A., Santini, A., 2008. Characterization of high purity lycopene from tomato wastes using a new pressurized extraction approach. J. Agric. Food Chem. 56, 6227–6231.

NASS-USDA, 2017. California Agricultural Statistics Review 2016–2017. https://www.nass.usda.gov/Statistics_by_State/California/Publications/Annual_Statistical_Reviews/2017/2016cas-all.pdf.

Oldfield, T.L., Achmon, Y., Perano, K.M., Dahlquist-Willard, R.M., VanderGheynst, J.S., Stapleton, J.J., Simmons, C.W., Holden, N.M., 2017. A life cycle assessment of biosolarization as a valorization pathway for tomato pomace utilization in California. J. Clean. Prod. 141, 146–156.

Pan, Z., Shih, C., McHugh, T.H., Hirschberg, E., 2008. Study of banana dehydration using sequential infrared radiation heating and freeze-drying. LWT Food Sci. Technol. 41, 1944–1951.

Pan, Z., Li, X., Khir, R., El-Mashad, H.M., Atungulu, G., McHugh, T.H., Delwiche, M., 2015. A pilot scale electrical infrared dry-peeling system for tomatoes: design and performance evaluation. Biosyst. Eng. 137, 1–8.

Pan, Z., El-Mashad, H.M., Li, X., Khir, R., Atungulu, G., Zhao, L., McHugh, T., Zhang, R., 2016. Demonstration tests of infrared peeling system powered by electrical emitters for tomato. Trans. ASABE 59 (4), 985–994.

Phinney, D.M., Frelka, J.C., Cooperstone, J.L., Schwartz, S.J., Heldman, D.R., 2017. Effect of solvent addition sequence on lycopene extraction efficiency from membrane neutralized caustic peeled tomato waste. Food Chem. 215, 354–361.

Poli, A., Strazzullo, G., De Giulio, A., Tommonaro, G., De Prisco, R., Nicolaus, B., Immirzi, B., and Malinconico, M. (2007). New bioproducts from solid waste of tomato processing industry. Proc. Xth IS on the Processing Tomato. B'Chir, A. and Colvine, S. (Eds.). Acta Hort. 758, ISHS.

Poojary, M.M., Passamonti, P., 2015. Extraction of lycopene from tomato processing waste: kinetics and modelling. Food Chem. 173, 943–950.

Rossini, G., Toscano, G., Duca, D., Corinaldesi, F., Pedretti, E.F., Giovanni, R., 2013. Analysis of the characteristics of the tomato manufacturing residues finalized to the energy recovery. Biomass Bioenergy 51, 177–182.

Ruiz Celma, A.R., Cuadros, F., López-Rodríguez, F., 2009. Characterization of industrial tomato by-products from infrared drying process. Food Bioprod. Process. 87 (4), 282–291.

Sabio, E., Álvarez-Murillo, A., Román, S., Ledesma, B., 2016. Conversion of tomato-peel waste into solid fuel by hydrothermal carbonization: influence of the processing variables. Waste Manag. 47, 122–132.

Sarkar, A., Kaul, P., 2014. Evaluation of tomato processing by-products: a comparative study in a pilot scale setup. J. Food Process Eng. 37, 299–307.

Saldana, M.D.A., Temelli, F., Guigard, S.E., Tomberli, B., Gray, C.G., 2010. Apparent solubility of lycopene and β-carotene in supercritical CO_2, CO_2 + ethanol and CO_2 + canola oil using dynamic extraction of tomato. J. Food Eng. 99, 1–8.

Sarada, R., Joseph, R., 1993. Biochemical changes during anaerobic digestion of tomato processing waste. Process Biochem. 28, 461–463.

Sarada, R., Joseph, R., 1994. Studies on factors influencing methane production from tomato processing waste. Bioresour. Technol. 47, 55–57.

Sayğılı, H., Güzel, F., 2016. High surface area mesoporous activated carbon from tomato processing solid waste by zinc chloride activation: process optimization, characterization and dyes adsorption. J. Clean. Prod. 113, 995–1004.

Sharma, S.K., Le Maguer, M., 1996a. Lycopene in tomatoes and tomato pulp fractions. Italian J. Food Sci. 8 (2), 107–113.

Sharma, S.K., Le Maguer, M., 1996b. Kinetics of lycopene degradation in tomato pulp solids under different processing and storage condition. Food Res. Int. 29 (3–4), 309–315.

Shao, D., Atungulu, G.G., Pan, Z., Yue, T., Zhang, A., Chen, X., 2013a. Separation methods and chemical and nutritional characteristics of tomato pomace. Trans. ASABE 56 (1), 261–268.

Shao, D., Yokoyama, W., Bartley, G., Pan, Z., Zhang, H., Zhang, A., 2013b. Plasma and hepatic cholesterol-lowering in hamsters by tomato pomace, tomato seed oil and defatted tomato seed supplemented in high fat diets. Food Chem. 139, 589–596.

Shao, D., Venkitasamy, C., Li, X., Pan, Z., Shi, J., Wang, B., Teh, H.E., McHugh, T.H., 2015. Thermal and storage characteristics of tomato seed oil. LWT Food Sci. Technol. 63, 191–197.

Shen, X., Xu, S., 2005. Supercritical CO_2 extraction of tomato seed oil. J. Food Tech. 3 (2), 226–231.

Shi, J., Le Maguer, M., 2000. Lycopene in tomatoes: chemical and physical properties affected by food processing. Crit. Rev. Food Sci. Nutr. 40 (1), 1–42.

Shi, J., Le Maguer, M., 2002. Lycopene in tomato: chemical and physical properties affected by food processing. Crit. Rev. Biotechnol. 20 (4), 293–334.

Sogi, D.S., Bawa, A.S., 1998. Studies on dehydration of tomato processing waste. Indian Food Packer 52 (2), 26–29.

Sogi, D.S., Garg, S.K., Bawa, A.S., 2002. Functional properties of seed meals and protein concentrates from tomato-processing waste. J. Food Sci. 67, 2997–3001.

Strati, I.F., Oreopoulou, V., 2011. Effect of extraction parameters on the carotenoid recovery from tomato waste. Int. J. Food Sci. Technol. 46, 23–29.

Strati, I.F., Oreopoulou, V., 2014. Recovery of carotenoids from tomato processing by-products – a review. Food Res. Int. 65 (C), 311–321.

Taungbodhitham, A.K., Jones, G.P., Wahlqvist, M.L., Briggs, D.R., 1998. Evaluation of extraction method for the analysis of carotenoids in fruits and vegetables. Food Chem. 63 (4), 577–584.

Tomatoland Information Services, 2011. World consumption: table tomato and processed tomato. http://www.tomatoland.com/documents/2248.pdf. (Accessed 10 July 2017).

Topal, U., Sasaki, M., Goto, M., Hayakawa, K., 2006. Extraction of lycopene from tomato skin with supercritical carbon dioxide: effect of operating conditions and solubility analysis. J. Agric. Food Chem. 54, 5604–5610.

Toscano, G., Pizzi, A., Foppa Pedretti, E., Rossini, G., Ciceri, G., Martignon, G., Duca, D., 2015. Torrefaction of tomato industry residues. Fuel 143, 89–97.

UC extension, 2012. 2012 Processing Tomato Variety Trails. http://vric.ucdavis.edu/pdf/TOMATO/ProcTomatoTrials2012YoloSolanoSac.pdf. (Accessed 10 July 2017).

USDA, 2017. National Nutrient Database for Standard Reference, Rel # 22. Available at http://www.nal.usda.gov/fnic/foodcomp/. (Accessed 11 July 2017).

Vidyarthi, S., 2017. Study and Modeling of a Novel Tomato Dry-Peeling Process Using Infrared Heating. (Ph.D. dissertation, UC Davis).

Weier, T.E., Stocking, R.C., Barbour, M.G., Rost, T.L., 1993. Botany: An Introduction to Plant Biology, sixth ed. John Wiley & Sons, Inc., USA.

Weiss, W.P., Frobose, D.L., Koch, M.E., 1997. Wet tomato pomace ensiled with corn plants for dairy cows. J. Dairy Sci. 80 (11), 2896–2900.

WPTC, 2017. WPTC World Production Estimate of Tomato for Processing. https://www.wptc.to/pdf/releases/WPTC%20World%20Production%20estimate%20as%20of%2031%20March%202017.pdf. (Accessed 1 May 2017).

Yamashoji, S., Onoda, E., 2016. Detoxification and function of immature tomato. Food Chem. 209, 171–176.

Zhang, R., El-Mashad, H.M., 2007. Biodiesel and biogas production from seafood processing byproducts. In: Shahidi, F. (Ed.), Maximizing the value of marine by-products. Woodhead Publishing, Abington Hall, Abington, Cambridge, CB1 6AH, England.

Zuknik, M.H., Nik Norulaini, N.A., Mohd Omar, A.K., 2012. Supercritical carbon dioxide extraction of lycopene: a review. J. Food Eng. 112, 253–262.

Zuorro, A., Fidaleo, M., Lavecchia, R., 2011. Enzyme-assisted extraction of lycopene from tomato processing waste. Enzym. Microb. Technol. 49, 567–573.

Zuorro, A., Lavecchia, R., Medici, F., Piga, L., 2013. Enzyme-assisted production of tomato seed oil enriched with lycopene from tomato pomace. Food Bioprocess Technol. 6, 3499–3509.

Further Reading

Germini, A., Paschke, A., Marchelli, R., 2007. Preliminary studies on the effect of processing on the IgE reactivity of tomato products. J. Sci. Food Agric. 87 (4), 660–667.

Kaur, D., Wani, A.A., Oberoi, D.P.S., Sogi, D.S., 2008. Effect of extraction conditions on lycopene extractions from tomato processing waste skin using response surface methodology. Food Chem. 108, 711–718.

Papaioannou, E.H., Karabelas, A.J., 2012. Lycopene recovery from tomato peel under mild conditions assisted by enzymatic pre-treatment and non-ionic surfactants. Acta Biochim. Pol. 59, 71–74.

Perretti, G., Troilo, A., Bravi, E., Marconi, O., Galgano, F., Fantozzi, P., 2013. Production of a lycopene-enriched fraction from tomato pomace using supercritical carbon dioxide. J. Supercrit. Fluids 82, 177–182.

Roy, B.C., Goto, M., Hirose, T., 1994. Extraction rates of oil from tomato seeds with supercritical carbon dioxide. J. Chem. Eng. Jpn. 27, 768–772.

Shao, D., Pan, Z., Yue, T., Atungulu, G.G., Zhang, A., Li, X., 2012. Study of optimal extraction conditions for achieving high yield and antioxidant activity of tomato seed oil. J. Food Sci. 77 (8), 202–208.

Udofa, A., 2014. Simulation of a Tomato Processing Plant (M.sc. thesis). Rensselaer Polytechnic Institute Hartford, CT.

Vasapollo, G., Longo, L., Rescio, L., Ciurlia, L., 2004. Innovative supercritical CO_2 extraction of lycopene from tomato in the presence of vegetable oil as CO-solvent. J. Supercrit. Fluids 29, 87–96.

Zhang, Y., Pan, Z., Ma, H., Li, Y., Venkitasamy, C., 2014. Umami taste amino acids produced by hydrolyzing extracted protein from tomato seed meal. Food Chem. April 29.

CHAPTER 6
Grapes

Chandrasekar Venkitasamy*, Liming Zhao†‡, Ruihong Zhang*, Zhongli Pan*
*Department of Biological and Agricultural Engineering, University of California - Davis, Davis, CA, United States, †Key Laboratory of Meat Processing of Sichuan, Chengdu University, Chengdu, China, ‡State Key Laboratory of Bioreactor Engineering, R&D Center of Separation and Extraction Technology in Fermentation Industry, East China University of Science and Technology, Shanghai, China

Chapter Outline

1 INTRODUCTION

Grapes are the berry fruit produced by plants from the genus *Vitis* which encompasses approximately 60 species of grapevines. Grapes were domesticated >6000 years ago. *Vitis vinifera* L. is the most commonly cultivated species, accounting for over 90% of the grape berries in the market. Grown in clusters of small round or elliptical berries, they can either be seedless or contain edible or nonedible seeds. Grapes are consumed both fresh or as processed products including wine, jam, juice, jelly, grape-seed extract, dried grapes, vinegar, and grape-seed oil. The reason for the prevalence of a broad range of processed

Integrated Processing Technologies for Food and Agricultural By-Products
https://doi.org/10.1016/B978-0-12-814138-0.00006-X

products is due to the extreme perishability of the fruit. As fresh fruit, grapes are very delicate and many are lost during harvesting and distribution. In developed countries, the table grape is one of the fruits with the highest input of technology (cooling, sulfuration, packing, cold storage) and manual practices (hand labor). Of all grapes, about 50%–75% are used to make wine, one third is consumed as fresh fruit, and the rest are dried, consumed as grape juice, or stored in the form of grape musts (FAO-OIV, 2016; García-Lomillo and González-SanJosé, 2017). Their use varies widely from country to country, often depending on the physical and politico-religious (e.g., wine prohibition) dictates of the region.

1.1 Grape Cultivation

Viticulture is the broad term encompassing the cultivation, protection, and harvest of grapes where the operations are outdoors. On the other hand, enology is the science dealing with wine and winemaking, including the fermentation of grapes into wine, which is mostly confined to the indoors. A vineyard is a plantation of grape-bearing vines grown for winemaking, raisins, table grapes, and nonalcoholic grape juice. Viticulture experienced one of the highest growths among agricultural commodities in terms of acreage and value over the past 30 years and is now a global multibillion dollar enterprise (Brostrom and Brostrom, 2009; Daane et al., 2018). This growth is related to factors such as increased international trade, improved global incomes, changing policies, technological innovations in production, storage, and transportation, by-product processing and utilization leading to development of novel and healthy products, and greater awareness of the health benefits of foods rich in antioxidants like grapes (Fig. 1).

Grapes are one of the world's most popular fruit crops, with approximately 77.44 million metric tons of production in 2016 from 7.1 million ha of land dedicated to its cultivation. China, Italy, and the United States are the leading grape producers (FAO, 2016). California is the principal grape producer in the United States; it produced 6.73 million metric tons in 2016, accounting for 87% of total grape production in the nation.

About 10,000 vine varieties are available in the world. Of all grapes, 13 varieties are grown in more than one-third of the world's vine area and 33 varieties in about 50% of the vine area. Some varieties which are grown in multiple countries are called international varieties. For example, Cabernet-Sauvignon is one of the most cultivated wine grapes in the world, covering >5% world vine area (OIV, 2017). Many countries in the world are specialized in wine production, such as Italy, France, Spain, and Argentina, while a few countries, such as China, India, and Turkey, are more focused on table and dried grapes.

1.2 Grape Varieties

Compared to wine grapes, table grapes usually have larger berries and firmer pulp and they are less prone to crushing and wilting during shipping. Table grapes form loose bunches and have thicker skin, making them easy to

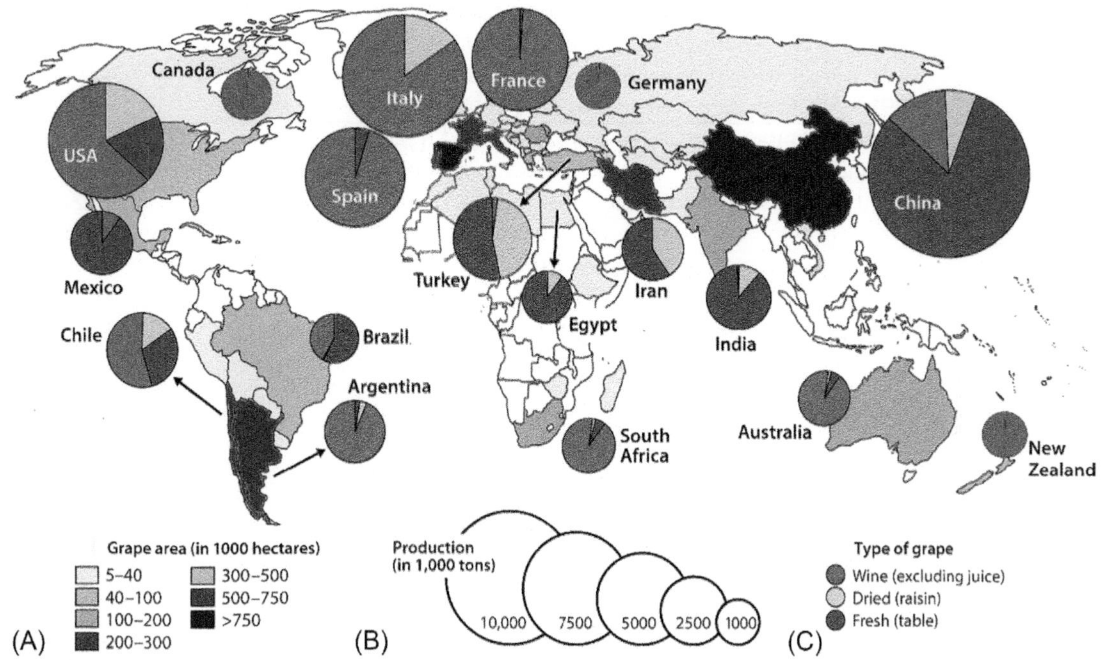

FIG. 1

Major grape production of the world showing (A) the land area planted to vineyards, (B) the grapefruit production from each of the 18 most productive countries, and (C) the percentage of the grape commodity (wine, dried fruit, and fresh fruit) in each of the 15 most productive countries (Daane et al., 2018).

eat (FAO-OIV, 2016). Dried grape varieties generally have small, seedless, and early-ripening berries that remain soft and not sticky. Another important characteristic of table grape varieties is the presence of aromatic compounds, which wine-producing varieties have little of. Aromas of wines originate from odorless precursors, acquiring their aromatic characteristics during the winemaking process.

1.2.1 TABLE GRAPES

Table grapes should be supplied fresh to consumers while dried grapes should have quality tolerance and maturity requirements. Consumer tastes and preferences for fresh grapes vary from region to region. In Europe, seedless mature yellow grapes with medium-sized bunches and well-developed colored berries that are crunchy, thin-skinned, and sweet are preferred. However, in emerging markets such as China, large, seeded, sweet-tasting berry varieties are preferred. Characteristics of selected table and dried grape varieties and the countries they are grown in are shown in Table 1.

Genetic research in table and dried grape varieties is focused toward the development of new cultivars to satisfy market demand and consumer tastes. The goal

Table 1 Characteristics of Selected Table and Dried Grape Varieties (FAO-OIV, 2016)

Variety	Characteristics	Countries	Fruits
Alphonse Lavallée	Very large black seeded berries, Crunchy skin	Argentina, Chile, Turkey, Peru	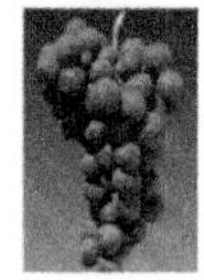
Crimson Seedless	Small to medium red, elliptic, seedless berries	Egypt, Italy, Peru, South Africa, United States of America	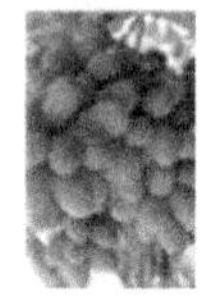
Dattier-de Beyrouth	Large white seeded berries, large cluster. Thick skin and firm pulp	Italy, Spain, Turkey	
Flame Seedless	Medium red seedless berries	Argentina, Chile, Egypt, Peru, South Africa, Argentina, United States of America	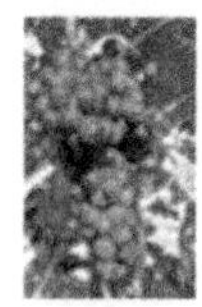
Muscat Hamburg	Medium- to large-sized black, seeded berries. Muscat flavors	China, mainland, France, Italy, Argentina, Chile, Turkey, Peru	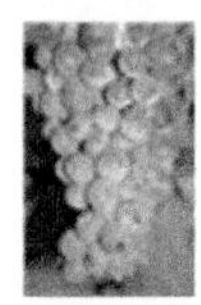
Italia	Very large seeded berries, large clusters	Argentina, China (mainland), Italy	
Muscat of Alexandria	Very large white, elliptic, seeded berries, large clusters. Firm pulp with intense muscat aromas	Algeria, Argentina, Greece, Morocco, South Africa, Spain	
Red Globe	Medium-sized red, round, seeded berries	Argentina, Australia, Chile, Egypt, Italy, Peru, South Africa, United States of America	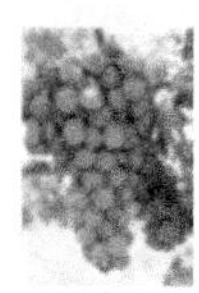

Table 1 Characteristics of Selected Table and Dried Grape Varieties (FAO-OIV, 2016)—cont'd

Variety	Characteristics	Countries	Fruits
Sugraon	Medium to large white, seedless berries, large clusters	Argentina, Australia, Egypt, Italy, Peru, South Africa, United States of America	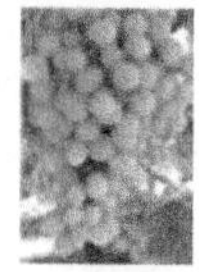
Sultanina	Small, white, seedless berries, large clusters. Thin skin and firm pulp	Argentina, Australia, Chile, China, mainland, Greece, India, Iran, South Africa, Egypt, Turkey, United States of America	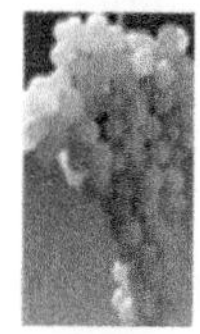
Victoria	Large white elliptic seeded berries. Firm pulp	Argentina, Italy	

of the research is to produce seedless varieties, which can adapt to various growing conditions and climates, are less prone to damage during picking, handling, and transportation and are suitable for food industry processing. Extending the maturity period and shelf life of grapes can extend marketability to a larger area where local cultivation is not possible. Different vinicultural methods and packing and storage methods have been developed to achieve this. Vineyard covering practices and greenhouses are used to accelerate or delay grape maturity, thus allowing harvests to extend for 6 months (Novello and Palma, 2008). Cane girdling, a practice consisting of removing a ring of bark from the trunk, is employed to hasten fruit maturity.

1.2.2 WINE GRAPES

Vitis vinifera, the grape variety that makes all quality wine, grows best in two relatively narrow bands between latitudes 30 and 50 degrees in the northern and southern hemispheres. In the United States, sultanina is the most cultivated variety, occupying 60,000 ha, or 14%, of the country's vine area. Chardonnay (10%), Cabernet Sauvignon (9%), and Concord (8%) are the next most popular varieties. Distribution of selected wine grape varieties cultivated in major grape growing countries and the area used to grow them are shown in Table 2.

During 2016, >7.67 million tons of grapes were grown commercially in the United States. California, with an ideal grape-growing climate, was the leading grape-growing state. California accounted for nearly 6.73 million tons, or 88%, of the grapes grown in the United States. Other top grape-growing states included Washington and New York (NASS, 2018). The United States had

Table 2 Distribution of Selected Wine Grape Varieties in 2015 (OIV, 2017)

Variety	Color	Area (ha)	Cultivated countries
Cabernet Sauvignon	Black	341,000	China, France, Chile, United States of America (USA), Australia, Spain, Argentina, Italy, and South Africa
Sultanina	White	273,000	Turkey, Iran, Iraq, Afghanistan, Pakistan, Uzbekistan, Turkmenistan, Tajikistan, USA, and South Africa
Merlot	Black	266,000	37 countries in the world
Tempranillo	Black	231,000	Grown in 17 countries, but 88% of area is in Spain
Airen	White	218,000	Spain
Chardonnay	White	210,000	Europe, America, and Oceania
Syrah	Black	190,000	France, Australia, Argentina, South Africa, the United States, and Chile
Red Globe	Black	159,000	China (91% of this variety's vineyard area). United States, Spain, Portugal, Italy, Turkey, Chile, Argentina, and South Africa
Garnacha Tinta/ Grenache Noir	Black	163,000	Spain and France
Sauvignon Blanc	White	123,000	All major wine producing countries
Pinot Noir/Blauer Burgunder	Black	112,000	France, Germany, Italy, Switzerland, Romania, Hungary, Spain, United States, New Zealand, Australia, Chile, Argentina, and South Africa
Trebbiano Toscano/Ugni Blanc	White	111,000	Italy, France, and Portugal

1,013,000 bearing acres producing an average of 7.57 tons/acre, valued at $6.26 billion in 2016 (NASS, 2018).

1.3 Botanical Description of Grapevines

Grapes are woody and grow on climbing vines. Grapevines use tendrils to attach themselves to fences or other tall-growing plants. Grapevines can extend their shoots by up to nearly a meter each year by spending most of their energy on growth in length rather than girth. Tendrils occur at nodes opposite of leaves and automatically begin to coil when they contact another object. Grapes are cultivated on a trellis, fence, or other structure for support.

Grape leaves vary in shape and size, depending on species and cultivar. Leaves are often large (8–10″ in width), sometimes deeply lobed as in many *V. vinifera* cultivars, or rounded with entire or serrated margins. Muscadine grapes have small, round, unlobed leaves with dentate margins. The buds are compound in grapes, meaning that they have multiple growing points or meristems. Flowers are small, indiscrete, and green, borne in racemose panicles opposite leaves at the base of the current season's growth. Species in Euvitis may have >100 flowers per inflorescence, whereas muscadine grapes have only 10–30 flowers per cluster.

Also, vinifera and concord grapes are perfect-flowered and self-fruitful, whereas some muscadines have only pistillate flowers. Fruits of grapes are true berries, small, round to oblong, and containing two to four seeds. The berries are often glaucous, having a fine layer of wax on the surface. Italian varieties and French-American hybrids may have 4–5 clusters of fruit per shoot and require cluster thinning to improve the development of quality and proper vine vigor. The thin skin of the grape is the source of the anthocyanin compounds that give rise to red, blue, purple, and black (dark purple) colored grapes. Green and yellow-skinned cultivars are often termed white grapes.

1.4 General Culturing Requirements

1.4.1 TEMPERATURE

Table grapes typically require a hot, dry climate, that is, warm days, cool nights, and low humidity. These conditions generally produce higher-quality grapes. The season at a particular site must be long enough to allow both the fruit and the vegetative parts of the vine to mature. The environment must also provide enough heat energy to ripen the fruit and vegetation. Adequate sunlight hours are necessary to ensure photosynthesis to mature the fruit and vine and to maintain the future productive potential for healthier vines and fruit sweetness. There must be very little rain during the ripening period to prevent various grape diseases, and winter must be long enough to ensure a period of dormancy for the vines. However, grapevines are sensitive to freezing temperatures. Damage to primary buds occurs at −18°C to −23°C, and trunks may be injured or killed below −23°C. Labrusca grapes are more cold resistant than vinifera or French-American hybrids but can be injured between −23°C and −29°C. Muscadine grapes are the least cold resistant, at risk of dying at temperatures below −18°C. Humidity is another limiting factor for the vinifera grape culture, due to disease susceptibility. Grapes cannot tolerate high RH or rain during harvest. Muscadines, however, grow much better in humid climates. The chilling requirement is highly variable among grape species; some grapes can grow in the tropics.

1.4.2 WATER

Grapes are intolerant of waterlogging and water stress. The volume of water required is determined by soil depth and soil bulk density. In areas with summer rainfall, the combined effect of these factors determines the volume of water which will be available to grapevines between rains. It is difficult to grow table grapes in areas receiving rain during the ripening period of the grapes, although this depends on the variety.

1.4.3 SOIL REQUIREMENTS

Table grapes can be grown in a wide variety of soil types. The most important characteristics of the soil are good internal drainage and adequate depth. Grapevines require deep, well-drained soil with a minimum of 75 cm to 1 m of permeable soil with no impeding layers (shallow bedrock, chemical or physical hardpans) for optimum vine growth. Although grapevines can be grown in various soil types, they grow best in a sandy loam soil with average fertility. Avoid growing grapes in soils that contain clay because it may cause poor drainage and salt accumulation. Grapes are fairly tolerant to a wide range of soils and pH, but grow best in a pH of 5.5–6.0.

1.4.4 *PROPAGATION*

The most common method of grape propagation is bench grafting, although rooted cuttings (where phylloxera is not a problem), T-budding, layering (difficult-to-root types like muscadine), and, to a limited extent, tissue culture are used in various situations.

Vitis vinifera was propagated on its own roots from the beginning of recorded history until about the 1870s. The grape phylloxera (Dactylosphaera vitifolii and Homoptera), also called the "grape root louse" (although it is actually an aphid), was brought from eastern North America to Europe in the 1860s, where it caused the most significant pest-related disaster in all of fruit culture. The search for resistant rootstocks led horticulturists to the native range of the phylloxera in eastern North America, where various species of American grapes had coexisted with the pest for millennia, and thus were resistant to it. Most grape rootstocks used today are numbered clonal selections of hybrids of *V. riparia*, *V. rupestris*, and *V. berlandieri* (Mencarelli et al., 2005).

1.5 Harvesting

Because the grape is a non-climacteric fruit, it should be harvested when fully ripened as taste and color will not improve after harvest. The harvesting period is determined by the variety, climatic conditions, total soluble solid (TSS), acidity and sugar acid ratio, and whether the grape is for the local or export market. The maturity standards of grapes are a minimum TSS of 160 Brix and sugar acid ratio of 20:1. Although the sugar content of the berries is considered as the indicator of their level of ripening, the ratio of sugar/acid is the correct index of ripening since this ratio indicates the taste of berries. Berries taste sour and less sweet when their acid content is higher. As harvest time approaches, berries should be sampled daily to determine sugar, acid, and pH levels. Physical appearance is considered primary criteria for choosing which grapes go to market; the best ones have a uniform size and appealing color. During the maturation of berries, the color of the peduncle undergoes changes. Grapes can be kept on the vine for several weeks after maturity. As long as no berries drop, it is best to leave the crop on the vines and pick the grapes as demanded by the market. Unless the weather is rainy, there will be no deterioration of fruit quality. If well protected by foliage, the grapes will withstand cold weather. Fruit quality can change under warm conditions. Possible damage to the crop by birds and bees must be considered in deciding whether or not the mature crop should be left on the vine.

Harvesting should be done by skilled workers wearing soft rubber gloves and using sharp secateurs/scissors for cutting. Grape clusters are cut from the vines with a sharp knife. Only attractive bunches fulfilling the minimum quality requirement should be harvested. The bunches should be held with one hand and cut upwards, away from your hands and arms to prevent injury. Hand shears are safer but are slower and more difficult to use. Do not jerk or pull clusters from the vine as this may crush some of the grapes. The bunch should always be held by the stem/peduncle. Rough handling results in loss of bloom (thin wax coating on berry surface), making the berries susceptible to decay. Careful handling of grapes during harvesting, transporting, cleaning, and packing is essential to prevent injury and abrasion. Green, immature, and diseased fruit should be discarded. Gently place the fruit into harvesting lugs or boxes and handle as little as possible. Take the container into the shade as soon as possible. Store the harvested grapes in a very cool, dry, well-ventilated place.

2 POSTHARVEST HANDLING

2.1 Sorting

Sorting of grapes is performed either manually by hand or electronically. Manual sorting requires several workers and is time-intensive. Initial sorting is conducted in the vineyard. The grapes are destalked and then put onto the sorting table (Fig. 2), which moves the berries using vibration. During sorting, leaves, stalks, snails, and uncolored berries are removed with the destemmer. The grapes are then crushed and put into the fermenter without leaves and stalks.

Electronic sorting is expensive but saves labor and time. Fruit is fed onto a horizontal conveyor belt and then dropped onto another horizontal conveyor belt.

FIG. 2
Manual sorting of grapes. Source: bkwine.com.

FIG. 3
Electronic sorter for grapes. Source: WECO, Woodland, CA.

Between the two conveyor belts are sensors and lasers that are connected to a computer. The computer has been programmed to distinguish desirable and undesirable qualities based on various parameters, such as coloring and size. If the computer recognizes that the fruit falling through the air does not meet the specifications, it is shot in the air by an air nozzle and lands onto another conveyor belt. The use of this technology makes it possible to sort individual berries according to differences in size (eliminating small green berries or large berries) and to eliminate berries that are not fully colored, along with stalks, leaves, pine needles, insects, and berries showing signs of botrytis, mealybug, bird damage, etc.

The electronic grape sorter manufactured by WECO, Woodland, CA is shown in Fig. 3. It uses advanced cameras, LEDs (light-emitting diodes), and software technology to identify any remaining stems, insects, unripe and damaged fruit, and raisins. Unwanted items are removed with precise blasts of air, increasing quality and throughput while reducing labor costs (AFF, 2012).

2.2 Packing

Table grapes should be packed and shipped immediately after harvest or stored properly when the harvested quantity exceeds demand. Table grapes are packed for protection during transportation and handling. Packaging is normally done in corrugated or solid fiberboard cartons. A layer of bubble pad or protective liner followed by a polyethylene lining is placed at the bottom of the carton to protect the grapes from bruising. Bunches from these weighed lots are placed in small, thin, and clean food-grade polyethylene pouches. One or two bunches weighing between 350 and 650g are placed in each pouch. No bunch weighing <150g is placed in a pouch. The grapes are then precooled to a temperature of 4°C and then a sulfur dioxide generating pad enclosed in absorbent tissue paper is placed over the grapes. The box is closed after covering the pouches with more polyethylene lining. During packing, care should be taken to avoid damage to the fruit from shattering, splitting, or bruising. Packing in multilayers within a crate should be avoided as it causes weakening of the pedicle attachment in the berries of the lower bunches due to pressure exerted by the bunches on top.

For raisin production, soil between the grape rows is smoothed and trays or papers used for field drying are placed before harvesting. Grapes are picked and placed into drying papers in single layers for drying. Dried raisins from paper trays are then collected and placed in storage. A dehydrator is used to dry the grapes for 12–24h to bring the raisins to the low moisture content required for storage. After drying, raisins are packed into boxes for sale.

For winemaking, harvested grapes are destemmed and crushed to extract the juice, which is fermented into wine. The requirements for the vinification process are decided based on initial quality evaluations. Sugar content and acidity are adjusted and sulfur dioxide is added to prevent oxidation and color deterioration.

2.3 Storage

2.3.1 *PRECOOLING*

The purpose of precooling is to reduce field heat. Retaining the freshness of grapes for a long time requires prompt removal of field heat after harvesting. The temperature in the packing house should be maintained at 18–20°C and the grapes should be transported to precooling units within 4–6h of harvest. The temperature of harvested grapes should be brought down to <4°C within 6–8h in the precooling chambers. A delay in bringing the grapes down to this temperature will reduce their keeping quality (Sharma, 2015).

2.3.2 *COLD STORAGE*

After precooling, the dual releasing sulfur dioxide pads (Grape guard) are placed with their coated surfaces facing downwards on the filled plastic pouches and covered with plastic sheet lining. The boxes are closed and shifted to cold storage rooms where the temperature and humidity are maintained at 0 ± 0.50°C and 93 ± 2%, respectively. A temperature of 0°C and humidity of 95% is optimal for maintaining freshness and preventing decay. Freezing injury to berries, pedicels, and rachis occur at −2°C, while at slightly higher temperatures decay may occur. Care should be taken to maintain the temperature and humidity strictly during storage and transit (Sharma, 2015). The arrangement of the boxes in cold storage is very important to ensure uniform cooling of all the berries in all the boxes.

2.3.3 TRANSPORT

Cooling must continue during the grapes' journey from the farm to the customer. During transportation, the temperature inside the box needs to be at the same low level as it was in cold storage. Berries should be transported from the farm to the airport or harbor using refrigerated trucks and stored in cold store at the airport. Cold store temperature should be maintained in the cargo aircraft to preserve the quality of grapes.

3 PROCESSING OF GRAPES

3.1 Raisins

Raisins are produced by drying the berries using solar energy or hot air. The use of sunshine is the most popular method for raisin production. Raisins were sun-dried from grapes as long ago as 1490 BCE (Mencarelli et al., 2005). Currently, grapes are handpicked from the vine, put on clean trays between the vine rows, and laid in the sun for 2 or 3 weeks to dry. When the moisture content reaches 15%, the trays are rolled carefully into bundles and baked in the sun for a few more days. Afterward, they are loaded on a conveyor belt and large stems are separated, placed in large wooden bins for moisture equalization, and shipped off to packing plants. At the packing site, the raisins are inspected for any imperfections and then passed through a series of conveyor belts and drums to remove remaining stems, chaffs, and lightweight fruit. The raisins are also sent through a vacuum air stream to catch any other undesirable materials. Finally, they are graded by size and thoroughly washed in pure water. Before packaging, the raisins are moved through a laser sorter, which uses bursts of air to knock any foreign material out. After final inspections, raisins are weighed and packed in a variety of convenient sizes.

Production of mycotoxins due to fungi is the major problem with raisin production by field drying and storage. Warm temperature (20–30°C) and high humidity must be avoided, and good ventilation through the raisins is required to prevent fungi formation. Solar dryers are used to reduce the drying time and control the drying conditions to avoid the formation of fungi and subsequent mycotoxin production. Raisins can be produced by using a perforated rotary drum dryer to pass hot air over the grapes. This is a faster method to produce raisins but requires more energy and careful control of temperature to avoid quality loss. Since the rate of moisture diffusion through the berries is influenced by the waxy cuticle of the grapes, pretreatments are performed to increase the drying rate. Dipping in hot water or the use of sulfur, caustic, and ethyl or methyl oleate emulsions are widely used as pretreatments. The pretreatments increase drying rates and improve quality. Surface treatment of grapes increases the drying rate by altering the waxy layer structure at the grape surface, thus reducing the internal resistance to water diffusion.

Raisins are used to make raisin juice and raisin paste through further processing. Raisins are leached with water several times to produce raisin juice, making it a pure extract of raisins (Mencarelli et al., 2005). The extracted liquid is evaporated in a vacuum pan to produce a self-preserving concentrate, which contains a minimum of 70% natural fruit soluble solids. Raisin juice is added to a variety of foods, including dairy, confectionery, and bakery items. Raisin paste is made completely from raisins, produced by extruding raisins through a fine mesh screen. Raisin paste can be used to add visual appeal and flavor. It is a stable ingredient that naturally sweetens fine confectionery fillings and soft-center candies. Raisin paste is also used in bakery items, such as bread, cookies, and pastries to inhibit mold growth, extend shelf life, and enhance flavor.

3.2 Wine Production

Wine is the product of partial or total alcoholic fermentation of fresh grapes, whether they are squeezed or not. This process has been performed since the age of the first settlements in the Tigris-Euphrates basin several thousand years ago. From this region, the vine *Vitis vinifera* has been carried to all the Mediterranean countries. Wine is produced and consumed on almost all continents. It contains healthy compounds such as flavonoids, which are useful for protection from cardiovascular disease, and tannins, which have antibacterial compounds and an astringent effect. Different kinds of wine are produced all over the world and the three main classifications (red, white, and dessert) determine the vinification process.

3.2.1 *RED WINE*

Red grapes for wine production must be picked at the right stage of ripening with the maximum content of flavonoids (anthocyanins which provide the red color) and non-flavonoids (phenolic acids which provide the brown color) as well as tannins localized mainly in the peel. The right equilibrium among these components is important for producing good quality wine, and the right sugar content is needed to produce wine. The stage of ripening for red grapes is easy to detect because the berries' peel changes from green to light red to deep red, and should be picked at the lattermost stage. The flavonoid and tannin contents are affected by the weather condition (high UV radiation, water stress, and thermal stress). To reach the right alcohol content, the sugar content (only fructose and glucose are found in grape berries), measured by a refractometer (RI = refractometer index in % or °Brix), should be above 19%. The taste of the berries can help determine when to harvest. Berries with a sugar content of over 17% are sweet and 19% sugar content provides a very sweet taste. At a sugar content of over 26%, there may be problems with the start of fermentation. However, if the value is too low, there may not be sufficient alcohol and the resulting wine will be very acidic. The conversion of sugar in alcohol is approximately 70%, which allows the potential alcohol content to be estimated by multiplying using the RI per 0.6. The acidity of the berries is another important characteristic and it is inversely correlated with the sugar content. In other words, acidity decreases when sugars increase and vice versa. The main acids in grapes are tartaric acid, malic acid, and citric acid. A good wine has acidity between 4 and 6 g/L in tartaric acid.

Picking: Grape pickers should avoid crushing the berries. Crushing the berries in the harvest containers provokes the start of acetic bacteria fermentation. Sometimes, potassium metabisulfite powder is applied over the grapes to avoid this anomalous fermentation, but this addition can kill the yeasts needed for alcoholic fermentation of the berries.

Unloading and destemming: Harvested grapes are unloaded in the crusher. The stems are removed to avoid the excessive salts which give an herbaceous taste to the wine. Parts of the stems can be left in the mass to provide acidity and tannins useful for wine stabilization.

Crushing: Crushing is a very important step for red grapes since it breaks the berries for extraction of must (fresh grape juice), and the fragmentation of the peel also allows for extraction of the pigments. Different kinds of crusher-destemmers are available to mechanically squeeze the berries. The typical machine is a roller type consisting of two fluted, horizontal rolls of rubber-coated steel and stainless steel. The rolls are adjustable and should be set so that the berries are thoroughly crushed without breaking the seeds or grinding the stems. For a smaller quantity of grapes, a simple screw type crusher and vertical presser (torque) can be used.

Maceration and fermentation: The objective of this operation is to transform sugars into alcohol and to permit the diffusion of the peel substances (pigments, pectins, acids, mineral ions) into the liquid. From the crusher, the juice and peels pumped to the steel container (fermentor) where fermentation and maceration will take place. The fermentor can be open on the top or sealed completely. In an open top fermentor, fermentation starts and continues easily due to the large availability of air for yeast. During fermentation, it is important to break the hat, or layer of pomace (peels, seeds, and residues of stems), which floats on the surface of the juice due to the carbon dioxide pressure created by the fermentation process. This practice is done with a special stick or by removing the liquid from the bottom of the fermentor and, with a pump, pouring the liquid on the hat with pressure. Other technologies are also used to break the hat and allow the pigments to diffuse from the solids to the liquid.

At the beginning of fermentation, sulfur dioxide is added (Table 4) to select the yeast, halt bacteria development, and favor pigment diffusion from the peel. During fermentation, a large amount of heat is released and the temperature of the mass can even reach 40°C depending on the environmental conditions. Temperature control is very important; the maximum temperature allowed is 35°C and should only be that high for a few hours. To reduce the temperature, wet the outside of the container. If it is too cold, the container must be warmed by heating the environment or heating the liquid with special tools such as infrared lamps and electrical resistance.

There is usually plenty of yeast in the peel (*Saccharomyces cerevisiae*) in red vinification and there is no need for an exogenous supply. On the other hand, in white vinification the yeast supply is common practice. During fermentation, a large amount of carbon dioxide is released; thus, the room must be ventilated to flush the CO_2 out. At the end of fermentation, the sugar content in the liquid should be below 1 g/L to avoid further fermentation (acetic fermentation) inside the bottle. By this time, the harvested juice will have completed the transformation into wine. This means that sugars have converted into alcohol and the liquid will have become red with light acidity and no sweetness.

Stabilization: When the liquid is removed from the fermentor, it is not completely clear. The clarity depends on the vinification process and the quality of grapes. Therefore, a secondary fermentation (malolactic) is performed to convert the malic acid into lactic acid, which reduces the acidity and gives a smoother taste. This fermentation process is adjusted based on the outside temperature. It is commonly induced by adding bacteria and heating the liquid slightly (30°C). During the secondary fermentation, sulfur dioxide concentration is controlled at around 20 mg/L. When this fermentation is accomplished, potassium metabisulfite is added to reach concentrations of 30 and 35 mg/L, respectively, for red and white wines.

Clarification: Proteins and pectins in the solution are removed before bottling the wine. Even tannins are sometimes removed to reduce the astringency. This process is called clarification and is carried out by adding protein compounds such as pure albumin (40 mL albumin/100 L wine), egg yolk (1 stirred egg yolk per 225 L of wine), gelatin (sold as liquid; 30 g gelatin/100 L wine), or fish glue (dilute 1 kg in 100 L of water and then take 2–3 g per 100 L of wine) to remove negatively charged substances like tannins.

After clarification, the red wine can be stored in a wooden barrel for a few months to a few years or in glass, plastic, and stainless steel containers. The pomace remaining in the fermentor is removed and undergoes a distillation process (grappa) or is pressed to obtain additional but lower quality wine.

Filtration and bottling: Filtration of wine before bottling is performed. However, for red wine it may be avoided as it removes pigments, reducing the color intensity and modifying the taste. Still, during shipping, the temperature changes continuously and filtration is necessary for stabilization. Several kinds of filtration can be done; one of the most commonly used methods involves fossil flour or cardboard to remove rough particles right before bottling. Vertical, synthetic septum filters are used to remove microorganisms. After filtration, the wine is bottled in a very well washed bottle and then capped with a cork or silicon cap. The wine can be kept in the bottle for several months before selling or it can be sold immediately.

3.2.2 WHITE WINE

White wine production differs from that of red wine in that the peel is removed from the flesh after crushing. The peel is removed to avoid the transfer of pigments, which are very susceptible to oxidation, thus producing an unpleasant color and flavor to the wine. In the past, white grape varieties were fermented with the peel as in red wine vinification but today only partial fermentation is done with the peel. The harvest and field transport processes are the same as those for red wine production. However, the crushing technique for white wine production differs by using a pneumatic pressure system. With this system, the pumping of air or nitrogen inside a plastic airtight balloon allows the squeezing of the grape berries against the wall of a cylindrical metal container with a perforated bottom. This causes the berries to release the flesh and the must while leaving the peels almost intact. The squeezing can be carried out with a simple torque using very light force. Afterward, the clear must is moved to the fermentor for fermentation without maceration.

In the fermentor, there is no formation of hat because there is only a small amount of pomace. All fermentation processes today occur without air contact to avoid oxidation at temperatures around or below 20°C. With this kind of vinification, the fermentation is started by adding yeast and nutrient compounds such as nitrogen (ammonia form), yeast ghosts, thiamine, and sulfur dioxide. Clarification and filtration are more or less the same as those of the red vinification. Usually, bentonite is used to clarify the wine. Today, most high-quality white wines are produced by barrique fermentation in which the must is pumped in a new barrique and left to ferment directly in the barrique. Using this method, there are precautions taken when carrying out the bâtonnage, which involves the mixing of the precipitate in the liquid mass during the primary and secondary fermentations and during the storage of the wine.

3.3 Grape Juice

The generalized flowchart for grape juice production is shown in Fig. 4, which has undergone a continuous change in several years. Several options are now available for pretreatment, juice extraction, and subsequent treatment after juice

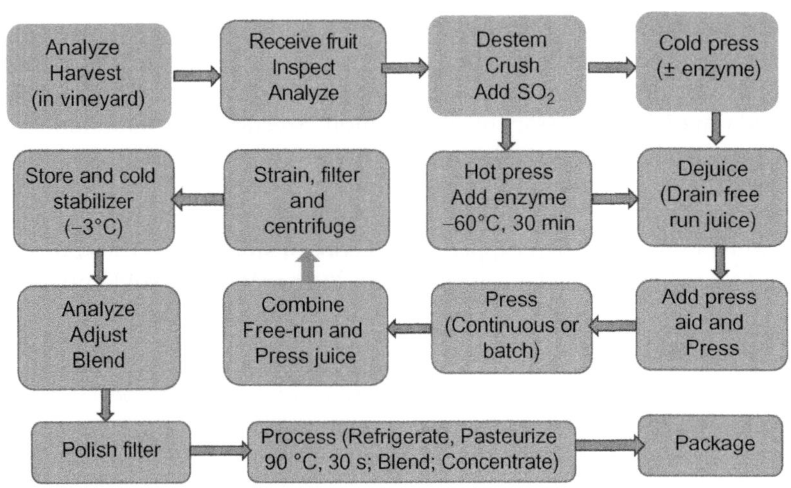

FIG. 4

Generalized flow chart for grape juice extraction (Bates et al., 2001).

extraction. The continuous hot pressing method is suitable for deeply pigmented grapes where maximum color extraction is desired. Cold pressing is suitable for light-colored grapes to maintain the color (Bates et al., 2001). There are two extraction processes for juice processing: the hot press and the cold press.

3.3.1 HOT PRESS

In hot press extraction, pectolytic enzyme is added to break down naturally occurring pectins and paper pulp. Rice hulls are used as press aids to facilitate extraction of juice. The hot-press method yields more juice with higher total solids and more non-sugar solids, tannins, pigments, and other substances than a cold-press juice operation. Hot pressing is performed at <65°C to preserve juice quality. As shown in the flow chart (Fig. 4), harvested grapes are dumped into a hopper and transported by augers or pumps to a rotary stemmer-crusher that separates the fruit from the stem and crushes the berries. The crushed berries are then pumped through a steam-jacketed vacuum preheater in which the pulp is heated to 60–63°C and passed into holding tanks. At this point, slow-moving agitators mix pectolytic enzyme and about 7 kg of purified paper pulp per 1000 kg of grapes. Enzymes take about 30–60 min to break down the pectin to make the grape pulp ready for pressing. The free run juice is separated in the de-juicer by a 40-mesh screen and the remaining pulp is pressed using a screw press. Later, the free run juice will be combined with the pressed juice and passed through the rotary vacuum filtration, pressure leaf filtration, or centrifugation. In a heat exchanger, the filtered juice is flash-heated at 80–85°C, rapidly cooled in another heat exchanger to −2.2°C and placed in tanks for rapid settling of argols for cold stabilization of juice. Seeding with bitartrate crystals and ion exchange methods can be used to accelerate the cold stabilization step. The final processing into a single-strength juice or concentrate can occur once the argols have settled and the juice is removed. Next, the juice is heated to 77°C in a heat exchanger and filled in bottles. The bottles are then closed with caps and pasteurized at 85°C for 3 min, cooled, and labeled. Before pasteurization treatment, the grape juice may be degassed. At times, the juice may be clarified with albumin, casein, or bentonite.

3.3.2 COLD PRESS

The cold-press method does not have the following steps: heating of the crushed berries to 60–63°C and holding in tanks with pectolytic enzymes. It is suitable for light-colored grape cultivars, which cannot be hot pressed because they lack skin pigment and yield a light green to yellow juice. Enzymes can be added to the cold-pressed juice to facilitate the clarification and filtration processes following cold stabilization. Sulfur dioxide can also be added to minimize enzymatic browning. Juice yields with the cold-press method are lower than those of the hot-press method.

The concentration of juice is an important operation in grape juice processing because the concentrated juice minimizes transportation and storage costs. The concentrate is diluted to make single strength grape juice or used to make jams, jellies, yogurt, frozen fruit desserts, cereals, cookies, and other bakery products.

4 GRAPE POMACE UTILIZATION

Grapes are one of the most extensively cultivated crops in the world, and about 75% of the total grape production is used to produce wine (García-Lomillo and González-SanJosé, 2017). In general, 20%–30% of the original grape weight is produced as pomace (Dwyer et al., 2014; Laufenberg et al., 2003), which constitutes the skin, stem, residual pulp, seed, small pieces of stalks, and yeast cells from the wine fermentation process (Christ and Burrit, 2013; Brenes et al., 2016; Beres et al., 2017). Seeds and skins represent 38%–52% and 5%–10% of grape pomace, respectively (Brenes et al., 2016). For a long time, wine pomace has been an undervalued product due to lack of alternative uses with economic benefits. Traditionally, wine pomace has been used as fertilizer or animal feed (Arvanitoyannis et al., 2006). Another use for it is to produce different types of "wine alcohol" (Silva et al., 2000), which are a part of well-appreciated and valorized distilled spirits, liquors, and liqueurs (Gonzalez-SanJos'e, 2014). Pomace has also been used as compost to increase the organic matter, nitrogen, and mineral contents of vineyard soils (Diaz et al., 2002). However, the application of pomace as fertilizer and animal feed has drawbacks due to the presence of antinutritive compounds that negatively affect crop yields and animal weight gain, and it fails to exploit the full potential market of this by-product (Dwyer et al., 2014). Only 3% of grape pomace produced is reused for animal feed (Dwyer et al., 2014; Brenes et al., 2016), waste-based compost (Santos et al., 2016), and thermal insulation in building construction (Muñoz et al., 2014). Incineration and discarding pomace in land fields are detrimental to the environment because the phenolic compounds decrease the pH of the pomace and increase resistance to biological degradation. In addition, inappropriate pomace disposal leads to surface and groundwater pollution, foul odor, fly, and pest attraction which may spread diseases, and oxygen depletion in soil and ground waters by tannins and other compounds (Beres et al., 2017; Christ and Burrit, 2013; Dwyer et al., 2014).

Many studies were performed to utilize pomace to extract phenolics and antioxidant fibers from grape skins (Chamorro et al., 2012; Duba et al., 2015; Beres et al., 2016) and oil from the seeds (Venkitasamy et al., 2014; Bail et al., 2008; Fernández et al., 2010; Fiori et al., 2014). There are several applications of grape pomace, including functional food [dietary fiber (DF) and polyphenols], food processing (biosurfactants), cosmetic (grape-seed oil and antioxidants), pharmaceutical, and supplements (grape pomace powder) (Rózek et al., 2010; Dwyer et al., 2014; Shinagawa et al., 2015).

4.1 Composition of Grape Pomace

The composition of grapes varies depending on climatic conditions and viticulture practices as well as factors such as variety, maturity, and sanitary conditions. The type of process and the conditions under which winemaking is carried out notably influence the composition of wine pomaces. Red wine pomace is a by-product that has been fermented, while white and rosé wine pomaces are removed before alcoholic fermentation and the fermentable sugars remain in

Table 3 **Basic Chemical Composition of Five Fresh (Wet) *V. vinifera* Grape Pomaces**

Grape Variety	Skin (%)	Seed (%)	Stem (%)	Total soluble solid content (%)
Morio Muscat	85.99	12.77	1.25	83.9
Muller Thurgau	90.67	7.84	1.49	72.4
Cab. Sauvignon	77.41	20.91	1.68	28.2
Pinot Noir	73.35	12.34	0.54	27.7
Merlot	83.18	14.98	1.84	22.2

Adapted from Jiang, Y., Simonsen, J., Zhao, Y., 2010. Compression-molded biocomposite boards from red and white wine grape pomaces. J. Appl. Polym. Sci. 119(5), 2834–2846.

their pomaces (García-Lomillo and González-SanJosé, 2017). Jiang et al. (2010) analyzed the compositions of various red and white pomaces using grape varieties from the United States. Table 3 shows the basic chemical compositions of grape pomaces from five different *Vitis vinifera* cultivars: two white (Morio Muscat and Muller Thurgau) and three red (Cabernet Sauvignon, Pinot Noir, and Merlot). The chemical composition is noticeably different for each grape variety with the largest variation found in the TSS content between red and white grapes. Jiang et al. (2010) calculated that grape pomace contains 7%–20% seeds; several studies have found varying results, ranging from 20% to 26% (Baydar and Akkurt, 2001) and as high as 38%–52% (Maier et al., 2009).

Deng et al. (2011) studied the chemical compositions of the grape pomace skins of the five varieties and reported that red pomace samples contained between 51% and 56% total DF, whereas the white grape pomace contained approximately half that (17%–28%), as shown in Table 4.

Table 4 **White and Red Grape Pomace Skin Compositions**

Composition (% DM)	Muller Thurgau (W)	Morio Muscat (W)	Cabernet Sauvignon (R)	Merlot (R)	Pinot Noir (R)
Soluble sugar	55.77	77.53	1.71	1.34	1.38
Total dietary fiber	28.01	17.28	53.21	51.09	56.31
Insoluble dietary fiber	27.29	16.44	52.40	49.59	54.59
Soluble dietary fiber	0.72	0.84	0.81	1.51	1.72
Condensed tannins	8.53	6.04	16.14	16.26	19.89

DM = dry matter; (W) = white wine pomace; (R) = red wine pomace.

Adapted from Deng, Q., Penner, M. H., Zhao, Y., 2011. Chemical composition of dietary fiber and polyphenols of five different varieties of wine grape pomace skins. Food Res. Int. 44(9), 2712–2720.

Table 5 Seed Oil Contents

Grape-seed cultivars	Oil content (% v/w)[a]	Degree of unsaturation (%)	Tocopherol (mg/kg)
Red	13.1–19.6	86.6–89.3	357–578
White	14.7–17.8	86.5–88.1	364–486

[a]Avg. (red and white) = 16.3%.

Adapted from Baydar, N. G., Akkurt, M., 2001. Oil content and oil quality properties of some grape seeds. Turk. J. Agric. For. 25(3), 163–168.

Baydar and Akkurt (2001) analyzed the oil contents in red and white pomace seeds and found them to be similar with an average of 16.3% (Table 5), of which 87% was unsaturated oil, which is beneficial to health when compared with saturated oil.

Linoleic acid (C18:2) is the predominant fatty acid (58%–75%) in grape-seed oil, followed by 12%–21% oleic acid (C18:1), and 7%–13% palmitic acid (C16:0) (Bordiga et al., 2015; Teh, 2016). The oil content in grape skin is minimal, ranging from 1.14% to 6.33% of dry matter (Deng et al., 2011).

4.2 Phenolics From Grape Pomace

Grape pomace is rich in polyphenols, which exert antioxidant properties, including resveratrol, procyanidins/condensed tannins, anthocyanins, flavanols, and flavanols/catechins. Phenolic compounds from grape pomace are effective free radical scavengers and have the potential to prevent cancer, cardiovascular diseases, cellular oxidative stress, diabetes, and pathogen proliferation. Total phenolic contents (TPC) in grape pomace ranged from 0.12 to 7.48 g gallic acid equivalents/100 g dry matter (DM) while the antioxidant activities ranged from 20 to 75 mM Trolox equivalents (Teh, 2016). Phenolic compounds are mostly distributed in grape seeds at 60%–70%, followed by skin at 28%–35%, and pulp at <10% (Bordiga et al., 2015; Negro et al., 2003; Shi et al., 2003). Epicatechin, catechin, and gallic acid are the major polyphenols in seeds, while procyanidins, epicatechin, epigallocatechin, gallic acid, and catechin are the main phenolics in grape peels (Cruz et al., 2004).

4.2.1 PHENOLICS EXTRACTION

The most widely reported extraction technique for polyphenols from grape pomace is solid-liquid extraction by mechanical agitation. The extraction efficiency depends on the solvent type, temperature, extraction time, and particle size, as well as the presence of interfering substances in the matrix (Ignat et al., 2011). Traditional extraction methods such as mechanical agitation are inefficient for extracting some phenolic compounds called "nonextractable phenolics," which are ester bound or trapped within proteins, polysaccharides, or cell walls (Saura-Calixto, 2012).

In previous years, novel methods were used to reduce extraction time and organic solvent consumption, aiming to increase process sustainability (Beres et al., 2017). Supercritical fluid extraction (SFE) is a novel technique used to extract phenolics from grape pomace. De Campos et al. (2008) performed SFE extraction of grape pomace by using pure CO_2 and CO_2 with ethanol as the cosolvent in concentrations of 10%, 15%, and 20% w/w. Accelerated Solvent Extraction, also known as pressurized fluid extraction or pressurized liquid extraction, of grape pomace was performed by Monrad et al. (2009). They found that a mixture of 50:50 ethanol/water (v/v) under a pressure of 6.8 MPa with one extraction cycle at 40–140°C was the best condition for the extraction of procyanidins from red grape pomace. Another nonconventional method that has been used for phenolic compound recovery from grape pomace is ultrasound-assisted extraction, which can achieve high yields in short periods of time (González-Centeno et al., 2015; González-Centeno et al., 2014; Drosou et al., 2015).

4.2.2 APPLICATION OF PHENOLICS

Grape pomace extracts can be applied in food, pharmaceutical, cosmetic, and other industries in the form of liquid extracts, concentrate, or powder (Prodanov et al., 2005; Beres et al., 2017). Grape pomace was used as a food protector to prevent lipid oxidation in fish-based products (Sánchez-Alonso et al., 2007a; Pazos et al., 2005), meat products (Guerra-Rivas et al., 2016), and as an antibacterial/antimicrobial agent against *Staphylococcus aureus, Bacillus cereus,* Campylobacter coli, *Escherichia coli* O157:H7, and Salmonella infantis (Katalinić et al., 2010). In the food industry, grape pomace can be used as a functional supplement to enrich beverages and as an ingredient of an osmotic solution to obtain dehydrated fruit with enhanced phenolic content (Rózek et al., 2010). It can also be incorporated into chitosan edible films to improve antioxidant content and shelf life. Grape pomace was used as a replacement for synthetic antioxidants in food products (Garrido et al., 2011) and to reduce acrylamide formation during Maillard reaction in chip dehydration (Zhu et al., 2009, 2015). Grape pomace has been found to have anticancer (Kaur et al., 2009), cholesterol lowering (Ferri et al., 2016), and antiaging properties (Wittenauer et al., 2015). Furthermore, there are cosmetic products with grape polyphenols, such as day and night creams, the face serum from Pure Super Grape (Marks and Spencer—United Kingdom), mattifying fluids, anti-wrinkle creams, and the protection fluid sold by Caudalíe (France), which declares its use of polyphenols from grape seeds.

4.3 Dietary Fiber From Grape Pomace

DF has been recognized for contributing almost no energy and having health benefits (Tungland and Meyer, 2002). It is the main component of dried wine pomace, with concentrations ranging between 43% and 75%. DF is mainly constituted of cell wall polysaccharides and lignin (García-Lomillo and González-SanJosé, 2017). Grape seeds are richer in fiber than the skins are, and red wine pomace is richer in fiber than white wine pomace is (Gul et al., 2013). Insoluble dietary fiber (IDF), especially acid insoluble lignin (Klason lignin), is the main component of DF in both red and white wine pomace (Saura-Calixto et al., 1991) and fiber contains tannins and proteins in considerable proportions (Arnous and Meyer, 2008).

The DF in grape pomace has the benefits of both fiber and antioxidants, which can help to prevent cancer and cardiovascular diseases (Zhu et al., 2015). Grape pomace has the characteristics of DF; for example, grape skin is a complex lignocellulosic material containing large amounts of hemicellulosic sugars that, after hydrolysis, produce solutions containing a wide variety of xylose and glucose monomers (Devesa-Rey et al., 2011).

4.3.1 *DIETARY FIBER EXTRACTION*

In general, the extraction of DF can be classified into three main types: conventional solvent extraction method, microwave or ultrasonic method, and enzyme-assisted extraction method (Zhu et al., 2015). Du et al. (2011). DF was extracted from grape pomace using hydrochloric acid as the solvent and reported that the best extraction efficiency was obtained with hydrochloric acid at 0.4 mol/L, a substrate solvent ratio of 1:12 (m:v), temperature under 75°C, and extraction time of 90 min to achieve a yield of 47% soluble fiber. Ferreira et al. (2013) investigated the soluble DF extraction from the white grape pomace of Chardonnay, and showed that a hot aqueous extraction at 90°C with a substrate/solvent ratio of 1:4 and no gridding of the grape pomace was able to extract various polysaccharides. Minjares-Fuentes et al. (2014) performed an ultrasound-assisted extraction of pectin from grape pomace using hydrochloric acid solution at pH 2.0 and obtained a maximum yield of 32% of pectin. Hot water extraction to recover simple sugars and polysaccharides from Pinot noir grape pomace was performed by Beres et al. (2016). The authors reported that the best conditions for sugar recovery were a temperature of 100°C, using the smallest particle size (approximately 300 nm), and a solute: solvent ratio of 1:12. Deng et al. (2011) studied the DF content in five varieties of wine grape pomace skins (Table 6) using the gravimetric-enzymatic method with sugar profiling by HPLC-ELSD. The IDF of grape pomace dominated with 98.5% of the total DF (Table 6). Insoluble DF composed of Klason lignin (7.9%–36.1% DM), neutral sugars (4.9%–14.6% DM), and uronic acid (3.6%–8.5% DM) weighed >95.5% of the total DF in all pomace skins of the five wine grape varieties.

4.3.2 *APPLICATION OF DIETARY FIBER*

The DF of grape pomace can be used in different food products, including bakery products, beverages, and meat products, and in the cosmetic and pharmaceutical industries, offering a range of health benefits. Grape pomace and seed flours

Table 6 Dietary Fiber Content in Five Varieties of Wine Grape Pomace Skins

Composition (% DM)	Muller Thurgau (W)	Morio Muscat (W)	Cabernet Sauvignon (R)	Merlot (R)	Pinot Noir (R)
Soluble sugar	55.77	77.53	1.71	1.34	1.38
Total dietary fiber	28.01	17.28	53.21	51.09	56.31
Insoluble dietary fiber	27.29	16.44	52.40	49.59	54.59
Soluble dietary fiber	0.72	0.84	0.81	1.51	1.72
Condensed tannins	8.53	6.04	16.14	16.26	19.89

DM = dry matter; (W) = white wine pomace; (R) = red wine pomace.

Adapted from Deng, Q., Penner, M. H., Zhao, Y., 2011. Chemical composition of dietary fiber and polyphenols of five different varieties of wine grape pomace skins. Food Res. Int. 44(9), 2712–2720; Dwyer, K., Hosseinian, F., Rod, M., 2014. The market potential of grape waste alternatives. J. Food Res. 3, 91–106.

have been applied in different products such as popsicles (Ishimoto et al., 2009), cereal bars biscuits and cookies (Mildner-Szkudlarz et al., 2013; Acun and Gül, 2014), extruded snacks (Bender et al., 2016), and muffins (Beres et al., 2017). These pomace-enhanced products are rich in fiber with antioxidant capacity potential and good acceptance by consumers. Pinot Noir grape fiber can be added to yogurt and salad dressings to increase their fiber and phenolic contents, and delay lipid oxidation during storage and extend shelf life (Tseng and Zhao, 2013). The grape pomace added to codfish and seafood was found to minimize color, flavor, and textural changes and to minimize lipid oxidation during freezing storage (Sánchez-Alonso et al., 2007a, 2007b, 2008). Addition of grape antioxidant DF in chicken breast hamburger patties (Sáyago-Ayerdi et al., 2009) and in fish muscles (Sánchez-Alonso et al., 2007b) improved the oxidative stability and the radical scavenging activity of the products. Moreover, grape fiber addition to red wine production removed up to 38% of the tannins generated during the process due to tannin's absorption ability of grape fiber (Guerrero et al., 2013).

The development of biodegradable products using white and red grape pomace fiber has been reported as a feasible process. For example, Ping et al. (2011) found that grape fiber can be used as an adhesive for wood particle boards. Due to the presence of soluble sugars in grape skins, they can be used as biodegradable packaging materials with excellent flexibility (Deng et al., 2011; Jiang et al., 2011). Grape pomace has great potential for being a part of environmentally friendly supporting materials due to its lignin, cellulose, and hemicellulose content (Deng et al., 2011; Zhu et al., 2015). Muñoz et al. (2014) used a mixture of 5% grape pomace and 95% clay to improve the thermal transmittance in construction.

4.4 Grape-Seed Oil

Grape-seed oil is already widely commercialized in the United States and other countries as it has been used for decades over several applications, particularly in cosmetics formulation and as a specialty salad or cooking oil (Venkitasamy et al., 2014; Baydar and Akkurt, 2001; Beveridge et al., 2005). Grape-seed oil is characterized by its light flavor with fruity touches, high smoke point (216°C), high digestibility, and a slight increase in viscosity when used for batch frying (Kinsella, 1974). Grape seeds have an oil content of 8%–15% (w/w) with high levels of unsaturated fatty acids, namely oleic and linoleic acids, which comprise >89% of the total oil composition (Choi et al., 2010b). Linoleic acid is an essential fatty acid, and when combined with its isomers, known as conjugated linoleic acids, they have biological effects (Beres et al., 2017).

4.4.1 *GRAPE-SEED OIL EXTRACTION*

Solid-liquid extraction using hexane as the solvent is the most common vegetable oil extraction from oleaginous seeds. The chemical (hexane) extraction method is detrimental to vital bioactive components in grape-seed oil and accelerates the oil's oxidation process. Therefore, other methods such as enzymatic-assisted aqueous, supercritical carbon dioxide, and ultrasound-assisted solvent extraction methods have been used for grape-seed oil extraction (Beres et al., 2017; Teh, 2016). Mechanical oil expellers have also been used to produce grape-seed oil since they are cheap, simple, and eco-friendly, and can be scaled up economically (Venkitasamy et al., 2014; Teh, 2016). The oil content and high lignin content in grape seeds are limitations to mechanical process extractions. Enzymatic pretreatment of seeds can improve the mechanical process as enzymes can hydrolyze seeds' cell walls, releasing bounded compounds and enhancing the oil yield (Azmir et al., 2013). Venkitasamy et al. (2014) used a laboratory screw press to extract oil from grape seeds and the effects of the press conditions, seed moisture, and particle sizes were studied. Grape-seed moisture content and screw diameter affected the oil extraction. Extracting whole grape seeds of 5.3% MC at 120 rpm using 10 mm die diameter was found to be the optimum condition which produced a grape-seed oil yield of 7.6% at an oil extraction rate of 0.57 $kg h^{-1}$ and seed residence time of 8.6 s. Rombaut et al. (2015) also used expeller pressing for grape-seed oil extraction at different preheating temperatures (90–120°C), screw rotation speeds (40–70 rpm), and die diameters (10–15 mm). Rombaut et al. (2014). Duba and Fiori (2015) extracted grape-seed oil with the supercritical extraction method and obtained higher oil yields than those of the pressing techniques but the phenolic content of oil decreased by >50%. Da Porto et al. (2013) compared grape-seed oil extraction using the Soxhlet and ultrasound assisted extraction methods. Compared with Soxhlet extraction, the ultrasound-assisted oil extraction produced higher oil yield with a lower solvent consumption and a shorter extraction time.

4.4.2 APPLICATIONS OF GRAPE-SEED OIL

Grape-seed oil has many advantages for human consumption owing to its high level of unsaturated fatty acids (Gómez et al., 1996). It has high linoleic acid content (important for prostaglandin synthesis, which has an influence on platelet aggregation and inflammatory processes), high vitamin E content (which helps to reduce the risk of arteriosclerosis), and low values of cholesterol. Therefore, its intake may be beneficial for preventing heart and circulatory problems (Oomah et al., 1998; Pardo et al., 2009). The beneficial effects of grape-seed oil include hepatoprotective, neuroprotective, and liver cholesterol reducing characteristics. Ismail et al. (2015, 2016) reported that grape-seed oil fed to liver and brain-injured rats had a protective effect on acute liver injury, which was attributed to the oil's potent antioxidant, anti-inflammatory, and antiapoptotic activities. The protective mechanisms of grape-seed oil against oxidative damage and inflammatory cascades in the brain proved a neuroprotective effect, which is related to its ability to scavenge free radicals, improve the activity of antioxidant enzymes, downregulate the gene expression levels of xanthine oxidase and inducible nitric oxide synthase, and suppress the inflammatory responses (Ismail et al., 2015). The grape-seed oil fed to rats showed good acceptability among the rats, a significant decrease in feed intake, and beneficial effects on liver cholesterol values (Asadi et al., 2010).

Choi et al. (2010a) used pre-emulsified grape-seed oil to replace pork back fat in frankfurter; which lowered the fat content closer to the target value of 20%, increased the gumminess and chewiness, and improved the textural properties. The incorporation of grape-seed oil (up to 10%) was also proposed to replace beef fat and improve the fatty acid profile of frankfurters and beef loin steaks (Jung et al., 2012; Özvural and Vural, 2014). Grape-seed oil with a high smoke point can be used as a potential biodiesel feedstock. In the cosmetic industry, grape-seed oil can be applied to moisturize the skin, aid in the healing of sunburns, and reduce *Acne vulgaris* due to the presence of linoleic acid (Vermaak et al., 2011).

5 SUMMARY

The grapes are one of the most valued fruits in the world, as it is consumed both fresh and in processed products such as wine, jam, juice, jelly, grape-seed extract, dried grapes, vinegar, and grape-seed oil. The end use varies among countries, depending on the religious and political interest of the region. Grapes use one of the highest inputs of technology and labor among all fruits. As fresh fruit, grapes are very delicate and extremely perishable and have a very high rate of loss during harvest and distribution. Therefore, processing into juice and raisins is performed to diminish waste and improve the marketability and profit from grape cultivation. About 50%–75% of grape production is processed to make wine and juice. Approximately 20%–30% of fresh grape produce is pomace, which consists of the remaining skin, seeds, and stalks. The pomace is a rich source of phenolics and antioxidants, but it is underutilized. Several studies have shown that the phenolics, DF, and grape-seed oil can be successfully extracted and used in animal feed and the pharmaceutical, cosmetic, and food industries to improve stability and nutritional characteristics. To make grape cultivation more sustainable, it is necessary to minimize resources involved in cultivation and processing and waste generation, and effectively use the by-products. This chapter provides an overview of grape production in the world, major varieties, growing conditions, postharvest handling, processing, and by-product utilization to minimize waste and achieve sustainability in the grape/wine processing industry.

References

Acun, S., Gül, H., 2014. Effects of grape pomace and grape seed flours on cookie quality. Qual. Assur. Saf. Crops Food 6, 81–88.

AFF, 2012. Grapes—Production Guideline. Department of Agriculture, Forestry and Fisheries. Department of South Africa. Available online at: http://www.nda.agric.za/docs/Brochures/grapesprod.pdf.

Arnous, A., Meyer, A.S., 2008. Comparison of methods for compositional characterization of grape (*Vitis vinifera* L.) and apple (*Malus domestica*) skins. Food Bioprod. Process. 86, 79–86.

Arvanitoyannis, I.S., Ladas, D., Mavromatis, A., 2006. Potential uses and applications of treated wine waste: a review. Int. J. Food Sci. Technol. 41, 475–487.

Asadi, F., Shahriari, A., Chahardah-Cheric, M., 2010. Effect of long-term optimal ingestion of canola oil, grape seed oil, corn oil and yogurt butter on serum, muscle and liver cholesterol status in rats. Food Chem. Toxicol. 48, 2454–2457.

Azmir, J., Zaidul, I.S.M., Rahman, M.M., Sharif, K.M., Mohamed, A., Sahena, F., Omar, A.K.M., 2013. Techniques for extraction of bioactive compounds from plant materials: a review. J. Food Eng. 117, 426–436.

Bail, S., Stuebiger, G., Krist, S., Unterweger, H., Buchbauer, G., 2008. Characterization of various grape seed oils by volatile compounds, triacylglycerol composition, total phenols and antioxidant capacity. Food Chem. 108, 1122–1132.

Bates, R.P., Morris, J.R., Crandall, P.G., 2001. Principles and Practices of Small- and Medium-Scale Fruit Juice Processing. Food & Agriculture Organization.

Baydar, N.G., Akkurt, M., 2001. Oil content and oil quality properties of some grape seeds. Turk. J. Agric. For. 25 (3), 163–168.

Bender, A.B.B., de Mello Luvielmo, M., Loureiro, B.B., Speroni, C.S., Boligon, A.A., da Silva, L.P., Penna, N.G., 2016. Obtention and characterization of grape skin flour and its use in an extruded snack. Braz. J. Food Technol. 19, 1–9.

Beres, C., Simas-Tosin, F.F., Cabezudo, I., Freitas, S.P., Iacomini, M., Mellinger-Silva, C., Cabral, L.M., 2016. Antioxidant dietary fibre recovery from Brazilian Pinot noir grape pomace. Food Chem. 201, 145–152.

Beres, C., Costa, G.N.S., Cabezudo, I., da Silva-James, N.K., Teles, A.S.C., Cruz, A.P.G., Mellinger-Silva, C., Tonon, R.V., Cabral, L.M.C., Freitas, S.P., 2017. Towards integral utilization of grape pomace from winemaking process: a review. Waste Manag. 68, 581–594.

Beveridge, T.H., Girard, B., Kopp, T., Drover, J.C., 2005. Yield and composition of grape seed oils extracted by supercritical carbon dioxide and petroleum ether: varietal effects. J. Agric. Food Chem. 53 (5), 1799–1804.

Bordiga, M., Travaglia, F., Locatelli, M., Arlorio, M., Coisson, J.D., 2015. Spent grape pomace as a still potential by-product. Int. J. Food Sci. Technol. 50 (9), 2022–2031.

Brenes, A., Viveros, A., Chamorro, S., Arija, I., 2016. Use of polyphenol-rich grape byproducts in monogastric nutrition. A review. Anim. Feed Sci. Technol. 211, 1–7.

Brostrom, G., Brostrom, J., 2009. The Business of Wine: An Encyclopedia. Greenwood Press, Westport, CT.

Chamorro, S., Viveros, A., Alvarez, I., Vega, E., Brenes, A., 2012. Changes in polyphenol and polysaccharide content of grape seed extract and grape pomace after enzymatic treatment. Food Chem. 133, 308–314.

Choi, Y., Choi, J., Han, D., Kim, H., Lee, M., Jeong, J., Chung, H., Kim, C., 2010a. Effects of replacing pork back fat with vegetable oils and rice bran fibre on the quality of reduced-fat frankfurters. Meat Sci. 84, 557–563.

Choi, Y., Choi, J., Han, D., Kim, H., Lee, M., Kim, H., Lee, J., Chung, H., Kim, C., et al., 2010b. Optimization of replacing pork back fat with grape seed oil and rice bran fiber for reduced-fat meat emulsion systems. Meat Sci. 84, 212–218.

Christ, K.L., Burrit, R.L., 2013. Critical environmental concerns in wine production: an integrative review. J. Clean. Prod. 53, 232–242.

Cruz, J.M., Dominguez, H., Parajo, J.C., 2004. Assessment of the production of antioxidants from winemaking waste solids. J. Agric. Food Chem. 52 (18), 5612–5620.

Da Porto, C., Porretto, E., Decorti, D., 2013. Comparison of ultrasound-assisted extraction with conventional extraction methods of oil and polyphenols from grape (*Vitis vinifera* L.) seeds. Ultrason. Sonochem. 20, 1076–1080.

Daane, K.M., Vincent, C., Isaacs, R., Ioriatti, C., 2018. Entomological opportunities and challenges for sustainable viticulture in a global market. Annu. Rev. Entomol. 2 (63), 193–214.

De Campos, L.M.A.S., Leimann, F.V., Pedrosa, R.C., Ferreira, S.R.S., 2008. Free radical scavenging of grape pomace extracts from Cabernet Sauvingnon (*Vitis vinifera*). Bioresour. Technol. 99, 8413–8420.

Deng, Q., Penner, M.H., Zhao, Y., 2011. Chemical composition of dietary fiber and polyphenols of five different varieties of wine grape pomace skins. Food Res. Int. 44 (9), 2712–2720.

Devesa-Rey, R., Vecino, X., Varela-Alende, J.L., Barral, M.T., Cruz, J.M., Moldes, A.B., 2011. Valorization of winery waste vs. the costs of not recycling. Waste Manag. 31 (11), 2327–2335.

Diaz, M.J., Madejon, E., L'opez, F., L'opez, R., Cabrera, F., 2002. Optimization of´ the rate vinasse/ grape marc for co-composting process. Process Biochem. 37, 1143–1150.

Drosou, C., Kyriakopoulou, K., Bimpilas, A., Tsimogiannis, D., Krokida, M., 2015. A comparative study on different extraction techniques to recover red grape pomace polyphenols from vinification byproducts. Ind. Crop. Prod. 75, 141–149.

Du, B., Li, F.Y., Fan, C.J., Zhu, F.M., 2011. Response surface methodology for optimization of extraction process for soluble dietary fibre from grape pomace with hydrochloric acid. Food Sci. 32, 128–134.

Duba, K.S., Fiori, L., 2015. Supercritical CO_2 extraction of grape seed oil: effect of process parameters on the extraction kinetics. J. Supercrit. Fluids 98, 33–43.

Duba, K.S., Casazza, A.A., Mohamed, H., Perego, P., Fiori, L., 2015. Extraction of polyphenols from grape skins and defatted grape seeds using subcritical water: experiments and modeling. Food Bioprod. Process. 94, 29–38.

Dwyer, K., Hosseinian, F., Rod, M., 2014. The market potential of grape waste alternatives. J. Food Res. 3, 91–106.

Fernández, C.M., Ramos, M.J., Pérez, Á., Rodríguez, J.F., 2010. Production of biodiesel from winery waste: extraction, refining and transesterification of grape seed oil. Bioresour. Technol. 101, 7019–7024.

Ferreira, C.S., de Pinho, M.N., Cabral, L.M.C., 2013. Solid-liquid extraction and concentration with processes of membrane technology of soluble fibers from wine grape pomace. Téc. Lisb, 1–9.

Ferri, M., Bin, S., Vallini, V., Fava, F., Michelini, E., Roda, A., Minnucci, G., Bucchi, G., Tassoni, A., 2016. Recovery of polyphenols from red grape pomace and assessment of their antioxidant and anti-cholesterol activities. New Biotechnol. 33, 338–344.

Fiori, L., Lavelli, V., Duba, K.S., Sri Harsha, P.S.C., Mohamed, H.B., Guella, G., 2014. Supercritical CO2 extraction of oil from seeds of six grape cultivars: modeling of mass transfer kinetics and evaluation of lipid profiles and tocol contents. J. Supercrit. Fluids 94, 71–80.

Food and Agricultural Organization and International Organization of Vine and Wine (FAO-OIV), 2016. FAO-OIV Focus 2016, Table and Dried Grapes. Available online at: www.oiv.int/public/medias/5116/booklet-fao-oiv-grapes-focus.pdf.

Food and Agriculture Organization of the United Nations (FAO). 2016. FAOSTAT Crop Information: Grapes. FAO.

García-Lomillo, J., González-SanJosé, M.L., 2017. Applications of wine pomace in the food industry: approaches and functions. Compr. Rev. Food Sci. Food Saf. 16 (1), 3–22.

Garrido, M.D., Auqui, M., Martí, N., Linares, M.B., 2011. Effect of two different red grape pomace extracts obtained under different extraction systems on meat quality of pork burgers. LWT Food Sci. Technol. 44, 2238–2243.

Gómez, A.M., López, C.P., De la Ossa, E.M., 1996. Recovery of grape seed oil by liquid and supercritical carbon dioxide extraction: a comparison with conventional solvent extraction. Short Communication. Chem. Eng. J. 61, 227–231.

González-Centeno, M.R., Knoerzer, K., Sabarez, H., Simal, S., Rosselló, C., Femenia, A., 2014. Effect of acoustic frequency and power density on the aqueous ultrasonic assisted extraction of grape pomace (*Vitis vinifera* L.)—a response surface approach. Ultrason. Sonochem. 21, 2176–2184.

González-Centeno, M.R., Comas-Serra, F., Femenia, A., Rosselló, C., Simal, S., 2015. Effect of power ultrasound application on aqueous extraction of phenolic compounds and antioxidant capacity from grape pomace (*Vitis vinifera* L.): experimental kinetics and modeling. Ultrason. Sonochem. 22, 506–514.

Gonzalez-SanJos'e, M.L., 2014. Liqueurs. In: Bamforth, C.W. (Ed.), The Oxford Handbook of Food Fermentations. Oxford University Press, Oxford, pp. 304–342.

Guerra-Rivas, C., Vieira, C., Rubio, B., Martínez, B., Gallardo, B., Mantecón, A.R., Lavín, P., Manso, T., 2016. Effects of grape pomace in growing lamb diets compared with vitamin E and grape seed extract on meat shelf life. Meat Sci. 116, 221–229.

Guerrero, R.L.F., Smith, P., Bindon, K.A., 2013. Application of insoluble fibers in the fining of wine phenolics. J. Agric. Food Chem. 61, 4424–4432.

Gul, H., Acun, S., Sen, H., Nayir, N., Turk, S., 2013. Antioxidant activity, total phenolics and some chemical properties of Öküzgözü and Narince grape pomace and grape seed flour. J. Food Agric. Environ. 11, 28–34.

Ignat, I., Volf, I., Popa, V.I., 2011. A critical review of methods for characterization of polyphenolic compounds in fruits and vegetables. Food Chem. 126, 1821–1835.

International Organization of Vine and Wine (OIV) (2017). OIV Focus 2017. Vine Varieties Distribution in the World. Available online at: http://www.oiv.int/public/medias/5336/infographie-focus-oiv-2017-new.pdf.

Ishimoto, E.Y., Capriles, V.D., Matias, A.G., Baccarin, A., Torres, E.A.F.S., 2009. Formulating functional popsicle with wine and grape juice by-products. Acta Hortic. 841, 603–606.

Ismail, A.F.M., Moawed, F.S.M., Mohamed, M.A., 2015. Protective mechanism of grape seed oil on carbon tetrachloride-induced brain damage in c-irradiated rats. J. Photochem. Photobiol. 153, 317–323.

Ismail, A.F.M., Salem, A.A.M., Eassawy, M.M.T., 2016. Hepatoprotective effect of grape seed oil against carbon tetrachloride induced oxidative stress in liver of Y-irradiated rat. J. Photochem. Photobiol. 160, 1–10.

Jiang, Y., Simonsen, J., Zhao, Y., 2010. Compression-molded biocomposite boards from red and white wine grape pomaces. J. Appl. Polym. Sci. 119 (5), 2834–2846.

Jiang, Y., Simonsen, J., Zhao, Y., 2011. Compression-molded biocomposite boards from red and white wine grape pomaces. J. Appl. Polym. Sci. 119, 2834–2846.

Jung, Y., Jung, S., Lee, H.J., Kang, M., Lee, S.K., Kim, Y.J., Jo, C., 2012. Effect of high pressure after the addition of vegetable oil on the safety and quality of beef loin. Korean J. Food Sci. Anim. Resour. 32, 68–76.

Katalinic, V., Mozina, S., Skroza, D., Generalic, I., Abramovic, H., Miloš, M., Miaden, B., 2010. Polyphenolic profile, antioxidant properties and antimicrobial activity of grape skin extracts of 14 *Vitis vinifera* varieties grown in Dalmatia (Croatia). Food Chem. 119, 715–723.

Kaur, M., Agarwal, C., Agarwal, R., 2009. Anticancer and cancer chemo preventive potential of grape seed extract and other grape-based products. J. Nutr. 139, 1806S–1812S.

Kinsella, J.E., 1974. Grapeseed oil: a rich source of linoleic acid. Food Technol. 28, 58–60.

Laufenberg, G., Kunz, B., Nystroem, M., 2003. Transformation of vegetable waste into value added products: (A) the upgrading concept; (B) practical implementations. Bioresour. Technol. 87, 167–198.

Maier, T., Schiber, A., Kammerer, D., Carle, R., 2009. Residues of grape (*Vitis vinifera* L.) seed oil production as a valuable source of phenolic antioxidants. Food Chem. 112 (3), 551–559.

Mencarelli, F., Bellincontro, A., DiRenzo, D., 2005. Grape: Post-Harvest Operations. FAO Publications, pp. 1–42.

Mildner-Szkudlarz, S., Bajerska, J., Zawirska-Wojtasiak, R., Górecka, D., 2013. White grape pomace as a source of dietary fibre and polyphenols and its effect on physical and nutraceutical characteristics of wheat biscuits. J. Sci. Food Agric. 93, 389–395.

Minjares-Fuentes, R., Femenia, A., Garau, M.C., Meza-Velázquez, J.A., Simal, S., Rosselló, C., 2014. Ultrasound-assisted extraction of pectins from grape pomace using citric acid: a response surface methodology approach. Carbohydr. Polym. 106, 179–189.

Monrad, J.K., Howard, L.R., King, J.W., Srinivas, K., Mauromoustakos, A., 2009. Subcritical solvent extraction of procyanidins from dried red grape pomace. J. Agric. Food Chem. 58, 4014–4021.

Muñoz, P., Morales, M.P., Mendívil, M.A., Juárez, M.C., Muñoz, L., 2014. Using of waste pomace from winery industry to improve thermal insulation of fired clay bricks. Eco-friendly way of building construction. Constr. Build. Mater. 71, 181–187.

NASS, 2018. Non-Citrus Fruits and Nuts 2017 Summary (June 2018). USDA, National Agricultural Statistics Service, Washington DC, USA. Available online at: http://usda.mannlib.cornell.edu/usda/current/NoncFruiNu/NoncFruiNu-06-26-2018.pdf.

Negro, C., Tommasi, L., Miceli, A., 2003. Phenolic compounds and antioxidant activity from red grape marc extracts. Bioresour. Technol. 87 (1), 41–44.

Novello, V., Palma, L., 2008. Growing grapes under cover. Acta Hortic. 785, 353–362.

Oomah, B.D., Liang, J., Godfrey, D., Mazza, G., 1998. Microwave heating of grape seed: effect on oil quality. J. Agric. Food Chem. 46 (10), 4017–4021. https://doi.org/10.1021/jf980412f.

Özvural, E.B., Vural, H., 2014. Which is the best grape seed additive for frankfurters: extract, oil or flour? J. Sci. Food Agric. 94, 792–797.

Pardo, J.E., Fernández, E., Rubio, M., Alvarruiz, A., Alonso, G.L., 2009. Characterization of grape seed oil from different grape varieties (*Vitis vinifera*). Eur. J. Lipid Sci. Technol. 111 (2), 188–193.

Pazos, M., Gallardo, J.M., Torres, J.L., Medina, I., 2005. Activity of grape polyphenols as inhibitors of the oxidation of fish lipids and frozen fish muscle. Food Chem. 92, 547–557.

Ping, L., Pizzi, A., Guo, Z.D., Brosse, N., 2011. Condensed tannins extraction from grape pomace: characterization and utilization as wood adhesives for wood particle board. Ind. Crop. Prod. 34, 907–914.

Prodanov, M.P., Domınguez, J.A., Blazquez, I., Salinas, M.R., Alonso, G.L., 2005. Some aspects of the quantitative/qualitative assessment of commercial anthocyanin-rich extracts. Food Chem. 90, 585–596.

Rombaut, N., Savoire, R., Thomasset, B., Belliard, T., Castello, J., Van Hecke, E., Lanoiselle, J.L., 2014. Grape seed oil extraction: interest of supercritical fluid extraction and gas-assisted mechanical extraction for enhancing polyphenol coextraction in oil. C. R. Chim. 17, 284–292.

Rombaut, N., Savoire, R., Thomasset, B., Castello, J., Van Hecke, E., Lanoisellé, J.L., 2015. Optimization of oil yield and oil total phenolic content during grape seed cold screw pressing. Ind. Crop. Prod. 63, 26–33.

Rózek, A., Achaerandio, I., Güell, C., López, F., Ferrando, M., 2010. Use of commercial grape phenolic extracts to supplement solid foodstuff. LWT Food Sci. Technol. 43, 623–631.

Sánchez-Alonso, I., Borderías, J., Larsson, K., Undeland, I., 2007a. Inhibition of hemoglobin-mediated oxidation of regular and lipid-fortified washed cod mince by a white grape dietary fibre. J. Agric. Food Chem. 55, 5299–5305.

Sánchez-Alonso, I., Jiménez-Escrig, A., Saura-Calixto, F., Borderías, A.J., 2007b. Effect of grape antioxidant dietary fibre on the prevention of lipid oxidation in miced fish: evaluation by different methodologies. Food Chem. 101, 372–378.

Sánchez-Alonso, I., Jiménez-Escrig, A., Saura-Calixto, F., Borderías, A.J., 2008. Antioxidant protection of white grape pomace on restructured fish products during frozen storage. LWT Food Sci. Technol. 41, 42–50.

Santos, F.T., Goufo, P., Santos, C., Botelho, D., Fonseca, J., Queirós, A., Costa, M.S.S.M., Trindade, H., 2016. Comparison of five agro-industrial waste-based composts as growing media lettuce: effect on yield, phenolic compounds and vitamin C. Food Chem. 209, 293–301.

Saura-Calixto, F., 2012. Concept and health-related properties of nonextractable polyphenols: the missing dietary polyphenols. J. Agric. Food Chem. 60, 11195–11200.

Saura-Calixto, F., Gofii, I., Mafias, E., Abia, R., 1991. Klason lignin, condensed ˜tannins and resistant protein as dietary fibre constituents: determination in grape pomaces. Food Chem. 39, 299–309.

Sáyago-Ayerdi, S.G., Brenes, A., Goñi, I., 2009. Effect of grape antioxidant dietary fiber on the lipid oxidation of raw and cooked Chicken hamburgers. LWT Food Sci. Technol. 42, 971–976.

Sharma, A.K., 2015. Postharvest Management in Grapes. Research Gate. https://doi.org/10.13140/RG.2.1.5025.9925.

Shi, J., Yu, J., Pohorly, J., Young, J.C., Bryan, M., Wu, Y., 2003. Optimization of the extraction of polyphenols from grape seed meal by aqueous ethanol solution. J. Food Agric. Environ. 1 (2), 42–47.

Shinagawa, F.B., Santana, F.C., Torres, L.R.O., Mancini-Filho, J., 2015. Grape seed oil: a potential functional food? Food Sci. Technol. 35, 399–406.

Silva, M.L., Macedo, A.C., Malcata, F.X., 2000. Review: steam distilled spirits from fermented grape pomace. Food Sci. Technol. Int. 6, 285–300.

Teh, H.E., 2016. Extraction and Characterization of Functional Components From Fruit and Vegetable Processing Waste. PhD Thesis Submitted to University of California, Davis. 234 p.

Tseng, A., Zhao, Y., 2013. Wine grape pomace as antioxidant dietary fibre for enhancing nutritional value and improving storability of yogurt and salad dressing. Food Chem. 138, 356–365.

Tungland, B.C., Meyer, D., 2002. Non-digestible oligo- and polysaccharides (dietary fiber): their physiology and role inhuman health and food. Compr. Rev. Food Sci. Food Saf. 3, 90–109.

Venkitasamy, C., Teh, H.E., Atungulu, G.G., McHugh, T.H., Pan, Z., 2014. Optimization of mechanical extraction conditions for producing grape seed oil. Trans. ASABE 57, 1699–1705.

Vermaak, I., Kamatou, G.P.P., Komane-Mofokeng, B., Viljoen, A.M., Beckett, K., 2011. African seed oils of commercial importance—cosmetic applications. S. Afr. J. Bot. 16, 41.

Wittenauer, J., Mäckle, S., Sußmann, D., Schweiggert-Weisz, U., Carle, R., 2015. Inhibitory effects of polyphenols from grape pomace extract on collagenase and elastase activity. Fitoterapia 101, 179–187.

Zhu, F., Cai, Y.Z., Ke, J., Corke, H., 2009. Evaluation of the effect of plant extracts and phenolic compounds on reduction of acrylamide in an asparagine/glucose model system by RP-HPLC-DAD. J. Sci. Food Agric. 89, 1674–1681.

Zhu, F., Du, B., Zheng, L., Li, J., 2015. Advance on the bioactivity and potential applications of dietary fibre from grape pomace. Food Chem. 186, 207–212.

CHAPTER 7
Blueberry and Cranberry

Giovana Bonat Celli*, Alisson Pacheco Kovalesk[†]
*Department of Food Science, Cornell University, Ithaca, NY, United States,
[†]School of Integrative Plant Science, Section of Horticulture, Cornell University, Geneva, NY, United States

Chapter Outline

1 INTRODUCTION

Blueberries and cranberries (*Vaccinium* spp.) are some of the main types of berries cultivated and consumed in North America. These fruits have a long history of use by Native Americans for the treatment of blood poisoning, urinary disorders, diarrhea, and diabetes (Howell, 2009). Recent studies have confirmed some of these folklore benefits, such as the prevention of urinary infections and maintenance of the genitourinary tract health by cranberry proanthocyanidins (Howell et al., 2005) and hypoglycemic properties of blueberry anthocyanins (Grace et al., 2009).

Cultivated blueberries and cranberries are shrubs native to the United States and Canada although their cultivation has expanded considerably in the last 20 years to other regions, such as Europe (Pliszka, 1997), Japan (Tamada, 2009), Chile (Caligari et al., 2009), and New Zealand (Miller et al., 2009). Worldwide, the United States is the main producer of blueberries, with almost 270,000 tons of fruit supplied in 2016 (USDA, 2017a). It is also responsible for >90% of the world production of cranberries (Skrovankova et al., 2015), with a forecasted production of >400,000 tons in 2017 (USDA, 2017b).

Integrated Processing Technologies for Food and Agricultural By-Products
https://doi.org/10.1016/B978-0-12-814138-0.00007-1

The majority of cranberries and a considerable amount of blueberries are processed into some form of industrialized product, thus resulting in considerable amounts of waste and by-products generated by the berry industry. In general, it is estimated that the conversion of fruits and vegetables into processed products generates up to 50% of by-product waste (e.g., peels, unripe fruits, etc.) (Virk and Sogi, 2004). In the case of berries, approximately 20% of the fruits remain as pomace or press cake—consisting of skins, seeds, and stems—during juice processing (Rohm et al., 2015). Although often destined for animal feed or used as a substrate for biogas or compost production, these by-products have a high nutritional and bioactive content. Opportunities still exist for the incorporation or extraction of valuable compounds from these by-products for the development of value-added products.

This chapter provides an overview of methods and strategies to convert blueberry and cranberry press cakes into value-added products. First, the production and processing of these berries are presented. Then, the characterization of the primary crops and their by-products is discussed, followed by relevant technologies used for value-added processing. Finally, a perspective of current opportunities and areas of future research is provided.

2 CROP PRODUCTION AND PROCESSING

2.1 Blueberry

Three blueberry species of economic importance are commonly grown in North America: *V. angustifolium* Aiton, or lowbush blueberry; *V. corymbosum* L., or northern highbush blueberry; and *V. ashei* Reade, or rabbiteye blueberry. The fourth type of blueberry, known as southern highbush, is a complex hybrid composed primarily of *V. corymbosum*. A summary is provided in Table 1. Lowbush blueberries are mainly produced in eastern Canada and the northeastern United States. They are machine harvested from managed or uncontrolled wild stands, and 99% of the production is sold for processing (Strik and Yarborough, 2005). Highbush and rabbiteye types are grown in subtropical and temperate regions worldwide (Boches et al., 2006), and are generally machine harvested for processed and hand harvested for fresh markets (Strik and Yarborough, 2005).

The total area harvested in the world in 2014 with blueberries was 95,000 ha, with a total yield of 525,000 tons. The United States and Canada are the largest producers, with 262,000 and 182,000 tons in 2014, respectively. The area planted with blueberries in the United States has been steadily increasing, with approximately 37,500 yielding hectares in 2016. The wild blueberry area has remained constant in the past 3 years, at around 9300 ha.

The common nomenclature refers to the relative stature of the plants in the case of lowbush and highbush blueberries; lowbush blueberries form colonies and are 15–60 cm tall, whereas the highbush plants can grow up to 4 m and are cane-forming species. Rabbiteye blueberries have a similar habit to highbush,

Table 1 Summary of Main Botanical and Growing Characteristics of Blueberries and Cranberries

Fruit	Characteristic
Blueberry	Main species: *V. angustifolium* Aiton (lowbush blueberry), *V. corymbosum* L. (northern highbush blueberry), and *V. ashei* Reade (rabbiteye blueberry) Locations: Canada and northeastern United States (lowbush) or subtropical and temperate regions worldwide (highbush and rabbiteye) Plant size: 15–60 cm (lowbush) to up to 4 m (highbush) or 6 m (rabbiteye) in height Fruit size: ~0.5 g (lowbush) to 1–3 g per fruit (highbush) Chilling requirement: 200–600 h (southern highbush) to 800–1500 chilling hours (northern highbush). Rabbiteye blueberries require 300–600 h Soil: acidic (pH 4.2–5.5), 250–300 ppm maximum salt content, low nutrient availability Recommended agricultural practice: nitrogen application, irrigation, annual pruning, pest and disease management, manual or mechanical harvesting
Cranberry	Main species and locations: *V. oxycoccos* L. (Europe, Asia, and some parts of North America) and *V. macrocarpon* Aiton (United States, Canada, and Chile) Plant size: up to 30 cm in height Fruit size: 8–10 mm (*V. oxycoccos*) to 10–20 mm long (*V. macrocarpon*) Chilling requirement: not well defined Soil: acidic (pH ~5.0), wetlands, nutrient-deficient Recommended agricultural practice: nitrogen supplementation, mechanical pruning, wet or dry harvest

with plants reaching 6 m in height. Southern and northern highbush blueberries differ primarily in their chilling requirement, where southern highbush was bred to have a lower chilling requirement and therefore adapt to regions with warm winters. Northern highbush requires 800–1500 chilling hours (number of hours below 7°C); while southern highbush requires 200–600 h. Rabbiteye blueberries are closer to southern highbush, requiring 300–600 h (Darnell, 2006).

In general, these fruits typically require acidic soils (between 4.2 and 5.5) (Hall et al., 1964) with maximum salt content in the range of 250–300 ppm (Holzapfel et al., 2004), and low nutrient availability. Although regular application of nitrogen by fertigation can improve production yield (Bryla and Strik, 2015; Ehret et al., 2014), soil pH still has a greater impact on plant growth than nutrient supplementation (Rosen et al., 1990).

Irrigation—either by drip or microjet system—also has a positive effect on fruit yield (Holzapfel et al., 2004), with up to 43% increase (Yarborough, 2004). This is due to the shallow, confined root system that can restrict water and nutrient uptake to superficial soil layers, which reduces yield and vegetative growth when the amount of precipitation is not enough to match the plant's requirement (Nunez et al., 2016). This water requirement varies considerably depending on several factors, including planting age (Bryla et al., 2011) and development phase (Mingeau et al., 2001), cultivar, and agricultural practice (e.g., plant spacing) (Bryla and Strik, 2007). In certain cases, overhead irrigation systems can be used to prevent spring frost (Strik, 2007). It is also recommended that shrubs are pruned annually (Kovaleski et al., 2015) to balance vegetative growth and fruit yield (Strik, 2007), which improves long-term sustainability (Strik et al., 2003). Moderate pruning may also result in increased yields for highbush and rabbiteye blueberries (Davies, 1983; Kovaleski et al., 2015). Another factor that can affect fruit yield and weight is the combined effect of pollination and pest and disease management, as demonstrated by Melathopoulos et al. (2014). Taking all these variables together, fruit yield can vary considerably, with the highest recorded yield of 20 MT/ha in northwest North America (Strik, 2007).

The different species of blueberries can also differ substantially in terms of fruit size and content of bioactive components. For instance, Kalt et al. (2001) showed that lowbush blueberries were significantly smaller (~0.5 g) and had higher concentrations of anthocyanins, total phenolic compounds, and antioxidant activity than highbush berries, whose size varied between 1 and 3 g per fruit. Kovaleski et al. (2015) reported anthocyanin content ranging from 70 to 100 mg L^{-1} in methanol extraction of southern highbush blueberries, which was affected by pruning intensity.

Blueberries can be harvested manually or mechanically. The majority of hand harvested and part of the mechanically harvested fruit is for fresh market, while processed fruit is mainly mechanically harvested. Historically, mechanized harvest of blueberries has been associated with a reduction of fruit quality, and research has been done to develop new equipment and cultivars that can withstand mechanical damage (Williamson and Cline, 2013). Casamali et al. (2016) reported a reduction in marketable yield of 40% in mechanically harvested southern highbush blueberries when compared to the hand harvested. Mechanical harvest also results in greater loss of quality during storage (Casamali et al., 2016).

2.2 Cranberry

Unlike the deciduous blueberry plants, cranberries are perennial evergreen dwarf shrubs (up to 30 cm in height) (Rodriguez-Saona et al., 2009). Two species are commonly cultivated: *V. oxycoccos* L. in Europe, Asia, and some parts of North America, and *V. macrocarpon* Aiton in the United States, Canada, and Chile. The total area planted with cranberries in the world was 56,000 ha, yielding 652,000 tons in 2014 (FAOSTAT, 2017). The United States is responsible for over half of the world production (380,000 tons), followed by Canada (176,000 tons), and Chile (82,000 tons). In the United States, the producer-owned company Ocean Spray largely controls (~70%) the fresh and processed cranberry market.

Cranberry shrubs grow in wetlands where they produce horizontal runners that cover the soil (DeMoranville, 2015). From the runners, vertical shoots called uprights grow and produce flowers in terminal mixed buds (Dale et al., 1994). Similar to blueberries, cranberries require acidic soil (pH ~5.0)

(Medappa and Dana, 1970) and are well adapted to nutrient-deficient soils; however, nitrogen supplementation has a significant impact on plant growth and fruit yield (Davenport, 1996), more than pH alone (Rosen et al., 1990). Native cultivars require approximately 25–70 kg/ha of nitrogen, <25 kg/ha of phosphorus, and 45–135 kg/ha of potassium (DeMoranville, 2015). Chilling requirements for cranberries are not well defined, although some information is available demonstrating that it needs >700 h for flowering to occur (Eady and Eaton, 1972). Pruning is mechanical and made using rotary rakes with blades in place of tines that cut the horizontal runners and promote regrowth (Roper and Vorsa, 1997).

Cranberries are cultivated in bogs on small farms (~17 ha) (Blake et al., 2007). The bogs are constructed using water-confining layers of soil and dikes that perch the water table (DeMoranville, 2006). This is important for the harvest when bogs are flooded with 30–60 cm of water to collect fruits that were previously knocked off the uprights. The cranberries float to the water surface due to a small internal air pocket that reduces fruit density (Miller, 2000). This process is known as *wet harvest* and is shown in Fig. 1. Flooding can also be used to prevent plant desiccation in winter in which case it can last for months (Kennedy, 2015). Wet harvesting can result in excessive mechanical injury, with a negative impact on postharvest handling (Forney, 2003). Alternatively, fruits can be *dry harvested*, either manually or mechanically, and these fruits are primarily destined to the fresh market (Stiles and Oudemans, 1999). After harvesting, fruits are processed (mostly concentrated) near growing regions to reduce shipment costs and avoid postharvest losses (Dale et al., 1994), generating a considerable amount of press cake.

FIG. 1

Cranberries harvested by the wet harvest method.

As with blueberries, the size of the fruits and the content of bioactive compounds vary between the two species of cranberries (Brown et al., 2012). *V. oxycoccos* yields small fruits (8–10 mm long), whereas *V. macrocarpon* fruit is relatively large (10–20 mm across) (Jacquemart, 1997). Jensen et al. (2002) showed that the European and American species have different concentrations of organic acids and iridoid glucosides that results in a significant impact on their taste profile.

3 PHYSICOCHEMICAL CHARACTERIZATION OF PRIMARY CROP AND BY-PRODUCT BIOMASS

Blueberries and cranberries can differ greatly in relation to their physicochemical attributes (e.g., pH, °Brix, shape, color, and firmness). Significant differences are observed throughout the fruit development, between cultivars (Wang et al., 2017), within each commercial harvest (Moggia et al., 2017), and during storage (Matiacevich et al., 2013). Forney et al. (2012) evaluated the differences between blueberries and cranberries in three degrees of maturity (unripe, intermediate, and fully ripe). Overall, an increase in primary sugars (i.e., glucose and fructose) was observed during fruit ripening, whereas total acid concentration (such as citric acid) decreased. Casamali et al. (2016) also observed an increase in total soluble solids (TSS) in blueberries during the progression of the harvest season, as well as a decrease in total titratable acidity (TTA). This resulted in a significant increase of the TSS:TTA ratio. Fruit size also decreases with the progression of the harvest season (Kovaleski et al., 2015). Forney et al. (2012) also noted that firmness decreased about 80% during blueberry ripening, whereas fully red cranberries were firmer than those at earlier development stages. Firmness is one of the most important attributes of the fruits for the fresh market, as it is a measure of freshness (Chiabrando et al., 2009). Blueberries are very susceptible to mechanical damage, which can reduce firmness during postharvest handling and result in the fruit of lower quality and reduced shelf life (Casamali et al., 2016; Xu et al., 2015).

Both fruits have been valued for their bioactive content and an overwhelming body of evidence has established a positive association between berry consumption and promotion of health (Blumberg et al., 2016; Whyte et al., 2016; Yang and Kortesniemi, 2015). These berries are rich sources of anthocyanins, a natural colorant with antioxidant health-promoting properties (Pojer et al., 2013). Kalt et al. (1999) showed that the anthocyanin content can range between 95 and 255 mg/100 g and 80–250 mg/100 g of fresh fruits with delphinidin as the main anthocyanidin, accounting for approximately 38% and 41% of the total anthocyanidins in lowbush and highbush blueberries, respectively. In the case of cranberries, the concentration of anthocyanins can increase by almost 26-fold during ripening, and the content almost doubled in overripe berries (Viskelis et al., 2009).

Unlike blueberries, which contain B-type procyanidins as (+)-catechin and (−)-epicatechin oligomers, cranberries have A-type procyanidins consisting mainly of (−)-epicatechin monomers and trace levels of the B-type (Prior et al., 2001). The main structural difference is that A-type procyanidins contain at least one double linkage, whereas in the B-type molecules are linked by a single bond. These compounds can interact with salivary proteins, contributing to the astringency sensation (Hofmann et al., 2006), and the A-type can inhibit the adherence of pathogenic microorganisms involved in urinary infections (Foo et al., 2000).

Blueberry and cranberry press cakes are also good sources of bioactive compounds. Skrede et al. (2000) found that ~1.5% of the total anthocyanin is lost during juice concentration due to the action of polyphenol oxidase, whereas 18% still remains in the press cake residue after pressing. Similarly, ~7% of flavonol glycosides and 3% of procyanidins were recovered in the press cake (Skrede et al., 2000). Viskelis et al. (2009) evaluated different cranberry cultivars and their press cake and concluded that the cakes had a higher concentration of anthocyanins than the whole berries. This is because the pigments are mostly concentrated on the berry skins, which are disposed in the press cake. Similarly, the concentration of phenolic compounds was higher in the press cakes (Viskelis et al., 2009). Blueberry and cranberry pomace are also sources of insoluble and soluble fibers (Gouw et al., 2017b), as shown by White et al. (2010a).

4 FROM BY-PRODUCTS TO VALUE-ADDED PRODUCTS

A considerable amount of research has investigated the conversion of blueberry and cranberry pomace and press cakes into value-added products. Different strategies have been used to extract bioactive compounds from these by-products, such as pulsed electric field (PEF) (Bobinaitė et al., 2015), ultrasound- and microwave-assisted techniques (Klavins et al., 2017; Raghavan and Richards, 2007), using supercritical CO_2 (Paes et al., 2014), among others summarized in Table 2. In certain cases, the bioactive compounds may be bound to the berry matrix, and some of the methods used thus far can disrupt the cell integrity, promoting the leakage of the compounds to the extraction medium. Procyanidins from cranberries are a good example. White et al. (2010b) optimized an extraction method based on alkaline hydrolysis to increase the yield up to 15 times in comparison to untreated samples. In a pilot-scale study, Harrison et al. (2013) investigated the maceration of cranberry pomace in ethanol followed by decantation with or without an additional press step. The combination of maceration, decantation, and pressing resulted in higher recovery of polyphenols and an extract with higher antioxidant and vasodilation properties in vivo (Harrison et al., 2013).

Preferably, these extracts need to be converted into a powder form to facilitate their storage, handling, and incorporation into various food products. However, a challenge arises when using organic solvents as extraction systems, as they require specific drying conditions to make the process safe. Another issue that can occur during the spray-drying process is the disruption of ternary phase

Table 2 Extraction Methods Used to Obtain Bioactive Compounds From Blueberry and Cranberry Pomace

Crop	Processing method	Results	Reference
Blueberry	Pulsed electric field (PEF) as a pretreatment to juice extraction	▪ Higher PEF intensity improved extractability of bioactive compounds from the press cake compared to untreated samples ▪ Resulting antioxidant activity of treated samples was higher	Bobinaitė et al. (2015)
	Supercritical fluid (SFE) and pressured liquid extractions (PLE)	▪ PLE extracts using 100% ethanol or ethanol + water had the highest antioxidant activities and phenolic contents, whereas acidified water yielded the highest anthocyanin content ▪ SFE with 5% water and 5% ethanol yield the highest concentrations of phenolic compounds, anthocyanins, and antioxidant activity	Paes et al. (2014)
	Thermal extraction	▪ Highest concentration of anthocyanins from the berry skins were obtained in samples treated at 50°C with 100 ppm SO_2 and 1% citric acid, whereas highest total phenolic content was obtained at 80°C	Lee and Wrolstad (2004)
	Thermal extraction	▪ Highest anthocyanin concentration was achieved when pomace was boiled at pH 1 for 15 min	Bener et al. (2013)
	Extrusion	▪ Mixtures containing 30% blueberry pomace and 70% sorghum flour were extruded at 160°C or 180°C using a screw speed of 150 or 200 rpm ▪ All conditions evaluated increased the concentrations of monomers, dimer, and trimers of procyanidins ▪ However, extrusion reduced the total anthocyanin content by 33%–42%	Khanal et al. (2009)
Cranberry	Ultrasound-(UAE) and microwave-assisted extraction (MAE)	▪ UAE resulted in higher yields of anthocyanins and polyphenols than MAE and Soxhlet extraction using acidified ethanol as the solvent system	Klavins et al. (2017)
	Solvent extraction (SE) and MAE	▪ Extracts were tested for their ability to inhibit lipid oxidation of mechanically separated turkey ▪ SE with 100% acetone or MAE with 100% ethanol were the most effective extracts	Raghavan and Richards, 2007

Table 2 Extraction Methods Used to Obtain Bioactive Compounds From Blueberry and Cranberry Pomace—cont'd

Crop	Processing method	Results	Reference
	Extrusion	▪ Pomace was extruded with corn starch at different ratios at 150°C, 170°C, and 190°C and with screw speeds of 150 or 200 rpm ▪ Percentage of pomace and temperature influenced anthocyanin retention, where the highest retention was observed at 150°C and 30% pomace ▪ Extrusion increase flavonol concentration by 30%–34% compared to control	White et al. (2010c)
	Extraction followed by film formation	▪ Aqueous extracts prepared from pomace were combined with (low or high methoxyl) pectin and sorbitol or glycerol ▪ Edible films formed from this mixture were bright red and presented cranberry flavor ▪ Different properties were obtained depending on the plasticizing agent used	Park and Zhao (2006)

equilibrium, when water, a solvent, and a wall material are used, such as addressed by Flores et al. (2014) using whey protein as the polymeric matrix.

In certain cases, berry pomace was incorporated directly into food products, without an extraction step. For instance, Mildner-Szkudlarz et al. (2016) incorporated cranberry pomace into muffins and investigated the effects of baking conditions on the recovery of bioactive compounds. They noted that baking at 180°C for 20 min resulted in muffins with better structure and texture and recovery of almost 20% of the anthocyanins from the pomace (Mildner-Szkudlarz et al., 2016). As another example, Šarić et al. (2016) used blueberry pomace powder as an ingredient in gluten-free cookies. Approximately 28.2% of blueberry pomace was used in combination with 1.8% raspberry pomace as substitutes of the flour mixture, resulting in cookies with similar protein and carbohydrate content but lower fat content than those containing gluten (Šarić et al., 2016).

The high sugar content of the pomaces makes them interesting materials for microbial conversion into value-added products. Zheng and Shetty (1998) showed the feasibility of using solid-state fermentation of cranberry pomace by economically relevant fungi, producing fungal inoculants such as those involved in pesticide or polymeric dye degradation. Fermentation of cranberry pomace can also increase the concentration of ellagic acid (Vattem and Shetty, 2002), a compound with anticarcinogenic properties (Ceci et al., 2016). Owing to their fiber content, the use of blueberry and cranberry pomace was also investigated in areas other than food technology. Gouw et al. (2017a) studied the use of these by-products with cellulose nanofiber to partially replace recycled newspaper on the production of molded pulp packaging boards, resulting in products with better or similar properties.

5 CONCLUSION AND FUTURE RESEARCH

Blueberry and cranberry pomace is in great availability from the processed fruit market, especially in the United States and Canada, where most of the production occurs. The pomace of these berries is a great source of fibers and bioactive components—such as anthocyanins and proanthocyanidins, which remain in part in the berry skins and seeds after juice pressing. Overall, the profile of these compounds differs greatly between species; blueberries contain mostly B-type procyanidins, while cranberries contain the A-type, which is associated with improvement of the genitourinary tract health.

In general, three main strategies are frequently described in the literature in terms of utilization of blueberry and cranberry pomace: (1) direct use in food products (e.g., baked goods, as natural colorants, etc.) with minimal processing, such as drying; (2) extraction to obtain high added value bioactive components; and (3) as substrate in solid-state fermentation to produce fungal inoculants or increase the concentration of certain components. Some examples of these strategies were covered in this chapter, although it was not the objective of the authors to provide an exhaustive review of the topic.

As most of the studies described in the literature are based on laboratory- and pilot-scale techniques for processing of pomace into value-added products, future research should focus on scaling up current protocols to meet market demand, produce high-quality value-added products, and create substantial economic impact while adhering to environmental and food regulations. An online search for suppliers or products that list blueberry and cranberry pomace as an ingredient or source of a bioactive compound provides little insight into the current use of these by-products by the food and pharmaceutical industries. More notably is their use as a component of pet formulas, which could indicate a potential issue in terms of consumer understanding and acceptance for human consumption. Overall, processors of these pomaces should focus on product quality and strategies to develop safe and affordable techniques.

References

Bener, M., Shen, Y., Apak, R., Finley, J.W., Xu, Z., 2013. Release and degradation of anthocyanins and phenolics from blueberry pomace during thermal acid hydrolysis and dry heating. J. Agric. Food Chem. 61, 6643–6649.

Blake, G., Sandler, H.A., Coli, W., Pober, D.M., Coggins, C., 2007. An assessment of grower perceptions and factors influencing adoption of IPM in commercial cranberry production. Renew. Agric. Food Syst. 22, 134–144.

Blumberg, J.B., Basu, A., Krueger, C.G., Lila, M.A., Neto, C.C., Novotny, J.A., Reed, J.D., Rodriguez-Mateos, A., Toner, C.D., 2016. Impact of cranberries on gut microbiota and cardiometabolic health: proceedings of the Cranberry health research conference 2015. Adv. Nutr. 7, 759s–770s.

Bobinaitė, R., Pataro, G., Lamanauskas, N., Šatkauskas, S., ViŠkelis, P., Ferrari, G., 2015. Application of pulsed electric field in the production of juice and extraction of bioactive compounds from blueberry fruits and their by-products. J. Food Sci. Technol. 52, 5898–5905.

Boches, P., Bassil, N.V., Rowland, L., 2006. Genetic diversity in the highbush blueberry evaluated with microsatellite markers. J. Am. Soc. Hortic. Sci. 131, 674–686.

Brown, P.N., Turi, C.E., Shipley, P.R., Murch, S.J., 2012. Comparisons of large (*Vaccinium macrocarpon* Ait.) and small (*Vaccinium oxycoccos* L., *Vaccinium vitis-idaea* L.) cranberry in British Columbia by phytochemical determination, antioxidant potential, and metabolomic profiling with chemometric analysis. Planta Med. 78, 630–640.

Bryla, D.R., Strik, B.C., 2007. Effects of cultivar and plant spacing on the seasonal water requirements of highbush blueberry. J. Am. Soc. Hortic. Sci. 132, 270–277.

Bryla, D.R., Strik, B.C., 2015. Nutrient requirements, leaf tissue standards, and new options for fertigation of northern highbush blueberry. HortTechnology 25, 464–470.

Bryla, D.R., Gartung, J.L., Strik, B.C., 2011. Evaluation of irrigation methods for highbush blueberry—I. Growth and water requirements of young plants. HortSci. 46, 95–101.

Caligari, P.D.S., Retamales, J.B., Lobos, G.A., 2009. Blueberry breeding for chilean conditions. Acta Hortic. 810, 125–128.

Casamali, B., Williamson, J.G., Kovaleski, A.P., Sargent, S.A., Darnell, R.L., 2016. Mechanical harvesting and postharvest storage of two southern highbush blueberry cultivars grafted onto *Vaccinium arboreum* rootstocks. HortSci. 51, 1503–1510.

Ceci, C., Tentori, L., Atzori, M.G., Lacal, P.M., Bonanno, E., Scimeca, M., Cicconi, R., Mattei, M., de Martino, M.G., Vespasiani, G., Miano, R., Graziani, G., 2016. Ellagic acid inhibits bladder cancer invasiveness and in vivo tumor growth. Nutrients 8.

Chiabrando, V., Giacalone, G., Rolle, L., 2009. Mechanical behaviour and quality traits of highbush blueberry during postharvest storage. J. Sci. Food Agric. 89, 989–992.

Dale, A., Hanson, E.J., Yarborough, D.E., McNicol, R.J., Stang, E.J., Brennan, R., Morris, J.R., Hergert, G.B., 1994. Mechanical harvesting of berry crops. In: Janick, J. (Ed.), Horticultural Reviews. In: vol. 16. John Wiley & Sons, Oxford, pp. 255–382.

Darnell, R.L., 2006. Blueberry botany/environmental physiology. In: Childers, N.F., Lyrene, P.M. (Eds.), Blueberries for Growers, Gardeners and Promoters. Dr Normal F. Childers Publications, Gainesville, pp. 5–13.

Davenport, J.R., 1996. The effect of nitrogen fertilizer rates and timing on cranberry yield and fruit quality. J. Am. Soc. Hortic. Sci. 121, 1089–1094.

Davies, F.S., 1983. Pruning, yield and morphology of 3 rabbiteye blueberry cultivars in Florida. Proc. Fla. State Hort. Soc. , vol. 96, 192–195.

DeMoranville, C.J., 2006. Cranberry best management practice adoption and conservation farm planning in Massachusetts. HortTechnology 16, 393–397.

DeMoranville, C.J., 2015. Cranberry nutrient management in southeastern Massachusetts: balancing crop production needs and water quality. HortTechnology 25, 471–476.

Eady, F.C., Eaton, G.W., 1972. Effects of chilling during dormancy on development of the terminal bud of the cranberry. Can. J. Plant Sci. 52, 273–279.

Ehret, D.L., Frey, B., Forge, T., Helmer, T., Bryla, D.R., Zebarth, B.J., 2014. Effects of nitrogen rate and application method on early production and fruit quality in highbush blueberry. Can. J. Plant Sci. 94, 1165–1179.

Flores, F.P., Singh, R.K., Kong, F., 2014. Physical and storage properties of spray-dried blueberry pomace extract with whey protein isolate as wall material. J. Food Eng. 137, 1–6.

Foo, L.Y., Lu, Y.R., Howell, A.B., Vorsa, N., 2000. The structure of cranberry proanthocyanidins which inhibit adherence of uropathogenic P-fimbriated *Escherichia coli* in vitro. Phytochemistry 54, 173–181.

Forney, C.F., 2003. Postharvest handling and storage of fresh cranberries. HortTechnology 13, 267–272.

Forney, C.F., Kalt, W., Jordan, M.A., Vinqvist-Tymchuk, M.R., Fillmore, S.A.E., 2012. Blueberry and cranberry fruit composition during development. J. Berry Res. 2, 169–177.

Gouw, V.P., Jung, J., Simonsen, J., Zhao, Y., 2017a. Fruit pomace as a source of alternative fibers and cellulose nanofiber as reinforcement agent to create molded pulp packaging boards. Compos. A: Appl. Sci. Manuf. 99, 48–57.

Gouw, V.P., Jung, J., Zhao, Y., 2017b. Functional properties, bioactive compounds, and in vitro gastrointestinal digestion study of dried fruit pomace powders as functional food ingredients. LWT Food Sci. Technol. 80, 136–144.

Grace, M.H., Ribnicky, D.M., Kuhn, P., Poulev, A., Logendra, S., Yousef, G.G., Raskin, I., Lila, M.A., 2009. Hypoglycemic activity of a novel anthocyanin-rich formulation from lowbush blueberry, *Vaccinium angustifolium* Aiton. Phytomedicine 16, 406–415.

Hall, I.V., Aalders, L.E., Townsend, L.R., 1964. The effects of soil pH on the mineral composition and growth of the lowbush blueberry. Can. J. Plant Sci. 44, 433–438.

Harrison, J.E., Oomah, B.D., Diarra, M.S., Ibarra-Alvarado, C., 2013. Bioactivities of pilot-scale extracted cranberry juice and pomace. J. Food Process. Preserv. 37, 356–365.

Hofmann, T., Glabasnia, A., Schwarz, B., Wisman, K.N., Gangwer, K.A., Hagerman, A.E., 2006. 'Protein binding and astringent taste of a polymeric procyanidin, 1,2,3,4,6-penta-O-galloyl-beta-D-glucopyranose, castalagin, and grandinin. J. Agric. Food Chem. 54, 9503–9509.

Holzapfel, E.A., Hepp, R.F., Marino, M.A., 2004. Effect of irrigation on fruit production in blueberry. Agric. Water Manag. 67, 173–184.

Howell, A.B., 2009. Update on health benefits of cranberry and blueberry. Acta Hortic. 810, 777–782.

Howell, A.B., Reed, J.D., Krueger, C.G., Winterbottom, R., Cunningham, D.G., Leahy, M., 2005. A-type cranberry proanthocyanidins and uropathogenic bacterial anti-adhesion activity. Phytochemistry 66, 2281–2291.

Jacquemart, A.L., 1997. *Vaccinium oxycoccos* L (*Oxycoccus palustris* Pers) and *Vaccinium microcarpum* (Turcz ex Rupr) Schmalh (*Oxycoccus microcarpus* Turcz ex Rupr). J. Ecol. 85, 381–396.

Jensen, H.D., Krogfelt, K.A., Cornett, C., Hansen, S.H., Christensen, S.B., 2002. Hydrophilic carboxylic acids and iridoid glycosides in the juice of American and European cranberries (*Vaccinium macrocarpon* and *V. oxycoccos*), lingonberries (*V. vitis-idaea*), and blueberries (*V. myrtillus*). J. Agric. Food Chem. 50, 6871–6874.

Kalt, W., McDonald, J.E., Ricker, R.D., Lu, X., 1999. Anthocyanin content and profile within and among blueberry species. Can. J. Plant Sci. 79, 617–623.

Kalt, W., Ryan, D.A.J., Duy, J.C., Prior, R.L., Ehlenfeldt, M.K., Vander Kloet, S.P., 2001. Interspecific variation in anthocyanins, phenolics, and antioxidant capacity among genotypes of highbush and lowbush blueberries (*Vaccinium* Section *cyanococcus* spp.). J. Agric. Food Chem. 49, 4761–4767.

Kennedy, C.D., 2015. Hydrologic and nutrient response of groundwater to flooding of cranberry farms in southeastern Massachusetts, USA. J. Hydrol. 525, 441–449.

Khanal, R.C., Howard, L.R., Brownmiller, C.R., Prior, R.L., 2009. Influence of extrusion processing on procyanidin composition and total anthocyanin contents of blueberry pomace. J. Food Sci. 74, H52–H58.

Klavins, L., Kviesis, J., Klavins, M., 2017. Comparison of methods of extraction of phenolic compounds from American cranberry (*Vaccinium macrocarpon* L.) press residues. Agron. Res. 15, 1316–1329.

Kovaleski, A.P., Williamson, J.G., Casamali, B., Darnell, R.L., 2015. Effects of timing and intensity of summer pruning on vegetative traits of two southern highbush blueberry cultivars. HortSci. 50, 68–73.

Lee, J., Wrolstad, R.E., 2004. Extraction of anthocyanins and polyphenolics from blueberry processing waste. J. Food Sci. 69, C564–C573.

Matiacevich, S., Cofré, D.C., Silva, P., Enrione, J., Osorio, F., 2013. Quality parameters of six cultivars of blueberry using computer vision. Int. J. Food Sci. 2013.

Medappa, K.C., Dana, M.N., 1970. The influence of pH, Ca, P and Fe on the growth and composition of the cranberry plant. Soil Sci. 109, 250–253.

Melathopoulos, A.P., Tyedmers, P., Cutler, G.C., 2014. Contextualising pollination benefits: effect of insecticide and fungicide use on fruit set and weight from bee pollination in lowbush blueberry. Ann. Appl. Biol. 165, 387–394.

Mildner-Szkudlarz, S., Bajerska, J., Górnaś, P., Segliņa, D., Pilarska, A., Jesionowski, T., 2016. Physical and bioactive properties of muffins enriched with raspberry and cranberry pomace powder: a promising application of fruit by-products rich in biocompounds. Plant Foods Hum. Nutr. 71, 165–173.

Miller, G.R., 2000. Crimson gold. Focus. Geogr. 46, 1–7.

Miller, S., Scalzo, J., Boldingh, H., 2009. Cranberry cultivars for the New Zealand industry. Acta Hortic. 810, 199–204.

Mingeau, M., Perrier, C., Ameglio, T., 2001. Evidence of drought-sensitive periods from flowering to maturity on highbush blueberry. Sci. Hortic. 89, 23–40.

Moggia, C., Graell, J., Lara, I., Gonzalez, G., Lobos, G.A., 2017. Firmness at harvest impacts postharvest fruit softening and internal browning development in mechanically damaged and non-damaged highbush blueberries (*Vaccinium corymbosum* L.). Front. Plant Sci. 8.

Nunez, G.H., Olmstead, J.W., Darnell, R.L., 2016. Towards marker assisted breeding for root architecture traits in southern highbush blueberry. J. Am. Soc. Hortic. Sci. 141, 414–424.

Paes, J., Dotta, R., Barbero, G.F., Martínez, J., 2014. Extraction of phenolic compounds and anthocyanins from blueberry (*Vaccinium myrtillus* L.) residues using supercritical CO_2 and pressurized liquids. J. Supercrit. Fluids 95, 8–16.

Park, S., Zhao, Y., 2006. Development and characterization of edible films from cranberry pomace extracts. J. Food Sci. 71, E95–E101.

Pliszka, K., 1997. Overview on *Vaccinium* production in Europe. Acta Hortic. 446, 49–52.

Pojer, E., Mattivi, F., Johnson, D., Stockley, C.S., 2013. The case for anthocyanin consumption to promote human health: a review. Compr. Rev. Food Sci. Food Saf. 12, 483–508.

Prior, R.L., Lazarus, S.A., Cao, G.H., Muccitelli, H., Hammerstone, J.F., 2001. Identification of procyanidins and anthocyanins in blueberries and cranberries (*Vaccinium* spp.) using high-performance liquid chromatography/mass spectrometry. J. Agric. Food Chem. 49, 1270–1276.

Raghavan, S., Richards, M.P., 2007. Comparison of solvent and microwave extracts of cranberry press cake on the inhibition of lipid oxidation in mechanically separated turkey. Food Chem. 102, 818–826.

Rodriguez-Saona, C.R., Rodriguez-Saona, L.E., Frost, C.J., 2009. Herbivore-induced volatiles in the perennial shrub, *Vaccinium corymbosum*, and their role in inter-branch signaling. J. Chem. Ecol. 35, 163–175.

Rohm, H., Brennan, C., Turner, C., Gunther, E., Campbell, G., Hernando, I., Struck, S., Kontogiorgos, V., 2015. Adding value to fruit processing waste: innovative ways to incorporate fibers from berry pomace in baked and extruded cereal-based foods—a SUSFOOD project. Foods 4, 690–697.

Roper, T.R., Vorsa, N., 1997. Cranberry: botany and horticulture. In: Janick, J. (Ed.), Horticultural Reviews. John Wiley & Sons, Oxford, pp. 215–249.

Rosen, C.J., Allan, D.L., Luby, J.J., 1990. Nitrogen form and solution ph influence growth and nutrition of two *Vaccinium* clones. J. Am. Soc. Hortic. Sci. 115, 83–89.

Šarić, B., Mišan, A., Mandić, A., Nedeljković, N., Pojić, M., Pestorić, M., Đilas, S., 2016. Valorisation of raspberry and blueberry pomace through the formulation of value-added gluten-free cookies. J. Food Sci. Technol. 53, 1140–1150.

Skrede, G., Wrolstad, R.E., Durst, R.W., 2000. Changes in anthocyanins and polyphenolics during juice processing of highbush blueberries (*Vaccinium corymbosum* L.). J. Food Sci. 65, 357–364.

Skrovankova, S., Sumczynski, D., Mlcek, J., Jurikova, T., Sochor, J., 2015. Bioactive compounds and antioxidant activity in different types of berries. Int. J. Mol. Sci. 16, 24673–24706.

Stiles, C.M., Oudemans, P.V., 1999. Distribution of cranberry fruit-rotting fungi in new jersey and evidence for nonspecific host resistance. Phytopathology 89, 218–225.

Strik, B.C., 2007. Horticultural practices of growing highbush blueberries in the ever-expanding U.S. and global scene. J. Am. Pomol. Soc. 61, 148–150.

Strik, B.C., Yarborough, D., 2005. Blueberry production trends in North America, 1992 to 2003, and predictions for growth'. HortTechnology 15, 391–398.

Strik, B., Buller, G., Hellman, E., 2003. Pruning severity affects yield, berry weight, and hand harvest efficiency of highbush blueberry. HortSci. 38, 196–199.

Tamada, T., 2009. Current trends of blueberry culture in Japan. Acta Hortic. 810, 109–115.

United States Department of Agriculture (USDA), 2017a. Noncitrus Fruits and Nuts—2016 Summary. National Agricultural Statistics Service, pp. 35–39. Available from: http://usda.mannlib.cornell.edu/usda/current/NoncFruiNu/NoncFruiNu-06-27-2017.pdf. (Accessed 1 October 2017).

United States Department of Agriculture (USDA), 2017b. Cranberries. National Agricultural Statistics Service (NASS), Agricultural Statistics Board. Available from, http://usda.mannlib.cornell.edu/usda/current/Cran/Cran-08-10-2017.pdf. (Accessed 1 October 2017).

Vattem, D.A., Shetty, K., 2002. Solid-state production of phenolic antioxidants from cranberry pomace by *Rhizopus oligosporus*. Food Biotechnol. 16, 189–210.

Virk, B.S., Sogi, D.S., 2004. Extraction and characterization of pectin from apple (*Malus pumila* Cv *Amri*) peel waste. Int. J. Food Prop. 7, 693–703.

Viskelis, P., Rubinskiene, M., Jasutiene, I., Sarkinas, A., Daubaras, R., Cesoniene, L., 2009. Anthocyanins, antioxidative, and antimicrobial properties of American cranberry (*Vaccinium macrocarpon* Ait.) and their press cakes. J. Food Sci. 74, C157–C161.

Wang, Y., Johnson-Cicalese, J., Singh, A.P., Vorsa, N., 2017. Characterization and quantification of flavonoids and organic acids over fruit development in American cranberry (*Vaccinium macrocarpon*) cultivars using HPLC and APCI-MS/MS. Plant Sci. 262, 91–102.

White, B.L., Howard, L.R., Prior, R.L., 2010a. Proximate and polyphenolic characterization of cranberry pomace. J. Agric. Food Chem. 58, 4030–4036.

White, B.L., Howard, L.R., Prior, R.L., 2010b. Release of bound procyanidins from cranberry pomace by alkaline hydrolysis. J. Agric. Food Chem. 58, 7572–7579.

White, B.L., Howard, L.R., Prior, R.L., 2010c. Polyphenolic composition and antioxidant capacity of extruded cranberry pomace. J. Agric. Food Chem. 58, 4037–4042.

Whyte, A.R., Schafer, G., Williams, C.M., 2016. Cognitive effects following acute wild blueberry supplementation in 7- to 10-year old children. Eur. J. Nutr. 55, 2151–2162.

Williamson, J.G., Cline, W.O., 2013. Mechanized harvest of southern highbush blueberries for the fresh market: an introduction and overview of the workshop proceedings. HortTechnology 23, 416–418.

Xu, R., Takeda, F., Krewer, G., Li, C.Y., 2015. Measure of mechanical impacts in commercial blueberry packing lines and potential damage to blueberry fruit. Postharvest Biol. Technol. 110, 103–113.

Yang, B., Kortesniemi, M., 2015. Clinical evidence on potential health benefits of berries. Curr. Opin. Food Sci. 2, 36–42.

Yarborough, D.E., 2004. Factors contributing to the increase in productivity in the wild blueberry industry. Small Fruits Rev. 3, 33–43.

Zheng, Z.X., Shetty, K., 1998. Cranberry processing waste for solid state fungal inoculant production. Process Biochem. 33, 323–329.

CHAPTER 8
Pomegranate

Chandrasekar Venkitasamy*, Liming Zhao†‡, Ruihong Zhang*, Zhongli Pan*
*Department of Biological and Agricultural Engineering, University of California - Davis, Davis, CA, United States, †Key Laboratory of Meat Processing of Sichuan, Chengdu University, Chengdu, China, ‡State Key Laboratory of Bioreactor Engineering, R&D Center of Separation and Extraction Technology in Fermentation Industry, East China University of Science and Technology, Shanghai, China

Chapter Outline

1 INTRODUCTION

Pomegranates (*Punica granatum* L.) belong to the family Punicaceae and have been grown since ancient times for their delicious fruits and as an ornamental garden plant for their red, orange, or occasionally creamy, yellow flowers.

Integrated Processing Technologies for Food and Agricultural By-Products
https://doi.org/10.1016/B978-0-12-814138-0.00008-3

The pomegranate is native to Iran in the Himalayan region and is grown in Mediterranean countries such as Iran, Spain, Egypt, India, China, and lately in the United States as well as in Near and Far East countries. It is also known as the Chinese apple, Apple of Carthage, or Apple with many seeds and is considered as the super fruit of the next generation (Dhinesh and Ramasamy, 2016). The estimated global cultivation area for pomegranates is around 300, 000 ha with a production of 3.0 million metric tons (NRCP, 2014). India ranks first in global pomegranate production with 0.81 million metric tons at a productivity of 7.40 t/ha (Dhinesh and Ramasamy, 2016). Spanish missionaries brought the pomegranate to the Americas in the 1500s (LaRue, 1977; Pareek et al., 2015). In the United States, pomegranates are grown where winters are mild, such as in California and Arizona, which enable the fruit to reach the quality necessary for successful commercial production. Recently, there has been a remarkable increase in the commercial farming of pomegranates globally, due to the potential health benefits attributed to its high antioxidant, anti-mutagenic, and antihypertension activities and the ability to reduce liver injury (Dhinesh and Ramasamy, 2016; Du et al., 1975; Tsuda et al., 1994; Lansky et al., 1998; Gil et al., 1996a).

Pomegranates are enjoyed for their sweetness, acidic juice, and chemopreventive medicinal properties. Fresh fruits are mainly consumed as dessert, while processed products such as bottled juice, syrups, and jelly are also popular. Regardless of its form, the fruit is delicious and nourishing. The pomegranate has been regarded as a medicinal food of great importance for therapeutic purposes, alleviating ailments such as colic, colitis-diarrhea, dysentery, leucorrhea, paralysis, and headache (Schubert et al., 1999; Sadeghi et al., 2009). It is also widely used in traditional Asian medicines both in Ayurvedic and Unani systems. Pomegranates are considered best for curing chronic stomach ailments and are also known for their anti-inflammatory and anti-atherosclerotic effect activity against osteoarthritis, prostate cancer, heart disease, and HIV-I (Malik et al., 2005; Sumner et al., 2005). The juice from pomegranates is one of nature's most powerful antioxidants as it has 3 times higher antioxidant activity than those of red wine and green tea (Gil et al., 2000). Pomegranate juice increases the body's resistance against infections, acts as a cooling beverage, and strengthens the functions of the kidney, liver, and heart. It is also used to cure leprosy patients. These therapeutic properties are due to the presence of betulinic and ursolic acids and different alkaloids such as pseudopelletierine, pelletierine, and other basic compounds. Pomegranate anthocyanins have scavenging activities, while its polyphenolic compounds elevate the antioxidant capacity of the human body (Singh et al., 1990; Dhinesh and Ramasamy, 2016).

All parts of the pomegranate tree including the roots, the reddish-brown bark, leaves, flowers, rinds, and seeds, have been used in medicine for thousands of years as they are rich sources of various chemical constituents (Table 1). The sweet varieties of pomegranate are considered a good laxative, while those which are between sweet and sour are a good cure for stomach inflammations and heart pain. Pomegranates have recently been found to boost activity of an enzyme which protects against cardiovascular risks. Its bark and rind are used for the

Table 1 Chemical Constituents of Different Parts of Pomegranate Plant (Dhinesh and Ramasamy, 2016)

Plant part	Constituents
Pomegranate Juice	Anthocyanins, glucose, ascorbic acid, ellagic acid, gallic acid, caffeic acid, catechin, Minerals, amino acids, quercetin, rutin
Pomegranate seed oil	95% punicic acid, ellagic acid, sterols
Pomegranate pericarp (peel, rind)	Phenolic punicalagins, gallic acid, catechin, flavones, flavonones, anthocyanidins
Pomegranate leaves	Tannins, flavone glycosides, luteolin, apigenin
Pomegranate flower	Gallic acid, ursolic acid, triterpenoids including maslinic, and asiatic acid
Pomegranate roots and bark	Ellagitannins, punicalin and punicalagin, piperidine alkaloids

control of dysentery and diarrhea. The rind is also used as dyeing material for cloth. The fruit juice can ferment easily and therefore it is commonly used in wine production. The citric acid and sodium citrate produced from the juice of wild pomegranates are used in the medical field. Furthermore, pomegranate seed (PS) oil has a potential for industrial use. Tannin that is available at all parts of the tree is successfully used alone or mixed with synthetic tannins for tanning leather (Koujalagi, 2012).

A well-ripened pomegranate fruit contains moisture (78%), protein (1.6%), carbohydrate (14.5%), fiber (5.1%), mineral matters (0.7%), and fat (0.1%). Phosphorus, oxalic acid, magnesium, calcium, and boron are present at the rates of 70, 14, 12, 10, and 0.3 mg/100 g, respectively. It also contains vitamins thiamine, riboflavin, nicotinic acid, and vitamin C at rates of 0.06, 0.1, 0.3, and 20 mg/100 g, respectively. The fruit has about 12%–16% digestible glucose and fructose. Acidity ranges between 1.5% and 3.0%. It provides energy of 65 cal/100 g (Koujalagi, 2012).

The pomegranate is a subtropical fruit. It can adapt to diverse types of soil and a wide range of climatic conditions and it can grow in altitudes of up to 1800 m above mean sea level. The fruit tree grows well in semiarid climates where cool winters and hot, dry summers prevail. However, the tree cannot produce sweet fruits unless the temperature is high for a sufficiently long period. Under tropical and subtropical climates, it behaves as an evergreen or partially deciduous tree. Under humid conditions, the sweetness of fruit is adversely affected. Therefore, pomegranate trees are hardy and can thrive well under drought conditions. The pomegranate starts fruiting in the third year after planting and continues for about 15 years. Economic yield is generally obtained after the third year of planting.

1.1 Cultivars

>500 cultivars of pomegranates have been named (IPGRI, 2001). A few cultivars of the same basic genotype are known by different names in different regions because the husk and aril color can vary markedly when the same genotype is

grown in different regions. Important traits used to identify the genotypic variations are fruit size, husk color (ranging from yellow to purple, with pink and red most common), aril color (ranging from white to red), seed hardness, maturity, juice content, acidity, sweetness, and stringency (Pareek et al., 2015). The primary cultivar of the United States, Wonderful, was discovered in Florida and brought to California in 1896 (California Rare Fruit Growers, 1997). It is also grown in Western Europe, Israel, and Chile. "Wonderful" is among the most deeply colored pomegranates in both of its husk and juice, with a rich flavor, good juice yield, sprightly acidity, and slight thirst-quenching astringency. It is considered to be among the best-tasting cultivars. Other commercial US cultivars include "Granada" (a "Wonderful" sport), "Early Wonderful" (also a "Wonderful" sport), and "Early Foothill." The Spanish cultivars "Mollar de Elche" and "Valenciana" are among the most widely marketed pomegranate cultivars in Western Europe (Costa and Melgarejo, 2000; Pareek et al., 2015). The "Valenciana" cultivar is harvested early (August) and has low risk from sun damage, pest attack, and bad weather, but it also has low yield, average to poor internal fruit quality, and small fruit size. The "Mollar" cultivar is harvested much later (end of September until mid-November) and displays more sun and split damage, but it has higher yield, excellent internal fruit quality, larger size, longer harvest period, and greater consumer acceptance. The prominent cultivars in India are "Bhagwa," "Ganesh," "Jodhpur Red," "Jalore Seedless," "Ruby," "Arkta," and "Mridula" (Pareek et al., 2015). Other major pomegranate cultivars in the main producing countries were listed by Pareek et al. (2015).

1.2 Fruit Maturity

Pomegranates are harvested when fruits reach a certain size and skin color. Pomegranates do not ripen once removed from the tree and should be picked when fully ripe to ensure the best eating quality for the consumer (Kader et al., 1984). The fruits should be harvested before they become overripe and crack open, especially under rainy conditions. The pomegranate fruit reaches full maturity within 4.5–6 months after full bloom, depending on climatic conditions (Ben-Arie et al., 1984). With experience, correct harvesting maturity of fruit can be determined by tapping the fruit and listening for a metallic sound (Mir et al., 2012). Titratable acidity (TA) and soluble solids content can be used as maturity indices to determine whether to harvest the pomegranates. Each pomegranate type requires a certain acid/soluble solids ratio at harvest. The acidity of pomegranates varies between 0.13% and 4.98% at harvest. It is $<1\%$ in sweet cultivars, 1%–2% in sweet–sour cultivars, and $>2\%$ in sour cultivars (Pareek et al., 2015). The total soluble solids (TSS) content of pomegranates varies between 8.3% and 20.5% at harvest. The maturity index (MI) using a ratio of TSS/TA is one of the most reliable indicators of pomegranate fruit maturity (Fawole and Opara, 2013a), although it depends on the cultivar and climatic conditions (Kulkarni and Aradhya, 2005; Schwartz et al., 2009). A science-based MI for pomegranate cultivars other than "Wonderful" is not currently available or well established. For "Wonderful," acidity should be lower than 1.85%, soluble sugar content >15%–17%, and sugar/acid ratio >18.5 (Sherafatian, 1994).

1.3 Pomegranate Fruit Quality

Pomegranate fruits have an irregular round shape with coriaceous rinds that vary in color from yellow, green, or pink to bright deep red, depending on the variety and stage of ripening (Holland et al., 2009). The fruit is a false berry, balausta type, which internally has multi-ovule chambers separated by membranous walls (septum) and a fleshy mesocarp. The chambers are filled with shiny red seeds encased in a succulent and edible red-pink pulp called arils. The arils develop from the outer epidermal cells of the seed and elongate to a very large extent in the radial direction (Fahan, 1976). The color of arils varies from white to deep red depending on the variety (Fig. 1).

Fruit quality depends largely on sugar and acid content of the juice (Dhinesh and Ramasamy, 2016). Pomegranate fruit quality depends on the following indices: freedom from internal and external decay (Kader, 2006); freedom from visible preharvest defects such as fruit cracking, sunburn, surface blackening, etc. which cause dark brown to black discoloration of the affected skin; and freedom from injury during harvesting and transportation without having surface abrasions, impact bruising, vibration injury, cuts, wounds, etc. Skin color should be characteristic to the cultivar. Dark red or pink-red color and a large fruit size are preferred. Aril color should be dark red and intense, with soft seeds or seedless. Flavor depends on sugar/acid ratio, and varies among cultivars. Soluble solids content above 17% and total phenolic content (TPC) below 0.25% are desirable for optimal levels of sweetness and astringency, respectively (Kader et al., 1984; Crisosto et al., 2000).

The edible portion of pomegranates is an excellent dietary source containing a significant proportion of organic acids, soluble solids, polysaccharides, vitamins, fatty acids, and mineral elements of nutritional significance as shown in Tables 2 and 3. The physicochemical properties of pomegranate fruit cultivars reported by several researchers show that they vary among the agroclimatic zones.

2 POSTHARVEST TECHNOLOGY OF POMEGRANATES

The pomegranate fruit takes 5–7 months to mature after blossom. The fruits are usually harvested from September to December in the United States. Due to the difficulties of separating the fruit from the tree, presently there is no machine or mechanical method of harvesting pomegranates and almost all harvesting is done by hand. The present method of harvesting pomegranates requires laborers to walk through the rows of trees, reach inside a tree to grasp an individual fruit, and then pull on the fruit to separate and detach it from the tree using clippers. However, this is a time consuming and costly process for the grower and potentially painful for the laborer as there are thorns and it is difficult to separate the stem from the laterals.

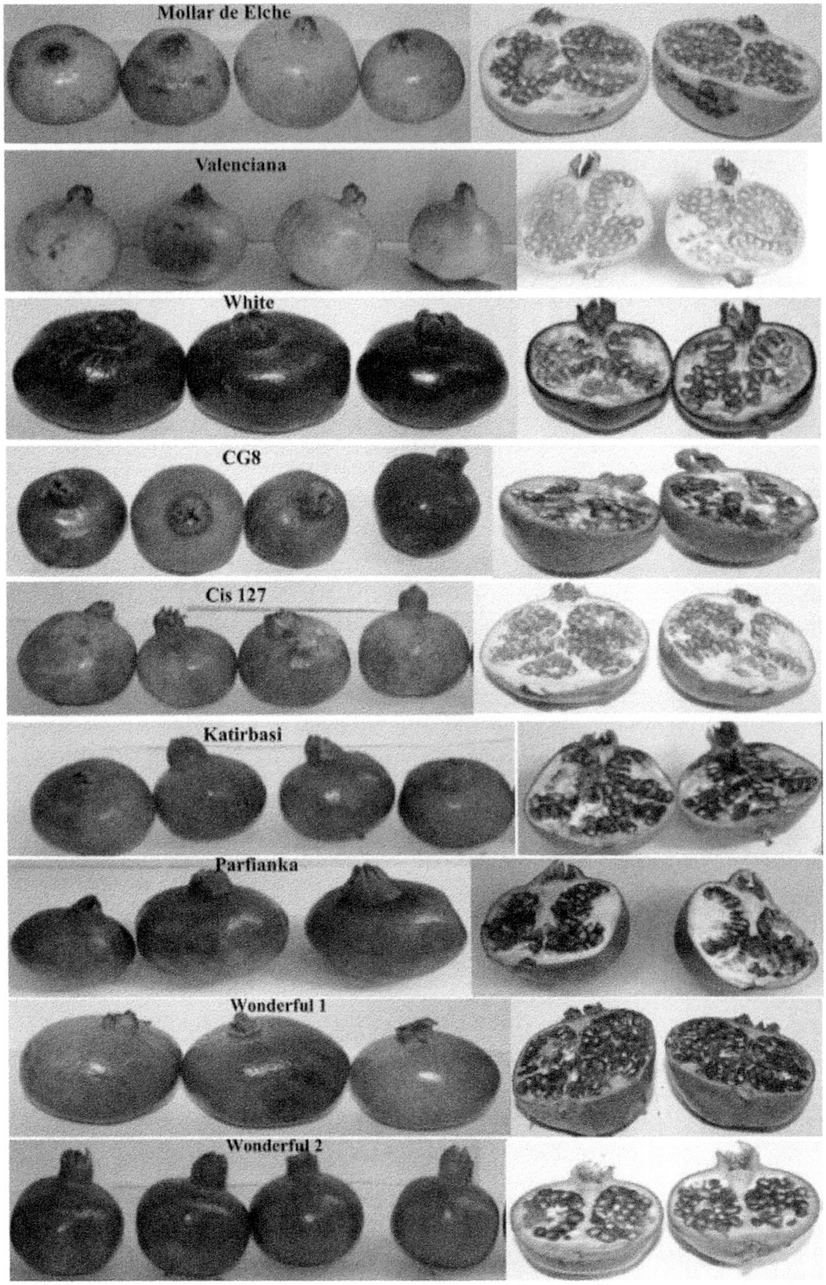

FIG. 1

Pomegranate fruits of nine cultivars grown in Spain with different quality characteristics. Source: Fernandes, L., Pereira, J.A., Lopéz-Cortés, I., Salazar, D.M., Ramalhosa, E., Casal, S. 2015. Fatty acid, vitamin E and sterols composition of seed oils from nine different pomegranate (Punica granatum L.) cultivars grown in Spain. J. Food Compos. Anal. 39, 13–22.

Table 2 Chemical Composition of Pomegranate Fruits (Dhinesh and Ramasamy, 2016; Ewaida, 1987)

Constituent	Value (dry weight basis)
Moisture (%)	19
Protein (%)	7.27
Total Sugars (%)	66.36
Ascorbic acid (mg/100 g)	72.73
Ash (%)	3.18
Acidity (%)	2.64

Table 3 Nutritional Composition of the Pomegranate (Dhinesh and Ramasamy, 2016; Ewaida, 1987)

Constituent	Unit	Value per 100 g of edible portion
Proximates		
Water	g	80.97
Energy	kcal	68
Protein	g	0.95
Total lipid	g	0.3
Ash	g	0.61
Carbohydrate, by difference	g	17.17
Total dietary fiber	g	0.6
Total sugars	g	16.57
Minerals		
Calcium	mg	3
Iron	mg	0.3
Magnesium	mg	3
Phosphorous	mg	8
Potassium	mg	259
Sodium	mg	3
Zinc	mg	0.12
Copper	mg	0.07
Selenium	mcg	0.6
Vitamins		
Vitamin C (Total ascorbic acid)	mg	6.1
Thiamin	mg	0.03
Riboflavin	mg	0.03
Niacin	mg	0.3
Pantothenic acid	mg	0.596
Vitamin B_6	mg	0.105
Total Folate	mcg	6
Vitamin A	IU	108
Vitamin E	mg	0.6
Vitamin K	mcg	4.6

After harvest, fruits are placed in picking bags for transfer to harvest bins which are transported to the packinghouse, where they are sorted to remove fruits with severe defects (such as scuffing, cuts, bruises, splitting, and decay). The remaining fruits are separated according to the magnitude of the physical defects. Pomegranates with slight or no defects are marketed fresh and those with moderate defects are processed into juice. The fruits are washed, air dried to remove surface moisture, fungicide treated, waxed, divided into several size categories, and packed in shipping containers. The fruits are held without movement in the shipping containers to reduce incidence and severity of scuffing and impact bruising during handling. Perforated plastic box liners are used to reduce water loss during postharvest handling of pomegranates. Packed fruits are forcibly air cooled to 7°C and kept at that temperature and 90%–95% relative humidity during storage and transport to retail centers (Kader, 2006).

Several postharvest conditions for the long-term storage of pomegranates were studied, including low temperature, delayed harvest, intermittent warming, controlled atmosphere (CA), and partial drying (Kader, 2006). Storage potential ranges from 3 to 4 months in air and from 4 to 6 months in a CA of 5% oxygen and 15% carbon dioxide (balance nitrogen). The use of CA storage, with 5% O_2 and 15% CO_2 at 7°C, extended the pomegranate postharvest life for up to 5 months. The optimal storage temperature of pomegranate ranges from 5°C to 8°C, depending on the variety and production area; 7°C is recommended for "Wonderful" pomegranates. In all cases, 90%–95% relative humidity should be maintained in the surrounding atmosphere. Using a storage temperature of 10°C prevents damage from cold. Pomegranates are very susceptible to water loss, which induces wrinkling of the pericarp. Storing the fruit with a coating or plastic lining or using waxes can reduce water loss particularly in conditions of low relative humidity. Long-time storage improves the flavor of the fruit (Table 4).

2.1 Pomegranate Fruit Processing

Sound pomegranate fruits are sold at a higher price in the market and are normally not used for processing. Fruit disorders such as sun burnt husks, splits and cracks, and husk scald on whole fruits reduce marketability and consumer acceptance. Harvested fruits with moderate defects are used for processing while the fruits without defects and with mild defects are marketed as fresh (Prasad et al., 2010). The major problem with pomegranates is cracking at maturity; about 20%–40% cracking has been reported and these cracked fruits neither look attractive nor fetch good prices for the growers. Pomegranate processing allows the use of low-quality fruits for the preparation of new products. Products like juice, juice concentrate, wine, isolated fresh arils, dried arils, rind powder, and other products have a great demand and a well-established market (Prasad et al., 2010; Adsule and Patil, 1995; Patil et al., 2002; Kleinberg, 2004; Seeram et al., 2006; Holland et al., 2009).

Table 4 Recommended Storage Requirements for Various Pomegranate Cultivars (Pareek et al., 2015)

Cultivar	Temperature (°C)	Relative humidity (%)	Storage period (months)	Reference
"Bhagwa"	5	>92	2–3	Fawole and Opara (2013b)
"Banati"	5	80–90	2	Al-Mughrabi et al. (1995)
"Hicaz"	6	85–90	5	Defilippi et al. (2006)
"Hicaz"	8–10	85–90	<2	Defilippi et al. (2006)
"Helow"	7	90–95	1	Al-Yahyai et al. (2009)
"Manfaloti"	5	80–90	2	Al-Mughrabi et al. (1995)
"Ruby"	5	>92	2–3	Fawole and Opara (2013b)
"Taeifi"	5	80–90	2	Al-Mughrabi et al. (1995)
"Wonderful"	5	95	2	Kader et al. (1984)
"Wonderful"	7.2	90–95	>2	Kader (2006).
"Mollar de Elche"	5–10	90–95	3	Mirdehghan et al. (2007)

2.2 Arils Separation and Processing

The demand of isolated fresh arils is increasing due to their high value, unique sensory characteristics, extended shelf life, and health benefits. Few juice manufacturers prefer the isolated arils because the juice is less bitter and tasty (Prasad et al., 2010). Each pomegranate contains several hundreds of arils completely held within the fruit. Manual processing for aril separation is performed by cutting the fruit into pieces by knife and then separating the arils by hand, which is very inefficient and highly labor intensive, time consuming, and tiring. Arils are so firmly attached to the rind, peel, and to each other that it is difficult to manual process arils into large quantities. Industrial aril production using machines created a new and innovative market for fresh arils, dried arils, juices, wines, and health and pharmaceutical products. Minimal processing of pomegranate arils mainly consists of washing with sanitizing agents to reduce the initial microbial load, pH modifications, use of antioxidants, modified atmosphere packaging, and temperature control. Few machines have been developed to efficiently extract intact arils on a commercial scale with a capacity to produce over 1 ton of arils per day (Rodov et al., 2005; Shmilovich et al., 2006).

Mechanical damage to the arils should be minimized during their extraction from the fruit, washing, drying to remove surface moisture, and packaging, since damaged arils are more susceptible to decay-causing fungi. Pre-extraction storage duration and post-extraction packaging and handling conditions of arils were found to affect the deterioration rate of pomegranate arils (Gil et al., 1996b, 1996c; Hess-Pierce and Kader, 1997, 2003). The respiration rates of arils were 1.5–3 and 3–6 mL CO_2/kg/h at 5°C and 7°C, respectively, and the ethylene production rate was 5–15 and 15–30 mL ethylene/kg/h at 5°C and 7°C, respectively. Arils can be stored for up to 14 days at 5°C from pomegranate fruits that are stored at 7°C for up to 3 months in air or up to 5 months in CA storage at 5% oxygen +15% carbon dioxide (balance nitrogen). Carbon dioxide-enriched atmospheres have a fungistatic effect and their optimal range for decay control without introducing off flavors in the arils is 15%–20% CO_2 added to either air or 5% O_2. Though the intact pomegranate fruits are chilling-sensitive, arils are low-temperature tolerant and can be stored at 0°C for up to 21 days, at 2°C for up to 18 days, or at 5°C for up to 14 days in marketable condition (Hess-Pierce and Kader, 2003).

The use of honey treatments preserves the fresh-like quality of arils and extends their shelf life. Dipping pomegranates in a honey solution (10% and 20%) for 5 min helped to store arils at 40°C for 10 days with a more brilliant aroma than those treated with water (Ergun and Ergun, 2009). It lowered the softening of the arils and extended the fresh-like quality of arils by delaying quality loss, microbial development, and pigment changes, thus providing a safe organic method. Dried arils, also known as "Anardhana," are prepared from sour, wild types and used as an acidulent in place of tamarind on dried green mangoes in the preparation of Indian styled curries, chutney, and other culinary preparations in North India. It is also used in the preparation of digestive candies and by the traditional systems of Ayurvedic and Unani medicine. The improved processing technique consists of pre-cleaning, mechanized extraction of arils, solar/sun drying, and packaging. After treating with sodium benzoate (600 ppm) for 10 min, arils are dehydrated in a drier at 45°C for 48 h to 10%–12% moisture content. It has an attractive brown color and can be stored for a long time in glass jars (Artes et al., 2000).

2.3 Pomegranate Juice Processing

Pomegranates contain 48%–52% of edible portions on the whole fruit basis, comprising about 78% juice and 22% seed (Dhumal et al., 2014), which varies by cultivar and growing location. The seeds and peels are available as by-products after juice extraction (Fig. 2). Teh (2016) studied the compositions of the juice, peel, and seed of five pomegranate varieties grown in California and reported that the pomegranate fruits contain 38%–50% juice, 39%–53% peel, and 8%–12% seed (Fig. 3).

The seeds along with the arils are crushed and juice is extracted and marketed as a fresh juice with an excellent flavor, attractive fragrance, delicious taste, and high nutritive and medicinal value.

Production of juice from pomegranate arils proved to be one of the important methods of value addition. The juice can then be processed into squash, syrup, nectar, jelly, concentrate, and other products. Pomegranate juice is also used to provide color to other products. It is a rich source of polyphenols. The phenolic constituents, such as anthocyanins, give the color. Flavonoids and some non-flavonoids are responsible for antioxidant properties, astringency, and bitterness of the juice (Gil et al., 2000). The high antioxidant content of pomegranate juice

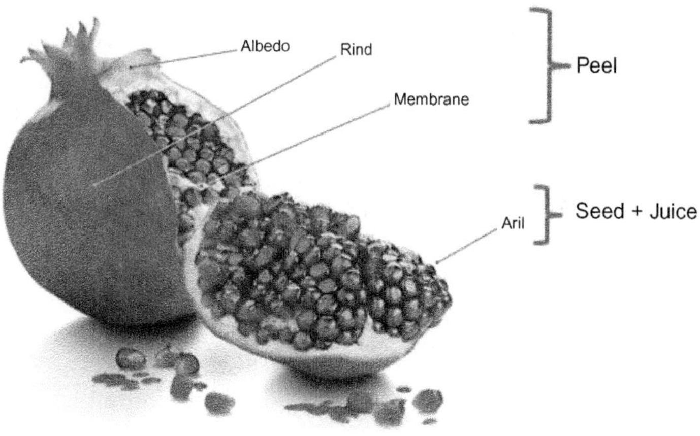

FIG. 2
Components of pomegranate fruit. Source: pomwonderful.com.

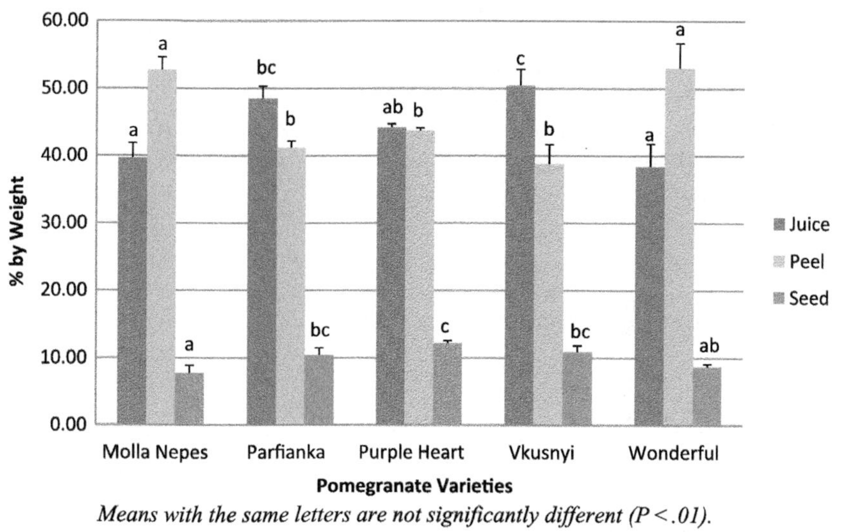

FIG. 3
Composition of pomegranate fruits of five varieties. Source: Teh, H.E. 2016. Extraction and Characterization of Functional Components From Fruit and Vegetable Processing Waste. PhD Thesis Submitted to University of California, Davis. 234 p.

makes it suitable for production of health supplements and nutraceuticals (Singh et al., 2002).

In earlier days, the fruits were quartered and crushed or the whole fruits were pressed in a hydraulic press and juice was strained out. As the juice extracted from the arils have improved purity, taste, and organoleptic properties and technology was successfully developed to separate arils, juice extraction from arils became popular. The various steps involved at the juice processing plant include receiving, washing, sorting to remove defective fruits, deseeding, separating the peels and membranes, extraction of juice, separation of solid particles, pasteurization, clarification, and ultrafiltration. Fruits received from the harvest are washed in the water tanks to remove dirt and soil particles and then conveyed on a sorting line where the defective fruits are manually handpicked by workers. The sorted fruits are sent to the deseeder, which loosens and separates the arils from the peels and membranes. Juice is extracted from the arils by a membrane press, which softly presses the arils to preserve the organoleptic properties of the juice. Afterwards, the remaining solid particles in the juice are removed in the vibroscreener. The resulting juice is pasteurized and pumped to treatment tanks, where enzymes and chemicals are added to increase the separation. On completion of the treatment for 3–4 h, the juice is sent to an ultrafiltration line for clarification.

Pomegranates have a brief harvest season and fresh pomegranate juice is susceptible to spoilage, which necessitates further processing to extend shelf life. Thermal processing is the most commonly used preservation technique. During thermal processing, many reactions, including pigment degradation, especially of carotenoids or anthocyanins, browning reactions such as the Maillard reaction, enzymatic browning, and oxidation of ascorbic acid take place, which induce adverse effects on sensory and nutritional values (Ibarz et al., 1999). Chen et al. (2013) studied the effect of a high hydrostatic pressure (HHP) (400 MPa/5 min) and high-temperature short-time (HTST) (110°C/8.6 s) treatment of pomegranate juice on microorganisms, pH, TSS, TA, color, total phenols, anthocyanins, antioxidant capacity, and shelf-life characteristics for 90 days at 4°C. They reported that HHP-treated and HTST-treated juices were microbiologically safe and had a greater retention of the original color, anthocyanins, and antioxidant capacity. In addition, increased total phenols were observed in HHP-treated samples immediately after processing. During storage, color change, anthocyanin content, total phenols, and antioxidant activity decreased.

3 BY-PRODUCTS UTILIZATION

Two main solid by-product streams are generated after extraction of juice from fruit and separation of seeds from juice: pomegranate peel (PP) and PS. The pomace is the term used to represent the by-product mixture containing PP and PS. PP and seed are excellent sources of several bioactive compounds such as phenolic compounds, tannins, flavonoids, sterols, fatty acids, dietary fibers, minerals, and vitamins. Pomegranate by-products have been used for the production of single-cell protein, industrial enzymes, and lovastatin with diversified market outlets and several economic and waste management benefits. PP extracts, seed oil, and pomegranate bagasse can be used in the fortification of food commodities generating the formation of functional novel products with diverse health benefits, increased quality, and longer shelf life (Charalampia and Koutelidakis, 2017). Viuda-Martos et al. (2012) studied the pomegranate pomace (bagasse) obtained after extracting juice from arils and whole fruits. The pomace obtained after juice triturated for 40 s in a vertical cutter to produce uniformly sized pieces increased the contact time during washing. Washing was done in hot water (1 L of water per kg of product) at 75°C for 10 min while constantly stirring the pomace. The product was then pressed to drain the water and lyophilized for 48 h before grinding to obtain a powder particle size of <0.417 mm. A flow chart of the processes is shown in Fig. 4.

The pomace from the whole fruit has lower protein, fat, and ash contents than those of the pomace obtained from arils. The protein, fat, and ash content of whole fruit pomace and aril pomace were 10.9 and 12.6 (g/100 g), 20.9 and 24.3 g/100 g, and 2.5 and 2.7 g/100 g, respectively. The total dietary fiber, insoluble dietary fiber (IDF), and soluble dietary fiber (SDF) contents is higher in

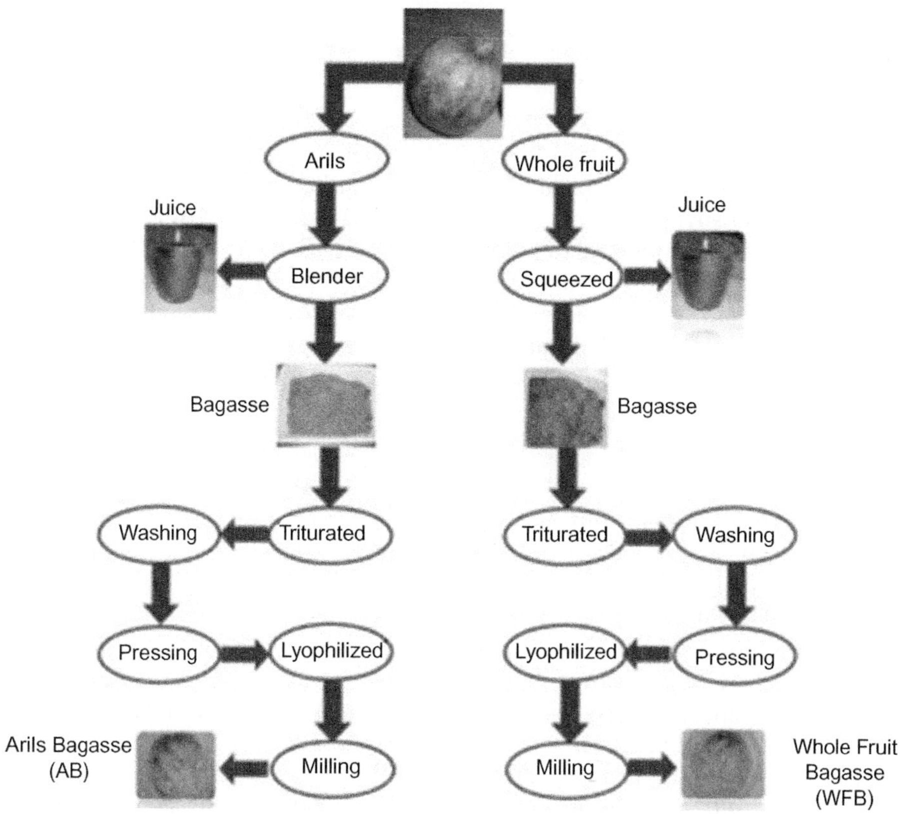

FIG. 4

Flowchart of processes to obtain pomegranate pomace (bagasse) from arils and whole fruits. Source: Viuda-Martos, M., Ruiz-Navajas, Y., Martin-Sanchez, A., Sanchez-Zapata, E., Fernandez-Lopez, J., Sendra, E., ... Perez-Alvarez, J. A. 2012. Chemical, physico-chemical and functional properties of pomegranate (Punica granatum L.) bagasses powder co-product. J. Food Eng. 110(2), 220–224.

Table 5 Average Chemical Composition of Pomegranate Peel and Seed Powder (Rowayshed et al., 2013; Charalampia and Koutelidakis, 2017)

Component	Pomegranate peel powder (%)	Pomegranate seed powder (%)
Moisture	13.7	5.8
Protein	3.1	13.7
Fat	1.8	29.6
Ash	3.3	1.5
Fiber	11.2	39.4
Carbohydrate	80.5	13.1
Total phenolic	27.9	0.25

pomace samples from whole fruits as the peel is included in the whole fruit pomace (Viuda-Martos et al., 2012). The chemical composition of PP and seeds are shown in Table 5.

Teh (2016) studied the physiochemical characteristics of peels and seeds from five novel pomegranate cultivars (Wonderful, Molla Nepes, Parfianka, Purple Heart, and Vkusnyi) grown in California, USA. Components of dried PP and seed determined through proximate analysis were reported. Carbohydrate was the principal constituent, making up 91% of PP, followed by 3.5% protein, 1.7% fat, and 3.7% ash or minerals. PS contained higher portions of protein and fat. They had 60%–72% carbohydrate, 11%–17% protein, 15%–21% fat, and 2%–3% mineral.

PP accounts for about 50% of the whole fruit. It is the nonedible portion comprised mainly of several bioactive compounds, such as hydrolysable tannins (pedunculagin, punicalin, punicalagin, and ellagic and gallic acids) at variant concentrations ranging from 27 to 172 g/kg (Fischer et al., 2013), flavonoids (catechins, anthocyanins, and other complex flavonoids), complex polysaccharides, and minerals (phosphorus, magnesium, calcium, potassium, and nitrogen) (Ismail et al., 2012). The primary polyphenols in PP are gallic acid (14.147%), proto catechuic acid (14.512%), chlorogenic acid (2.355%), vanillic acid (3.851%), coumarin (3.534%), caffeic acid (2.748%), oleuropein (0.590%), ferulic acid (1.857%), and quercetin (0.949%). PP also contains caffeine (6.420%) (Farag et al., 2014; Charalampia and Koutelidakis, 2017). PS account for about 10% of the edible part of pomegranate, ranging from 40 to 100 g/kg of fruit. They are rich sources of total lipids (12%–20%, w/w, of total seed), containing high concentrations of polyunsaturated ($n-3$) fatty acids. More specifically, the oil consists of up to 80% of conjugated octadecatrienoic fatty acids with a high percentage of *cis*, *trans*, *cis*-Δ-9, 11, 13 acids (e.g., punicic acid) (Jurenka, 2008). Fatty acid content reaches up to 95% of the seed oil, of which almost 99% is considered triacylglycerols. Also, PS oil contains sterols,

Table 6 Major Phytochemical Compounds Present in Pomegranate Juice and Pomegranate Seed Oil and Peel (Farag et al., 2014; Rowayshed et al., 2013; Liu et al., 2009)

Pomegranate juice	Anthocyanins; glucose; ascorbic acid; phenolics, such as ellagic acid, caffeic acid; catechin; epigallocatechin gallate; quercetin; rutin; amino acids and minerals
Pomegranate peel	Phenolic compounds like punicalagins, gallic acid, catechin, epigallocatechin gallate, quercetin, rutin, anthocyanidins and other flavonoids; dietary fibers; neutral sugars (xylose, aravinose)
Pomegranate seed oil	Punicic acid; ellagic acid; fatty acids; sterols (daucosterol, campesterol, stigmasterol and b-sito sterols); phenyl aliphatic glycosides; isoflavones and coumesterol

steroids, tocopherols, and cerebroside in minor quantities. The seed envelope consists of lignin and likely some antioxidant lignin derivatives as well as hydroxybenzoic/cinnamic acids and isoflavones (Fadavi et al., 2006; Charalampia and Koutelidakis, 2017).

Several scientific studies have confirmed pomegranate biological activities and medicinal effects of the edible part of the fruit, but very few data exist on the bioactivity of PP, seed, powder, and extracts. Major phytochemical compounds present in pomegranate juice and PS oil and peel are shown in Table 6. Due to these health benefits and their large annual production as by-products, PS and peels may have more beneficial applications in food industries instead of being only used as animal feed.

Effective separation and utilization of pomegranate juice by-product streams rich in valuable bioactive compounds could lead to the production of several products such as: (1) additives used in the industrial food sector to increase shelf life of products, (2) functional ingredients to enhance food quality and promote human health, and (3) single-cell protein and industrial enzymes by using proper pomegranate by-product streams based on fermentation media. Effective separation and utilization will strengthen the existing structure in pomegranates' juice production sector and also give rise to the development of new products while simultaneously solving waste management issues.

3.1 Phenolic Compounds

PPs and seeds are rich in bioactive compounds which are available as by-products of juice. PP contains higher antioxidant activity when compared with PS, as shown in Table 7. Li et al. (2006) reported that PP extracts contained 249.4 mg/g TPC while only 24.4 mg/g was found in the seed fractions. However, PSs are a rich source of tocopherols and phytosterols, containing 161.2–170.1 mg/100 g α-tocopherol and 80.2–92.8 mg/100 g γ-tocopherol (Pande and Akoh, 2009). Ellagitannins, especially punicalagin, are the most abundant polyphenols found in PP (Fischer et al., 2011). The compounds demonstrated antitumor and anti-inflammatory properties in vitro (Lu et al., 2007;

Table 7 Total Polyphenols, Flavonoids, Anthocyanins, and Hydrolysable Tannins Contents of Pomegranate Peels and Seeds (Elfalleh et al., 2012; Zaki et al., 2015)

By-product	Total polyphenol (mg/g GAE, db)[a]		Total flavonoids (mg/g RE, db)[b]		Total anthocyanins (mg/g CGE, db)[c]		Hydrolysable tannins (mg/g TAE, db)[d]	
	Water	Methanol	Water	Methanol	Water	Methanol	Water	Methanol
Peel	53.65	85.60	21.03	51.52	51.02	102.2	62.71	139.63
Seed	7.94	11.84	3.30	6.79	19.62	40.84	32.86	29.57

[a]mg gallic acid equivalents per g dry weight.
[b]mg tannic acid equivalents per g of dry weight.
[c]mg rutin equivalents per g dry weight.
[d]mg cyanidin-3-glucoside equivalents per g dry weight.

Seeram et al., 2007; Sharma et al., 2010). Pomegranate polyphenols such as punicalagin, punicalin, and ellagic and gallic acids were also shown to reduce macrophage oxidative stress and inhibit atherosclerosis progression (Aviram et al., 2008). Other phenolic compounds identified in PS, pulp, and peel included hydrolyzable tannins, caffeic acid, p-coumaric acid, ferulic acid, catechin, epicatechin, and quercetin (Elfalleh et al., 2011; Pande and Akoh, 2009). Remarkably high polyphenol contents in PP make it a potent natural antioxidant source. The peel extracts are used as food additives. For instance, sunflower oil fortified with PP extracts displayed delayed oxidation processes (Iqbal et al., 2008). PP extract contains vanillic acid as the main phenolic compound followed in order by syringic, caffeic acid, and p-coumaric and ferulic acid (Muhammad et al., 2015). Phenols, tannins, and flavonoids of PP extract showed an antibacterial and antifungal activity against several species (Khan and Hanee, 2011; Supayang et al., 2005). The antimicrobial mechanism of phenolic compounds is due to the reaction of phenolics with proteins and/or protein sulfhydryl components existing in the microbial cell membrane, forcing the precipitation of membrane protein and inhibition of enzymes (e.g., lycosyltransferases), therefore leading to microbial death (Naz et al., 2007; Supayang et al., 2005).

Cam et al. (2014) microencapsulated PP phenolics with maltodextrins and demonstrated that the encapsulated phenolics were effective in improving antioxidant activities in ice cream. Polyphenol extracts from pomegranate aril and peel could also inhibit the growth of foodborne pathogens, such as *Staphylococcus aureus*, *Escherichia coli*, and *Listeria monocytogenes* (Pagliarulo et al., 2016; Wu et al., 2016).

3.2 Extraction of Polyphenols

Several methods have been used to extract phenolic compounds from PP (Xi et al., 2017; Pan et al., 2011, 2012). Polyphenols can be extracted and purified from the peel using methanol, ethanol, water, pressurized water, or liquid chromatography systems (Cam and Hisil, 2010; Cam and Icyer, 2015; Qu et al., 2010; Rababah et al., 2010; Teh, 2016). The conventional methods include distillation and organic solvent extraction, which are tedious and time consuming and have a negative impact on environment in terms of emission of organic volatile compounds (Pan et al., 2012). Therefore, new and promising techniques such as pressurized water extraction (Cam and Hisil, 2010), instant controlled pressure drop (Ranjbar et al., 2016), supercritical fluid extraction (SFE) (Mushtaq et al., 2015), high-voltage electrical discharge (Xi et al., 2017), and pulsed ultrasound and ultrasound-assisted extraction (Malviya et al., 2014; Kazemi et al., 2016) have been introduced and show high extraction efficiency and low energy and solvent consumptions.

The extraction of peel and seed using green or chemical methods is economically beneficial for freely available, low-value pomegranate pomace. The commercial food-grade solvents used for the extraction of antioxidants are water, ethanol, and methanol or mixtures of them. Optimization of the extraction process such as the type of solvent, the particle size of the material, and the extraction temperature and time are required for economical production of food-grade antioxidants. Zaki et al. (2015) extracted phenolic and antioxidant compounds from peels of two pomegranate varieties using methanol and water concluded that methanol was the best extracting solvent for phenolic compounds. Malviya et al. (2014) investigated the antioxidant potential of PP of the Ganesh variety using either water, ethanol, methanol, or their mixtures, and found that the highest DPPH and ABTS inhibition activity was observed when methanol or a mixture with water (70% ethanol: 30% water) were used as extracting solvents. The pulsed ultrasound or ultrasonic-assisted extraction was found to be one of the most promising extraction techniques, better than the conventional ones, providing higher extraction yields, less energy, and fewer solvents and chemicals (Malviya et al., 2014). The enzyme-assisted SFE process doubled the recovery of crude extracts with an increased level of phenolic constituents, improved radical scavenging capacity, trolox equivalent antioxidant capacity, and inhibition of linoleic acid per oxidation (Tabaraki et al., 2012). TPC in the five cultivars grown in California ranged from 146.29 to 279.32, 6.61 to 13.20, 5.86 to 9.04, and 4.28 to 6.19 mg gallic acid equivalent/g dry sample for the peel soluble, peel insoluble, seed soluble, and seed insoluble fractions, respectively. Trolox equivalent antioxidant capabilities (TEAC) of the four fractions varied from 241.23 to 562.43, 17.40 to 26.51, 2.45 to 4.52, and 6.53 to 9.77 mg/g. About 87%–93% of the phenolic compounds were distributed in the peel soluble fractions. Polyphenols punicalagin, chlorogenic acid, gallic acid, catechin, epicatechin, and rutin were detected in the soluble peel extracts. The distribution and concentration of lipid, antioxidant, and phenolic compounds in the pomegranate tissues were significantly influenced by varietal characteristics (Teh, 2016).

3.3 Extraction of Pomegranate Seed Oil

PS account for 22% of the arils of pomegranates and contain high amounts of oil; the total lipid content on dry basis ranges from 66 to 193g per kg of fruit (Abbasi et al., 2008a, 2008b). The oil content of PS varies from 12% to 20% of the seed on a dry weight basis (Al-Maiman and Ahmad, 2002). Teh (2016) found that the oil content of PS varied from 15.23% to 21.32% depending on the variety (Fig. 5).

PS oil was reported to present biological properties (Eikani et al., 2012), such as antioxidant and eicosanoid enzyme inhibition properties (Qu et al., 2010), immune function and lipid metabolism (Yamasaki et al., 2006), estrogen content (Tong et al., 2006), skin photo-aging inhibition effect (Park et al., 2010), lipid peroxidation and activity of antioxidant enzymes (Melo et al., 2010), toxicological evaluation (Meerts et al., 2009), and protective effect against gentamicin-induced nephrotoxicity (Asadpour et al., 2010). PSs contain 4.4%–24.1% of oil (Fadavi et al., 2006; Kıralan et al., 2009; Teh, 2016). The oil is composed of 70.4%–86.0% punicic acid, 3.4%–6.6% linoleic acid, 3.2%–7.7% oleic acid, 2.1%–4.8% palmitic acid, and 1.4%–3.6% stearic acid (Abbasi et al., 2008b; Pande and Akoh, 2009; Schubert et al., 1999). In nature, PSs are the main source of punicic acid, which is shown to possess antioxidant, antitumor, and anti-inflammation properties. Several conjugated linolenic acid (CLnA) isomers of punicic acid, including α-eleostearic acid, catalpic acid, and β-eleostearic acid, are also present in PS oil, but they are often collectively reported as punicic acid due to difficulties in separating and identifying the individual peaks in gas chromatography (Sassano et al., 2009).

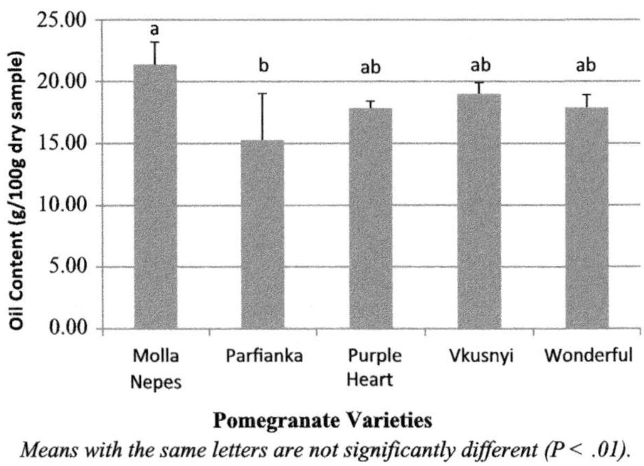

FIG. 5

Oil contents in the pomegranate seeds of different varieties. Source: Teh, H.E. 2016. Extraction and Characterization of Functional Components From Fruit and Vegetable Processing Waste. PhD Thesis Submitted to University of California, Davis. 234 p.

The antioxidant properties of the cold-pressed PS oil were found to be significantly superior to those of red wine, green tea, and the synthetic antioxidant butylated hydroxy anisole (BHA) (Schubert et al., 1999). PS oil consists of 65%–80% conjugated fatty acids, the most important of which is 9-trans, 11-cis, 13-trans, octadecatrienoic acid, also known as punicic acid. PS oil was reported to inhibit the upstream eicosanoid enzyme phospholipase A2 expressed by human prostate cancer cells (Abbasi et al., 2008a). In addition, PS oil can decrease the leukotriene production from arachidonic acid, which plays a major role in the occurrence of asthma in children, skin inflammation, and platelet aggregation associated with cardiovascular diseases. Furthermore, it has the non-steroidal estrogens daidzein (a glucoside), genistein (a glycone), and coumestrol (Moneam et al., 1988). PS oil has been reported to prevent skin cancer and act as chemopreventive agents (Abbasi et al., 2008b).

Different extraction methods including normal stirring, soxhlet, microwave irradiation, and ultrasonic irradiation hexane and petroleum benzene as solvents; the SFE method uses CO_2 as solvent and a cold press/extruder press to extract pomegranate oil (Teh, 2016; Abbasi et al., 2008a). PS oil from different methods of extraction with organic solvents (normal stirring, soxhlet, microwave irradiation, and ultrasonic irradiation) showed significant differences in the extraction yields, as shown in Table 8. However, no differences was found when a given method (e.g., microwave irradiation) was applied using different organic solvents. The oil yields extracted from various conditions of SFE were different and the extraction yield increased with an increase in pressure and a decrease in temperature. Application of modifiers with SFE resulted in an increase in the extraction yield in the order of hexane > ethanol > water. (Abbasi et al., 2008a, 2008b). Extraction yields from the various extraction conditions used in the SFE were lower than oil yields obtained from other extraction methods using organic solvents. No significant differences were observed in the fatty acid compositions of the extracted oils using organic solvents. However, the fatty acid

Table 8 Total Oil Recovery Applying Different Extraction Methods Using Petroleum Benzene and Hexane as Solvents (Micro Stands for Microwave) (Abbasi et al., 2008a)

	Solvent	
Extraction method	Petroleum benzene	Hexane
Blank	13.0 ± 0.0^{e}	12.7 ± 0.4^{e}
Stirring (4 h)	13.0 ± 0.0^{e}	13.0 ± 0.3^{e}
Ultrasound (45 min)	$15.7 \pm 0.1^{b,c}$	16.0 ± 0.5^{b}
Micro—200 W (10 min)	14.7 ± 0.4^{b}	$15.0 \pm 0.0^{c,d}$
Micro—800 W (10 min)	$15.6 \pm 0.3^{b,c,d}$	$15.8 \pm 0.3^{b,c}$
Soxhlet (6 h)	18.6 ± 0.2^{a}	18.7 ± 0.2^{a}

Two-factorial comparison among the data. Means with the same letters are not significantly different ($P < .01$).

Table 9 Total Phenolic Contents of the Extracted Oils Using Organic Solvents (mg/g) (Abbasi et al., 2008b)

	Solvent	
Extraction method	**Petroleum benzene**	**Hexane**
Stirring (4 h)	8.4 ± 0.6^a	8.5 ± 0.2^a
Ultrasound (45 min)	8.1 ± 0.1^a	8.2 ± 0.2^a
Microwave (200 W, 10 min)	8.0 ± 0.6^a	8.7 ± 0.5^a
Microwave (800 W, 10 min)	8.5 ± 0.2^a	9.0 ± 0.3^a
Soxhlet (6 h)	8.9 ± 0.6^a	9.0 ± 0.7^a

The letter *a* indicates that means with the same letters are not significantly different ($P = .0.01$).

compositions of the oils extracted under different conditions of the SFE system indicated significant differences among several fatty acids, including unsaturated fatty acids (Abbasi et al., 2008a).

The total amounts of phenolic compounds in the PS oil extracted using various extraction methods with the two organic solvents (petroleum benzene and hexane) are given in Table 9. No significant differences were found among the TPC of the extracted oils. Increasing the temperature in microwave extraction (due to an increase in microwave power) and using ultrasonic waves, both of which can improve solvent penetration and therefore the extraction yield, did not favor the extraction of phenolic compounds into the organic solvents. These results can be attributed to the somewhat polar nature of phenolic compounds, which cannot be dissolved in the nonpolar solvents used for the extraction of seed oil (Abbasi et al., 2008b). When using supercritical carbon dioxide as solvent, different operational conditions resulted in different amounts of phenolic compounds in the PS oil. Increases in pressure, temperature, and volume of the modifier resulted in a decrease in the TPC of the extracted oils. The TPC of the extracted oil from one of the SFE runs was several times greater than those in the extracted oils using organic solvents.

The physicochemical characteristics and shelf life of mechanically extracted (cold pressed) and solvent extracted pomegranate oil were studied by Teh (2016). She reported that PS oil is mainly comprised of unique CLnA that have an 18-carbon backbone structure and three uninterrupted double bonds. The chemical properties of PS oil extracted by the two methods are shown in Table 10.

Teh (2016) performed the hexane extraction of PS from five cultivars and reported that the total oil contents in seeds ranged from 15.23% to 21.32% (dry basis). In addition, seed oils were composed of 81%–84% punicic acid, 4%–5% palmitic acid, 3%–4% linoleic acid, 3%–4% oleic acid, 3% α-eleostearic acid, and 2%–3% stearic acid. Oil quality indicators were not significantly different among cultivars.

Table 10 Chemical Characteristics of Pomegranate Seed Oil Extracted Using Mechanical Press and Solvent Extraction (Teh, 2016)

	Pomegranate seed oil	
Chemical properties	**Hexane**	**Press**
Acid value (mg KOH/g)	0.93 ± 0.04^a	1.07 ± 0.04^b
Peroxide value (meq. O_2/kg)	3.90 ± 0.32^a	6.77 ± 0.37^b
Iodine value (g I_2/100 g)	147.32 ± 3.60^a	163.39 ± 5.25^b

The letter *a* and *b* indicates that means with the same letters are not significantly different ($P = .0.01$).

3.4 Integrated Utilization of Pomace to Produce Antioxidants, Oil, and Biogas

Qu et al. (2009, 2010) studied the feasibility of integrated extraction and anaerobic digestion process for the recovery of nutraceuticals and biogas from pomegranate pomace. The objectives of the earlier study include (1) to determine the yields and properties of antioxidants (total phenolics in terms of tannic acid equivalent) and oil extracted from various dry and wet constituents of PM, including peel, seeds, and mixture; (2) to examine the feasibility of converting the PM into biogas energy with different initial organic loadings before and after the extraction; and (3) to determine the kinetic parameters of biogas production using a modified Gompertz bacterial growth model. Water and petroleum ether were used as solvents in the extraction of antioxidants and oil, respectively. The process flow chart of the integrated extraction and anaerobic digestion of pomegranate pomace is shown in Fig. 6. The antioxidant extraction was performed using deionized water at the water to sample ratio of 50:1 g/g extraction temperature of 25±2°C, reaction times of 4 h for wet and dry peel and 8 h for wet and dry seeds and mixture, and using magnetic stirrer at a stirring speed of 1200 rpm (Qu et al., 2009). The liquid extract was separated from the residue by centrifugation at 3500 rpm for 20 min at 4°C and the extract yield, antioxidant yield, antioxidant content, and DPPH scavenging activity of antioxidants were determined and shown in Table 11.

Comparing the yield and antioxidant contents of wet and dried samples they found no significant effect on the extraction efficiency and functionality of the extract from either peel or seeds. It was concluded that the direct use of pomegranate pomace in wet or dried form for antioxidant (total phenolics in terms of tannic acid equivalent) production is feasible (Qu et al., 2009). The results showed that 1 ton of dry mass from wet PP can produce 508 kg of extract. The extract had an antioxidant content of 23.0%, which corresponded to an antioxidant yield of 106 kg per ton of peel on dry basis (db). The antioxidants extracted from all samples had DPPH scavenging activities ranging from 6.1 to 6.9 g/g as shown in Table 11. Thus, the peel was found to be a better source for producing

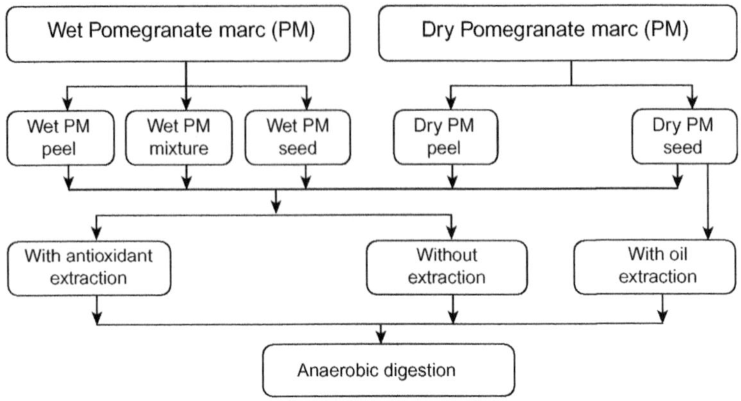

FIG. 6

Process flowchart of the integrated system of antioxidant extraction, oil extraction, and anaerobic digestion. Source: Qu, W., Pan, Z., Zhang, R., Ma, H., Zhu, B., Wang, Z., Atungulu, G., 2009. Integrated extraction and anaerobic digestion process for recovery of nutraceuticals and biogas from pomegranate marc. Trans. ASABE 52 (6), 1997–2006.

Table 11 Yields and Properties of Extracts From Wet and Dry PM Samples Using Water as Solvent (Qu et al., 2009)

Sample type	Extract yield (%)	Antioxidant Yield (%)	Antioxidant Content (%)	DPPH Scavenging Activity (g/g)
Wet peel	46.3[a]	10.6[a]	23.0[a]	6.6[a]
Dry peel	50.8[a]	10.2[a]	20.1[a]	6.5[a]
Wet seed	22.1[c]	1.1[c]	5.2[c]	6.1[a]
Dry seed	28.6[b,c]	1.4[c]	5.1[c]	6.8[a]
Wet peel and seed mixture	29.8[b]	4.1[b]	13.8[b]	6.9[a]

Different letters indicate significant difference at $P < .05$. Yield is the dry basis of the raw material, and DPPH scavenging activity is the amount of DPPH oxidized based on antioxidant weight.

antioxidant than the seeds and mixture as the peel had significantly higher extract yields, antioxidant yields, and antioxidant contents compared to the seeds and mixture.

Qu et al. (2010) reported that the yield and content of antioxidants increased with reduced particle size and increased water/sample ratio and temperature, but an antioxidant activity was low when extraction temperature was high. By considering the antioxidant activity and operation cost, the recommended extraction conditions were peel particle size of 0.2 mm, water/peel ratio of 50/1 (w/w), temperature of 25°C, and extraction time of 2 min, which gave the high antioxidant yield (11.5%) and content (22.9%), and DPPH scavenging activity of 6.2 g/g.

The oil from the dry PSs was extracted using petroleum ether with the petroleum ether to seed ratio of 10:3 $mL g^{-1}$, extraction temperature of $25 \pm 2°C$, reaction time of 60 min, and by using magnetic stirrer at a stirring speed of 1200 rpm. The oil solution was filtered through Whatman No. 1 filter paper. The residue from the extraction was re-extracted again using the earlier procedure. The oil solutions were dried to a constant weight by removing solvent in a rotary evaporator at 45°C. About 138 kg of oil was produced from 1 ton of dry seeds. The oil contained unsaturated fatty acids of linoleic (33%), docosahexaenoic (0.37%), and linolenic acids (0.1%). The oil also contained many saturated fatty acids of palmitic acid (14.9%) and stearic acid (11%) (Qu et al., 2009).

The anaerobic digestion tests of pomegranate pomace obtained after antioxidant and oil extraction were conducted at $35 \pm 2°C$ with a feedstock to microorganism ratio of 0.5 on volatile solid (VS) basis under two initial organic loadings of 3.0 and 5.0 g VS L-1. Compared to the low initial loading (3.0 g VS L-1), the high initial loading (5.0 g VS L-1) improved methane contents (55.1%–67.5%) but not biogas yield. The effect of antioxidant extraction from the PP on the biogas yield and methane content was studied by performing anaerobic digestion of pomegranate pomace after antioxidant and oil extraction (AE treated) and compared it with that of un-extracted raw pomace (untreated). Cumulative biogas yields and biogas production rates of untreated and AE treated wet PM samples during batch anaerobic digestion at $35 \pm 2°C$ and initial organic loading of 5.0 g VS L-1 is shown in Fig. 7.

>95% of the biogas production from untreated wet peel, seeds, and mixture was achieved within 8, 14, and 14 days, respectively, whereas the AE treated wet peel, seeds, and mixture required 13, 16, and 16 days, respectively, to attain 95% of the biogas yields due to their slow digestion (Fig. 6). After 20 days of digestion, the biogas yields from the untreated wet peel, seeds, and mixture were higher (396, 405, and 389 $mL g^{-1}$ VS, respectively) than the AE treated wet samples, which had corresponding biogas yields of 276, 298, and 292 $mL g^{-1}$ VS, respectively. The results confirmed that the antioxidant extraction process significantly reduced the cumulative biogas yields and biogas production rates, but increased the digestion time. This is because the extract removal reduced the utilizable soluble substrate for the inoculum, which resulted in reduced bacterial growth and biogas production. Both untreated and AE treated wet seeds had lower biogas production rates than wet peel because the seeds had more hardly degradable substances, such as lignin, cellulose, and hemicellulose compared to the peel. The biogas yield data from oil extracted PS and untreated seeds indicated that both biogas yield and biogas production rate were reduced by the oil extraction process.

Table 12 shows the methane yields from untreated and treated PM samples at digestion time of 20 days. The effects of drying and extraction on methane yields were similar to the effects on biogas production. The extraction process caused

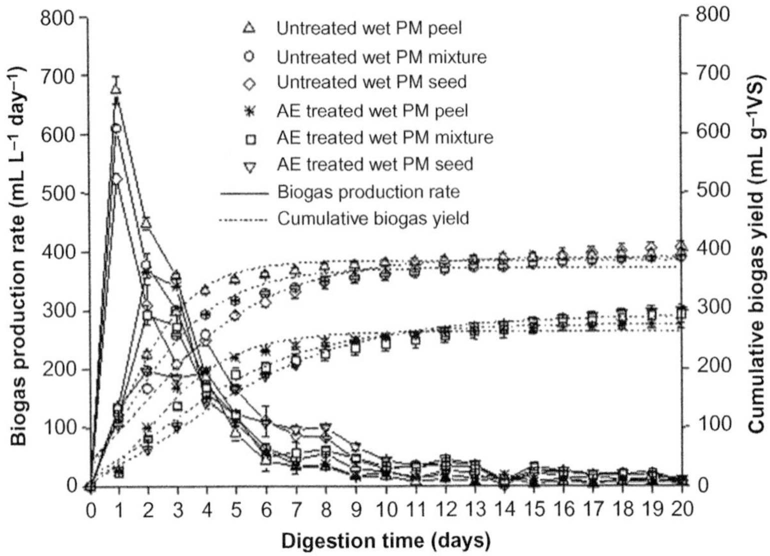

FIG. 7

Cumulative biogas yields and biogas production rates of untreated and AE treated wet PM samples during batch anaerobic digestion at 35 ± 2°C and initial organic loading of 5.0 g VS L-1. Source: Qu, W., Pan, Z., Zhang, R., Ma, H., Zhu, B., Wang, Z., Atungulu, G., 2009. Integrated extraction and anaerobic digestion process for recovery of nutraceuticals and biogas from pomegranate marc. Trans. ASABE 52 (6), 1997–2006.

Table 12 Methane Yields and VS Reductions of Untreated and Treated PM Samples Using Batch Anaerobic Digestion at Digestion Time of 20 days (Qu et al., 2009)

Sample type	Treatment condition	Methane yield (mL g^{-1} VS)	VS reduction (%)
Wet PM peel	Untreated	207	16.4
	AE treated	148	11.6
Dry PM peel	Untreated	213	17.6
	AE treated	140	13.1
Wet PM seed	Untreated	249	19.6
	AE treated	183	19.1
Dry PM seed	Untreated	276	22.7
	AE treated	222	18.5
	OE treated	200	20.5
Wet PM mixture	Untreated	221	16.3
	AE treated	161	13.2

the loss of biodegradable substrate for methanogenesis, which correspondingly reduced the methane yield.

The VS reductions of extracted PM peel, seeds, and mixture in wet and dry forms were 11.6%–13.1%, 18.5%–20.5%, and 13.2%, respectively, which were lower than those (16.4%–17.6%, 19.6%–22.7%, and 16.3%) of the untreated samples. The VS reduction had a positive relationship with biogas and methane production (Qu et al., 2009).

Qu et al. (2009, 2010) concluded that the drying process did not show any significant effect on the extraction efficiency and functionality of the extract—both PP and seed. This showed that the by-product from pomegranate juice processing can be directly used for antioxidant production (total phenolics in terms of tannic acid equivalent) or dried first when necessary. One ton of dry mass from wet PM peel can produce 508 kg of extract with an antioxidant content of 23.0%, which corresponded to an antioxidant yield of 106 kg per ton of PM peel on dry basis (db). One ton of dry mass from dry PM seeds can produce 138 kg of oil with high unsaturated fatty acids. With initial loading of 5.0 g VS L-1, the extracted residuals of peel, seeds, and mixture had low biogas yields of 276, 298, and 292 $mL g^{-1}$ VS, respectively, and methane yields of 148, 183, and 161 $mL g^{-1}$ VS, compared to the raw PM samples with biogas yields of 396, 405, and 389 $mL g^{-1}$ VS and methane yields of 207, 249, and 221 $mL g^{-1}$ VS, at digestion time of 20 days.

4 UTILIZATION OF POMEGRANATE POMACE

Pomegranate processing generates large volumes of pomace, which is either disposed of in landfills or used as animal feed. The peel is a rich source of phenolic compounds and antioxidants, while the oil from the seeds is a rich source of punicic acid with antioxidant, antitumor, and anti-inflammation properties. The potential uses of pomace are summarized in the following section.

4.1 Pomegranate Pomace as a Functional Food Ingredient

The pomegranate fruit is considered as a functional food as the different parts of the fruit have several compounds with functional properties which show medicinal effects. Viuda-Martos et al. (2010) compiled the various functional and medical effects of pomegranates which had been claimed by several researchers (Fig. 8). Pomegranate fruits can act as antioxidants, as antitumoral or antihepatotoxic agents, and improve cardiovascular health. They have been shown to have antimicrobial, anti-inflammatory, antiviral, and antidiabetic properties. They can also improve oral and skin health, help prevent Alzheimer's disease, and improve sperm quality and erectile dysfunction in male patients. However, these effects have not been solidly established through well-controlled clinical trials.

Pomegranate pomace and its extracts with a high bioactive property could be added to meat, seafood, drinks, and baked products to improve product quality,

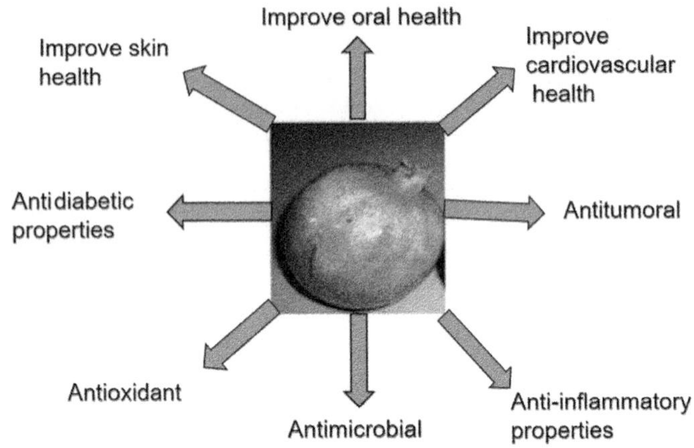

FIG. 8

Principal functional and medicinal effects of pomegranate. Source: Viuda-Martos M, Fernandez-Lopez, J., Perez-Alvarez, J.A. 2010. Pomegranate and its many functional components as related to human health: a review. Compr. Rev. Food Sci. Food Saf. 9, 635–654.

inhibit microbial growth, and delay deterioration during storage (Basiri et al., 2015; Devatkal et al., 2014; Kanatt et al., 2010; Mastrodi Salgado et al., 2012). Naveena et al. (2008) demonstrated that PP extract increased oxidative stability of grilled chicken burgers by several folds when compared to butylated hydroxytoluene, a commonly used synthetic antioxidant. Pomegranate extract addition in meat paté was found to inhibit L. *monocytogenes* growth by 5.1-log at refrigeration temperature (Hayrapetyan et al., 2012). Cookies and biscuits supplemented with PP powder had higher fiber, mineral, and antioxidant contents and could serve as healthier alternatives to conventional baked goods (Ismail et al., 2014; Srivastava et al., 2014).

4.2 Pomegranate Pomace as Dietary Supplements

A recent study demonstrated the use of bagasse from pomegranates as a source of total, soluble, and insoluble fiber. Pomegranate pomace contains 45.6–50.3 g/100 g total dietary fiber on dry basis. The IDF to SDF ratio is 1.5–1.7. Pomegranate fibers had good water-holding capability; they could hold 4.5–4.9 g water/g fiber. This excellent characteristic suggests that they could be used as bulking, thickening, or hydration agents in food formulation (Hasnaoui et al., 2014; Viuda-Martos et al., 2012). Polysaccharide fractions isolated and purified from pomegranates also carried several biological activities, such as antioxidant, antitumor, immunomodulation, and whitening properties (Joseph et al., 2012; Teh, 2016). The pomegranate bagasse powder coproduct exhibited a WHC of 4.86 times its own weight, which is similar to other fibrous residues such as sugar cane (4.98 g water/g product) and pear (5 g water/g product), indicating the potential to be used as a functional dietary ingredient. The oil-holding capacity of pomegranate bagasse was reported as 5.9 g oil/g fiber, showing that, even if pomegranate bagasse is added in products or used for frying processes, it would not retain much oil (Martos et al., 2012). A very important index regarding dietary fibers and their usage as a functional ingredients in food is the ratio of IDFs to SDF. The IDF to SDF ratio of pomegranate bagasse was approximately 1.6–1.7, values that indicate a balanced ratio of dietary fibers in pomegranate by-products (Martos et al., 2012). All the earlier characteristics show that PP could be used for the production of fiber-enriched products or even hygienic low fat products, using dietary fibers as fat replacers.

4.3 Enzymes and Single-Cell Protein Production

Uma et al. (2012) produced invertase with a specific activity of 197.5 U/mL using a fungal strain (*Cladosporium cladosporioides*) in submerged fermentations after 96 h of incubation at 30°C and a pH value of 4, using PP as a carbon source and yeast extract as a nitrogen source. Invertase production has also been evaluated using PP waste as fermentation feedstock by *Aspergillus flavus* bioconversions (Uma et al., 2010). Khan et al. (2010) evaluated the potential production of SCP through *Saccharomyces cerevisiae* fermentations using pomegranate rind as substrate and obtained 54.28% yield of crude protein from 100 g pomegranate rind.

4.4 Lovastatin From Pomegranate Seeds

Lovastatin is a fungal secondary metabolite which is used in the treatment of hypercholesterolemia, atherosclerosis, peripheral vascular disease, ischemic disease, bone fracture, and other diseases (Seraman et al., 2010). Naik and Lele (2012) evaluated lovastatin production with PS as substrate via solid state fermentation. PS enriched with 0.1% (w/v) potassium dihydrogen phosphate, 5% (w/v) glucose, and 60% (w/w) moisture were found to be the best substrate, resulting in the production of 4.2 ± 0.03 mg lovastatin/g dfm after 15 days of optimized solid-state fermentation at pH 5. A maximum lovastatin yield of 6.5 ± 0.07 mg/g dfm was achieved by mutation studies using ethyl methyl sulfonate (EMS), using PS as the fermentation substrate (Naik and Lele, 2012).

4.5 Pomegranate Pomace as Coloring Agent

PP is used as a natural and nontoxic coloring element for textile dyeing (Benli and Bahtiyari, 2015). Cotton fabric soaked with 4 times its volume of PP dye yielded cloth with yellow to red hues and supplementary antimicrobial properties (Davulcu et al., 2014). PP waste can also be made into absorbents though thermal activation. Pomegranate-based absorbents have the capacity of removing 65.7 mg/g aquatic pollutant phenols, 35.2 mg/g chromium ions, 19.23 mg/g Congo Red, and 7.3 mg/g copper ions from aqueous solution (Bhatnagar and Minocha, 2009; Guzel et al., 2014; Teh, 2016). Agricultural waste from pomegranate processing can also be used as growth substances for enzymes, antioxidants, and nutraceuticals production through biotechnological approaches. Catechin, gallic acid, and ellagic acid were some of the phenolic compounds that could be converted from polyphenols in PP and seeds by *Aspergillus niger* (Aguilar et al., 2008; Robledo et al., 2008; Teh, 2016). Tannase is an enzyme that catalyzes ester bond hydrolysis in tannins to produce gallic acids, and it has broad applications in the food industry. Pomegranate pomace is a suitable starting component for extracellular tannase production by fungi *A. Niger* and *Penicillium purpurogenum* (Jana et al., 2012; Srivastava and Kar, 2009). Laccase, lignin peroxidase, amylase, and other hydrolytic enzymes could also be produced by fungal fermentation of pomegranate waste. Fermentation of PP and pressed pomace with mesophilic inoculum for 100days could yield a total of 0.312 and 0.420L/g biome thane, respectively (Gunaseelan, 2004).

5 SUMMARY

Pomegranates are naturally available biological sources rich in nutrients and bioactive compounds. The most important of these compounds are anthocyanins, which exhibit strong chemopreventive activities such as anti-mutagenicity, anti-hypertension, anti-oxidative potential, and reduction of liver injury. Excellent flavor, nutritive and antioxidant values, and the medicinal properties of pomegranate fruits show their high potential for processing into value-added products with extended shelf lives. The by-products of pomegranates (peels and seeds) are also rich sources of several bioactive compounds, such as phenolic compounds, tannins, flavonoids, sterols, fatty acids, dietary fibers, minerals, and vitamins. These by-products can be used to produce single-cell proteins, industrial enzymes, and lovastatin with diversified market outlets and several economic and waste management benefits. Pomegranates can be utilized as minimally processed fresh arils, juice concentrates, jams, jellies, seeds in syrup, pomegranate spirits, PP powders, pomegranate rind powders, and PP extracts. A few studies reported the ingestion of pomegranate and its peel fractions in the form of pills, capsules, and gels as conventional treatment regimens against certain diseases in countries of the developing world. The recent clinical trials to explore the therapeutic potential of pomegranate products, particularly determining preventive efficacy of pomegranate extracts with cancer, cardiovascular diseases, inflammation, diabetes, and ultraviolet radiation-induced skin damage, show that pomegranates are a promising food with well-defined therapeutic benefits. However, these products are not yet popularized in the consumer market due to lack of the commercially viable processing technologies and scientifically established clinical trials to support the therapeutic claims. Therefore, future research should focus on the development of technologies for commercial production of novel food products from fruits, arils, peels, and seeds and conducting clinical trials to strengthen the therapeutic reputation of pomegranate food products. This would open new avenues for scientific research in the realm of food science and nutrition, and help people realize the potential of pomegranate fruit and its by-products.

References

Abbasi, H., Rezaei, K., Rashidi, L., 2008a. Extraction of essential oils from the seeds of pomegranate using organic solvents and supercritical CO2. J. Am. Oil Chem. Soc. 85, 83–89.

Abbasi, H., Rezaei, K., Emamdjomeh, Z., Mousavi, S.M.E., 2008b. Effect of various extraction conditions on the phenolic contents of pomegranate seed oil. Eur. J. Lipid Sci. Technol. 110, 435–440.

Adsule, R.N., Patil, N.B., 1995. Pomegranate. In: Salunkhe, D.K., Kadam, S.S. (Eds.), Handbook of Fruit Science and Technology. Marcel Dekker, New York, pp. 455–464.

Aguilar, C.N., Aguilera-Carbo, A., Robledo, A., Ventura, J., Belmares, R., Martinez, D., Contreras, J., 2008. Production of antioxidant nutraceuticals by solid-state cultures of pomegranate (*Punica granatum*) peel and creosote bush (*Larrea tridentata*) leaves. Food Technol. Biotechnol. 46 (2), 218–222.

Al-Maiman, S.A., Ahmad, D., 2002. Changes in physical and chemical properties during pomegranate (Punicagranatum L.) fruit maturation. Food Chem. 76, 437–441.

Al-Mughrabi, M.A., Bacha, M.A., Abdelrahman, A.O., 1995. Effects of storage temperature and duration on fruit quality of three pomegranate cultivars. J. King Saud Univ. 7, 239–248.

Al-Yahyai, R., Al-Said, F., Opara, L., 2009. Fruit growth characteristics of four pomegranate cultivars from northern Oman. Fruits 64, 335–341.

Artes, F., Tudela, J.A., Villaescua, R., 2000. Thermal postharvest treatment for improving pomegranate quality and shelf life. Postharvest Biol. Technol. 18, 245–251.

Asadpour, E., Boroushaki, M.T., Sadeghnia, H., 2010. Protective effect of pomegranate seed oil against gentamicin induced nephrotoxicity in rat. Toxicol. Lett. 196 (1), s232.

Aviram, M., Volkova, N., Coleman, R., Dreher, M., Reddy, M.K., Ferreira, D., Rosenblat, M., 2008. Pomegranate phenolics from the peels, arils, and flowers are antiatherogenic: studies in vivo in atherosclerotic apolipoprotein E-deficient mice and in vitro in cultured macrophages and lipoproteins. J. Agric. Food Chem. 56, 1148–1157.

Basiri, S., Shekarforoush, S.S., Aminlari, M., Akbari, S., 2015. The effect of pomegranate peel extract (PPE) on the polyphenol oxidase (PPO) and quality of Pacific white shrimp (*Litopenaeus vannamei*) during refrigerated storage (vol 60, pg 1025, 2015). LWT Food Sci. Technol. 63 (1), 798. https://doi.org/10.1016/j.lwt.2015.04.001.

Ben-Arie, R., Segal, N., Guelfat-Reich, S., 1984. Thematuration and ripening of the 'Wonderful' pomegranate. J. Am. Soc. Hortic. Sci. 109, 898–902.

Benli, H., Bahtiyari, M.I., 2015. Use of ultrasound in biopreparation and natural dyeing of cotton fabric in a single bath. Cellulose 22 (1), 867–877.

Bhatnagar, A., Minocha, A.K., 2009. Adsorptive removal of 2,4-dichlorophenol from water utilizing *Punica granatum* peel waste and stabilization with cement. J. Hazard. Mater. 168 (2–3), 1111–1117.

California Rare Fruit Growers. 1997. Pomegranate Fruit Facts. Available online at: http://www.crfg.org/pubs/ff/pomegranate.html.

Cam, M., Hisil, Y., 2010. Pressurised water extraction of polyphenols from pomegranate peels. Food Chem. 123 (3), 878–885.

Cam, M., Icyer, N.C., 2015. Phenolics of pomegranate peels: extraction optimization by central composite design and alpha glucosidase inhibition potentials. J. Food Sci. Technol. 52 (3), 1489–1497.

Cam, M., Icyer, N.C., Erdogan, F., 2014. Pomegranate peel phenolics: microencapsulation, storage stability and potential ingredient for functional food development. LWT Food Sci. Technol. 55 (1), 117–123.

Charalampia, D., Koutelidakis, A.E., 2017. From pomegranate processing by-products to innovative value added functional ingredients and bio-based products with several applications in food sector. BAOJ Biotech. 3, 025.

Chen, D., Xi, H., Guo, X., Qin, Z., Pang, X., Hu, X., Liao, X., Wu, J., 2013. Comparative study of quality of cloudy pomegranate juice treated by high hydrostatic pressure and high temperature short time. Innov. Food Sci. Emerg. Technol. 19, 85–94.

Crisosto, C.H., Mitcheam, E.J., Kader, A.A., 2000. Pomegranates, produce facts. Available from http://postharvest.ucdavis.edu/Commodity_Resources/Fact_Sheets/Datastores/Fruit_English/?uid=53&ds=798. Accessed 3 May 2019.

Costa, Y., Melgarejo, P., 2000. A study of the production cost of two pomegranate varieties grown in poor quality soils. Options Méditerr. Ser. A 42, 49–53.

Davulcu, A., Benli, H., Sen, Y., Bahtiyari, M.I., 2014. Dyeing of cotton with thyme and pomegranate peel. Cellulose 21 (6), 4671–4680.

Defilippi, B.G., Whitaker, B.D., Hess-Pierce, B.H., Kader, A.A., 2006. Development and control of scald on Wonderful pomegranates during long-term storage. Postharvest Biol. Technol. 41, 234–243.

Devatkal, S.K., Thorat, P., Manjunatha, M., 2014. Effect of vacuum packaging and pomegranate peel extract on quality aspects of ground goat meat and nuggets. J. Food Sci. Technol. 51 (10), 2685–2691. https://doi.org/10.1007/s13197-012-0753-5.

Dhinesh, K.V., Ramasamy, D., 2016. Pomegranate processing and value addition: review. J. Food Process. Technol. 7, 565. https://doi.org/10.4172/2157-7110.1000565.

Dhumal, S.S., Karale, A.R., Jadhav, S.B., Kad, V.P., 2014. Recent advances and the developments in the pomegranate processing and utilization: a review. J. Agric. Crop Sci. 1 (1), 1–17.

Du, C.T., Wang, P.L., Francis, F.J., 1975. Anthocyanins of pomegranate (*Punica granatum* L.). J. Food Sci. 40, 417–418.

Eikani, M.H., Golmohammad, F., Homami, S.S., 2012. Extraction of pomegranate (*Punica granatum* L.) seed oil using superheated hexane. Food Bioprod. Process. 90, 32–36.

Elfalleh, W., Tlili, N., Nasri, N., Yahia, Y., Hannachi, H., Chaira, N., … Ferchichi, A., 2011. Antioxidant capacities of phenolic compounds and tocopherols from Tunisian pomegranate (*Punica granatum*) fruits. J. Food Sci. 76 (5), C707–C713.

Elfalleh, W., Hannachi, H., Tlili, N., Yahia, Y., Nasri, N., et al., 2012. Total phenolic contents and antioxidant activities of pomegranate peel, seed, leaf and flower. J. Med. Plant Res. 6, 4724–4730.

Ergun, M., Ergun, N., 2009. Maintaining quality of minimally processed pomegranate arils by honey treatments. Br. Food J. 111 (4), 396–406.

Ewaida, E.H., 1987. Nutrient composition of Taifi pomegranate (*Punica granatum* L.) fragments and their suitability for the production of jam. Persian Gulf Sci. Res. Agric. Biol. Sci. 3, 367–378.

Fadavi, A., Barzegar, M., Azizi, H.M., 2006. Determination of fatty acids and total lipid content in oilseed of 25 pomegranates varieties grown in Iran. J. Food Compos. Anal. 19 (6–7), 676–680.

Fahan, A., 1976. The seed. In: Plant Anatomy. Hakkibutz hameuhad Publication, Jerusaleum, pp. 419–430.

Farag, R.S., Latif, M.S.A., Emam, S.S., Tawfeek, L.S., 2014. Phytochemical screening and polyphenol constituents of pomegranate peels and leave juices. Arch. Agron. Soil Sci. 1 (6), 86–93.

Fawole, O.A., Opara, U.L., 2013a. Developmental changes in maturity indices of pomegranate fruit: a descriptive review. Sci. Hortic. 159, 152–161.

Fawole, O.A., Opara, U.L., 2013b. Effects of storage temperature and duration on physiological responses of pomegranate fruit. Ind. Crop. Prod. 47, 300–309.

Fischer, U.A., Carle, R., Kammerer, D.R., 2011. Identification and quantification of phenolic compounds from pomegranate (*Punica granatum* L.) peel, mesocarp, aril and differently produced juices by HPLC-DAD-ESI/MSn. Food Chem. 127 (2), 807–821.

Fischer, U.A., Jaksch, A.V., Carle, R., Kammerer, D.R., 2013. Influence of origin source, different fruit tissue and juice extraction methods on anthocyanin, phenolic acid, hydrolysable tannin and isolariciresinol contents of pomegranate (*Punica granatum* L.) fruits and juices. Eur. Food Res. Technol. 237, 209–221.

Gil, M.I., Artes, F., Tomas-Barberan, F.A., 1996a. Minimal processing and modified atmosphere packaging effects on pigmentation of pomegranate seeds. J. Food Sci. 61, 161–164.

Gil, M.I., Martínez, J.A., Artés, F., 1996b. Minimally processed pomegranate seeds. Lebensm. Wiss. Technol. 29, 708.

Gil, M.I., Sánchez, R., Marin, J.G., Artés, F., 1996c. Quality changes in pomegranates during ripening and cold storage. Z. Lebensm. Unters. Forsch. 202, 481.

Gil, M.I., Tomas-Barberan, F.A., Hess-Pierce, B., Holcroft, D.M., Kader, A.A., 2000. Antioxidant activity of pomegranate juice and its relationship with phenolic composition and processing. J. Agric. Food Chem. 48, 4581–4589.

Gunaseelan, V.N., 2004. Biochemical methane potential of fruits and vegetable solid waste feedstocks. Biomass Bioenergy 26 (4), 389–399.

Guzel, F., Aksoy, O., Akkaya, G., 2014. Evaluation of Pomegranate (*Punica granatum* L.) pulps for the removal of copper(II) ions: kinetic, equilibrium, and desorption studies. J. Dispers. Sci. Technol. 35 (4), 482–493.

Hasnaoui, N., Wathelet, B., Jimenez-Araujo, A., 2014. Valorization of pomegranate peel from 12 cultivars: dietary fibre composition, antioxidant capacity and functional properties. Food Chem. 160, 196–203.

Hayrapetyan, H., Hazeleger, W.C., Beumer, R.R., 2012. Inhibition of *Listeria monocytogenes* by pomegranate (*Punica granatum*) peel extract in meat pate at different temperatures. Food Control 23 (1), 66–72. https://doi.org/10.1016/j.foodcont.2011.06.012.

Hess-Pierce, B.M., Kader, A.A., 1997. Carbon dioxide-enriched atmosphere extend postharvest life of pomegranate arils. In: Proceedings of CA-97 Program and Abstracts of University of California, USA, p. 135.

Hess-Pierce, B.M., Kader, A.A., 2003. Responses of 'Wonderful' pomegranates to controlled atmospheres. Acta Hortic. 600, 751.

Holland, D., Hatib, K., Bar-Ya'akov, I., 2009. Pomegranate: botany, horticulture, breeding. In: Janick, J. (Ed.), Horticultural Reviews. vol. 35. John Wiley and Sons, New Jersey, pp. 127–191.

Ibarz, A., Pagan, J., Garza, S., 1999. Kinetic models for colour changes in pear puree during heating at relatively high temperatures. J. Food Eng. 39 (4), 415–422.

IPGRI, 2001. Regional Report CWANA 1999–2000. International Plant Genetic Resources Institute, Rome, pp. 20–28.

Iqbal, S., Haleem, S., Akhtar, M., Zia-ul-Haq, M., Akbar, J., 2008. Efficiency of pomegranate peel extracts in stabilization of sunflower oil under accelerated conditions. Food Res. Int. 41 (2), 194–200. https://doi.org/10.1016/j.foodres.2007.11.005.

Ismail, T., Piero, S., Saeed, A., 2012. Pomegranate peel and fruit extracts: a review of potential anti-inflammatory and anti-infective effects. J. Ethnopharmacol. 143, 397–405. https://doi.org/10.1016/j.jep.2012.07.004.

Ismail, T., Akhtar, S., Riaz, M., Ismail, A., 2014. Effect of pomegranate peel supplementation on nutritional, organoleptic and stability properties of cookies. Int. J. Food Sci. Nutr. 65 (6), 661–666.

Jana, A., Maity, C., Halder, S.K., Mondal, K.C., Pati, B.R., Das Mohapatra, P.K., 2012. Tannase production by Penicillium purpurogenum PAF6 in solid state fermentation of tannin-rich plant residues following OVAT and RSM. Appl. Biochem. Biotechnol. 167 (5), 1254–1269. https://doi.org/10.1007/s12010-012-9547-5.

Joseph, M.M., Aravind, S.R., Varghese, S., Mini, S., Sreelekha, T.T., 2012. Evaluation of antioxidant, antitumor and immunomodulatory properties of polysaccharide isolated from fruit rind of *Punica granatum*. Mol. Med. Rep. 5 (2), 489–496.

Jurenka, J.S., 2008. Therapeutic applications of pomegranate (Punicagranatum L.): a review. Altern. Med. Rev. 13 (2), 128–144.

Kader, A.A., 2006. Postharvest biology and technology of pomegranates. In: Seeram, N.P., Schulman, R.N., Heber, D. (Eds.), Pomegranates: Ancient Roots to Modern Medicine. CRC Press, Boca Raton, FL, pp. 211–220.

Kader, A.A., Chordas, A., Elyatem, S., 1984. Responses of pomegranates to ethylene treatment and storage temperature. Calif. Agric. 38, 14–15.

Kanatt, S.R., Chander, R., Sharma, A., 2010. Antioxidant and antimicrobial activity of pomegranate peel extract improves the shelf life of chicken products. Int. J. Food Sci. Technol. 45 (2), 216–222. https://doi.org/10.1111/j.13652621.2009.02124.x.

Kazemi, M., Karim, R., Mirhosseini, H., Hamid, A.A., 2016. Optimization of pulsed ultrasound-assisted technique for extraction of phenolics from pomegranate peel of Malas variety: punicalagin and hydroxybenzoic acids. Food Chem. 206, 156–166.

Khan, J.A., Hanee, S., 2011. Antibacterial properties of *Punica granatum* peels. Int. J. Appl. Biol. Pharm. Technol. 2 (3), 23–27.

Khan, M., Khan, S.S., Zafar, A., Tanveer, A., 2010. Production of single cell protein from *Saccharomyces cerevisiae* by utilizing fruit wastes. Nanobiotech. Univ. 1 (2), 127–132.

Kıralan, M., Gölükcü, M., Tokgöz, H., 2009. Oil and conjugated linolenic acid contents of seeds from important pomegranate cultivars (*Punica granatum* L.) grown in Turkey. J. Am. Oil Chem. Soc. 86 (10), 985–990.

Kleinberg, A., 2004. Pomegranates, first ed. Ten Speed Press, Berkeley, California, pp. 7–109.

Koujalagi, C.B., 2012. An Economic Analysis of Production, Marketing and Export Performance of Pomegranate in Karnataka. PhD Thesis Submitted to University of Agricultural Sciences, Dharwad, Karnataka, India. pp. 1–110.

Kulkarni, A.P., Aradhya, S.M., 2005. Chemical changes and antioxidant activity in pomegranate arils during fruit development. Food Chem. 93, 319–324.

Lansky, E., Hubert, S., Neeman, I., 1998. Pharmacological and the rapeutical properties of pomegranates. In: Proceedings of the First International Symposium on Pomegranate, Spain, pp. 231–235.

LaRue, J. 1977. Growing pomegranates. DANR Leaflet 2459: available online at http://ucce.ucdavis.edu/files/programs/5419/Growing_Pomegranates_in_California.htm.

Li, Y., Guo, C., Yang, J., Wei, J., Xu, J., Cheng, S., 2006. Evaluation of antioxidant properties of pomegranate peel extract in comparison with pomegranate pulp extract. Food Chem. 96, 254–260.

Liu, G., Xu, X., Hao, Q., Gao, Y., 2009. Supercritical CO2 extraction optimization of pomegranate (*Punica granatum* L.) seed oil using response surface methodology. LWT Food Sci. Technol. 42 (9), 1491–1495.

Lu, J., Wei, Y., Yuan, Q., 2007. Preparative separation of punicalagin from pomegranate husk by high-speed countercurrent chromatography. J. Chromatogr. B 857 (1), 175–179.

Malik, A., Afaq, F., Arfaraz, S., Adhami, V.M., Syed, D.N., 2005. Pomegranate juice for chemoprevention and chemotherapy of prostate cancer. Proc. Natl. Acad. Sci. 102, 14813–14818.

Malviya, S., Jha, A.A., Hettiarachchy, N., 2014. Antioxidant and antibacterial potential of pomegranate peel extract. J. Food Sci. Technol. 51 (12), 4132–4137.

Martos, M.V., Navajas, Y.R., Sanchez, A.M., Zapata, E.S., Lopez, J.F., et al., 2012. Chemical, physico-chemical composition, functional properties of pomegranate (*Punica granatum* L) bagasses powder co-product. J. Food Eng. 10 (2), 220–224.

Mastrodi Salgado, J.M., Baroni Ferreira, T.R., de Oliveira Biazotto, F., dos Santos Dias, C.T., 2012. Increased antioxidant content in juice enriched with dried extract of pomegranate (*Punica granatum*) peel. Plant Foods Hum. Nutr. 67 (1), 39–43. https://doi.org/10.1007/s11130-011-0264-y.

Meerts, I.A.T.M., Verspeek-Rip, C.M., Buskens, C.A.F., Keizer, H.G., Bassaganya-Riera, J., Jouni, Z.E., van Huygevoort, A.H.B.M., van Otterdijk, F.M., van de Wart, E.J., 2009. Toxicological evaluation of pomegranate seed oil. Food Chem. Toxicol. 47, 1085–1092.

Melo, I.L.P., Carvalho, E.B.T., Silva, A.M.O., ManciniFilho, J., 2010. Effects of pomegranate seed oil on lipoperoxidation and activity of antioxidant enzymes in liver and brain of rats. Free Radic. Biol. Med. 49, s189.

Mir, M.M., Umar, I., Mir, S.A., Rehman, M.U., Rather, G.H., Banday, S.A., 2012. Quality evaluation of pomegranate crop—a review. Int. J. Agric. Biol. 14, 658–667.

Mirdehghan, S.H., Rahemi, M., Castillo, S., Martinez-Romero, D., Serrano, M., Valero, D., 2007. Prestorage application of polyamines by pressure or immersion improves shelf life of pomegranate stored at chilling temperature by increasing endogenous polyamine levels. Postharvest Biol. Technol. 44, 26–33.

Moneam, N.M.A., El Sharaky, A.S., Badreldin, M.M., 1988. Oestrogen content of pomegranate of seeds. J. Chromatogr. 1988 (438), 438–442.

Muhammad, M., Bushra, S., Farooq, A., Ahmad, A., Rizviet, S., et al., 2015. Enzyme-assisted supercritical fluid extraction phenolic antioxidants from pomegranate peel. J. Supercrit. Fluids 104, 122–131.

Mushtaq, M., Sultana, B., Anwar, F., Adnan, A., Rizvi, S.S.H., 2015. Enzyme assisted supercritical fluid extraction of phenolic antioxidants from pomegranate peel. J. Supercrit. Fluids 104, 122–131.

Naik, A., Lele, S.S., 2012. Solid state fermentation of pomegranate seed for lovastatin production. A bioprocessing approach. Adv. Biosci. Biotechnol. 3, 643–647.

Naveena, B.M., Sen, A.R., Vaithiyanathan, S., Babji, Y., Kondaiah, N., 2008. Comparative efficacy of pomegranate juice, pomegranate rind powder extract and BHT as antioxidants in cooked chicken patties. Meat Sci. 80 (4), 1304–1308. https://doi.org/10.1016/j.meatsci.2008.06.005.

Naz, S., Siddiqi, R., Ahmad, S., Rasool, S.A., Sayeed, A., 2007. Antibacterial activity directed isolation of compounds from *Punica granatum*. J. Food Sci. 72 (9), M341–M345.

NRCP, 2014. Pomegranate: Cultivation, Marketing and Utilization. Technical Bulletin No. NRCP/2014/…, ICAR-National Research Centre on Pomegranate, Solapur, India.

Pagliarulo, C., De Vito, V., Picariello, G., Colicchio, R., Pastore, G., Salvatore, P., Volpe, M.G., 2016. Inhibitory effect of pomegranate (*Punica granatum* L.) polyphenol extracts on the bacterial growth and survival of clinical isolates of pathogenic *Staphylococcus aureus* and *Escherichia coli*. Food Chem. 190, 824–831.

Pan, Z., Qu, W., Ma, H., Atungulu, G.G., McHugh, T.H., 2011. Continuous and pulsed ultrasound-assisted extractions of antioxidants from pomegranate peel. Ultrason. Sonochem. 18 (5), 1249–1257.

Pan, Z., Qu, W., Ma, H., Atungulu, G.G., McHugh, T.H., 2012. Continuous and pulsed ultrasound-assisted extractions of antioxidants from pomegranate peel. Ultrason. Sonochem. 19 (2), 365–372.

Pande, G., Akoh, C.C., 2009. Antioxidant capacity and lipid characterization of six Georgia grown pomegranate cultivars. J. Agric. Food Chem. 57 (20), 9427–9436.

Pareek, S., Valero, D., Serrano, M., 2015. Postharvest biology and technology of pomegranate. J. Sci. Food Agric. 95, 2360–2379. https://doi.org/10.1002/jsfa.7069.

Park, H.M., Moon, E., Kim, A.J., Kim, M.H., Lee, S., Lee, J.B., Park, Y.K., Jung, H., Kim, Y.B., Kim, S.Y., 2010. Extract of *Punica granatum* inhibits skin photoaging induced by UVB irradiation. Int. J. Dermatol. 49, 276–282.

Patil, A.V., Karale, A.R., Bose, T.K., 2002. Pomegranate. In: Bose, T.K., Metra, S.K., Sanyal, D. (Eds.), Fruits: Tropical and Subtropical. In: vol. 2. Naya Udyog, Bidhan Sarani, Calcutta, India, pp. 125–162.

Prasad, R.N., Chandra, R., Teixeira da Silva, J.A., 2010. Postharvest handling and processing of pomegranate. Fruit Veg. Cereal Sci. Biotechnol. 4 (Special Issue 2), 88–95.

Qu, W., Pan, Z., Zhang, R., Ma, H., Zhu, B., Wang, Z., Atungulu, G., 2009. Integrated extraction and anaerobic digestion process for recovery of nutraceuticals and biogas from pomegranate marc. Trans. ASABE 52 (6), 1997–2006.

Qu, W., Pan, Z., Ma, H., 2010. Extraction modeling and activities of antioxidants from pomegranate marc. J. Food Eng. 99 (1), 16–23.

Rababah, T.M., Banat, F., Rababah, A., Ereifej, K., Yang, W., 2010. Optimization of extraction conditions of total phenolics, antioxidant activities, and anthocyanin of oregano, thyme, terebinth, and pomegranate. J. Food Sci. 75 (7), C626–C632.

Ranjbar, N., Eikani, M.H., Javanmard, M., Golmohammad, F., 2016. Impact of instant controlled pressure drop on phenolic compounds extraction from pomegranate peel. Innov. Food Sci. Emerg. Technol. 37, 177–183.

Robledo, A., Aguilera-Carbo, A., Rodriguez, R., Martinez, J.L., Garza, Y., Aguilar, C.N., 2008. Ellagic acid production by Aspergillus Niger in solid state fermentation of pomegranate residues. J. Ind. Microbiol. Biotechnol. 35 (6), 507513. https://doi.org/10.1007/s10295-008-0309-x.

Rodov, V., Schimilovitch, Z., Ronen, B., Hoffman, A., Egozi, H., Porat, R., Goldman, G., Horev, B., Weiss, B., Vinokur, Y., Shomer, I., Holland, D., 2005. Mechanically separated pomegranate arils: a new lightly processed fresh product. In: Proceedings of the Fresh-Cut Expo Phoenix, 14–16 April, AZ, p. 11.

Rowayshed, G., Salama, A., Fadl, M.A., Hamza, S.A., Emad, A.M., 2013. Nutritional and chemical evaluation for pomegranate (*Punica granatum* L.) fruit peel and seeds powders by products. Middle East J. Appl. Sci. 3 (4), 169–179.

Sadeghi, N., Jannat, B., Oveisi, M.R., Hajimahmoodi, M., Photovat, M., et al., 2009. Antioxidant activity of Iranian pomegranate (*Punica granatum* L.) seed extracts. J. Agric. Sci. Technol. 11, 633–638.

Sassano, G., Sanderson, P., Franx, J., Groot, P., van Straalen, J., Bassaganya-Riera, J., 2009. Analysis of pomegranate seed oil for the presence of jacaric acid. J. Sci. Food Agric. 89 (6), 1046–1052.

Schubert, S.Y., Lansky, E.P., Neeman, I., 1999. Antioxidant and eicosanoid enzyme inhibition properties of pomegranate seed oil and fermented juice flavonoids. J. Ethnopharmacol. 66, 11–17.

Schwartz, E., Glazer, I., Bar-Ya'akov, I., Matityahu, I., Bar-Ilan, I., Holland, D., et al., 2009. Changes in chemical constituents during the maturation and ripening of two commercially important pomegranate accessions. Food Chem. 115, 965–973.

Seeram, N.P., Schullman, R.N., Heber, D., 2006. Commercialization of pomegranates: fresh fruit, beverages and botanical extracts. In: Seeram, N.P., Schullman, R.N., Heber, D. (Eds.), Pomegranates Ancient Roots to Modern Medicine. CRC Press Taylor and Francis Group, Boca Raton, Florida, pp. 187–196.

Seeram, N.P., Aronson, W.J., Zhang, Y., Henning, S.M., Moro, A., Lee, R.-P., … Heber, D., 2007. Pomegranate ellagitannin-derived metabolites inhibit prostate cancer growth and localize to the mouse prostate gland. J. Agric. Food Chem. 55 (19), 7732–7737.

Seraman, S., Rajendran, A., Thangavelu, V., 2010. Statistical optimization of anticholesterolemic drug lovastatin production by the red mold Monascus purpureus. Food Bioprod. Process. 88 (2–3), 266–276.

Sharma, M., Li, L., Celver, J., Killian, C., Kovoor, A., Seeram, N.P., 2010. Effects of ellagitannin extracts, ellagic acid, and their colonic metabolite, urolithin A, on Wnt signalling. J. Agric. Food Chem. 58, 3965–3969.

Sherafatian, D., 1994. The effect of harvesting date and temperature during storage on keeping quality of pomegranate. Seed Plant 10, 25–34.

Shmilovich, Z., Sarig, Y., Ronen, B., Horman, A., Egozi, H., Beres, H., Bar-Lev, E., Groz, F., 2006. Development of a method and system for extracting the seeds (arils) from pomegranate fruits: concept to commercial utilization. Acta Hortic. 818, 256–258.

Singh, R.P., Gupta, A.K., Bhatia, A.K., 1990. Utilization of wild pomegranate in north west Himalayas-status and problems. In: Proceedings of National Seminar on Production and Marketing of Indigenous Fruits New Delhi, pp. 100–107.

Singh, R.P., Murthy, K.N.C., Jayaprakasha, G.K., 2002. Studies on the antioxidant activity of pomegranate (*Punica granatum*) peel and seed extracts using in vitro models. J. Agric. Food Chem. 50 (1), 81–86.

Srivastava, A., Kar, R., 2009. Characterization and application of tannase produced by Aspergillus niger ITCC 6514.07 on pomegranate rind. Braz. J. Microbiol. 40 (4), 782–789.

Srivastava, P., Indrani, D., Singh, R.P., 2014. Effect of dried pomegranate (*Punica granatum*) peel powder (DPPP) on textural, organoleptic and nutritional characteristics of biscuits. Int. J. Food Sci. Nutr. 65 (7), 827–833.

Sumner, M.D., Elliott-Eller, M., Weidner, G., Daubenmier, J.J., Chew, M.H., Marlin, R., 2005. Effects of pomegranate juice consumption on myocardial perfusion in patients with coronary heart disease. Am. J. Cardiol. 96, 810–814.

Supayang, P.V., Treechada, S., Surasak, L., Thanomjit, S., Tetsuya, I., et al., 2005. Inhibitory effects of active compounds from Punicagranatum pericarp on verocytotoxin production by Enterohemorrhagic *Escherichia coli* O157:H7. J. Health Sci. 51 (5), 590–596.

Tabaraki, R., Heidarizadi, E., Benvidi, A., 2012. Optimization of ultrasound- assisted extraction of pomegranate (Punicagranatum L) peel antioxidants by response surface methodology. Sep. Purif. Technol. 98, 16–23.

Teh, H.E. 2016. Extraction and Characterization of Functional Components From Fruit and Vegetable Processing Waste. PhD Thesis Submitted to University of California, Davis. 234 p.

Tong, P., Kasuga, Y., Khoo, C.S., 2006. Liquid chromatographic mass spectrometric method for detection of estrogen in commercial oils and in fruit seed oils. J. Food Compos. Anal. 19, 150–156.

Tsuda, T., Watanabe, M., Ohshima, K., Norinobu, S., Choi, S., et al., 1994. Antioxidant activity of the anthocuanin pigments cyaniding 3-O-ß-D glucoside and cyaniding. J. Agric. Food Chem. 42, 2407–2410.

Uma, C., Gomathi, D., Muthulakshmi, C., Gopalakrishnan, V.K., 2010. Production, purification and characterization of invertase by Aspergillus flavus using fruit peel waste as substrate. Adv. Biol. Res. 4 (1), 31–36.

Uma, C., Gomathi, D., Ravikumar, G., Kalaiselvi, M., Palaniswamy, M., 2012. Production and properties of invertase from a Cladosporium cladosporioides in SmF using pomegranate peel waste as substrate. Asian Pac. J. Trop. Biomed. 2 (2), S605–S611.

Viuda-Martos, M., Fernandez-Lopez, J., Perez-Alvarez, J.A., 2010. Pomegranate and its many functional components as related to human health: a review. Compr. Rev. Food Sci. Food Saf. 9, 635–654.

Viuda-Martos, M., Ruiz-Navajas, Y., Martin-Sanchez, A., Sanchez-Zapata, E., Fernandez-Lopez, J., Sendra, E., … Perez-Alvarez, J.A., 2012. Chemical, physico-chemical and functional properties of pomegranate (*Punica granatum* L.) bagasses powder co-product. J. Food Eng. 110 (2), 220–224.

Wu, J., Jahncke, M.L., Eifert, J.D., O'Keefe, S.F., Welbaum, G.E., 2016. Pomegranate peel (*Punica granatum* L) extract and Chinese gall (Galla chinensis) extract inhibit *Vibrio parahaemolyticus* and *Listeria monocytogenes* on cooked shrimp and raw tuna. Food Control 59, 695–699.

Xi, J., He, L., Yan, L.G., 2017. Continuous extraction of phenolic compounds from pomegranate peel using high voltage electrical discharge. Food Chem. 230, 354–361.

Yamasaki, M., Kitagawa, T., Koyanagi, N., Chujo, H., Maeda, H., KohnoMurase, J., Imamura, J., Tachibana, H., Yamada, K., 2006. Dietary effect of pomegranate seed oil on immune function and lipid metabolism in mice. Nutrition 22, 54–59.

Zaki, S.A., Abdelatif, S.H., Abdelmohsen, N.R., Ismail, F.A., 2015. Phenolic compounds and antioxidant activities of pomegranate peels. Int. J. Food Eng. 1 (2), 73–76.

Further Reading

D'Aquino, S., Palma, A., Schirra, M., Continella, A., Tribulato, E., et al., 2010. Influence of film wrapping and fludioxonil application on quality of pomegranate fruit. Postharvest Biol. Technol. 55, 121–128.

Fernandes, L., Pereira, J.A., Lopéz-Cortés, I., Salazar, D.M., Ramalhosa, E., Casal, S., 2015. Fatty acid, vitamin E and sterols composition of seed oils from nine different pomegranate (*Punica granatum* L.) cultivars grown in Spain. J. Food Compos. Anal. 39, 13–22.

CHAPTER 9
Citrus

Yike Chen, Tyler J. Barzee, Ruihong Zhang, Zhongli Pan
Department of Biological and Agricultural Engineering, University of California - Davis, Davis, CA, United States

Chapter Outline

1 CITRUS INTRODUCTION

Citrus is an important agricultural crop among the fruit sector due to its high annual consumption and trade. It is widely accepted that citrus originated in southeast Asia and was spread to Africa thousands of years ago through trading and migration (Gmitter and Hu, 1990). Citrus was then introduced to southern Europe during the Roman Empire as a symbol of the noble society. During the Middle Ages, citrus adapted quickly to the climates of the Mediterranean regions and was brought to the Americas by Spanish explorers (Webber, 1967). Nowadays, citrus is widely cultivated in over 140 countries throughout the tropical and subtropical regions across the world (Baldwin, 1993; Ladaniya, 2008). The orange colored regions in Fig. 1 depict the major citrus growing areas in the world.

Citrus fruits are one of the most popular fruit commodities because of their unique, refreshing flavor, as well as their nutritional values. In addition to oranges, tangerines, lemons, limes, and grapefruit, other citrus hybrids are also commonly available in the market, such as kumquats, citrons, and pomelos. As shown in Fig. 2, the total world citrus production in 2017 accounted for the largest single category of fruit production with over 146 million tons, followed by

Integrated Processing Technologies for Food and Agricultural By-Products
https://doi.org/10.1016/B978-0-12-814138-0.00009-5

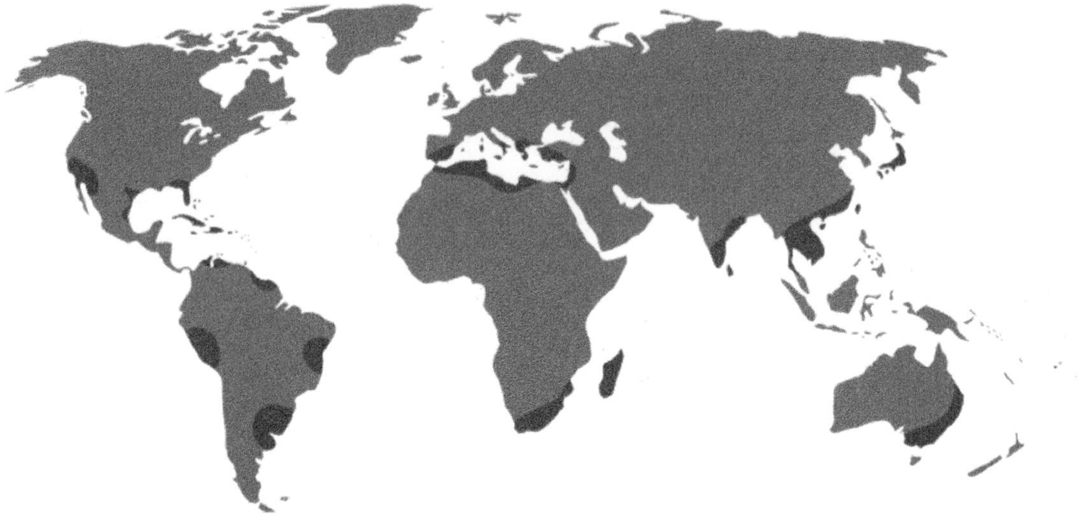

FIG. 1
Major producer regions for citrus fruits in the world (Wikimedia Commons, 2010).

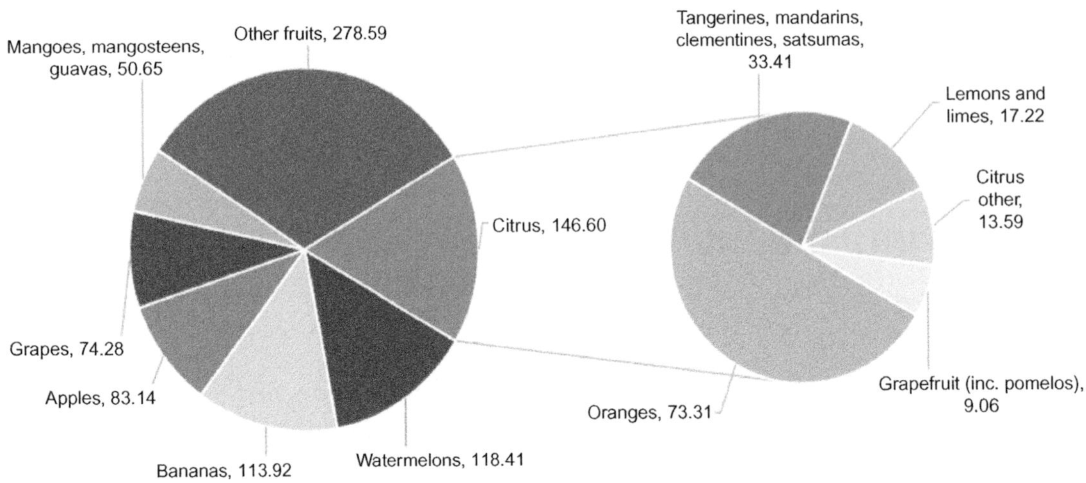

FIG. 2
Global fruit production by variety in 2017 (in million tons) (FAO, 2019).

watermelons, bananas, and apples. Oranges make up about half of the total citrus production worldwide, and one-third of which are processed into juice with the rest being marketed as whole fruit (FAO, 2017).

Brazil and the United States are the two major citrus producers in America. As shown in Table 1, the majority of the citrus harvested in these two countries are oranges. For decades, Brazil led the world in citrus production, especially oranges, because of its favorable weather conditions (Ladaniya, 2008). The

Table 1 Varieties of Citrus Fruits Production in China, Brazil, and the United States (FAO, 2017)

Variety	China		Brazil		United States	
	Production (MT[a])	Percentage[b] (%)	Production (MT[a])	Percentage[b] (%)	Production (MT[a])	Percentage[b] (%)
Oranges	8.69	22	17.50	88	4.62	66
Tangerines	18.19	46	0.97	5	0.94	13
Lemons and limes	2.35	6	1.29	7	0.80	11
Grapefruit	4.73	12	0.08	0	0.63	9
Citrus, other	5.35	14	0	0	0.01	0

[a]MT: million tons

[b]Percentage of each country's total citrus production.

annual citrus fruit production in Brazil is currently around 20 million tons with oranges accounting for 88% (17.5 million tons) of its total citrus production (FAO, 2019). It is estimated that over 60% of the oranges produced in Brazil are used for processing, making it the largest orange juice exporter in the world accounting for three-quarters of the total global export market (USDA, 2019). In the United States, Florida and California have been growing citrus for more than 250 years (Liu et al., 2012; Tobey and Wetherell, 1995). In the 1920s, railroad expansions allowed commercial orange farms and industrial-scale orange processing plants to be built in Florida and California, leading to a rapid growth of the citrus market in the US (Webber et al., 2014). Currently, Arizona, Texas, Hawaii, Louisiana, and Washington all produce citrus; but Florida and California remain the two leading citrus producing states in the US (USDA, 2019). Most of the citrus grown in Florida is used for juice processing. In 2017, 96% of the oranges, 54% of the grapefruit, and 28% of the tangerines produced in Florida were processed into citrus juice (Florida Department of Citrus, 2018). On the other hand, a high portion of California's citrus is packed for the fresh market. Around 80% of the oranges and grapefruit, 75% of the lemons, and 70% of the mandarins harvested in California were sent for direct consumption in 2017 (Babcock, 2018).

China's citrus production has been rapidly increasing, reaching around 40 million tons in 2017 (Fig. 3), with total citrus production about double that of Brazil, the next largest citrus producer. The main varieties of China's citrus products are tangerines, oranges, and grapefruits (Table 1). Around 90% of the citrus in China is marketed as fresh produce, leaving the rest 10% for canned fruit production (Shan, 2016; Spreen et al., 2012). Other citrus producers in Asia include India and Japan, with India surpassing the United States, Mexico, Argentina, and Europe in total citrus production in 2017 (Fig. 3).

Citrus produced in Mediterranean countries is currently targeted for fresh consumption. Less than 10% of the total citrus production is utilized for processing

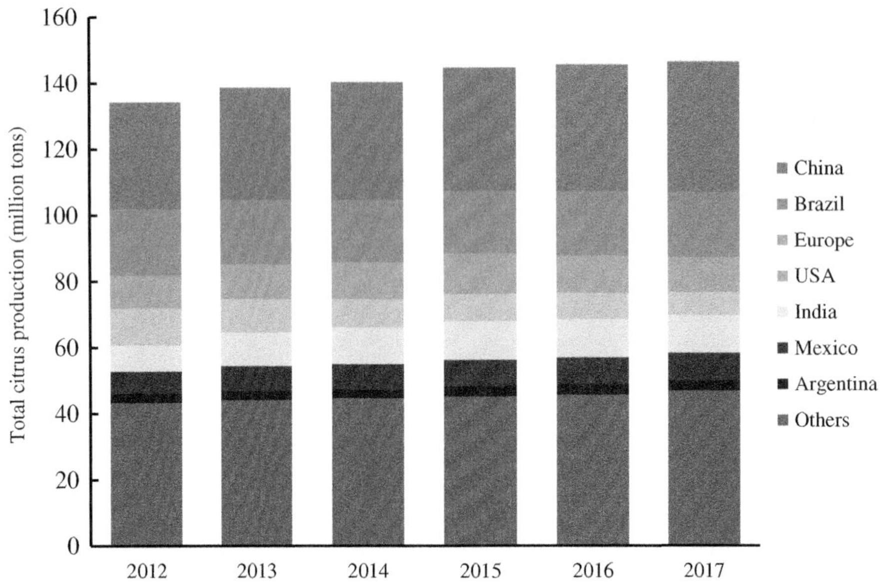

FIG. 3
Citrus production by country/region in the past 5 years (FAO, 2017).

(FAO, 2017). On average, the total annual citrus production in the Mediterranean area is 10 million tons. Spain, Portugal, and Morocco are the three main citrus producing countries in the Mediterranean due to their favorable weather for flowering and fruit ripening (USDA, 2019).

2 CITRUS HARVESTING

Depending on weather conditions, citrus fruits take a year or more to ripen. Pruning shears or pulling forces are required to harvest the citrus fruits before natural fruit detachment occurs (Ladaniya, 2008). Unlike many other fruits, citrus fruits cease ripening upon detachment from the tree (Kader, 1999). As a result, precise timing of the harvesting of citrus fruits is of great importance. Harvesting of citrus fruits is mainly accomplished through manual and mechanical means throughout the world. It was estimated that harvesting contributes 35%–45% of the total operational cost of citrus production (Sanders, 2005).

2.1 Manual Harvesting

Manual harvesting is a traditional and common method to harvest fruits. Although it is labor intensive, manual harvesting is particularly popular for fruits that have large time windows for optimal maturity or for fruits that are marketed for direct consumption (Sanders, 2005). Historically, manual harvesting has been the preferred method for achieving high-quality control and minimizing tree damage (Benkeblia et al., 2011). Also, manual harvesting is sometimes carried out as a clean-up operation either after the mechanical harvesting or at the end of the harvesting season to pick up the fruits that are left behind (Sanders, 2005).

The major cost concern for manual harvesting is associated with labor. Singerman et al. (2017) performed a survey of 15 harvesting companies that were responsible for harvesting 18% of the total citrus cultivation area in Florida (79,996 acres) to determine the costs associated with manually harvesting various citrus varieties. The average harvesting costs for Florida fresh citrus in 2017 were estimated to be $3.28, $2.65, and $4.46 per box for sweet oranges, grapefruit, and specialty citrus, respectively. The high costs for harvesting specialty citrus fruits, such as tangelos and tangerines, are due to the required extra caution and labor process. The procedure requires workers to clip, pick, and handle these specific fruits carefully due to their thin and easily damageable skin. Citrus in Spain is also harvested manually. Researchers estimated that manual harvesting costs accounted for 29% and 43% of the total direct production cost of oranges and mandarins in Andalucia, respectively (Moreno et al., 2015). Further, researchers from Brazil also showed the labor costs during harvesting represented 44% of the total citrus production cost (Costa and Camarotto, 2012).

Relatively low harvesting efficiency is a crucial problem for manually harvested citrus orchards. Workers normally need to climb up and down the trees with a ladder to pick the fruits that are then delivered to a central packaging facility. In certain cases, time spent on activities other than picking may account for more than 50% of the total working time (Ehsani and Udumala, 2010). Harvesting aids were developed to increase picking efficiency and reduce the time spent in non-picking activities during manual harvesting. For instance, multi-picker positioning platforms were designed to allow groups of pickers to work simultaneously on different positions of a citrus tree (Ehsani and Udumala, 2010). Several companies in the US developed and demonstrated their picking platforms for research in the 1970s. However, none of them proved to be economically feasible because the small improvement in picking rates often did not justify the capital cost of the equipment (Sanders, 2005). Therefore, manual harvesting of citrus is still limited by the comparatively low harvesting rates.

Moving forward, other issues related to decreased labor availability and worker safety concerns may work to increase the prevalence of mechanical citrus harvesting methods due to their ability to minimize on-site worker needs and safety concerns. Reduced harvesting costs are attainable by utilizing an appropriate and efficient mechanical harvesting system (Sanders, 2005).

2.2 Mechanical Harvesting

Mechanical harvesters were originally developed to minimize the production cost while increasing the harvest efficiency. Extensive research and development were initiated in Florida in 1994 to evaluate the available technologies and provide cost estimation by applying these mechanical harvesting methods. If citrus growers opt to use mechanical harvesters provided by third-party contractors, they can save at least 50% or more of the total harvesting costs, compared to the price of hiring manual labor. Predicted reduction in harvesting expenses can be contributed towards the increase in labor productivity, which is estimated to be 5–15 times more efficient than manual harvesting (Brown, 2005). Mechanical harvesting can be divided into four techniques based on their harvesting mechanisms: air shaking, trunk shaking, limb shaking, and canopy shaking (Sanders, 2005).

2.2.1 AIR SHAKING

Air shaking uses a blast of high-speed airflow to blow off the citrus fruits. Whitney (1968) first developed and evaluated the concept of the air shaking technique. The test was carried out over a 4-year period, and the overall fruit removal efficiencies were concluded to be lower than the acceptable level, within a range of 39.3%–84.7%, depending on the maturity and variety of the citrus fruits.

Later on, developments in understanding of plant physiology allowed for the development of chemical abscission technology, which improved fruit removal efficiency due to a strong dependency on the fruit bonding strength when used in combination with air shaking. Researchers found that chemicals, such as ascorbic acid, ethephon, and 2-chloroethyl phosphoric acid, could trigger the enzymatic abscission process of citrus fruits (Cooper and Henry, 1971; Li et al., 2011). Subsequent research conducted by Whitney (1977) concluded a 90%–95% of the fruit removal efficiency using the air shaking method with abscission chemicals. Although abscission chemicals could increase the fruit removal efficiency by reducing the fruit bonding force, the usage of these chemicals has been regulated due to potential food safety and health concerns (Sanders, 2005). For example, in the US, all the abscission chemicals need to be tested and registered through the US Environmental Protection Agency (Blanco and Roka, 2009).

2.2.2 TRUNK AND LIMB SHAKING

Trunk shakers grip the trunk of the citrus trees right above the ground and shake the whole tree using a vibrating machine. In the 1970s, trunk shaking technology was first demonstrated and studied in Florida for orange harvesting (Whitney and Wheaton, 1987). The key parameters that are crucial to the harvest efficiency include shaking frequency, amplitude, and vibration time. Research found that shaking frequencies between 6 and 10 Hz with a high amplitude of 50 mm gained a higher fruit removal efficiency than frequencies between 15 and 18 Hz combined with a low amplitude of 30 mm (Whitney et al., 2000). The field study showed that most of the fruits get detached within the first 2–3 s of shaking (Ortiz and Torregrosa, 2013).

Similar to trunk shaking, limb shaking applies forces directly to individual branches. Fruit removal efficiency during limb shaking was studied to determine the optimal shaking amplitude and frequency for lime fruit harvesting

(Loghavi and Mohseni, 2006). A 98.5% harvest efficiency was obtained at 80 mm amplitude and 10 Hz frequency with negligible leaf shattering.

Abscission chemicals were also evaluated to determine their effectiveness in enhancing the removal efficiency of trees subjected to both trunk and limb shaking. Generally speaking, abscission chemicals can increase the harvest efficiency by 5%–15% with a reduction in fruit bonding force of 50%–80% (Whitney et al., 2000). Moreno et al. (2015) studied the effect of trunk shakers on fruit detachment while using ethephon as an abscission chemical during a 3-year trial. The application of ethephon did not show significant effects on the fruits' quality and trees' health; however, the project demonstrated an increase in removal efficiency by an average of 13% relative to those without abscission chemicals.

2.2.3 *CANOPY SHAKING*

Unlike trunk and limb shaking, canopy shakers apply shear force by inserting vibrating horizontal rods into the tree canopy instead of gripping the tree body. The average fruit removal efficiency of canopy harvesting ranges from 80% to 90% without the use of abscission chemicals, which makes it superior to the other mechanical harvesting methods due to the cost of chemicals and related environmental and health concerns (Liu et al., 2017; Sanders, 2005). However, because of the physical contact between the rods and the tree canopy, canopy shaking can sometimes cause damage to the fruit and tree (Hannan and Burks, 2004). Canopy shaking is currently the most advanced and commercially available system among the four mechanical harvesting techniques reviewed. A substantial number of studies have been conducted to understand and optimize the abscission pattern of canopy harvesting (Ehsani and Udumala, 2010; Gupta et al., 2016; Savary et al., 2010). Additional research for canopy shakers, consisting of different rod shapes and materials, was performed to increase fruit removal efficiencies and decrease damage (Aragon-Rodriguez et al., 2019; Liu et al., 2017; Pu et al., 2018). Computer simulations have also been utilized to examine the effects of frequency and penetrating depth on the detachment of fruit. Finite element models were applied to find the optimal operation conditions (shaking frequency and amplitude) and design parameters (rod configuration and stiffness) (Gupta et al., 2016; Liu et al., 2018).

2.3 Robotic Harvesting

As mentioned above, different mechanical harvesting techniques have been developed to reduce labor dependency and increase harvesting efficiency. Although all mechanical harvesters are able to enhance the citrus harvesting efficiency rate, they lack the ability to differentiate the size and quality of the fruits. Also, mechanical harvesting could severely damage the fruits, making it less desirable by vendors in the fresh fruit market (Sanders, 2005). Robotic harvesting machines have been designed and tested to overcome the identified barriers of manual and mechanical harvesting techniques. The main task for robotic harvesting is to identify the fruit in a 3-D coordinate system and control the manipulator to retrieve the fruit from the tree (Mehta and Burks, 2014). The whole system can also be divided into three phases: including two-dimensional (2D) processing, three-dimensional (3D) processing, and path optimization (Plebe and Grasso, 2001). A machine vision system, consisting of a color CCD camera and a master computer, was developed to provide fast and accurate identification of matured fruits (Lü et al., 2014). This system was able to recognize 92.4% of the citrus fruit and the branches thicker than 5 pixels (about 5.9 mm) in diameter. Although robotic harvesting is a promising technology for the citrus industry, it still needs a lot of research and development, mainly because of its relatively low reliability and high cost (Ladaniya, 2008).

3 CITRUS PROCESSING

After harvesting, citrus fruits are either consumed as fresh or processed for juice. The majority of processed citrus fruits are oranges, accounting for around 80% of the total processed citrus (Fig. 4). Orange juice production utilizes approximately one-third of the annual world orange production (FAO, 2017). Brazil and the United States are the two leading countries for orange juice production (FAO, 2017). Depending on the processes used, the juice can be categorized as "Not From Concentrate" juice or "From Concentrate" juice. Both kinds of juice are known as "100% juice," which is also called "single-strength juice" (Berk, 2016). The first commercial canned orange juice facility was built in the early 20th century in California, and the juice industry has been thoroughly developed in order to meet the fast-growing demand of juice consumption (Berk, 2016).

Berk (2016) reviewed the citrus processing industry which can be further summarized in four major steps. The first step after harvesting is transporting fresh

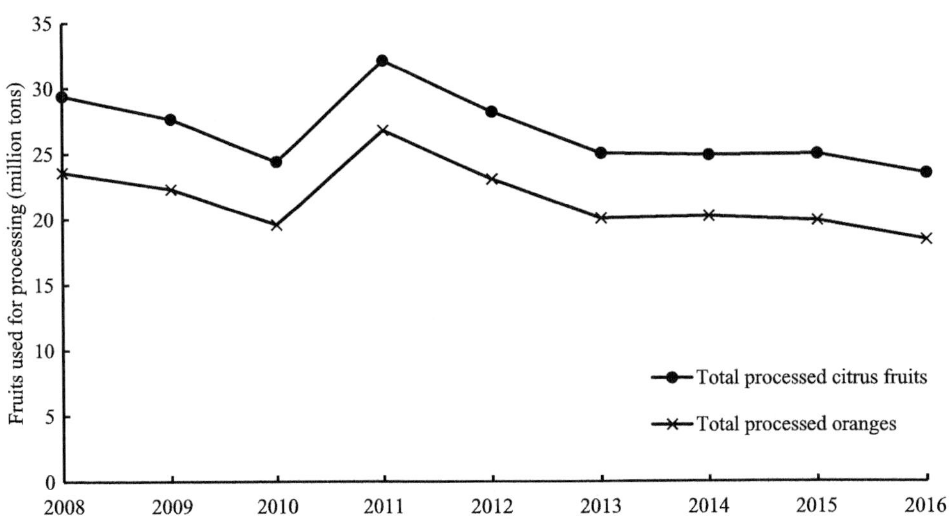

FIG. 4
The amount of citrus fruits and oranges being processed in the world (FAO, 2017).

fruit to the production facility. After arriving at the processing plant, fruits are graded, and debris is removed. The fruit is then washed in brush washers and sorted out by size. The pre-sorting step is essential for later juice extraction processes because different operating conditions are applied to fruits with different sizes (Crupi and Rispoli, 2002). The third step is the main step of juice production. This phase starts with juice extraction by pressing or reaming. Some machines cut the fruit into halves before the process of juice extraction, while others accept the whole fruit (Berk, 2016). After juice extraction, raw juice is chilled to a low temperature to reduce microbial activities and is screened to separate the pulp and other residual solids in the juice (Zema et al., 2018). Deaeration is also applied to strip oxygen from the liquid stream to prevent oxidation of vitamin C, degradation of aromatic compounds, and introduction of unfavorable flavors (Sandhu and Minhas, 2006). The juice is then packaged or further concentrated to produce juice concentrate. Concentrated juice can be stored for longer periods of time with less volume, however, there are concerns of it being "less healthy" than non-concentrated juice by consumers (Berk, 2016). Burdurlu et al. (2006) showed the vitamin C degradation in citrus juice concentrates followed a first-order kinetic model over 8-week storage. The final step of citrus juice processing is pasteurization and packaging. Sometimes, citric acid, enzymes, and vitamin C are added before pasteurization to enrich the flavor and stabilize the final product (Berk, 2016). The conventional industrial pasteurization practice for juice is to heat it to 90–99°C and hold for 15–30 s (Braddock, 1999). Besides thermal pasteurization, pulsed electric field (PEF) technology was studied as an alternative pasteurization method by many researchers (Cserhalmi et al., 2006; Yeom et al., 2000). Similar microbial and enzyme inactivation was achieved by the PEF method with a comparable or even better juice quality. Finally, the juice is filled into individual packages, labeled, and transported to market (Burdurlu et al., 2006). Fig. 5 shows the process diagram of a typical citrus juice processing plant.

4 CITRUS BY-PRODUCTS

After juice extraction, the residual solids are considered to be citrus by-products, which constitute around 50% of the whole fruit weight (Chavan et al., 2018). Citrus by-products contain the peel, pulp, rag, and seeds and are isolated by solid liquid separation during the juice production process (Fig. 6). Although citrus by-products are sometimes called citrus waste, they still contain large amounts of valuable compounds, such as fiber, protein, pectin, polyphenols and essential oils. The utilization of these by-products could potentially create more value-added products to the citrus processing industry. Table 2 shows the composition of typical orange waste after juice extraction.

Although the composition of other citrus by-products greatly depends on the juice extraction processes, the citrus variety, climate, and location, the proximate compositions are close to each other. Other types of citrus by-products also have

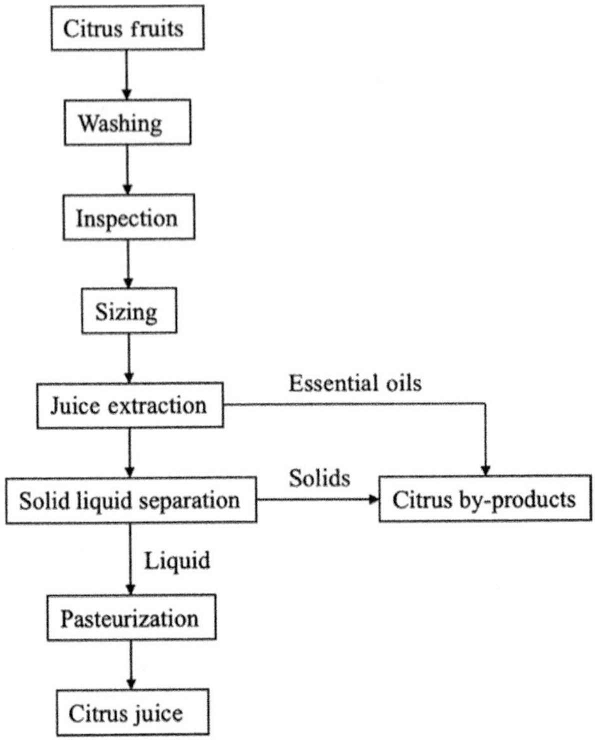

FIG. 5
Process diagram of citrus juice production.

FIG. 6
Orange by-products after juice extraction (left: orange peels; right: orange pomace) (Zhang et al., 2015).

Table 2 Composition of Orange By-Products (Pourbafrani et al., 2010)

Components	Composition (% dry basis)
Total solids	20±0.8
Sugar	
Glucose	8.10±0.46
Fructose	12.00±0.21
Sucrose	2.80±0.15
Pectin	25.00±1.20
Protein	6.07±0.10
Fiber	
Cellulose	22.00±1.95
Hemicellulose	11.09±0.21
Lignin	2.19±0.04
Limonene	3.78±0.30
Ash	3.73±0.20
Total	96.76±2.39

Table 3 Proximate Composition of Citrus By-Products (in % Dry Basis)

	Lemon peels	Grapefruit waste	Kinnow-mandarin waste
Protein	9.42	7.03	5.78
Fiber	15.18	9.48	14.38
Fat	6.26	n.d.[a]	n.d.[a]
Ash	4.98	4.03	3.23
Source	Janati et al. (2012)	Sinclair (1972)	Oberoi et al. (2011)

[a]n.d., not determined.

similar compositions to orange waste. Table 3 summarizes the compositions of lemon peels, grapefruit waste, and mandarin waste.

By knowing the compositions of the citrus by-products, technologies can be developed to convert these "waste" materials into value-added products. Some of the techniques, such as converting the citrus by-products to cattle feed and extracting essential oils, have already been applied in some of the citrus waste management practices. Others are still under research and are introduced here due to their potential future applications.

4.1 Animal Feed

Currently, most of the citrus by-products are added to animal feed as an ingredient in the fresh or dried form (Bampidis and Robinson, 2006). It is still the simplest and cheapest way to dispose of large amounts of waste produced from citrus processing facilities every day (Chavan et al., 2018). The fresh citrus by-products contain high carbohydrate and low lignin, which is ideal for the digestion system of ruminant animals (Oluremi et al., 2006). The fibrous content is believed to help improve the animal diet and meat quality (Zema et al., 2018). Bampidis and Robinson (2006) reviewed the applications of citrus by-products as ruminant feed. The main finding was that the citrus by-products, both fresh and dried, could support the growth and lactation of ruminant animals like other conventional feed materials. Fresh bergamot pulp was added to feed as a cereal substitute to test the lamb health and meat quality in Italy (Scerra et al., 2018). The results showed that no significant difference in meat quality was observed in lamb feed mixtures containing up to 35% bergamot, suggesting promise for this method in reducing feed cost without compromising animal health and meat quality.

In addition to ruminant animals, fresh orange pulp was also successfully fed as a substitute for alfalfa in guinea pig production in Spain (Mínguez and Calvo, 2018). One of the disadvantages of using citrus by-products as animal feed is their low nitrogen content, but this can be compensated by mixing with other feedstocks (Berk, 2016). The bitter taste emitted from the essential oils in citrus by-products, especially lemon by-products, can affect the palatability and maximum contribution potential of citrus by-products within the feed (Zema et al., 2018). Another disadvantage of using fresh citrus by-product is their high moisture content (80%, Table 2). The moist citrus by-products can support microbial growth and degradation of the material and excessive amounts of water could increase the costs associated with material handling and transportation. Therefore, citrus by-products are normally dried for feed applications (Arthington et al., 2002; Berk, 2016).

Fresh citrus by-products are pressed and dried to total solids up to 90% (Berk, 2016). Drying of the by-products offers benefits of more stable and uniform final products and also reduces the volume and the cost of hauling. The effect of enzymatic hydrolysis before drying on the digestibility of citrus by-products as dry feed substitute was studied (Tripodo et al., 2004). Pectolytic enzymes were proved to be able to enhance the digestibility of the citrus by-products. In addition, lime or calcium hydroxide is added during the drying process to raise the pH and help to break pectin-carbohydrate bonds to release water in the pulp (Chavan et al., 2018). Tayengwa and Mapiye (2018) reviewed the effects of a mixture of dried citrus and winery by-products as feed on ruminant animal's diet, health, production, and meat quality, which were characterized by different analyses, including the rumen digestibility, methane and ammonia emissions, parasite infection, oxidative stress, meat fatty acid composition, and sensory quality. They found that adding up to 15% of dry citrus and winery by-products to the regular feed could improve the ruminant animal's diet, health, production, and meat quality because of the nutritional compounds, such as pectin, essential oils, and polyphenols, contained in the citrus and winery processing by-products. In addition, dried citrus by-products were used as a feed substitute for lamb production. No significant differences were found in the nutrient intake and digestibility of the lambs that were fed for ages with up to 40% of dried citrus by-products (Sharif et al., 2018). Compared to sugar beet pulp, corn grain, and sorghum grain as a feed additive, citrus by-products are believed to be more favorable for ruminant animals due to the high content of degradable neutral detergent fiber, especially pectin (Cullen et al., 1986). Further, the essential oils contained in the citrus by-products were also claimed to have the potential to improve rumen fermentation efficiency and reduce the greenhouse gas (GHG) emissions from ruminant animals (Cobellis et al., 2016). Most of the researchers believe that the activities of the methanogens and hyper-ammonia-producing bacteria were inhibited by the presence of the essential oils, resulting in mitigation of methane and ammonia emissions (Bodas et al., 2012; Cobellis et al., 2016; Patra and Yu, 2012). No significant difference in terms of the nutrient intake and digestibility was found when using dried citrus by-products as a feed substitute for lamb production (Sharif et al., 2018).

Dried citrus by-products have also been fed to non-ruminant animals. Goliomytis et al. (2018) fed dried orange pulp to laying hens in order to study the antioxidant levels of the resulting egg yolk. The results showed the eggs from hens fed with orange pulp had a significant increase in the oxidative stability in the egg yolk within 10 days compared to the control with the basal diet. However, egg quantity and quality decreased due to the lower rate of feed intake and egg laying in hens that consumed the dried orange pulp. This might be due to the bitter taste from the essential oils in the pulp. Dried citrus pulp was added to a low crude protein pig feed to investigate the feasibility of using citrus by-products as a feed substitute for weanling pigs (Almeida et al., 2017). The results concluded that adding 7.5% of dried citrus pulp did not show significant differences in the growth of weanling pigs and gut morphology during the 28-day study period. Other technologies have also been utilized to treat citrus by-products. Enzymatic pretreatment was conducted on citrus by-products to modify the fibrous materials for food applications (Canela-Xandri et al., 2018). Essential oils in the orange peel were extracted and fed to silver catfish as a feed additive with significant improvements observed in its growth performances (specific growth rate and relative weight gain) as well as metabolic and antioxidant levels (Lopes et al., 2019).

4.2 Biofuel

Due to their high moisture content, citrus by-products are more suitable for biochemical conversions than thermochemical to produce biofuel. Citrus by-products contain large amounts of fibrous materials, such as cellulose, hemicellulose, and pectin. Biofuel production from cellulosic materials has been extensively studied in the past decade. Biogas and bioethanol production from waste materials has been utilized across a wide range of industries, and in many cases, has been shown to have positive effects on waste management and environmental impacts (Zhou et al., 2008). The estimated bioethanol and biomethane yields from citrus by-products range from 50 to 60 L/ton of waste and 300 to 600 mL CH_4/g VS, respectively (Pourbafrani et al., 2013; Ruiz and Flotats, 2014; Zhang et al., 2015).

Simultaneous saccharification and fermentation was used to produce bioethanol from mandarin by-products. The mandarin waste was first dried and ground to a homogenous product, and then subjected to a mixture of cellulases and pectinase solution, as well as the yeast (Oberoi et al., 2011). The final ethanol concentration was 42.8 g/L after 12-hfermentation, and the yield was around 70 L/ton of waste. The author concluded that the enzymatic hydrolysis could degrade most of the structural sugars while the limonene, an inhibitory essential oil, was volatilized during the autoclave pretreatment.

An economic model was developed to estimate the economic feasibility of converting citrus by-products to ethanol in Florida, USA. The whole process comprises limonene extraction, enzymatic hydrolysis, fermentation, distillation, and dewatering. The production cost of ethanol from citrus by-products was estimated to be $0.32/L, which was between the production cost values for corn ethanol ($0.26/L) and cellulosic ethanol ($0.36–$0.43/L) at the time of publication (Zhou et al., 2007). Although the production cost of bioethanol from citrus by-products was higher than corn ethanol, it could offer benefits with value-added products, such as limonene. It is worth noting that the cost of enzymes contributed around 48% of the total operational cost in the process. Also, other value-added coproducts could be generated together with bioethanol during the fermentation processes under certain conditions. For instance, Cypriano et al. (2018) developed a process combining enzymatic hydrolysis, precipitation, solid extraction, and fermentation to produce ethanol, hesperidin, and nanocellulose from the orange pulp.

A process diagram of an advanced biorefinery producing ethanol, methane, and limonene from citrus by-products is shown in Fig. 7. Life cycle assessment (LCA)

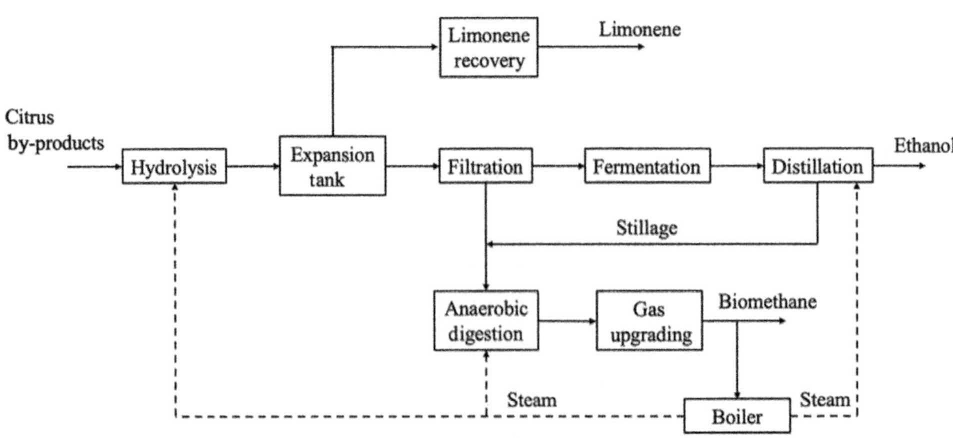

FIG. 7
Process diagram of a model citrus by-product biorefinery (Pourbafrani et al., 2013).

was used to estimate the possible reductions in GHG emissions associated with utilizing citrus by-products as feedstock for biofuel and limonene production (Pourbafrani et al., 2013). Ethanol and biomethane produced from the biorefinery were used to displace gasoline for light duty vehicles and natural gas for electricity generation, respectively. Limonene was considered to displace acetone as a solvent. Digestate post-anaerobic digestion was included in the assessment as a soil amendment. A range from 77% to 134% reduction in GHG emissions compared to the baseline scenario was achieved by the model simulation depending on the scale of the biorefinery.

One of the most important issues for biofuel production from citrus by-products is microbial inhibition by the limonene contained in most of the citrus peels. Limonene is a type of citrus essential oil that probably developed to protect the fruits from bacterial and viral infections (Schieber, 2017). Normally, limonene is extracted before any other fermentation or digestion process. However, a group of Japanese researchers isolated two yeasts, *Clostridiu beijerinckii* and *C. cellulovorans*, that demonstrated limonene tolerance up to 0.05% (*v*/v) (Tomita et al., 2019). These two yeasts could potentially be used for ethanol fermentation using citrus by-products without any feedstock pretreatment. Lukitawesa et al. (2018) developed a novel two stage anaerobic digestion system with filtration and recirculation to eliminate limonene inhibition in the digesters. The first reactor served as a hydrolysis tank and the effluent from the first stirred tank reactor was filtered with a rotary drum to separate the solids from the liquid stream. The liquid stream with decreased limonene content was then fed to a second up-flow anaerobic sludge bed reactor for gas production. Effluent from the second reactor was recirculated to compare with the control without recirculation. The results showed that the filtration step could successfully reduce the limonene inhibition, and the recirculation increased the biomethane yield from 66 to 113 mL CH_4/g VS to 160–203 mL CH_4/g VS.

The potential of biofuel production from citrus by-products was also studied in Iran (Taghizadeh-Alisaraei et al., 2017). The rapid growth of citrus production in Iran caused serious problems associated with the proper disposal of citrus waste, creating around 680 thousand tons of waste materials every year. It was shown that converting the solid waste into energy could ease some of the country's limitations in storage and disposal space and also produce 27 million liters of ethanol or 37 million m^3 of methane as renewable energy. However, a more detailed technical and economic analysis including transportation needs to be conducted to evaluate the accessibility of the feedstock, although the author claimed that the citrus production is centralized in several provinces in Iran.

Not only are biochemical conversions being studied, but also thermal chemical conversions of citrus by-products are currently under research. The feasibility of fluidized bed torrefaction was studied to convert orange peels into a high energy content solid biofuel (Brachi et al., 2019). Drying for 24 h at room temperature was first carried out to remove the moisture. The torrefaction process then converted the low value orange peels into biofuel with high energy density. The final solid biochar had a 50% increase in energy density compared with the raw orange peels and a 67% decrease in the oxygen to carbon ratio. Also, torrefaction (200–325°C) and pyrolysis (400–650°C) using orange and lemon peels were conducted in a horizontal fix-bed pyrolysis reactor (Volpe et al., 2015). Both torrefaction and pyrolysis increased the energy density and stability of the citrus by-products by reactive hemicellulose decomposition. At 500°C and 30 min, the highest energy densities were obtained for bio-chars from lemon and orange peels. The gross calorific values of the lemon and orange peel bio-char products were 31,200 and 31,800 J/g, respectively, a 70%–75% increase from the initial feedstock materials.

4.3 Other Value-Added Products

Animal feed and biofuel production are the two most promising ways to manage citrus by-products. However, there are other value-added products that are abundant and can be extracted from these waste materials. Technologies such as microwave-assisted extraction (MAE), ultrasound-assisted extraction (UAE), pulse electric field (PEF), and enzymatic hydrolysis, have been applied in the citrus waste management practices to extract and separate value-added products in the past decade (Chavan et al., 2018). MAE is a promising technology for citrus by-product utilization because microwaves can enhance the mass and heat transfer by penetrating the pores between the plant cells and decreasing the heat up time compared to conventional heating (Lopez-Avila, 2000). Similarly, ultrasound could intensify the mass transfer process by causing the collapse of cavitation bubbles which further increase the diffusion rates of the products (Khan et al., 2010). In addition, PEF-assisted extraction has also been widely utilized in the food industry. The treatment introduces electric strength on the plant cells to destruct the cytoplasmic membranes so that the valuable products can become more permeable to the solvent (Luengo et al., 2013). Furthermore, enzymatic hydrolysis is sometimes used as a pretreatment method to improve the accessibility of the valuable compounds by degrading the structural carbohydrates in the plant cell wall. All of these extraction technologies are considered non-toxic and environmental-friendly methods while consuming less energy and chemicals (Canela-Xandri et al., 2018). Other novel technologies have also been studied and included below to show their potential application in citrus by-product extraction. However, research and engineering development are needed to make those processes more economically favorable for industrial applications.

4.3.1 PECTIN

Pectin is a gelatinous polysaccharide, which can be found in the plant cell wall and middle lamella. The function of pectin is to bind the plant cells together to provide both strength and flexibility (Satari et al., 2017). Traditionally, pectin is used in the food industry as a thickening and stabilizing agent, but researchers also found that pectin could be used as a fat substitute (Min et al., 2010). Normally, citrus peels contain the highest pectin among the citrus by-products (Satari et al., 2017). Conventional pectin extraction involves acid and heat, which requires high chemical and energy inputs. For example, researchers reported the use of 0.05- to 2-M sulfuric acid solutions to extract pectin from citrus peels at 80–100°C for 1 h (Marić et al., 2018).

Satari et al. (2017) developed a simultaneous pectin extraction and pretreatment method to convert citrus waste into pectin and soluble sugars. The process was optimized to pretreat a mixture of orange and grapefruit waste using dilute nitric acid and ethanol within a single-step process at pH 1.8, 80°C and 2 h. At the optimal condition, 58.7 g of glucose and pectin in total could be extracted from 100 g of dry raw material.

In addition, an innovative technology was developed to extract pectin from orange and lemon peels using only microwaves and water (Fidalgo et al., 2016). This acid free microwave extraction method could significantly reduce the requirements of energy and chemicals while producing similar yields as the conventional extraction method. Essential oils, mainly limonene, were also extracted from the feedstock in the oil phase after the microwave extraction.

Besides acids and microwave-assisted processes, ultrasound has also been tested for pectin extraction. Ultrasound assisted heating extraction (UAHE) was carried out to evaluate the efficiency of pectin extraction from the grapefruit peel. The results showed that the UAHE could increase the yield of pectin by14.08%, and reduce the retention time and temperature by 39.53% and 20°C, respectively (Xu et al., 2014).

Further, Dominiak et al. (2014) studied six commercial enzymes on the pectin extraction yield of lime peels. At pH 4.8 and 50°C, pectin yields ranged from 15% to 32% depending on the types of the enzyme. Because of the undefined composition of different citrus by-products, the lack of a broad-spectrum enzyme mixture could be the drawback for the wide application of enzymatic extraction of pectin.

4.3.2 PHENOLIC COMPOUNDS

High amounts of phenolic compounds are also found in citrus by-products, especially flavonoids. Citrus by-products contain around 0.67%–19.62% dry basis of phenolic compounds mainly in the peels (M'hiri et al., 2017). Phenolic compounds have been widely used in food, cosmetic, and pharmaceutical products because of their antioxidant activity. M'hiri et al. (2017) reviewed the major phenolic compounds contained in the citrus by-products and their industrial applications. Different industrial extraction approaches are summarized in the review article to provide options for biorefineries with different citrus by-product treatment processes (Putnik et al., 2017). Conventional solvent extraction methods normally involve the use of organic solvents, such as ethanol and methanol (Garcia-Castello et al., 2015).

Flavonoids from grapefruit solid wastes were extracted by conventional ethanol extraction and UAE methods (Garcia-Castello et al., 2015). Response surface analysis was carried out to identify the optimal operating parameters and predict the highest yields. The optimal conditions for the conventional ethanol extraction method were at a temperature of 69°C, an ethanol concentration of 30%, and a retention time of 190 min. The UAE achieved the highest extraction yield with a higher ethanol concentration of 40% at a lower temperature (25°C) and a shorter processing time (55 min) compared to conventional ethanol extraction. Also, UAE increased the total polyphenol yield and total antioxidant activity by 30% and 38%, respectively, compared to the conventional method.

A process combining high-voltage electrical discharge (HVED) with enzymatic hydrolysis was developed to extract carbohydrates and polyphenols from orange peels (El Kantar et al., 2018). With an HVED energy input of 222 kJ/kg of substrate and 12 fungal β-Glucanase Units/g of substrate, 50 and 0.7 g of reducing sugars and polyphenols, respectively, were extracted from 100 g of dry orange peel. In addition, the effect of PEF treatment on polyphenol extraction from orange peels was investigated (Peiró et al., 2019). The samples treated with PEF at 7 kV/cm for 60 μs (20 pulses of 3 μs) almost doubled the antioxidant activity compared to the untreated control. The same technology was also used for polyphenol extraction from lemon by-products (Luengo et al., 2013). An electric field intensity of 7 kV/cm and duration of 900 μs (30 pulses of 30 μs) were determined to result in the highest yield of polyphenols.

A new type of liquid solvent, deep eutectic solvents (DES) are composed of strong hydrogen bond donors and acceptors. Similar to the ionic liquid, DES can be used to extract metal ions or organic compounds in the liquid or solid phase due to their higher partitioning coefficient (Dai et al., 2013). DES are also less volatile and toxic in contrast to organic solvents used in preparing the ionic liquid. Further, they are non-flammable and biodegradable which can lead to beneficial extraction solvents for various crucial bioproducts. The possibility of extracting the phenolic compounds from orange peels using choline chloride-based DES with glycerol or ethylene was investigated (Ozturk et al., 2018). DES extraction had higher polyphenol yields than the conventional ethanol (30%, *w*/w) extraction. The highest yield of 3.61-mg gallic acid equivalent per gram of orange peel was achieved at DES to water ratio of 1:10, temperature of 333.15 K, extraction time of 100 min and solid to solvent ratio of 1:10.

Supercritical carbon dioxide (CO_2) extraction is another advanced technology. It extracts the products by their different diffusion coefficients in the gas and supercritical phases of CO_2 (Clifford and Williams, 2000). Supercritical CO_2 was used to extract the antioxidant compounds from orange leaves, which contains high amounts of phenolic and flavonoid compounds (Montes et al., 2019). The leaves were first dried and ground to particle size below 5 mm. Rapid expansion supercritical solution (RESS) experiments were then conducted to study the extraction efficiency and determine the optimal operating temperature (40–100°C) and pressure (80–300 bar). The results showed that high temperatures (100°C) and high pressures (larger than 275 bar) were required to form powdered products after the RESS extraction. Further, supercritical antisolvent extraction (SAE) was also carried out in the same study to extract and produce antioxidant nanoparticles from liquid orange leaf extracts produced by the conventional ethanol extraction. The conventional ethanol extraction method was used to produce the liquid orange leaf extracts as the substrate for SAE needs to be in the liquid form. High pressure (120 bar) and low temperature (40°C) were recommended to acquire nanoparticles in the powder form. In addition, the initial extract concentration played an important role in the SAE process as indicated by the experiments. The extract concentration of 20 mg/mL was recommended for the formation of nanoparticle powders.

4.3.3 ESSENTIAL OILS

Citrus essential oils (CEOs) are volatile aromatic compounds usually present in the peels of the citrus fruits at a concentration between 0.6% and 3.8% dry basis (Forgács et al., 2012; M'hiri et al., 2017). CEOs have been reported to offer various benefits to human health and wellness due to their antimicrobial properties (Negro et al., 2016). CEOs comprise at least 20–60 different varieties of compounds. The predominant essential oil detected in the citrus peels is limonene (Schieber, 2017). The traditional method for CEO extraction is using a mechanical cold press. The liquid emulsion obtained after cold pressing is then centrifuged to obtain CEOs in the oil phase (Zema et al., 2018). Hydro-distillation is another common way to extract essential oils from citrus by-products (Ruiz and Flotats, 2014).

Steam explosion is normally used to pretreat the biomass before biofuel production to destruct the cell wall structure. However, it can also be used to extract CEOs such as limonene that can be collected from the condensate of the vapor (Forgács et al., 2012). Research showed that injection of steam at 60 bar, 150°C for 20 min could reduce the limonene concentration in the citrus by-products by 94%. The final yield of the limonene in the condensate was 3.5% dry matter, which was 92% of the theoretical yield (3.8%) of the citrus by-product used in the research.

Ultrasound-assisted hydro-distillation was used to extract essential oils from bergamot waste (Xing et al., 2019). A final oil concentration of 0.48% was achieved using 180-W ultrasound power for 60 min, which was 118% higher than using the conventional hydro-distillation method.

Different microwave-assisted extraction methods were compared based on the yield and quality of the essential oils, the extraction time, and the energy consumption (Razzaghi et al., 2019). The six extraction methods included two conventional methods: steam distillation and steam hydro-diffusion, two microwave-based approaches: solvent free microwave extraction and microwave hydro-diffusion and gravity, and two microwave steam combined processes: microwave steam distillation and microwave steam diffusion. Among the six techniques, microwave treatment combined with steam extraction methods were concluded to be the most efficient for essential oil extraction from orange peels, followed by methods using only microwaves. The conventional extraction methods were reported to be the least efficient among those tested.

Supercritical CO_2 extraction was also studied to extract and separate limonene from citrus by-products. Yasumoto et al. (2015) studied the supercritical CO_2 separation on a mixture of D-limonene (65%) and canola oil (35%) solution, representing the seed of Yuzu. Yuzu is a popular citrus fruit in northeast Asia with an estimated production amount of 20,000 tons per year in Japan (Sprunger et al., 2018). At 30°C, 10 MPa, and solvent to feed ratio of 6, the highest D-limonene recovery (>99% w/w) was acquired by supercritical CO_2 extraction. Despite the high yield and purity, this process is still far from the industrial application because of its relatively high equipment capital investments and operational expenses (Zema et al., 2018).

5 SUMMARY AND FUTURE RESEARCH

Citrus fruits are one of the most consumed fruit commodities in the world. However, the citrus production industry generates a large number of by-products (peel, pulp, rag, and seeds) from juice processing facilities. A plethora of research has been conducted to investigate the feasibilities of different approaches to utilize or convert these "waste" materials into value-added products. The utilization of citrus by-products as animal feed (both raw and dried) is the most cost-effective and efficient method; nonetheless, it is constrained by the high transportation costs used to deliver products from the citrus processing facilities to the adjacent animal farms. Additionally, animal production is becoming more centralized and efficient; therefore, the consumption of citrus by-products as animal feed is expected to decrease in the near future. Finding alternative technologies for the efficient management of citrus by-products is a crucial component in the citrus industry. Biofuel production is a promising alternative citrus waste management practice. Both biochemical and thermochemical conversion techniques have been concluded to be feasible in producing biofuel from citrus by-products. However, research must be conducted to evaluate and optimize the operational conditions for biofuel production technologies. Essential oils are extracted and separated before initiating other downstream processes within the biochemical conversion process. This is because the essential oils are believed to be the inhibitors for enzyme facilitated biochemical reactions (Schieber, 2017). Other than the essential oil extraction, pretreatment processes are usually necessary for achieving a high biofuel production efficiency. Research on effective pretreatment methods is of great importance for the different types of citrus by-products due to their unique characteristics. Technical and economic analysis, and LCA are always recommended in order to evaluate the economic feasibility and environmental profit of biofuel refineries using citrus by-products as the feedstock. Several emerging technologies, including microwave-assisted extraction, ultrasound-assisted extraction, pulse electric field, and enzymatic hydrolysis have been studied in both laboratory and pilot scales to extract value-added products, such as pectin, essential oils, and phenolic compounds, from the citrus by-products. Currently, these technologies are not widely used in the industry because of their comparatively high capital requirements and operating costs. Research and development are needed to justify the costs and benefits for applying these innovative methods on an industrial scale.

References

Almeida, V.V., Nuñez, A.J.C., Schinckel, A.P., Alvarenga, P.V.A., Castelini, F.R., Silva-Guillen, Y.V., Thomaz, M.C., 2017. Interactive effect of dietary protein and dried citrus pulp levels on growth performance, small intestinal morphology, and hindgut fermentation of weanling pigs. J. Anim. Sci. 95, 257–269.

Aragon-Rodriguez, F., Castro-Garcia, S., Sola-Guirado, R.R., Gil-Ribes, J.A., 2019. Fruit abscission pattern of 'Valencia' orange with canopy shaker system. Sci. Hortic. (Amsterdam) 246, 916–920.

Arthington, J.D., Kunkle, W.E., Martin, A.M., 2002. Citrus pulp for cattle. Vet. Clin. North Am. Food Anim. Pract. 18, 317–326.

Babcock, B., 2018. Economic impact of California's citrus industry. Citrograph 9, 36–39.

Baldwin, E.A., 1993. Citrus fruit. In: Biochemistry of Fruit Ripening. Springer, Dordrecht, pp. 107–149.

Bampidis, V.A., Robinson, P.H., 2006. Citrus by-products as ruminant feeds: a review. Anim. Feed Sci. Technol. 128, 175–217.

Benkeblia, N., Tennant, D.P.F., Jawandha, S.K., Gill, P.S., 2011. Preharvest and harvest factors influencing the postharvest quality of tropical and subtropical fruits. Postharvest Biol. Technol. Trop. Subtrop. Fruits, 112–142e.

Berk, Z., 2016. Citrus Fruit Processing. Academic Press, London, UK.

Blanco, G., Roka, F., 2009. Cost/benefit analysis of abscission registration for citrus mechanical harvesting. In: Southern Agricultural Economics Association Annual Meeting, Atlanta, Georgia.

Bodas, R., Prieto, N., García-González, R., Andrés, S., Giráldez, F.J., 2012. Manipulation of rumen fermentation and methane production with plant secondary metabolites. Anim. Feed Sci. Technol. 176, 78–93.

Brachi, P., Chirone, R., Miccio, F., Miccio, M., Ruoppolo, G., 2019. Valorization of orange peel residues via fluidized bed torrefaction: comparison between different bed materials. Combust. Sci. Technol. 1–15.

Braddock, R., 1999. Handbook of Citrus By-Products and Processing Technology. John Wiley & Sons, New York, NY.

Brown, G.K., 2005. New mechanical harvesters for the Florida citrus juice industry. HortTechnology 15, 69–72.

Burdurlu, H.S., Koca, N., Karadeniz, F., 2006. Degradation of vitamin C in citrus juice concentrates during storage. J. Food Eng. 74, 211–216.

Canela-Xandri, A., Balcells, M., Villorbina, G., Cubero, M.Á., Canela-Garayoa, R., 2018. Effect of enzymatic treatments on dietary fruit fibre properties. Biocatal. Biotransform. 36, 172–179.

Chavan, P., Singh, A.K., Kaur, G., 2018. Recent progress in the utilization of industrial waste and by-products of citrus fruits: a review. J. Food Process Eng. 41.

Clifford, A.A., Williams, J.R., 2000. Introduction to supercritical fluids and their applications. In: Supercritical Fluid Methods and Protocols. Humana Press, New Jersey, pp. 1–16.

Cobellis, G., Trabalza-Marinucci, M., Yu, Z., 2016. Critical evaluation of essential oils as rumen modifiers in ruminant nutrition: a review. Sci. Total Environ. 545–546, 556–568.

Cooper, W.C., Henry, W.H., 1971. Abscission chemicals in relation to citrus fruit harvest. J. Agric. Food Chem. 19, 559–563.

Costa, S.E.A., Camarotto, J.A., 2012. An ergonomics approach to citrus harvest mechanization. Work 41, 5027–5032.

Crupi, F., Rispoli, G., 2002. Citrus juices technology. In: Citrus. The Genus Citrus. Taylor & Francis, London, pp. 77–113.

Cserhalmi, Z., Sass-Kiss, Á., Tóth-Markus, M., Lechner, N., 2006. Study of pulsed electric field treated citrus juices. Innov. Food Sci. Emerg. Technol. 7, 49–54.

Cullen, A.J., Harmon, D.L., Nagaraja, T.G., 1986. In vitro fermentation of sugars, grains, and by-product feeds in relation to initiation of ruminal lactate production. J. Dairy Sci. 69, 2616–2621.

Cypriano, D.Z., da Silva, L.L., Tasic, L., 2018. High value-added products from the orange juice industry waste. Waste Manag. 79, 71–78.

Dai, Y., van Spronsen, J., Witkamp, G.-J., Verpoorte, R., Choi, Y.H., 2013. Natural deep eutectic solvents as new potential media for green technology. Anal. Chim. Acta 766, 61–68.

Dominiak, M., Søndergaard, K.M., Wichmann, J., Vidal-Melgosa, S., Willats, W.G.T., Meyer, A.S., Mikkelsen, J.D., 2014. Application of enzymes for efficient extraction, modification, and development of functional properties of lime pectin. Food Hydrocoll. 40, 273–282.

Ehsani, R., Udumala, S., 2010. Mechanical harvesting of citrus—an overview. Resour. Mag. 17, 4–6.

El Kantar, S., Boussetta, N., Rajha, H.N., Maroun, R.G., Louka, N., Vorobiev, E., 2018. High voltage electrical discharges combined with enzymatic hydrolysis for extraction of polyphenols and fermentable sugars from orange peels. Food Res. Int. 107, 755–762.

FAO, 2017. Citrus Fruit Fresh and Processed Statistical Bulletin 2016. FAO, Rome.

FAO, 2019. FAOSTAT [WWW Document]. Food Agric. Organ. United Nations. http://www.fao.org/faostat/en/#data/QC. (Accessed 13 April 2019).

Fidalgo, A., Ciriminna, R., Carnaroglio, D., Tamburino, A., Cravotto, G., Grillo, G., Ilharco, L.M., Pagliaro, M., 2016. Eco-friendly extraction of pectin and essential oils from orange and lemon peels. ACS Sustain. Chem. Eng. 4, 2243–2251.

Florida Department of Citrus, 2018. 2016–17 Annual Processor's Statistics Report.

Forgács, G., Pourbafrani, M., Niklasson, C., Taherzadeh, M.J., Hováth, I.S., 2012. Methane production from citrus wastes: process development and cost estimation. J. Chem. Technol. Biotechnol. 87, 250–255.

Garcia-Castello, E.M., Rodriguez-Lopez, A.D., Mayor, L., Ballesteros, R., Conidi, C., Cassano, A., 2015. Optimization of conventional and ultrasound assisted extraction of flavonoids from grapefruit (*Citrus paradisi* L.) solid wastes. LWT—Food Sci. Technol. 64, 1114–1122.

Gmitter, F.G., Hu, X., 1990. The possible role of Yunnan, China, in the origin of contemporary *Citrus* species (Rutaceae). Econ. Bot. 44, 267–277.

Goliomytis, M., Kostaki, A., Avgoulas, G., Lantzouraki, D.Z., Siapi, E., Zoumpoulakis, P., Simitzis, P., Deligeorgis, S.-G., 2018. Dietary supplementation with orange pulp (*Citrus sinensis*) improves egg yolk oxidative stability in laying hens. Anim. Feed Sci. Technol. 244, 28–35.

Gupta, S.K., Ehsani, R., Kim, N.H., 2016. Optimization of a citrus canopy shaker harvesting system: mechanistic tree damage and fruit detachment models. Trans. ASABE 59, 761–776.

Hannan, M.W., Burks, T.F., 2004. Current developments in automated citrus harvesting. In: ASAE/CSAE Annual International Meeting, Ottawa, pp. 1–10.

Janati, S.S.F., Beheshti, H.R., Feizy, J., Fahim, N.K., 2012. Chemical composition of lemon (*Citrus limon*) and peels its considerations as animal food. GIDA J. Food 37, 267–271.

Kader, A.A., 1999. Fruit maturity, ripening, and quality relationships. Acta Hortic. 203–208.

Khan, M.K., Abert-Vian, M., Fabiano-Tixier, A.-S., Dangles, O., Chemat, F., 2010. Ultrasound-assisted extraction of polyphenols (flavanone glycosides) from orange (*Citrus sinensis* L.) peel. Food Chem. 119, 851–858.

Ladaniya, M.S., 2008. Commercial fresh citrus cultivars and producing countries. In: Citrus Fruit: Biology, Technology and Evaluation. Academic Press, San Diego, CA, pp. 13–65.

Li, P., Lee, S., Hsu, H.-Y., 2011. Review on fruit harvesting method for potential use of automatic fruit harvesting systems. Proc. Eng. 23, 351–366.

Liu, Y., Heying, E., Tanumihardjo, S.A., 2012. History, global distribution, and nutritional importance of citrus fruits. Compr. Rev. Food Sci. Food Saf. 11, 530–545.

Liu, T.-H., Ehsani, R., Toudeshki, A., Zou, X.-J., Wang, H.-J., 2017. Experimental study of vibrational acceleration spread and comparison using three citrus canopy shaker shaking tines. Shock. Vib. 2017, 1–9.

Liu, T.-H., Luo, G., Ehsani, R., Toudeshki, A., Zou, X.-J., Wang, H.-J., 2018. Simulation study on the effects of tine-shaking frequency and penetrating depth on fruit detachment for citrus canopy-shaker harvesting. Comput. Electron. Agric. 148, 54–62.

Loghavi, M., Mohseni, S., 2006. The effect of shaking frequency and amplitude on detachment of lime fruit. Iran Agric. Res. 24, 27–37.

Lopes, J.M., de Freitas Souza, C., Saccol, E.M.H., Pavanato, M.A., Antoniazzi, A., Rovani, M.T., Heinzmann, B.M., Baldisserotto, B., 2019. Citrus x aurantium essential oil as feed additive improved growth performance, survival, metabolic, and oxidative parameters of silver catfish (*Rhamdia quelen*). Aquac. Nutr. 25, 310–318.

Lopez-Avila, V., 2000. Microwave-assisted extraction. In: Encyclopedia of Separation Science. Academic Press, San Diego, CA, pp. 1389–1398.

Lü, Q., Cai, J., Liu, B., Deng, L., Zhang, Y., 2014. Identification of fruit and branch in natural scenes for citrus harvesting robot using machine vision and support vector machine. Int. J. Agric. Biol. Eng. 7, 115–121.

Luengo, E., Álvarez, I., Raso, J., 2013. Improving the pressing extraction of polyphenols of orange peel by pulsed electric fields. Innov. Food Sci. Emerg. Technol. 17, 79–84.

Lukitawesa, Wikandari, R., Millati, R., Taherzadeh, M.J., Niklasson, C., 2018. Effect of effluent recirculation on biogas production using two-stage anaerobic digestion of citrus waste. Molecules 23, 3380.

M'hiri, N., Ioannou, I., Ghoul, M., Mihoubi Boudhrioua, N., 2017. Phytochemical characteristics of citrus peel and effect of conventional and nonconventional processing on phenolic compounds: a review. Food Rev. Int. 33, 587–619.

Marić, M., Grassino, A.N., Zhu, Z., Barba, F.J., Brnčić, M., Brnčić, S.R., 2018. An overview of the traditional and innovative approaches for pectin extraction from plant food wastes and by-products: ultrasound-, microwaves-, and enzyme-assisted extraction. Trends Food Sci. Technol. 76, 28–37.

Mehta, S.S., Burks, T.F., 2014. Vision-based control of robotic manipulator for citrus harvesting. Comput. Electron. Agric. 102, 146–158.

Min, B., Bae, I.Y., Lee, H.G., Yoo, S.H., Lee, S., 2010. Utilization of pectin-enriched materials from apple pomace as a fat replacer in a model food system. Bioresour. Technol. 101, 5414–5418.

Mínguez, C., Calvo, A., 2018. Effect of supplementation with fresh orange pulp (*Citrus sinensis*) on mortality, growth performance, slaughter traits and sensory characteristics in meat guinea pigs. Meat Sci. 145, 51–54.

Montes, A., Hanke, F., Williamson, D., Guamán-Balcázar, M.C., Valor, D., Pereyra, C., Teipel, U., Martínez de la Ossa, E., 2019. Precipitation of powerful antioxidant nanoparticles from orange leaves by means of supercritical CO_2. J. CO_2 Util. 31, 235–243.

Moreno, R., Torregrosa, A., Moltó, E., Chueca, P., 2015. Effect of harvesting with a trunk shaker and an abscission chemical on fruit detachment and defoliation of citrus grown under Mediterranean conditions. Spanish J. Agric. Res. 13.

Negro, V., Mancini, G., Ruggeri, B., Fino, D., 2016. Citrus waste as feedstock for bio-based products recovery: review on limonene case study and energy valorization. Bioresour. Technol. 214, 806–815.

Oberoi, H.S., Vadlani, P.V., Nanjundaswamy, A., Bansal, S., Singh, S., Kaur, S., Babbar, N., 2011. Enhanced ethanol production from Kinnow mandarin (*Citrus reticulata*) waste via a statistically optimized simultaneous saccharification and fermentation process. Bioresour. Technol. 102, 1593–1601.

Oluremi, O.I.A., Ojighen, V.O., Ejembi, E.H., 2006. The nutritive potentials of sweet orange (*Citrus sinensis*) rind in broiler production. Int. J. Poult. Sci. 5, 613–617.

Ortiz, C., Torregrosa, A., 2013. Determining adequate vibration frequency, amplitude, and time for mechanical harvesting of fresh mandarins. Trans. ASABE 56, 15–22.

Ozturk, B., Parkinson, C., Gonzalez-Miquel, M., 2018. Extraction of polyphenolic antioxidants from orange peel waste using deep eutectic solvents. Sep. Purif. Technol. 206, 1–13.

Patra, A.K., Yu, Z., 2012. Effects of essential oils on methane production and fermentation by, and abundance and diversity of, rumen microbial populations. Appl. Environ. Microbiol. 78, 4271–4280.

Peiró, S., Luengo, E., Segovia, F., Raso, J., Almajano, M.P., 2019. Improving polyphenol extraction from lemon residues by pulsed electric fields. Waste Biomass Valoriz. 10, 889–897.

Plebe, A., Grasso, G., 2001. Localization of spherical fruits for robotic harvesting. Mach. Vis. Appl. 13, 70–79.

Pourbafrani, M., Forgács, G., Horváth, I.S., Niklasson, C., Taherzadeh, M.J., 2010. Production of biofuels, limonene and pectin from citrus wastes. Bioresour. Technol. 101, 4246–4250.

Pourbafrani, M., McKechnie, J., MacLean, H.L., Saville, B.A., 2013. Life cycle greenhouse gas impacts of ethanol, biomethane and limonene production from citrus waste. Environ. Res. Lett. 8.

Pu, Y., Toudeshki, A., Ehsani, R., Yang, F., 2018. Design and evaluation of a two-section canopy shaker with variable frequency for mechanical harvesting of citrus. Int. J. Agric. Biol. Eng. 11, 77–87.

Putnik, P., Bursać Kovačević, D., Režek Jambrak, A., Barba, F., Cravotto, G., Binello, A., Lorenzo, J., Shpigelman, A., 2017. Innovative "green" and novel strategies for the extraction of bioactive added value compounds from citrus wastes—a review. Molecules 22, 680.

Razzaghi, S.E., Arabhosseini, A., Turk, M., Soubrat, T., Cendres, A., Kianmehr, M.H., Perino, S., Chemat, F., 2019. Operational efficiencies of six microwave based extraction methods for orange peel oil. J. Food Eng. 241, 26–32.

Ruiz, B., Flotats, X., 2014. Citrus essential oils and their influence on the anaerobic digestion process: an overview. Waste Manag. 34, 2063–2079.

Sanders, K.F., 2005. Orange harvesting systems review. Biosyst. Eng. 90, 115–125.

Sandhu, K.S., Minhas, K.S., 2006. Oranges and citrus juices. In: Handbook of Fruits and Fruit Processing. vol. 60. Blackwell Publishing, Oxford, UK, p. 309.

Satari, B., Palhed, J., Karimi, K., Lundin, M., Taherzadeh, M.J., Zamani, A., 2017. Process optimization for citrus waste biorefinery via simultaneous pectin extraction and pretreatment. BioResources 12, 1706–1722.

Savary, S.K.J.U., Ehsani, R., Schueller, J.K., Rajaraman, B.P., 2010. Simulation study of citrus tree canopy motion during harvesting using a canopy shaker. Trans. ASABE 53, 1373–1381.

Scerra, M., Foti, F., Caparra, P., Cilione, C., Violi, L., Fiammingo, G., D'Agui', G., Chies, L., 2018. Effects of feeding fresh bergamot (*Citrus bergamia* Risso) pulp at up to 35% of dietary dry matter on growth performance and meat quality from lambs. Small Rumin. Res. 169, 160–166.

Schieber, A., 2017. Side streams of plant food processing as a source of valuable compounds: selected examples. Annu. Rev. Food Sci. Technol. 8, 97–112.

Shan, Y., 2016. Overview of the canned citrus industry. In: Canned Citrus Processing. Science Press, Beijing, China, pp. 1–6.

Sharif, M., Ashraf, M.S., Mushtaq, N., Nawaz, H., Mustafa, M.I., Ahmad, F., Younas, M., Javaid, A., 2018. Influence of varying levels of dried citrus pulp on nutrient intake, growth performance and economic efficiency in lambs. J. Appl. Anim. Res. 46, 264–268.

Sinclair, W., 1972. The Grapefruit: Its Composition, Physiology and Products. UCANR Publications, Davis, CA.

Singerman, A., Burani-Arouca, M., Futch, S.H., Ranieri, R., 2017. Harvesting Charges for Florida Citrus, 2016/17.

Spreen, T.H., Gao, Z., Gmitter, F., Norberg, R., 2012. An overview of the citrus industry of China. Proc. Florida State Hortic. Soc. 125, 119–121.

Sprunger, A., Marmillod, I., Kosińska-Cagnazzo, A., Andlauer, W., 2018. Bioactive compounds of juice and peels of Yuzu fruits cultivated in Switzerland. Chim. Int. J. Chem. 72, 728–732.

Taghizadeh-Alisaraei, A., Hosseini, S.H., Ghobadian, B., Motevali, A., 2017. Biofuel production from citrus wastes: a feasibility study in Iran. Renew. Sust. Energ. Rev. 69, 1100–1112.

Tayengwa, T., Mapiye, C., 2018. Citrus and winery wastes: promising dietary supplements for sustainable ruminant animal nutrition, health, production, and meat quality. Sustainability 10, 3718.

Tobey, R., Wetherell, C., 1995. The citrus industry and the revolution of corporate capitalism in Southern California, 1887–1944. Calif. Hist. 74, 6–21.

Tomita, H., Okazaki, F., Tamaru, Y., 2019. Direct IBE fermentation from mandarin orange wastes by combination of *Clostridium cellulovorans* and *Clostridium beijerinckii*. AMB Express 9(1).

Tripodo, M.M., Lanuzza, F., Micali, G., Coppolino, R., Nucita, F., 2004. Citrus waste recovery: a new environmentally friendly procedure to obtain animal feed. Bioresour. Technol. 91, 111–115.

USDA, 2019. Citrus: World Markets and Trade.

Volpe, M., Panno, D., Volpe, R., Messineo, A., 2015. Upgrade of citrus waste as a biofuel via slow pyrolysis. J. Anal. Appl. Pyrolysis 115, 66–76.

Webber, H.J., 1967. History and Development of the Citrus Industry. University of California, Division of Agricultural Sciences, Berkeley, CA.

Webber, H.J., Barker, R., Ferguson, L., 2014. History and development of the California citrus industry. In: Ferguson, L., Grafton-Cardwell, E.E. (Eds.), Citrus Production Manual. UCANR Publications, Davis, CA, pp. 3–11.

Whitney, J.D., 1968. Citrus Fruit Removal With an Air Harvester Concept. vol. 81. Florida State Horticultural Society, Miami, FL, pp. 43–48.

Whitney, J.D., 1977. Design and performance of an air shaker for citrus fruit removal. Trans. Am. Soc. Agric. Eng. 20, 52–56.

Whitney, J.D., Wheaton, T.A., 1987. Shakers affect Florida orange fruit yields and harvesting efficiency. Appl. Eng. Agric. 3, 20–24.

Whitney, J.D., Hartmond, U., Kender, W.J., Salyani, M., 2000. Orange removal with trunk shakers and abscission chemicals. Appl. Eng. Agric. 16, 367.

Wikimedia Commons, 2010. The world's major producer regions for citrus fruits [WWW Document]. https://commons.wikimedia.org/wiki/File:Hauptanbaugebiete-Zitrusfrüchte.svg. (Accessed 10 April 2019).

Xing, C., Qin, C., Li, X., Zhang, F., Linhardt, R.J., Sun, P., Zhang, A., 2019. Chemical composition and biological activities of essential oil isolated by HS-SPME and UAHD from fruits of bergamot. LWT 104, 38–44.

Xu, Y., Zhang, L., Bailina, Y., Ge, Z., Ding, T., Ye, X., Liu, D., 2014. Effects of ultrasound and/or heating on the extraction of pectin from grapefruit peel. J. Food Eng. 126, 72–81.

Yasumoto, S., Quitain, A.T., Sasaki, M., Iwai, H., Tanaka, M., Hoshino, M., 2015. Supercritical CO_2-mediated countercurrent separation of essential oil and seed oil. J. Supercrit. Fluids 104, 104–111.

Yeom, H.W., Streaker, C.B., Zhang, Q.H., Min, D.B., 2000. Effects of pulsed electric fields on the quality of orange juice and comparison with heat pasteurization. Agric. Food Chem. 48, 4597–4605.

Zema, D.A., Calabrò, P.S., Folino, A., Tamburino, V., Zappia, G., Zimbone, S.M., 2018. Valorisation of citrus processing waste: a review. Waste Manag. 80, 252–273.

Zhang, R., Zhang, M., Aramrueang, N., 2015. Characterization of Food Processing Waste as Feedstock for Anaerobic Digesters. University of California, Davis.

Zhou, W., Widmer, W., Grohmann, K., 2007. Economic analysis of ethanol production from citrus peel waste. Proc. Florida State Hortic. Soc. 120, 310–315.

Zhou, W., Widmer, W., Grohmann, K., 2008. Developments in ethanol production from citrus peel waste. Proc. Florida State Hortic. Soc. 121, 307–310.

SECTION 3

Vegetables and Root Crops

CHAPTER 10
Leafy Vegetables

Natthiporn Aramrueang*, Suvaluk Asavasanti†‡, Aphinya Khanunthong§
*Department of Biotechnology, Faculty of Science, Mahidol University, Bangkok, Thailand, †Food Technology & Engineering Laboratory, Pilot Plant Development & Training Institute, KMUTT, Bangkok, Thailand, ‡Food Security and Process Innovation Research Group, Food Engineering Department, Faculty of Engineering, KMUTT, Bangkok, Thailand, §Office of Agricultural Economics, Ministry of Agriculture and Cooperatives, Bangkok, Thailand

Chapter Outline

1 INTRODUCTION

Leafy vegetables, grown and consumed in almost every single continent, are present in different types and varieties due to the distinct geographical area. Lettuce has five distinctive types, each with its own varieties and requiring different growing and harvesting practices. Some leafy varieties are widely adapted and can be grown in many different temperatures and geographical areas.

According to Food and Agriculture Organization (FAO), the world's total harvested area for primary vegetables increased by threefold between 1961 and 2014 (FAO, 2017). Harvested area for the main leafy vegetables (cauliflowers and broccoli, spinach, lettuce and chicory, and cabbages, and other brassicas) had been steadily increasing during the past five decades from 83% (cabbages and other brassicas) to 488% (cauliflowers and broccoli), or by an average of

Integrated Processing Technologies for Food and Agricultural By-Products
https://doi.org/10.1016/B978-0-12-814138-0.00010-1

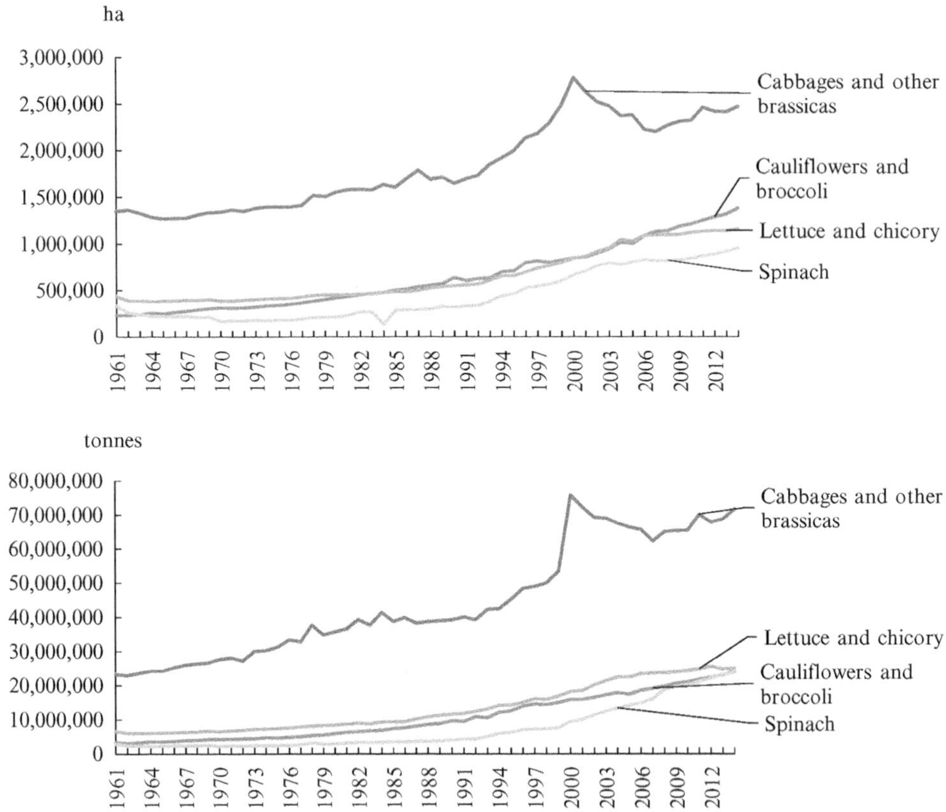

FIG. 1

(A) The world's total harvested area (ha), and (B) the world's total production (tons) during the 1961–2014 (FAO, 2017).

230% overall (Fig. 1). Meanwhile, the increases in the quantity of crop production noticeably surpassed the increases in harvested area.

According to FAO, China dominates the world's production of lettuce, cauliflowers, broccoli, cabbage, and spinach. The United States and India are also the main producers for these leafy vegetables. The top two countries (China and the United States) produce more than half of the world's production for lettuce and chicory, cabbages, cauliflowers, broccoli, and spinach (Fig. 2). Cabbage is the leading leafy vegetable produced in the world, and almost half of the world's production comes from China. Lettuce is the most produced leafy green in the United States, however, still more than half of its total production comes from China. Just over one-third of cauliflower and broccoli production comes from China and India, and almost all of the world's spinach is produced in China (approximately 91% in 2014). The growing demand for healthy foods is a significant driving factor in an increasing demand for leafy vegetables, however fresh consumption and cooking of leafy greens is a ubiquitous and longstanding practice worldwide.

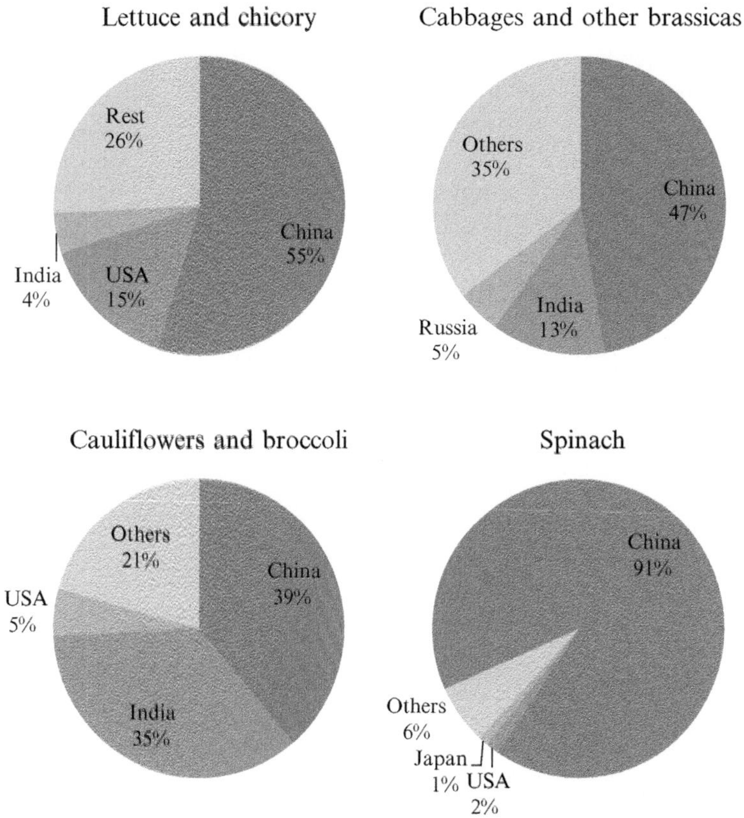

FIG. 2
The world's major producers of primary leafy vegetables in 2014. Source: FAO database; note that the database consists of countries with available data or able to compute by using FAO methodology; not all countries are included.

In the United States, the market for fresh leafy vegetables is led by lettuce, followed by cabbage, broccoli, and spinach. Leafy vegetables comprise one-fifth of the US vegetable market (Table 1), yet account for more than one-third of fresh market produce. Lettuce is primarily distributed for fresh produce, while cabbage, broccoli, and spinach partially go into processed foods. Leafy vegetables are mostly utilized for human consumption. Generally, most of its leafy body is consumable. Therefore, by-products could potentially be from nonedible or parts remaining during harvesting, or that wasted or spoiled during postharvest processing and storage, in bringing the product to market. In the consumer market, varieties with extended shelf life are preferred, while other short-lived varieties may be grown and distributed locally as they are easily perishable.

The US Department of Agriculture publishes grade standards for each specific type of vegetable, which include specific metrics to grade produce quality such as decay and damage (e.g., discoloration, freezing, scars) (USDA, 2004).

Table 1 The US Harvested Area for Leafy Vegetable (acres), Total Production (cwt), Utilized Production (cwt), Fresh Market Utilized Production (cwt), and Harvested but not Sold Production (cwt) in 2016

	Acres harvested	Production	Utilized production	Utilized production (fresh market)	Harvested production not sold
Lettuce	281,700 (11%)	89,811,000 (12%)	89,811,000 (12%)	89,811,000 (24%)	
-Head	126,300	47,601,000	47,601,000	47,601,000	
-Romaine	96,200	28,946,000	28,946,000	28,946,000	
-Leaf	59,200	13,264,000	13,264,000	13,264,000	
Cabbage	57,300 (2%)	22,665,000 (3%)	22,658,900 (3%)	18,449,000 (5%)	6100
Broccoli	131,300 (5%)	22,465,000 (3%)	22,465,000 (3%)	21,938,000 (6%)	
Spinach	47,500 (2%)	7,135,000 (1%)	7,135,000 (1%)	5,566,000 (1%)	
Total principle leafy vegetable (4)	517,800 (20%)	142,076,000 (18%)	142,069,900 (18%)	135,764,000 (36%)	
Total US principle vegetable[a] (22)	2,565,900 (100%)	780,271,000 (100%)	775,016,200 (100%)	375,305,200 (100%)	

[a]Total US principal vegetables includes artichokes, asparagus, beans (lima, snap), broccoli, cabbage, cantaloupes, carrots, cauliflower, celery, corn (sweet), cucumbers, garlic, honeydews, lettuce (head, leaf, romaine), onions, pea (green), peppers (bell, chili), pumpkins, spinach, squash, tomatoes, and watermelons (USDA, 2017).

However, by-products, which are not attached to the marketable produces, would not be graded or shall be graded as "No Established US Grade" (NOG). For example, intentionally detached romaine leaves from the bottom are graded as NOG.

2 CHEMICAL COMPOSITION OF LEAFY VEGETABLES AND BY-PRODUCTS

In the food industry, the nutritional importance of vegetables is their complex chemical composition, providing primary nutrients (carbohydrates, proteins, and lipids), a good source of fibers, a high content of minerals (Ca, Fe, Mg, P, K, Zn, Cu, Mn, and Se) and vitamins (A, B complex, C, E, K, carotene, lutein, and zeaxanthin), and other phytochemicals. The complex chemicals in vegetables have important effects on human metabolism because they provide nutritional and energy content, regulate metabolism, hydrate due to their high water content, and stimulate muscular and skeletal systems, internal glands, and enzyme activity. In addition to the primary nutrients used in food industry, vegetables, and their by-products contain a variety of organic phytochemicals

Table 2 Percentage of Refuse (%wb) and Refuse Parts for Leafy Vegetables (USDA, 2016)

Leafy vegetables	Scientific name	Refuse	Refuse parts
Cabbage	*Brassica oleracea* (Capitata Group)	20	Outer leaves and core
Chinese cabbage (Pe-tsai)	*Brassica rapa* (Pekinensis Group)	7	Outer leaves and root base
Chinese cabbage (pak-choi)	*Brassica rapa* subsp. *chinensis*	12	Base and damaged leaves
Cauliflower	*Brassica oleracea* (Botrytis Group)	61	Leaf stalks, cores and trimmings
Broccoli	*Brassica oleracea* var. *italica*	39	Leaves and tough stalks with trimmings
Broccoli, leaves	*Brassica oleracea* var. *italica*		
Broccoli, stalks	*Brassica oleracea* var. *italica*		
Lettuce, romaine	*Lactuca sativa* var. *logifolia*	6	Core
Lettuce, green leaf	*Lactuca sativa* var. *crispa*	36	Outer leaves
Lettuce, red leaf	*Lactuca sativa* var. *crispa*	20	Core and damaged outer leaves
Lettuce, butterhead	*Lactuca sativa* var. *capitata*	26	Outer leaves and core
Lettuce, iceberg	*Lactuca sativa* var. *capitata*	5	Core
Spinach	*Spinacia oleracea*	28	Large stems and roots

that can be extracted as a raw materials for other applications having medicinal, cosmetic, dermatological, and commercial importance. Vegetables derived from different parts of a plant, such as leaves, root, stem, bulb, fruit, flower, seeds, vary in characteristic and composition. In general, leafy green vegetables are rich in fibers, protein, vitamins A and C, Mg, chlorophyll, and carotenoids.

The five most produced leafy vegetables selected for investigating their characteristics and compositions in this chapter are cabbage, cauliflower, broccoli, lettuce, and spinach. Cruciferous leafy greens or the members of the genus *Brassica*, including cabbage, broccoli, kale, mustard greens, and collard greens, are high in nutrients and contain glucosinolates, which are naturally pungent components and contribute to health-promoting properties. Table 2 presents the percentage of refuse (inedible) for the selected leafy vegetables, and their common refused by-products. Cauliflower has the highest refuse rate at 61% among the common leafy vegetables studied. The percentage of refuse for broccoli and lettuce (green leaf) are 39% and 36%, respectively. In general, the refuse parts from leafy vegetables are the outer leaves, core, damaged leaves, stems, roots, and trimmings.

Most fresh vegetables and fruits are high in water content, which generally range from 75% to 95% by fresh weight (Butnariu and Butu, 2015). Leafy vegetables tend to have higher water contents, typically >90% (wb, wet basis) as illustrated in Table 3. Lettuce (in all varieties studied) and Chinese cabbage (pak-choi) contain high water content in the range of 95%–96% wb, whereas broccoli contains the lowest water content around 89% wb.

Table 3 Proximate Composition (% dry basis, db) of Leafy Vegetables and Their By-Products

	DM[a]	Carbs[b]	Fiber[c]	Sugars	Protein	Lipid	Ash
Primary crops							
Cabbage	7.8	74.2	32.0	40.9	16.4	1.3	8.2
Chinese cabbage (Pe-tsai)	5.6	57.6	21.4	25.1	21.4	3.6	17.5
Chinese cabbage (pak-choi)	4.7	46.6	21.4	25.2	32.1	4.3	17.1
Cauliflower	7.9	62.7	25.2	24.1	24.2	3.5	9.6
Broccoli	10.7	62.1	24.3	15.9	26.4	3.5	8.1
Lettuce, romaine	5.4	61.0	39.0	22.1	22.8	5.6	10.8
Lettuce, green leaf	5.0	57.2	25.9	15.5	27.1	3.0	12.4
Lettuce, red leaf	4.4	51.8	20.6	11.0	30.5	5.0	12.6
Lettuce, butterhead	4.4	51.0	25.2	21.5	30.9	5.0	13.0
Lettuce, iceberg	4.4	68.1	27.5	45.2	20.6	3.2	8.3
Spinach	8.6	42.2	25.6	4.9	33.3	4.5	20.0
By-products							
Cabbage, leaves[d]	10.0	61.7	33.7[e]	20.6	19.9	2.6	15.8
Cauliflower, leaves[d]	13.0	65.1	27.5[e]	18.6	17.0	4.2	13.7
Broccoli, leaves	9.3	54.4	24.7	15.9	32.0	3.8	9.9
Broccoli, stalks	9.3	56.3	N/A[f]	N/A[f]	32.0	3.8	9.9

[f]data not available (USDA, 2016).
[a]Dry matter (% wb).
[b]carbohydrate, calculated by difference.
[c]total dietary fiber.
[d]Wadhwa and Bakshi (2013).
[e]natural detergent fiber (NDF).

Protein from nutrient-dense leafy green vegetables accounts for approximately 30% of its calories and provides all the essential amino acids. The protein content of the selected leafy greens ranges from 16% to 33% db which is considerably higher than that db in dried legumes (20%–34%) (e.g., lentils, dried beans, and soybeans). In addition, spinach, lettuce (read leaf and butterhead), broccoli leaves and stalks, and Chinese cabbage (pak-choi) have >30% db protein. The amino acid profiles for the selected leafy vegetables are presented in Table 4. The main amino acids in the leafy green are glutamic, aspartic, lysine, and leucine. In leafy vegetables, lipid contents are low ranging from 1.3%–5.6% db and most of the lipids are in the form of saturated fatty acids, monounsaturated fatty acids, polyunsaturated fatty acids, and phytosterols, making them a low-calorie protein source.

Leafy vegetables are important sources of both digestible and fibrous indigestible carbohydrates. The digestible carbohydrates are present largely in the form of soluble sugars and starches for energy sources while indigestible fibers are important for normal digestion. Starch is an insoluble compound, often encapsulated in rigid cellulose, rendering them inaccessible to digestive enzymes. Starch is a common storage form of carbohydrate found mostly in roots and legumes. Leafy vegetables are very low in starch content, while its sugar content varies by species

Table 4 Amino Acid Composition (% db) of Leafy Vegetables and Their By-Products

	Trp	Thr	Ile	Leu	Lys	Met	Cys	Phe	Tyr	Val	Arg	His	Ala	Asp	Glu	Gly	Pro	Ser
Cabbage	0.14	0.45	0.38	0.52	0.56	0.15	0.14	0.41	0.24	0.54	0.96	0.28	0.54	1.56	3.76	0.38	0.61	0.68
Chinese cabbage (Pe-tsai)	0.21	0.70	1.21	1.25	1.27	0.12	0.23	0.62	0.41	0.94	1.19	0.37	1.23	1.53	5.13	0.62	0.45	0.68
Chinese cabbage (pak-choi)	0.32	1.05	1.82	1.88	1.90	0.19	0.36	0.94	0.62	1.41	1.79	0.56	1.84	2.31	7.69	0.92	0.66	1.03
Cauliflower	0.25	0.96	0.90	1.34	2.74	0.25	0.25	0.82	0.64	1.58	1.08	0.71	1.46	2.23	3.24	0.90	0.90	1.08
Broccoli	0.31	0.82	0.74	1.21	1.26	0.36	0.26	1.09	0.47	1.17	1.79	0.55	0.97	3.04	5.07	0.83	1.03	1.13
Broccoli, leaves	0.31	0.98	1.17	1.41	1.51	0.37	0.21	0.90	0.68	1.37	1.56	0.54	1.27	2.29	4.03	1.02	1.22	1.07
Broccoli, stalks	0.31	0.98	1.17	1.41	1.51	0.37	0.21	0.90	0.68	1.37	1.56	0.54	1.27	2.29	4.03	1.02	1.22	1.07
Lettuce, romaine	0.19	0.80	0.83	1.41	1.19	0.28	0.11	1.21	0.46	1.02	1.00	0.39	1.04	2.58	3.30	0.91	0.83	0.93
Lettuce, green leaf	0.18	1.18	1.67	1.57	1.67	0.32	0.32	1.10	0.64	1.39	1.41	0.44	1.12	2.83	3.63	1.14	0.96	0.78
Lettuce, red leaf	0.50	1.10	0.87	1.61	1.03	0.37	0.21	1.54	0.67	1.10	0.94	0.44	1.17	3.21	3.56	1.10	0.87	1.01
Lettuce, butterhead	0.30	0.94	0.89	1.62	1.28	0.32	0.21	1.21	0.43	1.24	1.17	0.39	1.24	3.25	4.71	1.01	0.85	1.05
Lettuce, iceberg	0.21	0.57	0.41	0.57	0.55	0.11	0.11	0.53	0.16	0.55	0.34	0.21	0.57	2.87	4.45	0.34	0.23	0.57
Spinach	0.45	1.42	1.71	2.59	2.02	0.62	0.41	1.50	1.26	1.87	1.88	0.74	1.65	2.79	3.99	1.56	1.30	1.21

Description of amino acid abbreviations: Ala, Alanine; Arg, Arginine; Asp, Aspartic acid; Cys, Cystine; Glu, Glutamic acid; Gly, Glycine; His, Histidine; Ile, Isoleucine; Leu, Leucine; Lys, Lysine; Met, Methionine; Phe, Phenylalanine; Pro, Proline; Ser, Serine; Thr, Threonine; Trp, Tryptophan; Tyr, Tyrosine; Val, Valine (USDA, 2016).

in the range of 5%–45% db. Among the selected leafy vegetables, the ones with higher sugar content are cabbage (41% db) and iceberg lettuce (45% db), while spinach contains the lowest sugar content (5% db). Primary sugars found in leafy vegetables are glucose and fructose. Other trace sugars found in certain leafy greens include sucrose, lactose, maltose, and galactose. Dietary fibers are resistant to human digestion, either derived from plant cell wall (cellulose, hemicelluloses, pectin, and lignin) or not (such as gums, algal polysaccharides, and mucilages) (Trowell et al., 1976). Fiber contents in the leafy greens range from 21% to 39% db, mostly in the form of cellulose. High-fiber leafy greens are romaine lettuce (39% db) and cabbage (32% db).

Leafy vegetables are good sources of minerals, mainly K, Ca, Mg, P, and Fe as illustrated in Table 5. They also contain trace minerals such as Zn, Mn, Cu, and Se. Important sources of K are spinach, lettuce, and Chinese cabbage. High Ca contents are found in Chinese cabbage and spinach. Since Mg is an essential component of chlorophyll, great sources of Mg are the dark green vegetables such as spinach, lettuce, and broccoli. These vegetables also contain high amounts of Zn. High Mn-containing vegetables are spinach and lettuce.

Table 5 Macro (% db) and Micro Mineral ([a]ppm, db) Content of Leafy Vegetable and Their By-Products

	K	Ca	Mg	P	Fe	Na	Zn[a]	Mn[a]	Cu[a]	Se[a]
Primary crops										
Cabbage	2.17	0.51	0.15	0.33	0.01	0.23	23.02	20.46	2.43	0.04
Chinese cabbage (Pe-tsai)	4.24	1.37	0.23	0.52	0.01	0.16	41.00	33.87	6.42	0.11
Chinese cabbage (pak-choi)	5.38	2.24	0.41	0.79	0.02	1.39	40.60	33.97	4.49	0.11
Cauliflower	3.77	0.28	0.19	0.55	0.01	0.38	34.05	19.55	4.92	0.08
Broccoli	2.95	0.44	0.20	0.62	0.01	0.31	38.32	19.63	4.58	0.23
Lettuce, romaine	4.58	0.61	0.26	0.56	0.02	0.15	42.67	28.76	8.91	0.07
Lettuce, green leaf	3.86	0.72	0.26	0.58	0.02	0.56	35.86	49.80	5.78	0.12
Lettuce, red leaf	4.29	0.76	0.28	0.64	0.03	0.57	45.87	46.56	6.42	0.34
Lettuce, butterhead	5.45	0.80	0.30	0.76	0.03	0.11	45.77	40.96	3.66	0.14
Lettuce, iceberg	3.23	0.41	0.16	0.46	0.01	0.23	34.40	28.67	5.73	0.02
Spinach	6.49	1.15	0.92	0.57	0.03	0.92	61.63	104.30	15.12	0.12
By-products										
Cabbage, leaves[b]	0.44	2.38	0.68	0.23	0.09	0.43	48.30	54.60	9.40	–[c]
Cauliflower, leaves[b]	0.60	2.17	0.44	0.34	0.04	0.39	40.80	40.80	4.00	–[c]
Broccoli, leaves	3.49	0.52	0.27	0.71	0.01	0.29	42.96	24.60	4.83	0.32
Broccoli, stalks	3.49	0.52	0.27	0.71	0.01	0.29	42.96	24.60	4.83	0.32

[a]ppm.
[b]Wadhwa and Bakshi (2013).
[c]data not available (USDA, 2016).

Leafy vegetables are also important sources of certain vitamins, especially vitamins C, A, β-carotene, lutein, and zeaxanthin (Table 6). The amount of vitamins in vegetables depends on the type, growth development and maturation, soil, nutrient management, and mode of conservation. Vitamin C is an essential nutrient required by human body and a powerful antioxidant for the development and maintenance of blood vessels, scar tissue, and cartilage. Vitamin C is required for generating ATP, dopamine, peptide hormones, and tyrosine. Dark leafy vegetables tend to have a higher vitamin C or ascorbic acid content than lighter ones. Thus, a greater amount of vitamin C is found in broccoli, cauliflower, and cabbage.

Leafy vegetables are rich in vitamin B. Thiamin (B1) is found in spinach, broccoli, lettuce, and cabbage. Among the selected leafy vegetables, a higher content of riboflavin (B2) is found in spinach and broccoli. Niacin (B3) is found in a higher amount in spinach, broccoli, and cauliflower. Broccoli and cauliflower are also a good source of pantothenic acid (B5). Pyridoxine (B6) content is higher in spinach, broccoli, and cabbage. Folate (B9) is found in a higher amount in spinach and romaine lettuce. β-carotene, vitamin A (retinol activity equivalents), lutein and zeaxanthin, and vitamin K are often found in leafy greens, with the highest content in spinach and lettuce. Chlorophyll is most abundant in dark green leafy vegetables since plant photosynthesis takes place primarily in leaves. Spinach is listed as the top source of chlorophyll with 23.7 mg of chlorophyll in 1 cup of raw spinach (Busch, 2017). Dark leafy greens (spinach and lettuce) are high in β-carotene but their leaves contain so much chlorophyll that the green pigment hides the orange pigment of β-carotene. Lutein and zeaxanthin are xanthophyll, one of the two major divisions of the carotenoid group mostly found in the dark greens.

3 PROCESSING TECHNOLOGY OF BY-PRODUCTS FOR THE PRODUCTION OF VALUE-ADDED COMPOUNDS

3.1 Nutraceuticals

Recently, there is a blurring line between food and medicine. In 1989, the term "nutraceutical" was defined by Dr. Stephen DeFelice, the founder and chairman of the Foundation for Innovation in Medicine, by combining the terms "nutrition" and "pharmaceutical." It refers to any substance that may be considered a food or part of a food and provides medical or health benefits, encompassing prevention and treatment of diseases (Rajasekaran et al., 2008).

Due to growing trend of health-seeking consumers, the market size of "nutraceuticals" or "functional foods" have seen rapid growth. In 2008, over 470 nutraceutical and functional food products were commercially available with documented health benefits. In 2016, the value of nutraceutical market was 205.39 billion USD, and is expected to reach 297 billion USD by the end of 2022 (MordorIntelligence, 2017). The leading nations in the global nutraceutical markets are the United States, Japan, Israel, and Germany.

Table 6 Vitamins (Value Per 100 g wb) of Leafy Vegetables and Their By-Products

	C (mg)	Thia (mg)	Ribo (mg)	Niac (mg)	Pant (mg)	B6 (mg)	Fol (μg)	Chol (mg)	A (μg)	βC (μg)	αC (μg)	L & Z (μg)	E (mg)	K (μg)
Primary crops														
Cabbage	36.60	0.06	0.04	0.23	0.21	0.12	43.00	10.70	5.0	42.0	33.0	30.0	0.15	76.0
Chinese cabbage (Pe-tsai)	27.00	0.04	0.05	0.40	0.11	0.23	79.00	7.60	16.0	190.0	1.0	48.0	0.12	42.9
Chinese cabbage (pak-choi)	45.00	0.04	0.07	0.50	0.09	0.19	66.00	6.40	223.0	2681.0	1.0	40.0	0.09	45.5
Cauliflower	48.20	0.05	0.06	0.51	0.67	0.18	57.00	44.30	–	–	–	1.0	0.08	15.5
Broccoli	89.20	0.07	0.12	0.64	0.57	0.18	63.00	18.70	31.0	361.0	25.0	1403.0	0.78	101.6
Lettuce, cos or romaine	4.00	0.07	0.07	0.31	0.14	0.07	136.00	9.90	436.0	5226.0	–	2312.0	0.13	102.5
Lettuce, green leaf	9.20	0.07	0.08	0.38	0.13	0.09	38.00	13.60	370.0	4443.0	–	1730.0	0.22	126.3
Lettuce, red leaf	3.70	0.06	0.08	0.32	0.14	0.10	36.00	11.80	375.0	4495.0	–	1724.0	0.15	140.3
Lettuce, butterhead	3.70	0.06	0.06	0.36	0.15	0.08	73.00	8.40	166.0	1987.0	–	1223.0	0.18	102.3
Lettuce, iceberg	2.80	0.04	0.03	0.12	0.09	0.04	29.00	6.70	25.0	299.0	4.0	277.0	0.18	24.1
Spinach	28.10	0.08	0.19	0.72	0.07	0.20	194.00	19.30	469.0	5626.0	–	12,198.0	2.03	482.9
By-products														
Broccoli, leaves	93.20	0.07	0.12	0.64	0.54	0.16	71.00	–	–	–	–	–	–	–
Broccoli, stalks	93.20	0.07	0.12	0.64	0.54	0.16	71.00	0.00	20.0	–	–	–	–	–

Description of vitamin abbreviations: A, vitamin A (retinol activity equivalents); B6, vitamin B-6; C, vitamin C (total ascorbic acid); Chol, choline; E, vitamin E (alpha-tocopherol); Fol, folate; K, Vitamin K (phylloquinone); L, lutein; Niac, niacin; Pant, pantothenic acid; Ribo, riboflavin; Thai, thiamin; Z, zeaxanthin; αC, alpha carotene; βC, beta (USDA, 2016).

Extract from leafy vegetables contains significant amount of bioactive compounds and is also claimed to provide protection against many diseases. Extracts from spinach, lettuce, and broccoli contain β-carotene (derivative of carotenoid), the important precursor of vitamin A, which has antioxidant properties and help in preventing cancer and other diseases (McMillan et al., 2002). Dark leafy vegetables, such as kale, collards, pak-choi, spinach, cabbage, and lettuce, contain lutein and zeaxanthin, which are derivatives of carotenoid and reported to act as vision improving agents (AOA, 2017; Rajasekaran et al., 2008). In addition, broccoli, cauliflower, cabbage, pak-choi, kale, and collards also contain glucosinolates, which are powerful activators of liver detoxification enzymes and have been associated with lower risk of lung and colorectal cancer (Fenwick and Heaney, 1983; Higdon et al., 2007). Spinach was reported to possess antitumor activity due to its curcumin (diferuloylmethane) content (Thanopoulou et al., 2006).

In addition, leafy vegetables also contain significant amounts of dietary fiber; cauliflower, broccoli, lettuce, and spinach have approximately 20%–25% db of dietary fiber (Butnariu and Butu, 2015), while cabbage contains approximately 40%–43% db of dietary (Jongaroontaprangsee et al., 2007). Thus, leafy vegetable by-products have a great potential to be transformed into dietary fibers. Dietary fibers appear to have many health benefits including the potential to lower the risk of developing coronary heart disease, stroke, obesity, diabetes, hypertension, and certain gastrointestinal diseases (Anderson et al., 2009). The major characteristics of the commercialized dietary fiber are: total dietary fiber content higher than 50% wb, moisture content lower than 9% wb, low lipid content, low calorie (lower than 8.36 kJ/g), and neutral flavor and taste (Larrauri, 1999).

Besides soybeans, green vegetables have long been recognized as the cheapest and the most abundant sources of protein from plants. Leafy vegetables contain approximately 27% db of protein, ranging from 16.4% in cabbage to 33.7% in spinach (Butnariu and Butu, 2015). Recently, the production and utilization of leaf protein extracts have been extensively studied in Europe, United States, and Asia due to global demand toward alternative protein sources that contribute to lower energy consumption and generate lesser CO_2 emission than the animal-based protein production (Aletor et al., 2002; Sari et al., 2015).

3.2 Natural Food Colorants

Besides use as nutraceuticals, leafy vegetable by-products have been used to produce natural coloring agents. In 2015, the global market size of natural food color was estimated at 1.32 billion USD and kept on increasing. Due to the rapid growth of the food industry and health awareness related to the use of synthetic color, the demand for natural food color in Asia Pacific region is expected to gain the highest growth rate of over 7% by revenue from 2016 to 2025 (GRV, 2017; Vayupharp and Laksanalamai, 2015). US-FDA and Codex have approved the use of four important classes of natural pigments including chlorophylls, carotenoids, betalains and anthocyanins, and their extracts as food additives (Francis, 1996).

Since these pigments cannot be synthesized by humans and animals (Žnidarčič et al., 2011), consumption of fruits and vegetables or its extracts is recommended for healthy diets. Leafy vegetables contain three types of photosynthetic pigments, that is, chlorophylls, carotenoids, and anthocyanins (Humphrey, 2004; Kimura and Rodriguez-Amaya, 2002; Larsen and Christensen, 2005). Besides giving an appealing color to foods, these pigments have been shown to impose several health benefits including the prevention of various diseases associated with oxidative stress such as cancer, cardiovascular diseases, and other chronic diseases (Sangeetha and Baskaran, 2010). More details about the basic chemical structure of each pigment, its natural sources, and color stability are described in the following.

3.2.1 *CHLOROPHYLLS*

Chlorophylls are commonly found oil-soluble pigments responsible for the green color of plants. The structure of chlorophylls is closed ring tetrapyroles chelated with magnesium atom in the center (Marquez and Sinnecker, 2008) as shown in Fig. 3. There are five forms of chlorophylls found in plants and photosynthetic organism, but in plant kingdom only two major forms are commonly found, that is, chlorophylls *a* and *b*. The difference between these chlorophylls is chemical compound at position 7. Chlorophyll *a* is composed of $-CH_3$ while chlorophyll *b* is composed of $-CHO$. The difference in chemical composition leads to the difference in color; chlorophyll *a* exhibits blue-green, while chlorophyll *b* appears yellow-green color (Ngamwonglumlert et al., 2017).

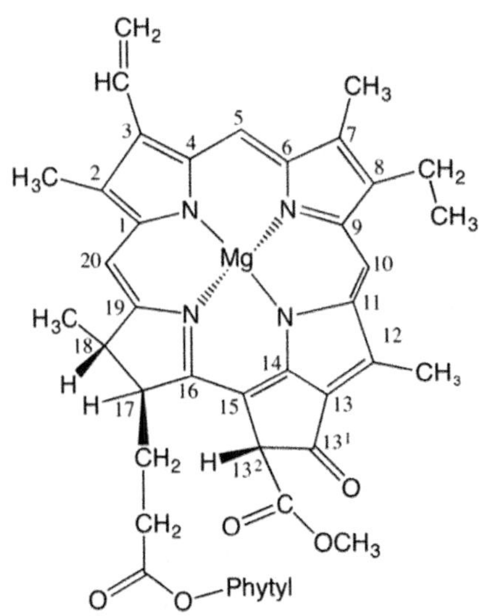

FIG. 3
Basic structure of chlorophylls pigment (Marquez and Sinnecker, 2008).

These two forms of chlorophyll coexist in plants in an approximate ratio of 3:1 with chlorophyll *a* being predominant (Humphrey, 2004).

Leafy vegetables that possess high amounts of chlorophylls are spinach, lettuce, and broccoli, respectively (Larsen and Christensen, 2005). Chlorophylls are generally confined to the chloroplasts and held within phospholipid membrane from which they can be extracted using proper solvents and extraction methods. General production process of commercial chlorophyll products by solvent extraction is described in Humphrey's study (Humphrey, 2004). Chlorophyll degradation and color change can easily occur when chlorophylls are exposed to heat, light, oxygen, acids, and enzymes. The loss of central magnesium atom in the chlorophylls structure is the major cause of the structural transformation from native chlorophyll to pheophytin, which exhibits olive brown color (Marquez and Sinnecker, 2008).

3.2.2 *CAROTENOIDS*

Carotenoids are lipid-soluble and widely found pigments in yellow, orange, and red fruits and vegetables; several of which have vitamin A activity. In nature, >600 types of carotenoids have been identified and they are categorized into three groups: carotenes, xanthophylls (yellow-orange pigments), and lycopene (red pigment). Carotenes are composed of carbon and hydrogen, while xanthophylls are composed of carbon, hydrogen, and oxygen. Lycopene has two additional double bonds compared with β-carotene (Khoo et al., 2011). The structure of major carotenoids is as shown in Fig. 4.

Regardless of the color and appearance, green leafy vegetables are reported to be a good source of carotenoids; carotenoids coexist with chlorophylls in many vegetables such as broccoli, lettuce, and spinach in the form of β-carotene, lutein, and zeaxanthin. In addition, raw red cabbage is regarded as a good source of lycopene (Khoo et al., 2011; Larsen and Christensen, 2005). Carotenoids are localized in chloroplasts and chromoplasts of plant cells. In chloroplasts, the carotenoids are mainly associated with proteins and act as photosynthesis aid, whereas in chromoplasts, they exist in crystalline form or as oily droplets (Bartley and Scolnik, 1995). Carotenoids are generally degraded by oxidation and isomerization reactions leading to fading of red and yellow colors of plant materials. These reactions can be stimulated by light, heat, peroxide, metal ions, and enzymes. Most carotenoids in plants are *trans*-isomer and isomerization converts *trans*-isomers to *cis*-isomers during cooking and further thermal processing (Khoo et al., 2011).

3.2.3 *ANTHOCYANINS*

Anthocyanins are water-soluble pigments found in many flowers, fruits, and vegetables (e.g., butterfly pea, rosella, grape, berries, red cabbage, and purple cauliflower). The shade of anthocyanins can vary from orange to red to blue depending on the pH of food matrix (Vayupharp and Laksanalamai, 2015) and its structure (Ngamwonglumlert et al., 2017). The basic structure of anthocyanins is glycosides of anthocyanidins, which is referred to as anthocyanins

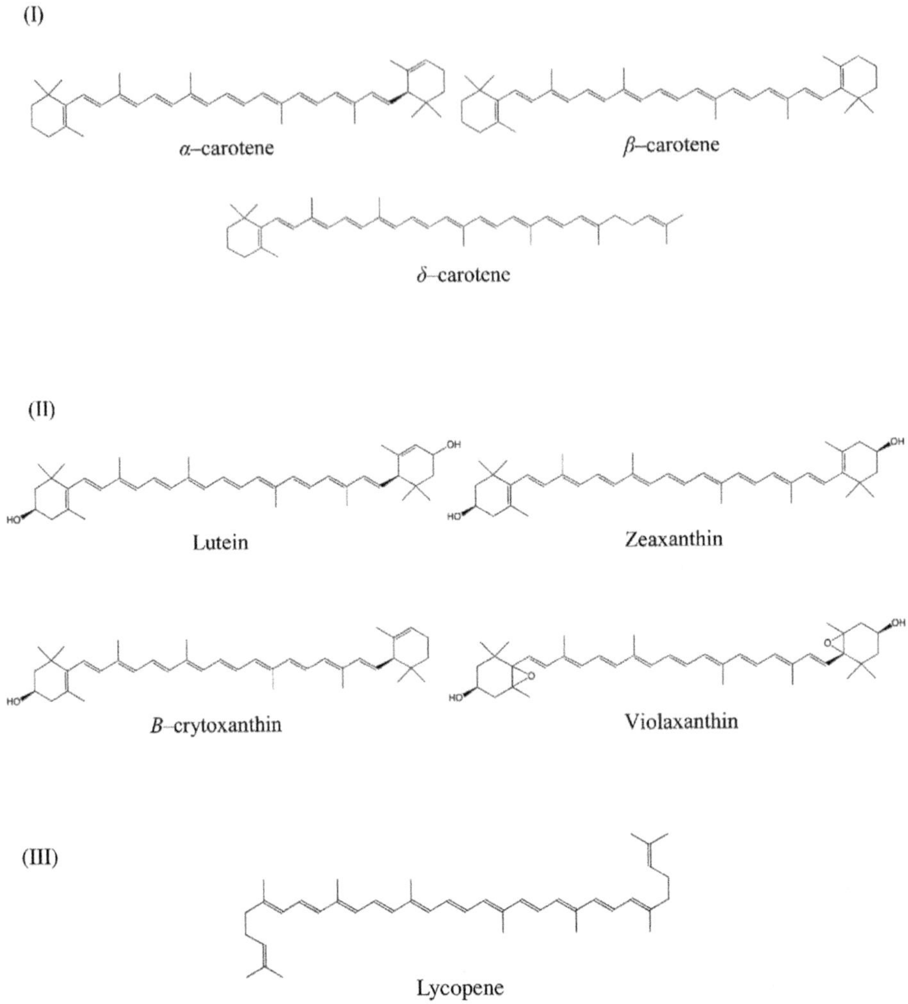

FIG. 4

Structures of major carotenes (I), xanthophylls (II), lycopene (III). Adapted from Ngamwonglumlert, Luxsika, Devahastin, Sakamon, Chiewchan, Naphaporn, 2017. Natural colorants: pigment stability and extraction yield enhancement via utilization of appropriate pretreatment and extraction methods. Crit. Rev. Food Sci. Nutr. 57 (15), 3243–3259.

without sugar molecules. Although there is >23 groups of anthocyanins (Zhao et al., 2015), the commonly found anthocyanins in nature are based on six anthocyanidins (Tanaka et al., 2008; Wrolstad, 2004): pelargonidin (plg), cyanidin (cyd), delphindin (dpd), penodin (pnd), petunidin (ptd), and malvidin (mvd), as shown in Fig. 5. Pigments that contain a larger number of methoxyl groups exhibit more redness, whereas those having a larger number of hydroxyl groups appear more bluish (Delgado-Vargas and Paredes-López, 2002).

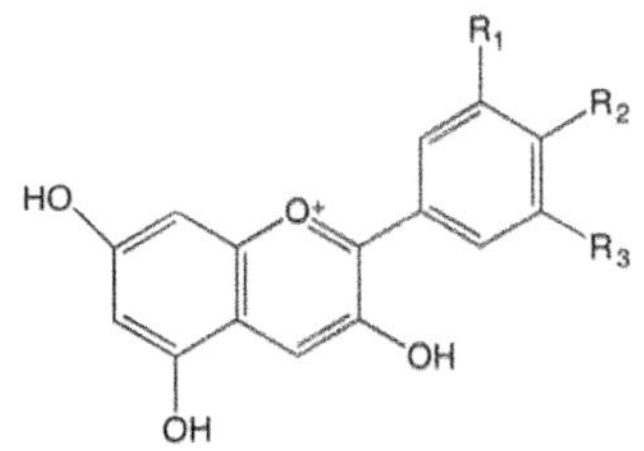

Anthocyanidin	Substitution pattern			Color
	R1	R2	R3	
Pelargonidin (plg)	H	OH	H	Orange
Cyanidin (cyd)	OH	OH	H	Magenta, Crimson
Delphinidin (dpd)	OH	OH	OH	Purple, Mauve, Blue
Peonidin (pnd)	OCH_3	OH	H	Magenta
Petunidin (ptd)	OCH_3	OH	OH	Purple
Malvidin (mvd)	OCH_3	OH	OCH_3	Purple

FIG. 5
Structures of six important anthocyanidins in nature (Ngamwonglumlert et al., 2017).

Anthocyanins are synthesized in cytosol and localized in vacuole of flowers and fruits, and also found in leaves, stems, and roots. They are mostly concentrated in outer cell layers such as the epidermis and peripheral mesophyll cells (Tanaka et al., 2008). Nature, traditional agricultural methods, and plant breeding have produced various uncommon crops containing anthocyanins, including purple or red broccoli, cabbage, cauliflower, carrots, and corn. Anthocyanin-based food colorants are commercially manufactured from horticultural crops (e.g., red cabbage, red shiso) and processing wastes (Wrolstad, 2004). Among red vegetables, red cabbage has generally been used for the production of natural dye for food application due to its high anthocyanin content (40–188 mg Cy 3-glcE/100 g of fresh weight) (Mizgier et al., 2016). Anthocyanins are highly sensitive to light, heat, pH, and are susceptible to oxidation and enzymatic reactions.

3.3 Extraction of Bioactive Compounds and Natural Coloring Agents

The basic extraction method of nonprotein chemical compounds from leafy vegetables is hydrodistillation with organic solvents or water. However, it requires long extraction time and a large amount of solvent. Protein extracts from leafy vegetables are usually obtained from mechanical size reduction of the leaves, followed by heat coagulation of the proteins, centrifugation, and drying. Elevated temperatures are used during protein coagulation and drying causes protein denaturation; the proteins loose most of their functional (e.g., foaming, emulsifying), but not their nutritional properties.

The main drive for novel extraction techniques is the concern of food industry toward the stability of natural color (Ngamwonglumlert et al., 2017), the functional properties of protein (Sari et al., 2015), and the degradation of bioactive compounds during extraction (Tanongkankit et al., 2010). Advanced extraction methods, such as supercritical fluid extraction (SFE), pressurized liquid extraction (PLE), microwave-assisted extraction (MAE), ultrasound-assisted extraction (UAE), pulsed electric field (PEF) extraction, and enzyme-assisted extraction (EAE), have emerged as an alternative choice since they consume less solvent, require shorter extraction time, and are more environmentally friendly (Cheok et al., 2014). Combined extraction technique may also be employed to increase extraction yield, improve extract quality, and may enhance generation of secondary plant metabolites. The secondary plant metabolites that exhibit pharmacological or toxicological effects in human or animals are considered as bioactive compounds (Azmir et al., 2013).

3.3.1 CONVENTIONAL EXTRACTION METHODS

Most of the conventional extraction methods are based on the ability to extract different solvents and simple application of heat and/or mixing. Three most commonly applied methods are Soxhlet extraction, maceration, and hydrodistillation.

3.3.1.1 Soxhlet Extraction

This method was initially designed by German chemist Franz Ritter Von Soxhlet in 1879 for lipid extraction from a solid material. Nowadays, it is often used as a reference method to determine an extraction yield of lipid from plant materials. The principle of Soxhlet extraction is using a solvent vapor to dissolve oil-base compounds inside the plant sample housed in a thimble. The vapor is then condensed to separate the compounds of interest from the solvent. Although Soxhlet extraction is simple, inexpensive, and easy to operate, it requires a large amount of solvent, long extraction time, and leads to degradation of heat-labile bioactive compounds (De Castro and Garcıa-Ayuso, 1998; Ngamwonglumlert et al., 2017).

3.3.1.2 Maceration

Maceration is a simple extraction method that can be conducted without any additional heat source. A plant sample is normally ground to increase an exposure of bioactive compounds to the solvent and mixed with the solvent of choice. The mixture is left in an extraction vessel with occasional agitation. After the process is finished, the liquid is separated from the solid by mechanical press or centrifugation. Since this method is generally performed at room temperature, it requires long extraction time. Moreover, a large amount of solvent is also needed to repeat the extraction until there is no more interesting compounds left in the sample (Azmir et al., 2013). One of the major drawbacks of this method is that the extract is obtained in the crude form and further purification is needed.

3.3.1.3 Hydrodistillation

Hydrodistillation has long been used for the extraction of essential oils and bioactive compounds from plant materials. Hydrodistillation can be performed in three different ways: water distillation, water and steam distillation, and direct steam distillation (Azmir et al., 2013). Three main physicochemical processes involved in hydrodistillation are hydro-diffusion, hydrolysis, and decomposition by heat. Hot water and steam act as the main media to free bioactive compounds in the plant matrix and carry them along. Indirect cooling condenses the vapor mixture and make separation of oil and bioactive compounds from the water happen. The essential oils and oil-based bioactive compounds are generally dried over anhydrous sodium sulfate. Since hydrodistillation is often conducted at temperature above boiling point of water, some volatile components, natural pigments, and heat-labile bioactive compounds may be lost.

3.3.2 NOVEL EXTRACTION METHODS

3.3.2.1 Enzyme-Assisted Extraction

Most of the bioactive compounds in plant are either dispersed in cytoplasm enclosed by cell membranes and cell wells or entrapped in the polysaccharide-lignin network by hydrogen or hydrophobic bonding making them not easily accessible by solvent used in routine extraction process. The addition of specific enzymes such as pectinase, cellulose, α-amylase, and hemicellulose during extraction facilitates the release of cellular constituents and enhance extraction recovery by breaking the cell walls and membranes and hydrolyzing polysaccharides and lipid complexes, including membranes (Azmir et al., 2013). Enzymes have been derived from fungi, bacteria, animal organs, or fruit/vegetable extracts. They have been used particularly for the treatment of plant materials prior to conventional extraction. The EAE has been considered as a novel and eco-friendly alternative to release bounded compounds and increase overall yield. The advantages of EAE are its nonthermal nature, lower solvent and energy consumption, lower toxicity, and its effectiveness in aqueous solution (Puri et al., 2012).

To effectively use enzymes for extraction enhancement, it is critical to understand their catalytic property and mode of action, optimal operating conditions, and which enzyme or enzyme combination is suitable for each plant material. Factors influencing efficacy of EAE are enzyme composition and concentration, particle size of plant materials, solid to water ratio, and hydrolysis time (Niranjan and Hanmoungjai, 2004).

3.3.2.2 Supercritical Fluid Extraction

In SFE, a specific gas is exposed to pressure and temperature beyond its critical point, above which distinctive gas and liquid phases do not exist. Supercritical fluid possesses liquid-like density and solvation power, and gas-like properties of diffusion, viscosity, and surface tension. These properties make it an excellent solvent to extract flavor and bioactive compounds from plant materials in a short time with higher yields at low temperature. As the pressure decreases, their solvation power increases (Sowbhagya and Chitra, 2010). The SFE has many advantages over other extraction methods such as the absence of solvent residues and the rich top flavor notes of the extracts. Selective extraction of a single component can be achieved by using optimal process condition. Main factors influencing the extraction efficiency are temperature, pressure, particle size and moisture content of raw material, extraction time, gas flow rate, and solvent-to-raw material ratio (Azmir et al., 2013).

An ideal solvent for SFE is carbon dioxide (CO_2) because its critical temperature (31 °C) is close to room temperature, and its critical pressure (74 bars) is relatively lower making it possible to operate at moderate pressures. The SFE is commonly operated in a temperature range of 30–60°C and a pressure range of 8–40 MPa (Ngamwonglumlert et al., 2017). The only disadvantage of CO_2 is its low polarity, which makes it unsuitable for the extraction of polar substances, which are common in pharmaceuticals and drug samples. This limitation has been successfully overcome by the use of chemical modifier like dichloromethane (CH_2Cl_2). Usually a small amount of modifier can significantly enhance the polarity of carbon dioxide (Hawthorne et al., 1994). Although SFE is quite efficient in extracting heat-labile bioactive compounds and consume small amount of solvent, this extraction method requires high capital and operating costs because of the high pressure required for the operation.

3.3.2.3 Pressurized Liquid Extraction

The concept of PLE is the application of high pressure to remain solvent liquid beyond their normal boiling point. The common range of pressure and temperature used are 10.3–13.8 MPa and 40–200°C, respectively. High pressure forces the solvent into the pores of sample matrix allowing better contact between the solvent and compounds to be extracted. High temperature, on the other hand, results in better diffusion of solvent into the sample matrix and also helps disrupt plant cells, as well as, increase solubility and mass transfer rate. Thus, PLE can significantly reduce extraction time and amount of solvent required.

The PLE can effectively extract both water- and oil-based compounds, depending on the solvent used. For extraction of polar compounds, PLE is considered to be an alternative to SFE (Azmir et al., 2013). However, PLE is not suitable for the extraction of heat-labile compounds since the sample is subjected to high temperature. Similar to SFE, one of the major drawbacks of PLE is high capital and operating costs due to the use of high pressure during extraction.

The PLE has been successfully used for the extraction of bioactive compounds from various plant materials. For example, extraction of flavonoids from spinach by PLE using an ethanol and water mixture (70:30) as a solvent at 50–150°C were more effective than using water solvent alone at 50–130°C (Howard and Pandjaitan, 2008).

3.3.2.4 Ultrasound-Assisted Extraction

The principle of UAE is to use bubble cavitation created by ultrasound waves (20 kHz to 100 MHz) to rupture plant cell walls and intact plant materials, and thus, enhance the extraction efficiency. Cavitation is a phenomenon caused by the passage of ultrasound wave through liquid medium creating alternative compression and expansion cycles of the bubbles. As the bubbles become too large to contain by its surface tension, they collapse resulting in massive shearing force and large amount of energy converted from kinetic energy of motion to heating of the bubble content. According to Suslick and Doktycz (1990), the bubbles have temperature of about 5000 K, pressure of about 1000 atm and, heating and cooling rate above 10^{10} K/s. Based on this principle, only liquid and liquid-containing solid materials have cavitation effect. The UAE can effectively facilitate organic and inorganic compound leaching from plant matrix.

Extraction efficiency is greatly influenced by moisture content of sample, milling degree, particle size, and the type of solvent used. In addition, temperature, pressure, frequency, and time of sonication are the key process parameters for the action of ultrasound. The UAE has also been incorporated with other techniques to enhance the efficiency of a conventional system.

The advantages of UAE include the ability to extract at lower temperatures due to the absence of external heat input, low solvent consumption, and reduction in extraction time. Ultrasound energy can also facilitate faster energy transfer, more effective mixing, lower thermal gradients and extraction temperature, selective extraction, and reduced equipment size.

3.3.2.5 Microwave-Assisted Extraction

The MAE is a novel method for extracting soluble compounds into fluids (both polar and nonpolar) from plant materials using microwave energy. Microwaves are electromagnetic fields in the frequency range from 300 MHz to 300 GHz. Rapid heating is generated by ionic conduction and dipole rotation mechanisms, which results in rapid expansion of cell structure and rupture of plant cell walls and membranes. These phenomena lead to enhance mass transport and extraction rate. The advantages of MAE are high extraction yield, short process time, small temperature gradient, small solvent consumption, and small extraction unit (Azmir et al., 2013). However, it is not suitable for the extraction of heat-labile compounds. Recently, MAE combined with vacuum has been proposed to make it possible to operate at lower temperature (Ngamwonglumlert et al., 2017). The factors affecting MAE efficiency are solvent type and concentration, extraction time, and microwave power.

3.3.2.6 Pulsed Electric Field Extraction

Pulsed electric field (PEF) pretreatment has been used for improving many mass transfer processes, for example, drying, diffusion, pressing, and extraction. The application of short and high-voltage electric field can nonthermally rupture biological membranes by electroporation effect (Asavasanti et al., 2011). During

PEF treatment, molecules separate according to their charges and accumulate on the opposite sides of lipid bilayer membranes. Once the transmembrane exceeds a critical value of approximately 1 V, pores are formed in the weak area of the membrane and cause a dramatic increase in the electrical conductivity and cell membrane permeability (Azmir et al., 2013). Thus, PEF can be used to increase extraction yield and decrease extraction time.

The factors affecting PEF efficiency include electric field strength applied, specific energy input, pulse number, frequency, treatment temperature, and properties of the samples to be treated. The advantages of PEF are short treatment time and little to no heat generation making possible the extraction of heat-sensitive compounds. However, nonpolar compounds do not benefit from PEF treatment since this technique requires solvent with good electrical conductivity (Ngamwonglumlert et al., 2017). In addition, PEF requires high-power supply generator and special design treatment chamber making it expensive and difficult for industrial-scale application.

3.4 Dietary Fiber Production Processes

Dietary fiber powder is generally made from fruit and vegetable by-products. Outer leaves of leafy vegetables, such as cabbage, lettuce, and broccoli, are generally discarded and used as animal feed and fertilizer. Due to its high dietary fiber and antioxidant contents, many researchers have explored different methods to produce dietary fiber powder from these by-products (Femenia et al., 1998; Larrauri, 1999; Tanongkankit et al., 2010). Extensive review on the utilization of food processing by-products as dietary, functional, and novel fiber was investigated by Sharma et al. (2016).

A basic process of dietary fiber powder production starts from washing the raw material and cutting it into small pieces. The following treatments have been performed to improve the functionality of insoluble fiber, which is the main component of some commercial products (Larrauri, 1999):

- Blanching and dipping in chemicals to inhibit various enzymes.
- Partial delignification of lignocellulose by alkaline hydrogen peroxide treatment or acid hydrolysis to improve water-holding capacity and swelling property.
- Enzymatic treatment to improve sensory properties and enhance soluble fiber.
- Encapsulation with soluble fiber to produce a product with better textural properties.
- Extrusion.

The pretreated raw material is then dried to the desired moisture content and finally ground into fine powder.

3.5 Anaerobic Digestion for Biogas Production

Most of the leafy vegetable wastes generated are disposed in municipal landfill or dumping sites, which cause many environmental problems. Leafy vegetables deteriorates easily, and can generate foul smells, greenhouse gases (GHGs), and leachate. The anaerobic digestion of leafy wastes offers a double advantage as it produces biogas energy and simultaneously treats the organic wastes, reducing the disposal in landfills (Appels et al., 2011). Considering the high moisture and organic content, leafy wastes are better suited for anaerobic digestion rather than direct combustion technologies like incineration.

Anaerobic digestion is a biofuel conversion technology for energy recovery from organic material to produce biogas fuel, which contains mainly methane, carbon dioxide, and trace amounts of hydrogen, hydrogen sulfide, ammonia, and water vapor. Through anaerobic digestion, biogas is produced, and nutrients in the influent are mineralized and remain in the digester effluent. Biogas, containing 50%–70% methane generated from anaerobic digestion, can be used directly as a combustion fuel for the generation of electricity and heat. The biogas can be processed and purified to compressed natural gas (CNG) for transportation fuel. The effluent obtained after digestion is biologically stabilized and rich in nutrients, which can be used for soil amendment or organic fertilizer.

When composition of the organic matter is known, theoretical methane and ammonia yields produced from anaerobic digestion can be estimated using Buswell's equation as follows (Buswell and Mueller, 1952). However, this theoretical approach does not take into account the needs for cell maintenances and anabolism.

$$C_aH_bO_cN_d + \left(\frac{4a-b-2c+3d}{4}\right)H_2O \rightarrow \left(\frac{4a+b-2c-3d}{8}\right)CH_4 + \left(\frac{4a-b+2c+3d}{8}\right)CO_2 + dNH_3 \tag{1}$$

According to the composition of substrates, different methane yield can be estimated from carbohydrates, proteins, and lipids for 0.415, 0.851, and 1.014 L CH_4/gVS, respectively (Angelidaki and Sanders, 2004). The anaerobic digestion process has been widely used for solid waste treatment, and proved to be an efficient energy conversion process with high net energy yield. In addition, anaerobic digestion has been emphasized as an approach to reduce the external energy demand and to recycle fertilizer (Lardon et al., 2009; Sialve et al., 2009).

3.5.1 *ANAEROBIC DIGESTION OF LEAFY VEGETABLES*

Biomethane production and biodegradability of typical leafy vegetables through anaerobic digestion was comprehensively explored by Yan et al. (2017). The established data on biogas production and the kinetic model describing anaerobic digestion process from leafy vegetables provide useful information for agro-industrial application of vegetable wastes using anaerobic digestion technology. Experimental and theoretical methane yield based on the organic composition

(VFA, protein, lipid, carbohydrate, and lignin) from Yan's study are presented in Table 7.

Anaerobic digestion of leafy by-products was studied for the biomethane potential, and digester performance in a continuous system (Zuo et al., 2015). Different digester designs for anaerobic digestion of vegetables and by-products were described by Bouallagui et al. (2005).

3.5.2 *ANAEROBIC DIGESTION OF LEAFY VEGETABLE BY-PRODUCTS*

In addition to the digestion of waste from the primary leafy vegetables, by-product wastes, such as, leaves, stem, and roots, could be used as a feedstock to anaerobic digestion, either as a single feedstock or co-digest with other organic materials. Wastes from leafy vegetables are produced mostly during agricultural production and harvesting.

Cabbage leaves, stem, and waste from food processing are the major source of cabbage by-products. According to Gunaseelan (2004), the methane yield of cabbage leaves and stems were 309 and 291 mL/g volatile solids (VS), respectively (Gunaseelan, 2004). Mesophilic anaerobic digestion of cabbage waste showed better performance in terms of biogas yield compared with thermophilic anaerobic digestion. The digestion of cabbage leaves (*Brassica oleracea* var. *capitata*) at mesophilic condition showed 43% increase in biogas yield than that at thermophilic digestion (Liu et al., 2009), making it more suitable for digestion at mesophilic condition as studied.

Similarly, anaerobic digestion of Chinese cabbage waste generated from a kimchi (fermented cabbage) production during trimming processes showed that mesophilic conditions obtained a higher biogas yield than thermophilic conditions

Table 7 Methane Yield and Biodegradability of Common Leafy Vegetables (Yan et al., 2017)

Common name	Scientific name	Methane yield ($mL\,g_{VS}^{-1}$)	
		Experiment	Theoretical[a]
Cabbage	*Brassica oleracea L.* var. *capitata L.*	204	414
Chinese cabbage	*Brassica pekinensis (Lour.) Rupr.*	129	449
Cauliflower	*Brassica oleracea L.* var. *botrytis L.*	250	464
Broccoli	*Brassica oleracea L.* var. *italic Planch.*	247	408
Lettuce	*Lactuca sativa L.*	199	404
Lettuce, romaine	*Lactuca sativa L.* var. *ramosa Hort.*	244	471
Spinach	*Spinacia oleracea L.*	158	386

[a]Theoretical methane yield based on organic composition.

(Kafle et al., 2014). Cauliflower (*Brassica oleracea* var. *botrytis*) and its by-products, leaves and stems, produced methane yield of 0.190 and 0.331 mL/g VS, respectively (Gunaseelan, 2004). Coriander plant, also known as cilantro or Chinese parsley, is also easy to digest. Anaerobic digestion of by-products from coriander plants (*Coriandrum sativum*) obtained methane yields of 0.325 mL/g VS for leaves, 0.309 mL/g VS for stems, 0.283 mL/g VS for roots, and 0.322 mL/g VS for the whole plant (Gunaseelan, 2004).

In addition to single type of feedstock, it is well established that mixtures of substrates containing both nitrogen- and carbon-rich substrates should be used in a proper proportion to optimize biomethane production. Previous studies have showed improvement in anaerobic co-digestion of vegetable wastes with other feedstocks, both on laboratory-scale and pilot-scale productions (Bouallagui et al., 2003; Scano et al., 2014; Sitorus et al., 2013). Co-digestion is suitable for mixed stream of leafy wastes with other common organic feedstocks such as market wastes, organic fraction of municipal organic waste, and manure.

3.5.3 *ANAEROBIC DIGESTION OF LEAFY VEGETABLE BY-PRODUCTS FOR FERTILIZER PRODUCTION*

Digestate from anaerobic digestion can be recycled and used as fertilizer for growing vegetables for sustainable agriculture. Complex organic nitrogen compounds from digested feedstocks are mineralized to ammonia in the digester. The digester microorganisms use a part of the ammonia nitrogen for growth. Other parts of ammonia form struvite and ammonium carbonate, and trace amounts are volatilized in the biogas ($<1\%$) (Möller and Müller, 2012). Similarly, organic phosphorus as a constituent of adenylates, nucleic acids, and phospholipids from the leafy feedstocks entering anaerobic digestion can be recycled. The degradation processes during anaerobic digestion has been shown to improve phosphorus plant availability (Massé et al., 2011). Other nutrients are also found in anaerobic digestate including potassium, sulfur, organic metals (calcium, magnesium), and micronutrients. The use of digestate from anaerobic digestion has been studied for growing a variety of leafy vegetables, including lettuce (Liedl et al., 2004; Liedl et al., 2006; Liu et al., 2011), cabbage (Tokuda et al., 2010), and Chinese cabbage (Wang et al., 2007).

4 SUMMARY

A large variety of different leafy vegetables comprise one-fifth of the US vegetable market, and account for more than one-third of the fresh produce market. The production and harvest rate of leafy vegetables has been increasing over the past decades. The world's leading producers of leafy vegetables are China, United States, and India. The five most produced leafy vegetables discussed in this chapter are cabbage, cauliflower, broccoli, lettuce, and spinach.

The main by-products of leafy vegetables derived from trimming process and nonedible parts are outer and damaged leaves, core, stalk, and root base. Leafy vegetables have a high content of protein, dietary fibers, vitamins, and minerals. For the most common leafy vegetables studied, protein content ranges from 16% to 30% db with a higher amount in spinach, Chinese cabbage (pak-choi), and lettuce. Dietary fiber content ranges from 21% to 39% db, which is found in a higher level in romaine lettuce and cabbage. In addition, leafy vegetables are a great source of vitamins, especially vitamins C, B complex, and A, and carotenoids (β-carotene, lutein, and zeaxanthin).

As leafy vegetables are composed of unique phytochemical compounds, nutraceuticals and natural food colorants are extracted for medical or health benefits. The market size of these pharmaceutical compounds and food colorants are growing, along with the development of extraction technology. To improve the drawbacks of conventional methods, several new advanced extraction methods have been developed, including EAE, SFE, PLE, UAE, MAE, and PEF. By-products of leafy vegetables or the biomass remained after the extraction of valuable compounds can be used as a feedstock for producing biogas energy and fertilizer through anaerobic digestion technology. Since leafy wastes are easily biodegradable, the results showed viable methane potentials. The high protein content in leafy waste can be recovered from nitrogen fertilizer during anaerobic digestion, along with other compounds such as phosphorus, potassium, and minerals.

The utilization of leafy vegetable by-products is often limited by the lack of flexible and scalable technology for collecting and processing the biomass, since the leafy greens vary largely in types, varieties, and quantity produced locally. In addition, the quality of leafy greens and their by-products vary by season, soil, growth stage, and nutrient management. Most leafy greens are produced locally in a smaller amount and easily perishable and their by-products may not be economically feasible to harvest, limiting the economies of scale of conversion for value-added products. Further development in conversion technologies toward flexibility and scalability are important considerations for this type of feedstock.

References

Aletor, O., Oshodi, A.A., Ipinmoroti, K., 2002. Chemical composition of common leafy vegetables and functional properties of their leaf protein concentrates. Food Chem. 78 (1), 63–68.

Anderson, J.W., Baird, P., Davis, R.H., Ferreri, S., Knudtson, M., Koraym, A., ... Williams, C.L., 2009. Health benefits of dietary fiber. Nutr. Rev. 67 (4), 188–205.

Angelidaki, I., Sanders, W., 2004. Assessment of the anaerobic biodegradability of macropollutants. Rev. Environ. Sci. Biotechnol. 3 (2), 117–129.

AOA, 2017. Lutein and Zeaxanthin - Eye-Friendly Nutrients. Retrieved 10 November 2017, from, American Optometric Association. https://www.aoa.org/patients-and-public/caring-for-your-vision/nutrition/lutein-and-zeaxanthin.

Appels, L., Lauwers, J., Degrève, J., Helsen, L., Lievens, B., Willems, K., ... Dewil, R., 2011. Anaerobic digestion in global bio-energy production: potential and research challenges. Renew. Sustain. Energy Rev. 15 (9), 4295–4301. https://doi.org/10.1016/j.rser.2011.07.121.

Asavasanti, S., Ristenpart, W., Stroeve, P., Barrett, D.M., 2011. Permeabilization of plant tissues by monopolar pulsed electric fields: effect of frequency. J. Food Sci. 76 (1), 98–111

Azmir, J., Zaidul, I.S.M., Rahman, M.M., Sharif, K.M., Mohamed, A., Sahena, F., ... Omar, A.K.M., 2013. Techniques for extraction of bioactive compounds from plant materials: a review. J. Food Eng. 117 (4), 426–436.

Bartley, G.E., Scolnik, P.A., 1995. Plant carotenoids: pigments for photoprotection, visual attraction, and human health. Plant Cell 7 (7), 1027.

Bouallagui, H., Ben Cheikh, R., Marouani, L., Hamdi, M., 2003. Mesophilic biogas production from fruit and vegetable waste in a tubular digester. Bioresour. Technol. 86 (1), 85–89. https://doi.org/10.1016/S0960-8524(02)00097-4.

Bouallagui, H., Touhami, Y., Ben Cheikh, R., Hamdi, M., 2005. Bioreactor performance in anaerobic digestion of fruit and vegetable wastes. Process Biochem. 40 (3), 989–995. https://doi.org/10.1016/j.procbio.2004.03.007.

Busch, S., 2017. Kind of Food Contains a High Amount of Chlorophyll. Retrieved 15 August 2017, from, San Francisco Chronicle (SFGATE). http://healthyeating.sfgate.com/kind-food-contains-high-amount-chlorophyll-11022.html.

Buswell, A.M., Mueller, H.F., 1952. Mechanism of methane fermentation. Ind. Eng. Chem. 44 (3), 550–552. https://doi.org/10.1021/ie50507a033.

Butnariu, M., Butu, A., 2015. Chemical composition of vegetables and their products. In: Handbook of Food Chemistry. Springer, pp. 627–692.

Cheok, C.Y., Salman, H.A.K., Sulaiman, R., 2014. Extraction and quantification of saponins: a review. Food Res. Int. 59, 16–40.

De Castro, M.D.L., Garcıa-Ayuso, L.E., 1998. Soxhlet extraction of solid materials: an outdated technique with a promising innovative future. Anal. Chim. Acta 369 (1), 1–10.

Delgado-Vargas, F., Paredes-López, O., 2002. Natural Colorants for Food and Nutraceutical Uses. CRC press.

FAO, 2017. Food and Agriculture Organization of the United Nations. FAOSTAT. Retrieved 17 November 2017, http://www.fao.org/faostat/en/#data/QC.

Femenia, A., Robertson, J.A., Waldron, K.W., Selvendran, R.R., 1998. Cauliflower (Brassica oleracea L), globe artichoke (Cynara scolymus) and chicory witloof (Cichorium intybus) processing by-products as sources of dietary fibre. J. Sci. Food Agric. 77 (4), 511–518.

Fenwick, G.R., Heaney, R.K., 1983. Glucosinolates and their breakdown products in cruciferous crops, foods and feedingstuffs. Food Chem. 11 (4), 249–271.

Francis, F.J., 1996. Safety of food colorants. In: Hendry, G.A.F., Houghton, J.D. (Eds.), Natural Food Colorants, second ed. Chapman & Hall, New York.

GRV, 2017. Natural Food Colors Market Estimates & Trend Analysis by Product (Curcumin, Carotenoids, Anthocyanin, Carmine, Chlorophyllin), by Application (Bakery & Confectionery, Beverages, Dairy & Frozen Products, Meat Products), and Segment Forecasts, 2014–2025. Retrieved 1 October 2017, from, Grand View Research. http://www.grandviewresearch.com/industry-analysis/natural-food-colors-market.

Gunaseelan, V.N., 2004. Biochemical methane potential of fruits and vegetable solid waste feedstocks. Biomass Bioenergy 26 (4), 389–399. https://doi.org/10.1016/j.biombioe.2003.08.006.

Hawthorne, S.B., Yang, Y., Miller, D.J., 1994. Extraction of organic pollutants from environmental solids with sub-and supercritical water. Anal. Chem. 66 (18), 2912–2920.

Higdon, J.V., Delage, B., Williams, D.E., Dashwood, R.H., 2007. Cruciferous vegetables and human cancer risk: epidemiologic evidence and mechanistic basis. Pharmacol. Res. 55 (3), 224–236.

Howard, L., Pandjaitan, N., 2008. Pressurized liquid extraction of flavonoids from spinach. J. Food Sci. 73 (3), 151–157.

Humphrey, A.M., 2004. Chlorophyll as a color and functional ingredient. J. Food Sci. 69(5).

Jongaroontaprangsee, S., Tritrong, W., Chokanaporn, W., Methacanon, P., Devahastin, S., Chiewchan, N., 2007. Effects of drying temperature and particle size on hydration properties of dietary fiber powder from lime and cabbage by-products. Int. J. Food Prop. 10 (4), 887–897.

Kafle, G.K., Bhattarai, S., Kim, S.H., Chen, L., 2014. Effect of feed to microbe ratios on anaerobic digestion of Chinese cabbage waste under mesophilic and thermophilic conditions: biogas potential and kinetic study. J. Environ. Manage. 133, 293–301. https://doi.org/10.1016/j.jenvman.2013.12.006.

Khoo, H.-E., Prasad, K.N., Kong, K.-W., Jiang, Y., Ismail, A., 2011. Carotenoids and their isomers: color pigments in fruits and vegetables. Molecules 16 (2), 1710–1738.

Kimura, M., Rodriguez-Amaya, D.B., 2002. A scheme for obtaining standards and HPLC quantification of leafy vegetable carotenoids. Food Chem. 78 (3), 389–398.

Lardon, L., Helias, A., Sialve, B., Steyer, J.-P., Bernard, O., 2009. Life-cycle assessment of biodiesel production from microalgae. Environ. Sci. Technol. 43 (17), 6475–6481.

Larrauri, J.A., 1999. New approaches in the preparation of high dietary fibre powders from fruit by-products. Trends Food Sci. Technol. 10 (1), 3–8.

Larsen, E., Christensen, L.P., 2005. Simple saponification method for the quantitative determination of carotenoids in green vegetables. J. Agric. Food Chem. 53 (17), 6598–6602.

Liedl, B.E., Cummins, M., Young, A., Williams, M.L., Chatfield, J.M., 2004. Hydroponic lettuce production using liquid effluent from poultry waste bioremediation as a nutrient source. In: Paper Presented at the VII International Symposium on Protected Cultivation in Mild Winter Climates: Production, Pest Management and Global Competition 659.

Liedl, B.E., Bombardiere, J., Chatfield, J.M., 2006. Fertilizer potential of liquid and solid effluent from thermophilic anaerobic digestion of poultry waste. Water Sci. Technol. 53 (8), 69–79.

Liu, R.H., Wang, Y.Y., Sun, C., 2009. Effects of temperature on anaerobic fermentation for biogas production from cabbage leaves. Nongye Jixie Xuebao/Trans. Chin. Soc. Agric. Mach. 40 (9), 116–121.

Liu, W.K., Yang, Q.C., Du, L.F., Cheng, R.F., Zhou, W.L., 2011. Nutrient supplementation increased growth and nitrate concentration of lettuce cultivated hydroponically with biogas slurry. Acta Agric. Scand. Sect. B Soil Plant Sci. 61 (5), 391–394.

Marquez, U.M.L., Sinnecker, P., 2008. Chlorophylls in foods: sources and stability food colorants. In: Socaciu, C. (Ed.), Chemical and Functional Properties. CRC Press, Boca Raton, pp. 195–212.

Massé, D.I., Talbot, G., Gilbert, Y., 2011. On farm biogas production: a method to reduce GHG emissions and develop more sustainable livestock operations. Anim. Feed Sci. Technol. 166, 436–445. https://doi.org/10.1016/j.anifeedsci.2011.04.075.

McMillan, D.C., Talwar, D., Sattar, N., Underwood, M., O'reilly, D.S.J., Mcardle, C., 2002. The relationship between reduced vitamin antioxidant concentrations and the systemic inflammatory response in patients with common solid tumours. Clin. Nutr. 21 (2), 161–164.

Mizgier, P., Kucharska, A.Z., Sokół-Łętowska, A., Kolniak-Ostek, J., Kidoń, M., Fecka, I., 2016. Characterization of phenolic compounds and antioxidant and anti-inflammatory properties of red cabbage and purple carrot extracts. J. Funct. Foods 21, 133–146.

Möller, K., Müller, T., 2012. Effects of anaerobic digestion on digestate nutrient availability and crop growth: a review. Eng. Life Sci. 12 (3), 242–257.

MordorIntelligence, 2017. Nutraceuticals Market—Growths, Trends and Forecasts (2017-2022) from Mordor Intelligence. https://www.mordorintelligence.com/industry-reports/global-nutraceuticals-market-industry?gclid=EAIaIQobChMIz5KuuezH1wIVDQ4rCh3qLwi3EAAYASAAEgLYM_D_BwE.

Ngamwonglumlert, L., Devahastin, S., Chiewchan, N., 2017. Natural colorants: pigment stability and extraction yield enhancement via utilization of appropriate pretreatment and extraction methods. Crit. Rev. Food Sci. Nutr. 57 (15), 3243–3259.

Niranjan, K., Hanmoungjai, P., 2004. Enzyme-aided aqueous extraction. In: Dunford, N.T., Dunford, H.B. (Eds.), Nutritionally Enhanced Edible Oil Processing. AOCS Publishing.

Puri, M., Sharma, D., Barrow, C.J., 2012. Enzyme-assisted extraction of bioactives from plants. Trends Biotechnol. 30 (1), 37–44.

Rajasekaran, A., Sivagnanam, G., Xavier, R., 2008. Nutraceuticals as therapeutic agents: a review. Res. J. Pharm. Tech. 1 (4), 328–340.

Sangeetha, R.K., Baskaran, V., 2010. Carotenoid composition and retinol equivalent in plants of nutritional and medicinal importance: efficacy of β-carotene from Chenopodiumalbum in retinol-deficient rats. Food Chem. 119 (4), 1584–1590.

Sari, Y.W., Mulder, W.J., Sanders, J.P.M., Bruins, M.E., 2015. Towards plant protein refinery: review on protein extraction using alkali and potential enzymatic assistance. Biotechnol. J. 10 (8), 1138–1157.

Scano, E.A., Asquer, C., Pistis, A., Ortu, L., Demontis, V., Cocco, D., 2014. Biogas from anaerobic digestion of fruit and vegetable wastes: experimental results on pilot-scale and preliminary performance evaluation of a full-scale power plant. Energ. Conver. Manage. 77, 22–30. https://doi.org/10.1016/j.enconman.2013.09.004.

Sharma, S.K., Bansal, S., Mangal, M., Dixit, A.K., Gupta, R.K., Mangal, A.K., 2016. Utilization of food processing by-products as dietary, functional, and novel fiber: a review. Crit. Rev. Food Sci. Nutr. 56 (10), 1647–1661.

Sialve, B., Bernet, N., Bernard, O., 2009. Anaerobic digestion of microalgae as a necessary step to make microalgal biodiesel sustainable. Biotechnol. Adv. 27 (4), 409–416.

Sitorus, B., Sukandar, Panjaitan, S.D., 2013. Biogas recovery from anaerobic digestion process of mixed fruit-vegetable wastes. Energy Procedia 32, 176–182. https://doi.org/10.1016/j.egypro.2013.05.023.

Sowbhagya, H.B., Chitra, V.N., 2010. Enzyme-assisted extraction of flavorings and colorants from plant materials. Crit. Rev. Food Sci. Nutr. 50 (2), 146–161.

Suslick, K.S., Doktycz, S.J., 1990. The effects of ultrasound on solids. In: Mason, T.J. (Ed.), Advances in Sonochemistry. vol. 1. JAI Press, New York, pp. 197–230.

Tanaka, Y., Sasaki, N., Ohmiya, A., 2008. Biosynthesis of plant pigments: anthocyanins, betalains and carotenoids. Plant J. 54 (4), 733–749.

Tanongkankit, Y., Chiewchan, N., Devahastin, S., 2010. Effect of processing on antioxidants and their activity in dietary fiber powder from cabbage outer leaves. Drying Technol. 28 (9), 1063–1071.

Thanopoulou, E., Baltayiannis, N., Lykogianni, V., 2006. Nutritional aspects regarding lung cancer chemoprevention. J. BUON 11 (1), 7–20.

Tokuda, S., Tanaka, Y., Higashio, H., Murakami, K., Aizawa, S., Uragami, A., Kunihisa, M., 2010. Method for application of anaerobically digested cattle slurry to the field cultivation of cabbage. Jpn. J. Soil Sci. Plant Nutr. 81 (2), 105–111.

Trowell, H., Southgate, D.A., Wolever, T.M., Leeds, A.R., Gassull, M.A., Jenkins, D.J., 1976. Letter: dietary fibre redefined. Lancet 1 (7966), 967.

USDA, 2004. Lettuce, Shipping Point and Market Inspection Instructions. Retrieved 19 November 2017, From United States Department of Agriculture, Agricultural Marketing Service, https://www.ams.usda.gov/grades-standards/lettuce-grades-and-standards.

USDA, 2016. USDA National Nutrient Database for Standard Reference. Retrieved 15 August 2017, from United States Department of Agriculture, Agricultural Research Service, https://ndb.nal.usda.gov/.

USDA. (2017). Retrieved 23 October 2017, from United States Department of Agriculture, National Agricultural Statistics Service https://www.nass.usda.gov/Statistics_by_Subject/index.php?sector=CROPS.

Vayupharp, B., Laksanalamai, V., 2015. Antioxidant properties and color stability of anthocyanin purified extracts from thai waxy purple corn cob. J. Food Nutr. Res. 3 (10), 629–636.

Wadhwa, M., Bakshi, M.P.S., 2013. Utilization of Fruit and Vegetable Wastes as Livestock Feed and as Substrates for Generation of Other Value-Added Products. 4RAP Publication.

Wang, Y.-y., Shen, F., Liu, R.-h., Wu, L.-j., 2007. Effects of anaerobic fermentation residue on the yield and quality of Chinese cabbage. Renew. Energy Resour. 5, 013.

Wrolstad, R.E., 2004. Anthocyanin pigments—Bioactivity and coloring properties. J. Food Sci. 69 (5), 419–425.

Yan, H., Zhao, C., Zhang, J., Zhang, R., Xue, C., Liu, G., Chen, C., 2017. Study on biomethane production and biodegradability of different leafy vegetables in anaerobic digestion. AMB Express 7 (1), 27.

Zhao, Y., Wu, Y., Wang, M., 2015. Bioactive substances of plant origin. In: Cheung, P.C.K., Mehta, B.M. (Eds.), Handbook of Food Chemistry. Springer-Verlag Berlin Heidelberg, New York, pp. 967–1039.

Žnidarčič, D., Ban, D., Šircelj, H., 2011. Carotenoid and chlorophyll composition of commonly consumed leafy vegetables in Mediterranean countries. Food Chem. 129 (3), 1164–1168.

Zuo, Z., Wu, S., Qi, X., Dong, R., 2015. Performance enhancement of leaf vegetable waste in two-stage anaerobic systems under high organic loading rate: Role of recirculation and hydraulic retention time. Appl. Energy 147, 279–286. https://doi.org/10.1016/j.apenergy.2015.03.001.

CHAPTER 11

Onion and Garlic

Hamed M. El Mashad*,†, Ruihong Zhang*, Zhongli Pan*

*Department of Biological and Agricultural Engineering, University of California - Davis, Davis, CA, United States, †Department of Agricultural Engineering, Faculty of Agriculture, Mansoura University, El-Mansoura, Egypt

Chapter Outline

1 INTRODUCTION

Onion and garlic are important vegetables that have been used for centuries in many countries against parasitic, fungal, bacterial, and viral infections. They also have several biological and medical benefits such as antioxidant, anticarcinogic, anticarcinogenic, antimutagenic, and antiasthmatic activities (Corzo-Martínez et al., 2007). Onions have low calories (50 kcal/100 g) and moderate amounts of protein, fat, glucose, sucrose, fiber, calcium, phosphorous, potassium, vitamin C, B6, and a flavonoid called quercetin (Mitra et al., 2012). It is used as a flavor for many foods and as an herbal medicine. Onion composition depends on cultivar, stage of maturation, environmental conditions during the cultivation conditions, storage condition and duration, and bulb parts (Abayomi and Terry, 2009; Benítez et al., 2011). According to Benkeblia (2005), garlic contains higher

Integrated Processing Technologies for Food and Agricultural By-Products
https://doi.org/10.1016/B978-0-12-814138-0.00011-3

concentrations of total phenols than onion. The highest and lowest radical scavenging activities were determined in garlic and green onion extracts, respectively. Different types of onion (green, red, yellow, purple) and garlic extracts have different polyphenol contents and antioxidant activities. Antioxidants are important to reduce the risk of cardiovascular and cancer diseases.

Onion and garlic are grown worldwide in temperate, subtropical, and tropical areas (Fritsch and Friesen, 2002). Figs. 1 and 2 show the world production, harvested areas, and yield of dry onion and garlic. As can be seen, the total area of onion harvesting slightly increased in the period of 2008–2015, while it was relatively constant for garlic. The total production of both crops increased over the presented years. The yield of onion slightly decreased in the period of 2011–2014, and that of garlic increased in the period of 2007–2014.

Onion and garlic can be consumed in their original forms or after processing, which produces significant amounts of waste. In addition to the processing wastes, worthlessly marketed onion and garlic can also be produced. The majority of worthlessly marketed onion and garlic and their processing wastes are currently disposed of in landfills or used as soil amendments causing negative environmental impacts and plant phytotoxicity. These materials are not good feeds for animal and can cause a rapid growth of *Sclerotium cepivorum* (white rot), which cause disposal problems in landfills (Roldán et al., 2008). Moreover, due to the high moisture content of these materials, incineration is not a good option to dispose of them. Therefore, they need to be disposed of in environmental-friendly and cost-effective ways.

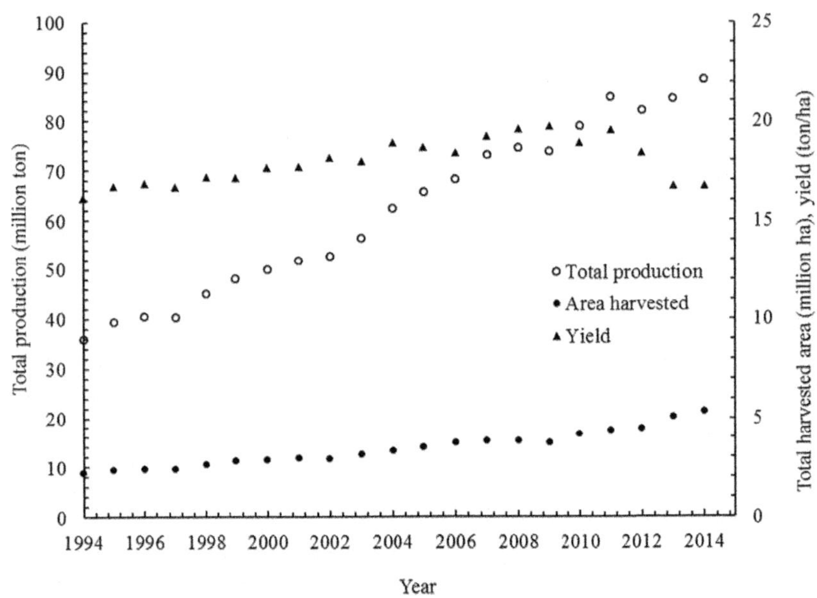

FIG. 1
Total world production; harvested areas, and yield of dry onion in the period of 1994–2014 (FAO, 2017).

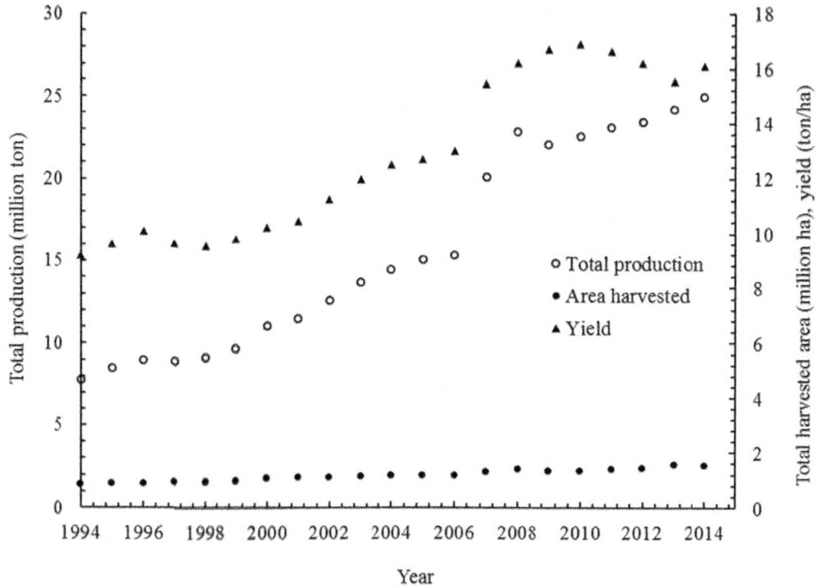

FIG. 2
Total world production, harvested areas, and yield of garlic in the period 1994–2014 (FAO, 2017).

2 MORPHOLOGY AND COMPOSITION OF ONION AND GARLIC

Botanically, onion (*Allium cepa* L.) and garlic (*Allium sativum* L.) are two of the edible species of the genus *Allium* that are cultivated in temperate to subtropical regions (Fritsch and Friesen, 2002). Shrestha (2007) reviewed the anatomy of the onion. Onions are a perennial herb that are composed of roots, a stem, and leaves. Roots are the part that is present underground. The onion stem is also underground, and it develops into a bulb. The bulb consists of a reduced stem and axillary buds surrounded by inner fleshy scale leaves and outer dry scales (skin). Depending on the variety and growth stage, onion bulbs have different shapes such as spherical, globular, flat, and conical. Onion color also varies from white, yellow, or red. Leaves are hollow, green, parallel-veined foliage with a fleshy sheathing base arising from the underground stem (Ranjitkar, 2003). Mann (1952) mentioned that the structure of garlic is similar to that of onion. However, the two plants differ in the storage structure of the bulb and the shape of the leaf blades. The dry garlic bulb is a storage structure that can be separated into small parts called cloves. Filaree Garlic Farm (2012) anatomically described garlic as composed of stem, pseudo-stem, leaves, cloves, and bulb. The true stem of garlic is underground, while the pseudo-stem is above ground with overlapping sheaths of leaves. Leaves are the part that appears above the ground and they are only the blade parts of the foliage leaves. Cloves are composed of different kinds of leaf primordia that make up the skin and flesh of the clove. The cloves are densely packed, elongated side bulbs that are the main economic organ.

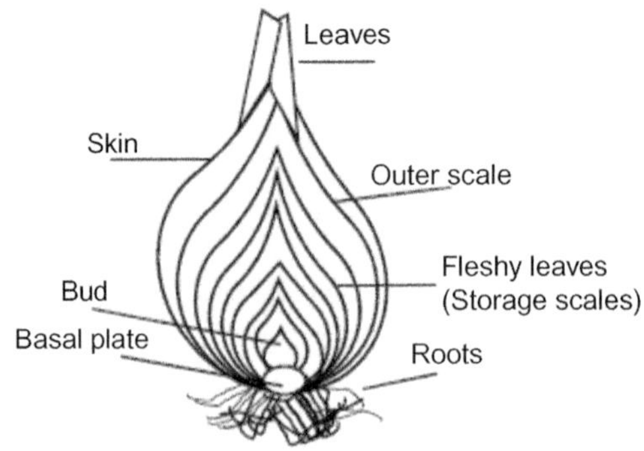

FIG. 3
Major parts of onion (e.g., Sharma et al., 2016).

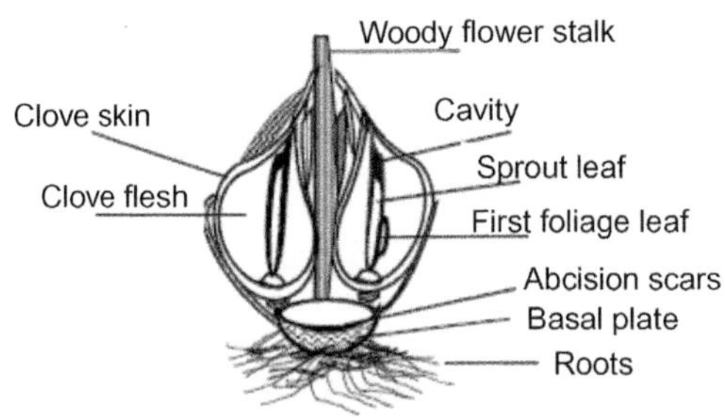

FIG. 4
Major parts of garlic (Filaree Garlic Farm, 2012).

Fresh leaves and pseudo-stems can also be consumed by humans (Fritsch and Friesen, 2002). The garlic bulb is a survival trait that stores energy for future growth. Depending on the variety, genetic traits, and cultivar type, garlic bulbs size ranges from 4.5 to 7.6 cm and contain 4–60 cloves of various shapes and sizes (The Herb Society of America, 2006). Figs. 3 and 4 show the main parts of onion and garlic.

3 CULTIVATION OF ONION AND GARLIC

Onion and garlic are perennial crops. Yet, they are cultivated as annual crops. Bosch Serra and Currah (2002) stated that onions are cultivated in all kinds of soils. The most suitable soils are loam and sandy loams with good structure and fair contents of organic matter. The authors presented detailed descriptions of onion cultivation and crop management that are summarized herein. Bulb and salad onions are cultivated using direct sowing in humid temperate climates. Cultivation and growing nursery beds are also applied where labor costs are low. Seedlings can be transplanted after reaching the three to four leaf stage. Onion are traditionally cultivated using synthetic fertilizers. Integrated crop management (ICM) systems are also applied to reduce resource inputs by avoiding their wastage. The ICM system can also reduce environmental pollution, production costs, and risks to human health. In addition to the traditional and ICM cultivation methods, the authors mentioned that onion can be organically cultivated without using synthetic chemical fertilizers and pesticides.

The irrigation requirements for onion depend on climate, soil conditions, and irrigation system. To optimize onion yield, soil water tension (SWT) criteria from 10 to 35 centibar was stated by Shock et al. (2010). The SWT is a measure of how strongly water is held by the soil. For a study conducted in India during the cultivation season of January to May, Kumar et al. (2007) found that the increase of onion yield from 20 to 35 ton/ha was not proportional to the increase in the amount of irrigation water in the range of 250–450 mm. Brown (2000) stated that onions require from 23 to 68 kg N per acre depending on the plant population, losses due to its leaching, nitrogen carryover from pervious crops, crop residue of the precedent crop, and mineralizable N released from soil organic matter. The phosphorus needed for onion can reach up to 145 kg of P_2O_5 per acre depending on the amount of soil content of phosphorus and lime, and the application of fumigation. In addition, to maximize onion yield, sulfur and micronutrients are also needed.

The Herb Society of America (2006) reviewed the conditions for garlic cultivation. Garlic can be cultivated in a variety of soils with a pH ranging from 4.5 to 8.3. However, light, well-drained loam soils, with a pH ranging from 6.5 to 6.7 are preferable. Garlic is generally planted in the fall. It can also be planted in the winter and spring, and the planting time depends on the location. Most garlics do not produce a fertile seed; therefore, garlic is propagated using cloves. Irrigation is important for achieving a high yield of garlic. The amount of irrigation water depends on the location and precipitation rate. For garlic cultivated in California during the spring, water requirement was estimated to be approximately 425 mm including 108 mm of rainfall (Ayars, 2008). There was no significant effect of irrigation methods (simulated furrow irrigation, surface drip irrigation, and subsurface drip irrigation) on garlic yield. Garlic needs moderate amounts of nitrogen depending on the soil and previous crop. It may not need phosphorus (De La Cruz Medina and García (2007)).

4 HARVESTING AND STORAGE OF ONION AND GARLIC

Mechanized harvesting of onion and garlic is the common practice in temperate climates. However, hand harvesting is still a common practice in many parts of the world. Harvest timing affects the yield and storability. Onion is mechanically harvested by specialized equipment. Onion harvesting and postharvesting processes include undercutting; curing using dry air for 2–3 days; clipping roots and tops, bagging it in burlap sacks; transporting it to warehouses; drying; grading; packaging for retails; and shipping to processing plants or markets (Marín, 2009; Timm et al., 1991). Curing is important prior to peeling and processing. It strengthens and seals the neck and outer layer of skin. It can be carried out for 48–72 h to remove excess moisture in the field or by forced ventilation in storage. Gubb and MacTavish (2002) reviewed several methods for curing and storing onion including field storage, ventilation with forced ambient air, ventilation with heated air, and stores with controlled-temperature facilities. Onions can be stored for 9 months with refrigerated and controlled atmosphere storage (Adam et al., 2000).

Garlic is harvested by pulling from the ground. Then, it is cured in the shade or indoors in the presence of good air circulation to prevent sunburns and the growth of mold. Fresh garlic can be stored for up to 3 months in a standard warehouse, 6 months in cold storage, or 1 year in a controlled atmosphere storage (Wiczkowski, 2011). Although most garlic is stored in the form of bulbs, processed, prepeeled, and chopped garlic products are also stored. Hong and Kim (2001) found that storing chopped garlic under high hydrostatic pressure (600 MPa, 1 min) combined with citric and/or ascorbic acids retarded its browning that resulted from nonenzymatic autoxidation of phenolic compounds. These storage conditions did not inhibit the production of a green pigment in dormancy-terminated bulbs.

5 ONION AND GARLIC PROCESSING AND PRODUCTION OF ITS WASTE

Onion and garlic are consumed in their fresh forms or after processing. However, onion products are less abundant than those of garlic (Corzo-Martínez et al., 2007). The characteristics of onion and garlic products depend on the variety and processing parameters and conditions (Corzo-Martínez et al., 2007). Processed onion is usually presented as dehydrated onion in the form of flakes and powder, canned onion pickle, and essential oils. Onion processing includes washing, top removing, peeling, coring, cutting, drying, and grinding. Similarly, garlic processing includes bulb breaking, peeling, dehydration, and grinding (Kallel and Ellouz Chaabouni, 2017). Lund and Lienau (2017) described the steps involved in the industrial dehydration of onion, including washing, slicing, dehydration, milling, and packaging. Industrial processing of frozen, cut, and freeze dried onions requires the separation of the top and bottom of the onion bulbs and also the two outer fleshy scales together with the brown skin (Benítez et al., 2011). Onion can be peeled using knife or steam. Steam peeling increases dry matter, decreases fructose, and minimally changes nonstructural carbohydrates (Kaack et al., 2004).

Approximately 38% of the fresh weight of processed onion is not suitable for consumption and discarded as waste (French et al., 2010). In addition, approximately 10% of harvested onions are worthlessly marketed, meaning that they do not meet the quality standards for marketing (Horiuchi et al., 2004). These onions are either disposed of or cheaply dumped. Onion processing waste includes damaged and unwanted bulbs (undersized, malformed, damaged, and diseased bulbs), onion skins, the two outer fleshy scales, roots, and the top and bottom of the bulbs (Benítez et al., 2011). Dry outer skin and stems represent 10% of the wet weight of onion (Roberts and Kidd, 2005). Skins in white and red onion ranged from 3% to 5% the total weight (Alexander and Sulebele, 1973). Onion damages such as internal bruising, cuts, and puncture damages can occur during mechanical packing. Timm et al. (1991) mentioned that bruising can result in tissue discoloration, decay, or a reduction in shelf life. Bruising can be reduced by improving handling operations using lower elevations for onion transportation, adding soft cushions on hard surfaces, and limiting transportation velocity. Onion processing wastes and worthlessly marketed onions can be used to produce valuable products such as antioxidants, chemicals, compost, and bioenergy.

Garlic can be processed into dehydrated garlic as an ingredient for food processing and other products such as dehydrated powder, essential oil, oil macerate, powder, and aged garlic extract (Wiczkowski, 2011). The major processes involved in the formation of garlic products include drying (low temperature or freeze drying), distillation, maceration in oil, hydroalcoholic short extraction, and hydroalchoholic long maceration (Staba et al., 2001; Wiczkowski, 2011). Garlic harvesting and processing wastes include the husk, stem, peels, straw, and unwanted bulbs. Garlic skins range from 8% to 10% of the total weight (Alexander and Sulebele, 1973). These wastes can be used for water purification (e.g., removal of heavy metals and dyes) and production of bioenergy and antioxidants (Kallel et al., 2014; Kallel and Ellouz Chaabouni, 2017).

6 CHARACTERIZATION OF ONION AND GARLIC PROCESSING WASTE

Different parts of onions contain different concentrations of components and elements. Ko et al. (2011) presented the chemical composition of onion skin. Carbohydrates, protein, lipids, and ash were 47.6%, 4.0%, 5.2%, and 7.1% (d.w.), respectively. Quercetin, quercetin-40-glucoside, kempherol, isorhamnetin, and rutin contents were 17.6, 3.7, 0.9, 0.7, and 0.9 mg/g onion skin, respectively. About 80% of the flavonoids found in onion are quercetin monoglucoside and quercetin diglucoside. Anthocyanins are concentrated in red onion skin, but minor amounts of them are found in the edible portion (Rhodes and Price, 1996). The dry outer scales of onions contain large amounts of quercetin, quercetin glucoside (3), and their oxidative products (Ly et al., 2005). Quercetin is a flavonol that is used as an antioxidant and nutritional supplement. It is yellow powder with a specific smell and a bitter taste (Ko et al., 2011). Benítez et al. (2011) determined the minerals, fiber, and antioxidant contents in Recas and Figueres cultivars of onions (Tables 1 and 2). They found that the brown skin and top and bottom of onions could be potentially used as functional food ingredient because they are rich in insoluble dietary fiber, total phenolics and flavonoids that have high antioxidant activity. Brown skins also

Table 1 Proximate Analysis and Concentration of Minerals of Different Parts of Recas and Figueres Cultivars of Onions (Benítez et al., 2011)

Components	Whole onion	Inner scales	Outer scales	Top-bottom	Brown skin
Dry matter (DM, %)	8.7–8.8	6.9–8.3	6.3–7.5	13.2–18.0	50.8–51.9
Crude protein (% of DM)	10.1–11.8	11.6–15.3	8.3–9.3	15.0–15.6	2.3–2.4
K (mg/g DM)	14.1–15.8	15.7–15.9	12.8–13.9	11.1–13.0	4.2–7.3
Ca (mg/g DM)	6.6[a]	1.8–3.5	4.2–7.8	10.2–16.5	22.1–30.7
Mg (mg/g DM)	1.0–1.3	0.6–0.7	0.7–0.8	1.4–1.5	1.1–1.5
Fe (μg/g DM)	53.2–100.3	19.6–43.6	19.6[a]	426.7–888.9	119.8–419.4
Zn (μg/g DM)	19.6–23.9	16.2–28.4	18.6–26.8	50.4–53.8	14.9–21.6
Mn (μg/g DM)	10.2–13.7	6.5–11.5	6.6–8.1	20.2–28.8	8.3–25.3
Se (μg/g DM)	0.15–0.34	0.13–0.93	0.06–0.18	0.03–0.43	0.05–0.90
Total ash (% of DM)	4.7–4.9	4.4–4.9	4.6–5.6	8.2–8.6	9.3–10.6

[a]Similar concentration in both cultivars.

Table 2 Total Phenols (mg GAE/g DM), Flavonoids (mg QE/g DM), Flavonols (mg/g DM), and Antioxidant Capacity (μmol Fe^{2+}/g DM) in Whole and Different Parts of Recas Cultivar of Onion (Benítez et al., 2011).

Components	Whole onion	Inner scales	Outer scales	Top-bottom	Brown skin
Total phenolics	17.30	9.40	19.70	30.50	52.70
Total flavonoids	10.30	7.00	19.50	25.90	43.10
Total flavonols	8.84	6.19	19.27	15.29	7.89
Quercetin 4′-glucoside	4.02	2.00	7.37	6.35	5.16
Quercetin 3,4′-glucoside	3.10	3.70	9.49	5.90	0.30
Quercetin	0.91	0.02	0.59	1.21	1.61
Quercetin 3-glucoside	0.16	0.10	0.42	0.40	0.31
Isorhamnetin 3,4′-diglucoside	0.12	0.12	0.37	0.57	0.19
Isorhamnetin 4′-diglucoside	0.53	0.25	1.03	0.86	0.32
Antioxidant activity (FRAP[a])	83.50	28.70	105.10	156.10	227.80

[a]FRAP: Ferric reducing ability assay.

contain high concentrations of quercetin aglycone and calcium. The tops and bottoms contain high concentrations of magnesium, iron, zinc, and manganese. The differences in mineral contents are attributed to the differences in the mobility of different elements. Herrmann (1976) mentioned that, unlike onions with white skins, onion with red skins contained high concentrations of flavones and flavonols. Quercetin is present in a large fraction in the free form, as aglycone, ranging from 67% to 86% of total quercetin, and in a smaller portion as spiraeoside (quercetin-4′-glucoside). Suh et al. (1999) isolated two flavonols from onion skin: 3, 5, 7, 3′-OH, and 3(5), 7, 3′, 4′-OH of flavonol. The first was composed of quercetin and glucose, and the second contained only quercetin. Both flavonols exhibited potent radical scavenging activity against 1, 1-diphenyl-2-picrylhydrazyl (DPPH), superoxide anion ($O_2\cdot-$), and hydroxyl (OH) radicals. Ramos et al. (2006) isolated quercetin-derived oxidation products and lunularin-4-O-β-D-glucoside from a water extract of onion skin, together with 17 other known compounds. Among the isolated compounds were 2-(3,4-dihydroxyphenyl)-4,6-dihydroxy-2-methoxybenzofuran-3-one, and 3-(quercetin-8-yl)-2,3-epoxyflavanone, which have antibacterial activities against *Helicobacter pylori,* and Methicillin-resistant *Staphylococcus aureus* (MRSA) and *H. pylori* strains, respectively.

Jaime et al. (2002) presented the compositions of different parts of three varieties (Sturon, Hysam, and Grano de Oro) of onions (Table 3). The brown skin and outer leaves of onions contained high concentrations of insoluble dietary fiber

Table 3 Composition of Three Cultivars (Sturon, Hysam, and Grano de Oro) of Onions (Jaime et al., 2002)

	Whole onion	Inner scales	Outer scales	Top	Bottom	Skin
Dry matter (DM, %)	10.8–17.0	8.8–16.6	9.7–14.3	17.7–25.7	17.5–26.8	60.3–82.9
Total nitrogen (g/kg DM)	15.8–17.3	16.3–18.1	15.0–15.6	8.2–10.7	21.2–35.0	2.6–5.4
Klason lignin (g/kg DM)	8.2–13.1	6.4–8.6	6.4–15.0	23.3–41.5	53.7–72.0	15.6–42.3
Insoluble dietary fiber (IDF, % of DM)	9.2–18.2	5.7–6.9	9.9–24.4	38.2–48.7	26.2–32.8	59.3–66.6
Soluble dietary fiber (SDF, % of DM)	4.4–6.7	3.5–5.2	4.2–5.1	3.4–4.7	7.4–10.2	1.7–4.4
Total dietary fiber (% of DM)	14.9–31.0	9.3–11.6	14.5–28.6	42.7–53.4	34.7–40.8	63.7–68.3
Total sugars in IDF (g/kg DM)	65.7–181.2	50.6–63.7	78.5–210.7	269.0–364.8	152.2–282.3	388.5–603.2
Total sugars in SDF (g/kg DM)	33.1–46.7	40.3–52.6	26.8–39.3	34.6–37.6	47.6–60.4	20.2–34.9
Total ash (g/kg DM)	34.3–55.1	32.9–57.0	31.7–53.5	69.9–96.3	96.5–107.7	68.9–85.0

that makes these parts good food product supplements. They also contained lower nitrogen than fleshy tissues. The top, bottom, and skin of these three cultivars contained high concentrations of lignin. The outer leaves had a low concentration of pectic galactose, and the brown skin contained small amounts of galactose. Total pectic substances accounted for 12% and 9.8% of the dry weight of white and red onion, respectively (Alexander and Sulebele, 1973). Benítez et al. (2013) characterized the paste, bagasse, and juice that were produced from onion waste. The bagasse was rich [361–453 mg/g dry weight (d.w.)] in dietary fiber, the paste contained a high concentration (5.6 mg/g, d.w.) of alkyl or alkenyl cystein sulfoxide; and the juice contained a high concentration (205–221 mg/g, d.w.) of fructans.

Garlic skins contain 7.8% protein, 9.9% ash, 27% pectin, 11.4% combined rhamnose, and 5.6% galactose (Abdel-Fattah and Edrees, 1972). Kallel et al. (2014) presented the chemical composition of garlic skin (Table 4). As it can be seen from Table 4, the skin contains high concentrations of dietary fibers, calcium, and potassium. The skins also contain phenolic and flavonoid compounds. The major phenolic compounds included caffeic, *p*-coumaric, ferulic, and *di*-ferulic acids (Kallel et al., 2014). Ichikawa et al. (2003) found that garlic skin extract had a strong 1, DPPH radical scavenging activity. They isolated and identified six antioxidants from garlic skins extract including N-trans-coumaroyloctopamine, N-trans-feruloyloctopamine, guaiacylglycerol-β-ferulic acid ether, guaiacylglycerol-β-caffeic acid ether, trans-coumaric acid, and trans-ferulic acid. Total pectic substances accounted for 11.2% of the dry weight of garlic skin (Alexander and Sulebele, 1973).

Table 4 Chemical Composition of Garlic Skin (Kallel et al., 2014)

Component	Values
Dry matter (DM, %)	96.5
Lignin (%, DM)	8.3
Insoluble dietary fiber (IDF, % of DM)	58.1
Soluble dietary fiber (SDF, % of DM)	4.2
Total dietary fiber (% of DM)	62.3
Protein (% of DM)	8.4
Lipids (g/kg DM)	0.9
Total ash (% DM)	16.7
K (mg/g DM)	701.3
Ca (mg/g DM)	1662.1
Mg (mg/g DM)	196.0
Na (mg/DM)	203.2
Fe (mg/g DM)	9.0
Mn (mg/g DM)	1.7
Zn (mg/g DM)	1.1
Cu (mg/g DM)	0.9

7 VALUABLE PRODUCTS FROM ONION AND GARLIC PROCESSING WASTE

7.1 Biosorbents

Onion and garlic wastes can be used directly or after processing as biosorbents and bioremediation agents. Saka et al. (2011) studied the effect preboiled treated onion skins (PTOS) and formaldehyde-treated onion skins (FTOS) on the adsorption of Pb^{2+} from aqueous solutions. Experiments were carried out to determine effect of the initial concentration of Pb^{2+}, agitation rate, pH, and temperature on the adsorption efficiency. The highest absorption efficiency of Pb^{2+} was obtained at a pH of 6. At a dose of 750 mg/L of both sorbents, the obtained highest removal efficiencies were 84.8% and 93.5% with PTOS and FTOS, respectively. Saka et al. (2012) studied the feasibility of using microwave heated and PTOS for the removal of methylene blue from aqueous solutions. They studied the effects of dye concentration, adsorbent dose, and pH on the removal efficiency of methylene blue. The removal efficiency increased with the increase of initial dye concentrations, contact time, and adsorbent dosage. Maximum removal efficiencies of 86.3% and 94.1% were obtained, respectively, with microwaved and PTOS at a pH of 10.0 and adsorbent dose of 750 mg/L. By applying the Langmuir isotherm equation, adsorption capacities of 142.7 and 55.6 mg/g were determined with microwave treated and preboiled onion skins.

The potential of garlic peels to remove methylene blue from aqueous solutions was evaluated by Hameed and Ahmad (2009). The peels were boiled for 30 min, dried, and then ground before being used in the experiments. The adsorption of methylene blue at equilibrium increased with a pH increase from 4 to 6. Adsorption was not changed by increasing the pH from 6 to 12. Adsorption isotherms were well described by the Freundlich model. The monolayer adsorption capacity increased with the temperature increase. A maximum adsorption capacity of 143 mg/g was determined at 323 K. Zhao et al. (2017) applied garlic peels as a sorbent medium to remove quinolone antibiotics that are excreted by humans and animals, causing negative effects on aquatic ecosystems and human health. Results showed that pretreated garlic peels with nitric acids had significantly (ca. 10 times) improved adsorption capacity than that of the untreated peels. The high affinity of the peels pretreated with nitric acid to quinolone antibiotics was attributed to the increase in the total amount of oxygen acidic groups. Asfaram et al. (2014) investigated the adsorption effect of garlic peels on the removal of Direct Red 12B. They studied the effect of pH, concentration of the dye, amount of adsorbent, particle size of peels, contact time, and temperature on the adsorption efficiency. Results showed that the Langmuir model described the equilibrium concentration of the dye in the solution. Adsorption efficiency increased with the decrease in particle size ($\leq$500 μm). Using garlic peels with particle sizes of up to 177 μm at an initial concentration of the dye of 50 mg/L resulted in a maximum removal efficiency of >99% of the dye at a pH of 2.0 and temperature of 55°C after 25 min of contact time.

Muthamilselvi et al. (2016) optimized the parameters (initial pH, adsorbent dosage, agitation speed, and contact time) that affect phenol adsorption onto garlic peels. Results showed that >80% of phenol could be removed from an aqueous solution containing a phenol content of 50 mg/L using an adsorbent dosage of 2.1 g/L, an agitation speed of 135 rpm, and a contact time of 7 h at an initial pH of 2. Kinetic results showed that phenol removal by garlic peels followed the pseudo-second-order kinetic model. Liu et al. (2014) studied the effect of initial pH ranging from 1 to 7, initial metal concentration ranging from 1 to 200 mg/L, contact time ranging from 1 to 120 min, and dosage of nonmercerized and mercerized garlic peels ranging from 1 to 100 mg on the removal of Pb^{2+} from aqueous solution. The adsorption capacity increased with the increase of pH from 1 to 3 reaching 7 and 9 mg/g of nonmercerized and mercerized garlic peels, respectively. With further increase in the pH, no increase in the adsorption was determined due to the formation of complexes between Pb^{2+} and functional groups of the garlic peels. The adsorption capacity increased with the increase of initial Pb^{2+} concentration in the range from 1 to 50 mg/L using the nontreated garlic peels and from 1 to 100 mg/L using mercerized peels. Adsorption equilibrium was achieved after 40 min for both adsorbents. For solutions containing 150 mg/L of Pb^{2+}, the removal efficiency increased with the increase of absorbent dosage before reaching 60 and 80 mg of the mercerized and nonmercerized peels, respectively. For both adsorbents, the equilibrium sorption of Pb^{2+} was well described by the Langmuir adsorption isotherm, and the adsorbent-adsorbate kinetics exhibited the pseudo-second-order model. Chowdhury et al. (2012) studied the effect of onion and garlic skins on the removal of Cu^{2+} from aqueous solution. The Cu^{2+} adsorption on the onion and garlic skins was spontaneous and endothermic. The adsorption capacity increased with an increase in the initial pH from 3.0 to 5.3. At an initial pH of 5.3 and temperature of 303 K, the maximum adsorption capacities were 66.7 and 76.9 mg/g for garlic and onion skins, respectively.

7.2 Bioenergy

Onion and garlic wastes have been used to produce energy via biochemical and thermochemical processes. Onion processing waste is characterized by a high C/N ratio of 52 (Pellejero et al., 2015). Therefore, mixing it with nitrogen-rich material is important for successful biochemical processing (e.g., anaerobic digestion and composting) of it. Anaerobic digestion is a biochemical process used to produce methane-rich biogas from feedstocks with high moisture contents. After upgrading, biogas can be used as a fuel for transportation, and as electricity and heat generation in combined heat and power generators. It can also be used for electricity generation in fuel cells. The performance of the anaerobic digestion depends on the characteristics of the feedstock and operational parameters of the process such as hydraulic retention time, organic loading rate, and temperature. Several designs of anaerobic digestion systems are installed to treat food wastes (Zhang and El-Mashad, 2017). Romano et al. (2004) studied the anaerobic digestion of whole onion solids in batch and two-phase anaerobic digestion systems without adjusting its C/N ratio. It was found that the digestion was limited by the hydrolysis of onion solids. The highest biogas and methane yields were in the ranges of 0.57–0.66 and 0.32–0.35 L/g volatile solids [VS] fed, respectively. It was found that when operating the two-phase system at an organic loading rate of 2.0 g [VS]/L/day, the pH decreased to 6.0 in the second biogasification reactor stage, due to the accumulation of volatile fatty acids. Lime or other alkalinity enhancing chemicals were necessary to raise the pH to the proper value (6.8–7.8) for a successful anaerobic digestion. For successful anaerobic digestion of onion juice and wastewater from onion processing in an anaerobic mixed biofilm reactor, Romano and Zhang (2008) used sludge, a cheap material, from an aerobic treatment process to adjust the alkalinity and fortify the deficiency of nitrogen and other essential elements in onion juice and onion processing wastewater. Average biogas and methane yields were 0.62 and 0.37 L/g VS, respectively.

Gills Onions, an onion processor in Oxnard, California, installed an anaerobic digester to treat onion processing waste and produce biogas. Onion wastes are first ground and then the extracted onion juice is treated in an upflow anaerobic sludge blanket (UASB) reactor to produce biogas. The system treats about 113 tons of onion waste (peels, tops, and tails) per day. The biogas produced is treated to remove sulfur, moisture, and particulates. It is then fed to two 300 kW molten carbonate fuel cells that do not require removal of CO_2. The electricity produced from the system supplies approximately 50% of the electricity needs of the company (Greer, 2010). The main challenge for the project is the removal of sulfur from the biogas to meet the sulfur content requirement of 0.1 ppm or less. In 2012, it was reported that the company saved $800,000 in electricity and labor costs associated with the disposal of onion remains. The fibrous solids that remain after juice extraction are compressed to produce cattle feed. Prior to installing the system, onion processing wastes were managed through composting and field application.

There are several thermochemical processes that can be applied to treat vegetables (e.g., onion and garlic) processing wastes. They include direct combustion (i.e., incineration), pyrolysis, gasification, and hydrothermal degradation. In the thermochemical processes, high temperatures are applied either at ambient or high pressures to break the bonds of organic matter and produce heat, synthesized gas, hydrocarbon liquid fuels (biooil), and biochar or ash. The product types of the thermophilic processes depend on their configuration and operational conditions. Compared with biochemical processes, thermochemical processes, except for hydrothermal processes, require lower moisture ($<40\%$). The operational concepts of different thermochemical processes, and the parameters (e.g., processing temperature and heating rates) affecting their performance were reviewed by Zhang and El-Mashad (2017). Park et al. (2012) carried out pyrolysis experiments on garlic stems at different temperatures (400–600°C). At a pyrolysis temperature of 500°C, the maximum yields of bio-oil, synthesis gas, and biochar were approximately 40%, 36%, and 24%, respectively. The water content and high heating value of the produced bio-oil were 57.1% and 7.25 MJ/kg, respectively.

7.3 Ethanol, Vinegar, and Acetic Acid

Horiuchi et al. (1999) used worthlessly marketed onions to produce vinegar that was rich in amino acids, organic acids, and minerals. The juice of Kurenai, a red onion cultivar, was fermented batchwise at 25°C for 40 h using yeast to produce alcohol. Then, acetic acid fermentation was conducted at 30°C for 200 h using *Acetobacter aceti* under aerobic conditions. The vinegar yield was 30 g/L of juice. The produced vinegar had higher potassium and magnesium and lower sodium contents than that in malt and rice vinegars. It also contained higher concentrations of malic, succinic, and pyroglutami acids than other commercial vinegars. Horiuchi et al. (2000b) developed a two-step fermentation system for the production of vinegar from onion juice at 30°C. The first step was batch fermentation of the juice to produce ethanol using immobilized *Saccharomyces cerevisiae* strain IR-2, and the second step was acetic acid fermentation using *A. aceti* with a vertical charcoal pellet bioreactor. Aseptic air was supplied from the bottom of the second reactor. Acetic acid yield was 38 g/L, which represented 86% of the theoretical yield calculated based on the sugar content of onion juice. Horiuchi et al. (2000a) applied a repeated batch process to produce alcohol from worthlessly marketed red onion using flocculating *S. cerevisiae* IR-2. A final ethanol concentration of 31 g/L could be obtained from each batch fermentation at 4 h retention time. The produced alcohol was red in color. It can be used as an alcoholic beverage because it contains various amino acids, minerals, and organic acids. The red color was attributed to the presence of anthocyanin at a concentration of 50 mg/L, which is significantly lower than that (2000 mg/L) in red wine. González-Sáiz et al. (2008) fermented onion alcohol in semicontinuous fermentation at 30°C for the production of vinegar using *A. aceti*. A maximum acetic acid concentration of 24 g/L was obtained after 90 h of fermentation.

7.4 Compost

Compost is a valuable soil amendment and contains micro and macro-fertilizer elements. Pellejero et al. (2015) studied the composting of three mixtures of onion processing waste with alfalfa and cattle manure. The first mixture was prepared by mixing one part of onion waste with 0.5 parts manure and 0.3 parts alfalfa. The second mixture was composed of equal amounts of onion waste and manure. The third mixture was composed of one part of onion waste and 0.3 parts of alfalfa. The initial C/N ratio of all mixtures was 30 and the initial pH ranged from 8 to 8.5. A mature compost could be obtained after 180 days. The pH and C/N of the produced compost ranged from 7.3 to 7.8 and from 10 to 11, respectively. The second mixture achieved the highest organic matter reduction.

Horiuchi et al. (2004) investigated the production of vinegar and compost from worthlessly marketed onions in an integrated system. Onion juice was extracted and fermented in a two-step fermentation system. Flocculating yeast *S. cerevisiae* IR-2 and *Acetobacter pasteurianus* no. 1 were used to produce alcohol and acetic

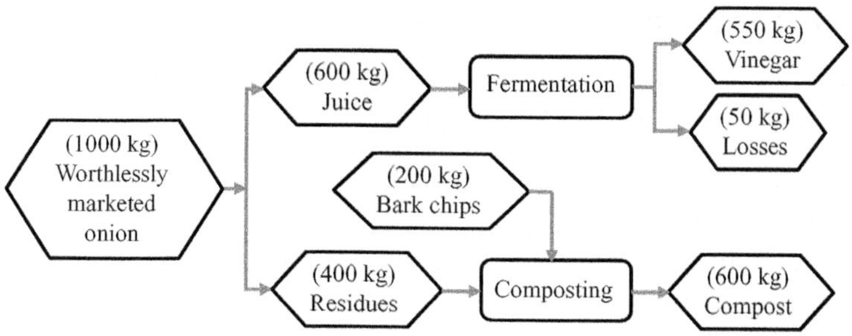

FIG. 5
A mass balance for the production of vinegar compost from worthless marketed onion (Horiuchi et al., 2004).

acid in the first and second steps, respectively. Residues of juice extraction represented 40% of the fresh weight of onion. They were converted to compost, using an in-vessel composter, within 1 week. Due to the high moisture content (86%, w.b.) of onion residues, bark chips were used to adjust the moisture content of the compost mixture to the suitable value (61%, w.b.) needed for successful compost. A mass balance of the integrated system for producing vinegar and compost is shown in Fig. 5. The produced compost had a pH of 7 and contained high concentrations of potassium and nitrogen. It also contained lower concentrations of heavy metals than the compost produced from sewage sludge or industrial wastes.

7.5 Enzymes

Solid onion waste was used to produce peroxidase-active crude extract as an alternative source of inexpensive peroxidases (Osman et al., 2008). A maximum yield of 3.5 mg of protein per 100 g fresh onion solid wastes, with a specific activity of 0.114 units/mg, could be obtained at the optimum temperature and pH of 30–40°C and 4–5, respectively. Crude onion peroxidase extract was used as a biocatalyst for the synthesis of the natural aurones, using 2′,3,4,4′,6′-pentahydroxy-chalcone as a substrate (Moussouni et al., 2010). Aurones represent a minor category of flavonoids. Takahama et al. (2001) mentioned that molecular oxygen was reduced to H_2O_2 by autooxidation when onion brown scales were suspended in water. 3,4-Dihydroxybenzoic acid was preferentially oxidized to produce oxidation products such as quinone. El Agha et al. (2008, 2009) investigated the utilization of peroxidase extracted from onion solid by-products for the oxidation of ferulic and caffeic acids, which can be found in food wastes. During the treatment of ferulic acid, the enzyme was active in the pH range of 4–6 and temperature range of 20–40°C. The optimal pH and temperature were 4 and 30°C, respectively. During the treatment of caffeic acid, the optimal enzyme activity was obtained at pH in the range of 5–6 and temperatures in the range of 20–30°C. The results showed that the produced peroxidase from onion waste could be an effective method for the bioremediation of olive processing wastewater.

7.6 Antibacterial and Antioxidant Production

Onion and garlic wastes are rich in antibacterial and antioxidant compounds. Furusawa et al. (2003) mentioned that the inedible brown scale of onion contains flavonoids that are bioactive compounds in the inhibition of platelet aggregation in human. Flavoniodal components such as quercetin dimers, quercetin, and querecetin-4′-glucoside were identified in the MeOH extricated brown scale of onions. The antiaggregatory and disaggregatory effects on human platelets were attributed to the interaction with membrane lipids that modifies membrane fluidity. The authors mentioned that the modification of membrane fluidity is also a reason for microbial growth-inhibition, cancer prevention, and hepatoprotection. Ifesan et al. (2014) indicated that garlic peel extracts have antioxidant activities similar to garlic bulbs. The authors studied the effect of garlic peel extracts on the survival of microorganisms. Results showed that the extracts reduced the number of *Staphylococcus aureus*, *Escherichia coli*, and *Proteus vulgaris* that were inoculated on cooked beef. Compared with the nontreated beef samples, a reduction of at least one log was obtained after the storage of the treated beef for 9 days at 4°C. The antibacterial effect was attributed to the interaction of sulfur compounds, allicin, with sulfur (thiol) groups of microbial enzymes such as trypsin and other proteases (Bakri and Douglas, 2005). Kallel et al. (2014) extracted phenolic compounds from garlic skins using water, ethanol, a methanol/water mixture 50/50 (v/v), and an ethanol/water mixture 50/50 (v/v) for 45 min at room temperature. The highest yields of total phenols and total flavonoids of 25 [gallic acid equivalents] per gram dry garlic skin and 0.62 [quercetin equivalent] per gram dry garlic skin, respectively, were obtained with 50/50 methanol/water (v/v). Methanol was a better solvent for extracting phenolic compounds than others because of its polarity and good solubility for phenolic components. All extracts exhibited antimicrobial activity against Gram-positive bacteria. The extracts with methanol and its mixture with water inhibited the growth of *Pseudomonas aeruginosa*, and *Klebsiella pneumonia* was inhibited with the 50/50 methanol/water and absolute methanol extracts (Kallel et al., 2014).

Roldán et al. (2008) characterized juice, past, and bagasse created from the by-products of two Spanish onion cultivars (Figueres and Recas). They also studied the effect of freezing (at −70°C), pasteurization (100°C for 11–17 min), and sterilization (115°C for 17–31 min) on the bioactive composition, antioxidant activity, and polyphenol oxidase (PPO) enzyme inhibition capacity of these by-products. The pasteurized paste produced from the Recas cultivar had the best characteristics, such as a remarkable antioxidant activity, high concentrations of total phenols and quercetin, and antibrowning effects, making it a good antioxidant. Similarly, Benítez et al. (2013) studied the effect of pasteurization (100°C for 11–12 min) and sterilization (115°C for 17–31 min) on bioactive compounds in paste, bagasse, and juice that were produced from the waste of the two onion cultivars. Results showed that pasteurization and sterilization improved the soluble/insoluble fiber ratio. While pasteurization did not affect the total dietary fiber content; sterilization decreased the total dietary fiber by

8%. Pasteurization and sterilization caused fructans losses by 36% and 59%, respectively. Due to thermal hydrolysis, losses of 30% and 64% in alkyl or alkenyl cystein sulfoxide could be determined in pasteurized paste and bagasse, respectively. Khiari et al. (2009) developed a sustainable process for the recovery of polyphenol from onion solid wastes using acidified water/ethanol mixtures, which are cheap and nontoxic solvents. High recovery yields were obtained with a water/ethanol mixture containing 60% ethanol and 0.1% HCl or 1% acetic acid. HCl was preferred due to its low cost. Extraction yield increased with the increase in extraction time, up to 6h. However, increasing the extraction temperature from 40°C to 60°C decreased the polyphenol recovery. The decrease of the recovery with the temperature increase might be attributed to the chemical and enzymatic degradation and/or the thermal decomposition of polyphenols (Moure et al., 2001).

Ultrasound and microwave were applied to increase the efficiency of the extraction of quercetin. Jang et al. (2013) extracted quercetin from solid onion waste using ultrasound-assisted ethanol extraction. Results showed that ethanol concentration, in the range of 40%–80% (v/v), and extraction temperature, in the range of 40–60°C, significantly affected the recovery efficiency of quercetin. The recovery efficiency was not significantly affected by pH, in the range of 2–6, liquid/solid ratio, in the range of 30–60mL/g, or extraction time, in the range of 15–35min. An optimal yield of 11.08mg per g dry weight of onion solid waste was obtained at an ethanol concentration of 59% and extraction temperature of 49°C. Without the application of ultrasound, the recovery of quercetin was half of that with the application of ultrasound. The authors attributed this to the positive effect of ultrasound agitation on the penetration of solvent into onion tissues and the disruption of cell walls. Kumar et al. (2014) applied microwave-assisted extraction and ultrasound-assisted extraction to extract quercetin from edible outer layers and the apical trimmings of onion (Bhima Red variety). Using microwave-assisted extraction and water as a solvent, the extraction yield of quercetin increased with the increase of pH from 2.71 to 6.25. A higher yield of quercetin (209mg/100g fresh weight) was obtained with microwave-assisted extraction at the pH of 6.2 and extraction time of 150s than that (175mg/100g fresh weight) obtained with ultrasonic extraction for 40min. This was attributed to the increase in effective swelling of plant tissues, which increased the contact between the solvent and onion particles, and the degradation of quercetin 3,4′ diglucosides to quercetin under the microwave irradiation. No data were presented for the temperature in both systems of extraction.

Super- and subcritical water was applied to extract quercetin from onion wastes. Ko et al. (2011) mentioned that subcritical water is an excellent alternative to organic solvents for the extraction of flavonol quercetin from onion skin because it has a temperature-dependent selectivity, environmental acceptability, high recovery efficiency, and low cost. They studied the effect of extraction temperature (100–190°C), extraction time (5–30 min), pressure (90–131bar), and mixture ratio of onion skin and diatomaceous earth (0.5:3.5–2:2). Diatomaceous earth, composed of remains of diatoms and probably algae, was used to assist in the extraction process by acting as an inert material to prevent the aggregation of the sample. A maximum yield of quercetin (16.3 mg/g onion skin) was obtained at an extraction temperature of 165°C, extraction time of 15 min, and a mixture ratio of 1.5:2.5 for onion skin and diatomaceous earth. The extraction efficiency with subcritical water yielded over four-, six-, and eightfolds greater than those obtained using boiling methanol, ethanol, and water extraction methods, respectively. Salak et al. (2013) found that free quercetin yield increased with the increase of supercritical water temperature up to 140°C. It then decreased to zero when the temperature increased to 220°C. The decrease in quercetin was attributed to its thermal decomposition at high temperatures.

Turner et al. (2006) developed a new method for the extraction of quercetin from yellow onion waste using subcritical water conditions catalyzed by thermostable enzymes. Two β-glucosidase originating from the hyperthermophilic bacterium *Thermotoga neapolitana* were used. *T. neapolitana* optimally grows at temperatures above 80°C. The performance of the new method was compared with the conventional extraction method at 80°C and ambient pressure using methanol, HCl, and ascorbic acid extraction. Quercetin yields of 13.31 and 12.55 mg per gram of dry weight of onion waste were obtained, after 40 and 150 min extraction in the new and conventional methods, respectively. Supercritical ethanol-modified carbon dioxide was used to extract quercetin from skins of yellow and red varieties of onions (Martino and Guyer, 2004). At a pressure of 393 bar, a temperature of 40°C, an ethanol molar concentration of 7.6%, and an extraction time of 2.5 h, free quercetin yields of 0.024 and 0.020 were obtained from the red and yellow verities, respectively. No significant difference was found between red and yellow varieties.

7.7 Dietary Fiber, Reducing Sugars and Cellulose

Dietary fibers have beneficial health effects such as reducing the risk of heart disease and cancer. Benítez et al. (2012) determined dietary fiber content in the unprocessed waste and by-products (bagasse, juice, and paste) of onion processing of two cultivars (Recas and Figueres). Results showed that insoluble dietary fiber was the main fiber fraction in all unprocessed and processed wastes from both cultivars, while soluble dietary fiber greatly varied among the cultivars and by-products. Bagasse, the solid by-product after juice extraction, was the best by-product because it had the highest fiber content and balanced soluble-to-insoluble fiber ratio. In addition, it was rich in uronic acids and polysaccharides. Inclusion of bagasse in a high-fat and high-cholesterol diet reduced serum lipid and total cholesterol in rats. Insoluble fiber contributes to intestinal regulation, and the soluble fraction can reduce the adsorption of blood cholesterol and glucose in the intestine (Scheneeman, 1987; Benítez et al., 2012). Several treatments are being applied to improve the functionality of insoluble fiber in fruit by-products, including partial delignification of lignocellulose, extrusion, enzymatic modification of fiber, and encapsulation with soluble fibers (Larrauri, 1999).

Soluble sugars (mono- and disaccharides) are major components in onions and their waste (Table 3). Onion skins are used to produce sugars for bioethanol production. A glucose yield of 0.31 g/g dry onion waste was obtained after enzymatic hydrolysis using 0.72, 0.16, and 1.0 mg of cellulase, pectinase, and xylanase, respectively, per gram of onion skin waste (Choi et al., 2015). Salak et al. (2013) studied the effect of temperature of the subcritical water, in the range 100–220°C, for the hydrolysis and decomposition of onion bulbs and skins for 5 min to increase the efficiency of the extraction of soluble sugars. The decomposition of onion bulbs increased sharply with temperature up to 167°C. Thereafter, slightly lower decomposition rates were determined. The decomposition of skin linearly increased with the temperature in the range of 100–220°C. A total soluble sugar value of >65% and 7% (of dry matter) was obtained from the bulbs and the skins at temperatures of 167–168°C. Reddy and Rhim (2014) isolated cellulose microfibers and nanocrystals from garlic skin fibers using alkali treatment and acid hydrolysis. The produced cellulose nanocrystals were spherical in shape with a size of 58–96 nm. The degree of crystallinity of the nanocrystals was higher than those of the crude fiber and the microfibers. The thermal stability of the cellulose nanocrystals was lower than that of the native cellulose microfibers due to the introduction of sulfate groups. The bionanocomposites can be used in food packaging and biomedical applications.

8 SUMMARY

Onion and garlic are two important crops that are cultivated under a wide range of weather conditions and soil types. Processing of these crops produce a significant amount of waste. Onion processing wastes include damaged and unwanted bulbs (undersized, malformed, damaged, and diseased bulbs), skins, roots, and tops and bottoms of the bulbs. Garlic processing wastes include peels, straw, and unwanted bulbs. These wastes and worthlessly marketed onions can cause negative environmental problems if they are not disposed of in environmental-friendly and cost-effective ways. Currently, most of these materials are land applied. These materials are rich in micro and macronutrients, soluble and insoluble fibers, antioxidants, and antibacterial compounds. Efficient and economic reuse of onion and garlic waste requires a continuous supply of the wastes, and established efficient systems for the production of value-added compounds.

Onion and garlic wastes can be used directly or after processing as biosorbents of Pb^{2+}, quinolone antibiotics, phenol, Cu^{2+}, and methylene blue from aqueous solutions. Biochemical (e.g., anaerobic digestion) and thermochemical (e.g., gasification) processes are applied to produce bioenergy from biomass materials. Anaerobic digestion is applied to treat onion processing wastewater and onion juice. The biogas produced from this process is cleaned to remove impurities and is then used as a fuel for transportation and generation of heat and electricity. A full-scale anaerobic digestion system is installed at an onion processor in California. The produced biogas powers two 300 kW fuel cells that produce about 50% of the electricity needed by the company. Thermochemical processes can be applied to produce energy from onion and garlic wastes with high solid contents. The liquid and gas products of thermochemical processes can be used as energy sources. The solid products from these processes (i.e., biochar) can be applied as a soil amendment or as a biosorbent. However, there is little information on the thermophilic processing of onion and garlic wastes. More research is needed to optimize the operation parameters of gasification, hydrolysis, and hydrothermal processing of onion and garlic wastes.

Worthless onions are used to produce alcohol and vinegar that are rich in amino acids, organic acids, and minerals. These onions and other onion wastes are also used to produce a high-quality compost. Furthermore, onion processing waste can be used to produce peroxidase-active extracts, which can be used as a biocatalyst for the synthesis of natural aurones, and as effective agents for the bioremediation of food processing wastewater. Onion processing wastes are rich in antibacterial and antioxidant compounds such as flavonoids, including quercetin dimers, quercetin, and querecetin-4′-glucoside. Flavonoids have antiaggregatory and disaggregatory effects on human platelets. Garlic processing wastes have antioxidant activities that have antibacterial effects. Extraction efficiency of antioxidants from onion and garlic wastes depends on the extraction solvent type, pH, temperature, extraction time, and application of ultrasound and microwave. Methanol is a better solvent for the extraction of phenolic compounds from garlic skin at room temperature than ethanol and water are. Subcritical water is more efficient for extraction of quercetin from onion skin than boiling methanol, ethanol, and water extraction methods. Onion and garlic processing wastes can be used to produce dietary fibers, sugars, cellulose, microfibers, and nanocrystals.

References

Abayomi, L.A., Terry, L.A., 2009. Implications of spatial and temporal changes in concentration of pyruvate and glucose in onion (Allium cepa L.) bulbs during controlled atmosphere storage. J. Sci. Food Agric. 89, 683–687.

Abdel-Fattah, A.F., Edrees, M., 1972. A study on the composition of garlic skins and the structural features of the isolated pectic acid. J. Sci. Food Agric. 23, 871–877.

Adam, E., Muhlbauer, W., Esper, A., Wolf, W., Spiess, W., 2000. Quality changes of onion (Allium cepa L.) as affected by the drying process. Nahrung 44, 32–37.

Alexander, M.M., Sulebele, G.A., 1973. Pectic substances in onion and garlic skins. J. Sci. Food Agric. 24, 611–615.

Asfaram, A., Fathi, M.R., Khodadoust, S., Naraki, M., 2014. Removal of direct red 12B by garlic peel as a cheap adsorbent: kinetics, thermodynamic and equilibrium isotherms study of removal. Spectrochim. Acta A Mol. Biomol. Spectrosc. 127, 415–421.

Ayars, J.E., 2008. Water requirement of irrigated garlic. Trans. ASABE 51 (5), 1683–1688.

Bakri, I.M., Douglas, C.W.I., 2005. Inhibitory effect of garlic extract on oral bacteria. Arch. Oral Biol. 50, 645–651.

Benítez, V., Mollá, E., Martín-Cabrejas, M.A., Aguilera, Y., López-Andréu, F.J., Cools, K., Terry, L.A., Esteban, R.M., 2011. Characterization of industrial onion wastes (Allium cepa L.): dietary fibre and bioactive compounds. Plant Foods Hum. Nutr. 66, 48–57.

Benítez, V., Mollá, E., Martín-Cabrejas, M.A., Aguilera, Y., López-Andréu, F.J., Esteban, R.M., 2012. Onion (*Allium cepa L.*) by-products as source of dietary Wber: physicochemical properties and effect on serum lipid levels in high-fat fed rats. Eur. Food Res. Technol. 234, 617–625.

Benítez, V., Mollá, E., Martín-Cabrejas, M.A., Aguilera, Y., Lòpez-Andréu, F.J., Terry, L.A., Esteban, R.M., 2013. The impact of pasteurization and sterilization on bioactive compounds of onion by-products. Food Bioprocess. Technol. 6, 1979–1989.

Benkeblia, N., 2005. Free-radical scavenging capacity and antioxidant properties of some selected onions (Allium cepa L.) and garlic (Allium sativum L.) extracts. Braz. Arch. Biol. Technol. 48, 753–759.

Bosch Serra, A.D., Currah, L., 2002. Agronomy of onions. In: Rabinowitch, H.D., Currah, L. (Eds.), Allium Crop Science: Recent Advances. CABI Publishing, Wallingford, UK.

Brown, B., 2000. Southern Idaho Fertilizer Guide: Onions. Idaho Cooperative Extension System. Agricultural Experiment Station. CIS 1081.

Choi, I.S., Cho, E.J., Moon, J.H., Bae, H.J., 2015. Onion skin waste as a valorization resource for the by-products quercetin and biosugar. Food Chem. 188, 537–542.

Chowdhury, A., Bhowal, A., Datta, S., 2012. Equilibrium, thermodynamic and kinetic studies for removal of copper (ii) from aqueous solution by onion and garlic skin. Water 4, 37–51.

Corzo-Martínez, M., Corzo, N., Villamiel, M., 2007. Biological properties of onions and garlic. Trends Food Sci. Technol. 18, 609–625.

De La Cruz Medina, J., García, H.S., 2007. Post-Harvest Operations. In: Instituto Tecnologico de Veracruz. http://www.itver.edu.mx.

El Agha, A., Makris, D.P., Kefalas, P., 2008. Peroxidase-active cell-free extract from onion solid wastes: biocatalytic properties and putative pathway of ferulic acid oxidation. J. Biosci. Bioeng. 106 (3), 279–285.

El Agha, A., Abbeddou, S., Makris, D.P., Kefalas, P., 2009. Biocatalytic properties of a peroxidase-active cell-free extract from onion solid wastes: caffeic acid oxidation. Biodegradation 20, 143–153.

FAO, 2017. http://www.fao.org/faostat/en/#data/QC. (Accessed 10 July 2017).

Filaree Garlic Farm, 2012. Filaree Garlic Farm's 2012 Catalogue. www.filareefarm.com. (Accessed 24 July 2017).

French, L., Hamman, L., Katz, S., Kozaki, Y., Frew, J., 2010. Zero Waste Strategies for Gills Onions Sustainable Innovation and Waste Management. http://www.esm.ucsb.edu/research/documents/onions_proposal.pdf. (Accessed 26 July 2017).

Fritsch, R.M., Friesen, N., 2002. Evolution, domestication and taxonomy. In: Rabinowitch, H.D., Currah, L. (Eds.), Allium Crop Science: Recent Advances. CABI publishing, Wallingford, UK.

Furusawa, M., Tsuchiya, H., Nagayama, M., Tanaka, T., Nakaya, K., Iinuma, M., 2003. Anti-platelet and membrane-rigidifying flavonoids in brownish scale of onion. J. Health Sci. 49, 475–480.

González-Sáiz, J.M., Esteban-Díez, I., Sánchez-Gallardo, C., Pizarro, C., 2008. Monitoring of substrate and product concentrations in acetic fermentation processes for onion vinegar production by NIR spectroscopy: value addition to worthless onions. Anal. Bioanal. Chem. 391, 2937–2947.

Greer, D., 2010. Onion grower invests in digester and fuel cells. Biocycle 51 (3), 23.

Gubb, I.R., MacTavish, H.S., 2002. Onion pre- and postharvest considerations. In: Rabinowitch, H.D., Currah, L. (Eds.), Allium Crop Science: Recent Advances. CABI Publishing, Wallingford, UK.

Hameed, B.H., Ahmad, A.A., 2009. Batch adsorption of methylene blue from aqueous solution by garlic peel, an agricultural waste biomass. J. Hazard. Mater. 164, 870–875.

Herrmann, K., 1976. Flavonols and flavones in food plants: a review. J. Food Technol. 11, 433–448.

Hong, S.I., Kim, D.M., 2001. Storage quality of chopped garlic as influenced by organic acids and high-pressure treatment. J. Sci. Food Agric. 81 (4), 397–403.

Horiuchi, J.I., Kanno, T., Kobayashi, M., 1999. New vinegar production from onions. J. Biosci. Bioeng. 88, 107–109.

Horiuchi, J.I., Yamauchi, N., Osugi, M., Kanno, T., Kobayashi, M., Kuriyama, H., 2000a. Onion alcohol production by repeated batch process using a flocculating yeast. Bioresour. Technol. 75, 153–156.

Horiuchi, J.I., Kanno, T., Kobayashi, M., 2000b. Effective onion vinegar production fermentation system by a two-step. J. Biosci. Bioeng. 90 (3), 289–293.

Horiuchi, J.I., Tada, K., Kobayashi, M., Kanno, T., Ebie, K., 2004. Biological approach for effective utilization of worthless onions—vinegar production and composting. Resour. Conserv. 40, 97–109.

Ichikawa, M., Ryu, K., Yoshida, J., Ide, N., Kodera, Y., Sasaoka, T., Rosen, R.T., 2003. Identification of six phenylpropanoids from garlic skin as major antioxidants. J. Agric. Food Chem. 51, 7313–7317.

Ifesan, B.O.T., Fadipe, E.A., Ifesan, B.T., 2014. Investigation of antioxidant and antimicrobial properties of garlic peel extract (Allium sativum) and its use as natural food additive in cooked beef. J. Sci. Res. Rep. 3 (5), 711–721.

Jaime, L., Mollá, E., Fernández, A., Martín-Cabrejas, M.A., López-Andréu, F.J., Esteban, R.M., 2002. Structural carbohydrate differences and potential source of dietary fiber of onion (Allium cepa l.) tissues. J. Agric. Food Chem. 50, 122–128.

Jang, M., Asnin, L., Nile, S.H., Keum, Y.S., Kim, H.Y., Park, S.W., 2013. Ultrasound-assisted extraction of quercetin from onion solid wastes. Int. J. Food Sci. Technol. 48, 246–252.

Kaack, K., Christensen, L.P., Hansen, S.L., Grevsen, K., 2004. Non-structural carbohydrates in processed soft fried onion (Allium cepa L.). Eur. Food Res. Technol. 218, 372–379.

Kallel, F., Ellouz Chaabouni, S., 2017. Perspective of garlic processing wastes as low- cost substrates for production of high-added value products: a review. Environ. Prog. Sustain. Energy. https://doi.org/10.1002/ep.12649.

Kallel, F., Driss, D., Chaari, F., Belghith, L., Bouaziz, F., Ghorbel, R., Ellouz Chaabouni, S., 2014. Garlic (Allium sativum L.) husk waste as a potential source of phenolic compounds: influence of extracting solvents on its antimicrobial and antioxidant properties. Ind. Crop Prod. 62, 34–41.

Khiari, Z., Makris, D.P., Kefalas, P., 2009. An investigation on the recovery of antioxidant phenolics from onion solid wastes employing water/ethanol-based solvent systems. Food Bioprocess. Technol. 2, 337–343.

Ko, M.J., Cheigh, C.I., Cho, S.W., Chung, M.S., 2011. Subcritical water extraction of flavonol quercetin from onion skin. J. Food Eng. 102, 327–333.

Kumar, S., Imtiyaz, M., Kumar, A., Singh, R., 2007. Response of onion (*Allium cepa L.*) to different levels of irrigation water. Agric. Water Manag. 89 (1–2), 161–166.

Kumar, B., Smita, K., Kumar, B., Cumbal, L., Rosero, G., 2014. Microwave-assisted extraction and solid-phase separation of quercetin from solid onion (*Allium cepa L.*). Sep. Sci. Technol. 49, 2502–2509.

Larrauri, J.A., 1999. New approaches in the preparation of high dietary fibre powders from fruit by-products. Trends Food Sci. Technol. 10, 3–8.

Liu, W., Liu, Y., Tao, Y., Yu, Y., Jiang, H., Lian, H., 2014. Comparative study of adsorption of Pb(II) on native garlic peel and mercerized garlic peel. Environ. Sci. Pollut. Res. 21, 2054–2063.

Lund, J.W., Lienau, P.J., 2017. Onion Dehydration. http://www.oit.edu/docs/default-source/geoheat-center-documents/publications/industrial/tp86.pdf?sfvrsn=2. (Accessed 21 August 2017).

Ly, T.N., Hazama, C., Shimoyamada, M., Ando, H., Kato, K., Yamauchi, R., 2005. Antioxidative compounds from the outer scales of onion. J. Agric. Food Chem. 53, 8183–8189.

Mann, L.K., 1952. Anatomy of the garlic bulb and factors affecting bulb development. Hilgardia 21 (8), 195–231.

Marín, M.E.R., 2009. Biological Activity and Nutritional Properties of Processed Onion Products (Ph. D. Thesis). Universidad Autónoma de Madrid, Spain.

Martino, K.G., Guyer, D., 2004. Supercritical fluid extraction of quercetin from onion skins. J. Food Process Eng. 27, 17–28.

Mitra, J., Shrivastava, S.L., Rao, P.S., 2012. Onion dehydration: a review. J. Food Sci. Technol. 49 (3), 267–277.

Moure, A., Cruz, J.M., Franco, D., Domínguez, J.M., Sineiro, J., Domínguez, H., et al., 2001. Natural antioxidants from residual sources. Food Chem. 72, 145–171.

Moussouni, S., Detsi, A., Majdalani, M., Makris, D.P., Kefalas, P., 2010. Crude peroxidase from onion solid waste as a tool for organic synthesis. Part I: cyclization of 2′,3,4,4′,6′-pentahydroxy-chalcone into aureusidin. Tetrahedron Lett. 51, 4076–4078.

Muthamilselvi, P., Karthikeyan, R., Kumar, B.S.M., 2016. Adsorption of phenol onto garlic peel: optimization, kinetics, isotherm, and thermodynamic studies. Desalin. Water Treat. 57 (5), 2089–2103.

Osman, A., Makris, D.P., Kefalas, P., 2008. Investigation on biocatalytic properties of a peroxidase-active homogenate from onion solid wastes: an insight into quercetin oxidation mechanism. Process Biochem. 43, 861–867.

Park, Y.K., Yoo, M.L., Lee, H.W., Park, S.H., Jung, S.-C., Park, S.-S., Kim, S.-C., 2012. Effects of operation conditions on pyrolysis characteristics of agricultural residues. Renew. Energy 42, 125–130.

Pellejero, G., Miglierina, A., Aschkar, G., Jiménez-Ballesta, R., 2015. Composting onion (Allium cepa) wastes with alfalfa (Medicago sativa L.) and cattle manure assessment. Agric. Sci. 6, 445–455.

Ramos, F.A., Takaishi, Y., Shirotori, M., Kawaguchi, Y., Tsuchiya, K., Shibata, H., Higuti, T., Tadokoro, T., Takeuchi, M., 2006. Antibacterial and antioxidant activities of quercetin oxidation products from yellow onion (Allium cepa) skin. J. Agric. Food Chem. 54, 3551–3557.

Ranjitkar, H.D., 2003. A Handbook of Practical Botany. Arun Kumar Ranjitkar, Kathmandu. Cited by Shrestha (2007).

Reddy, J.P., Rhim, J.W., 2014. Isolation and characterization of cellulose nanocrystals from garlic skin. Mater. Lett. 129, 20–23.

Rhodes, M.J.C., Price, K.R., 1996. Analytical problems in the study of flavonoid compounds in onions. Food Chem. 57, 113–117.

Roberts, J.S., Kidd, D.R., 2005. Lactic acid fermentation of onions. Lebensm. Wiss. Technol. 38, 185–190.

Roldán, E., Sánchez-Moreno, C., Ancos, B., Cano, M.P., 2008. Characterization of onion (Allium cepa L.) by-products as food ingredients with antioxidant and antibrowning properties. Food Chem. 108, 907–916.

Romano, R.T., Zhang, R.H., 2008. Co-digestion of onion juice and wastewater sludge using an anaerobic mixed biofilm reactor. Bioresour. Technol. 99, 631–637.

Romano, R., Zhang, R., Hartman, K., 2004. Anaerobic digestion of onion wastes using a continuous two-phase anaerobic solids digestion system. In: Paper Presented at ASAE/CSAE 2004 Annual International Meeting. August 1–4, 2004. Ottawa, Ontario, Canada. ASAE Paper Number 047070.

Saka, C., Şahin, Ö., Demir, H., Kahyaoğlu, M., 2011. Removal of lead (II) from aqueous solutions using pre-boiled and formaldehyde-treated onion skins as a new adsorbent. Sep. Sci. Technol. 46, 507–517.

Saka, C., Şahin, Ö., Çelik, M.S., 2012. The removal of methylene blue from aqueous solutions by using microwave heating and pre-boiling treated onion skins as a new adsorbent. Energy Sources, Part A 34, 1577–1590.

Salak, F., Daneshvar, S., Abedi, J., Furukawa, K., 2013. Adding value to onion (Allium cepa L.) waste by subcritical water treatment. Fuel Process. Technol. 112, 86–92.

Scheneeman, B.O., 1987. Soluble vs insoluble fiber-different physiological responses. Food Technol. 41, 81–82.

Sharma, K., Mahato, N., Nile, S.H., Lee, E.T., Lee, Y.R., 2016. Economical and environmentally-friendly approaches for usage of onion (*Allium cepa* L.) waste. Food Funct 7, 3354–3369.

Shock, C.C., Feibert, E., Jensen, L., Klauzer, J., 2010. Successful Onion Irrigation Scheduling. Malheur Experiment Station, Oregon State University .SR 1097.

Shrestha, H., 2007. A Plant Monograph on Onion (*Allium cepa* L.). http://www.uiennieuws.nl/kennis/docs/A%20Plant%20Monograpgh%20on%20Onion%20(Allium%20cepa%20L.)%202007.pdf . (Accessed 5 October 2017).

Staba, E.J., Lash, L., Staba, J.E., 2001. A commentary on the effects of garlic extraction and formulation on product composition. J. Nutr. 131, 1118S–1119S.

Suh, H.J., Lee, J.M., Cho, J.S., Kim, Y.S., Chung, S.H., 1999. Radical scavenging compounds in onion skin. Food Res. Int. 32, 659–664.

Takahama, U., Oniki, T., Hirota, S., 2001. Phenolic components of brown scales of onion bulbs produce hydrogen peroxide by autooxidation. J. Plant Res. 114, 395–402.

The Herb Society of America, 2006. Garlic: An Herb Society of America Guide. The Herb Society of America, Kirtland, OH.

Timm, E.J., Brown, G.K., Brook, R.C., Schulte, N.L., Burton, C.L., 1991. Impact bruise estimates for onion packing lines. Appl. Eng. Agric. 7, 571–576.

Turner, C., Turner, P., Jacobson, G., Almgren, K., Waldeback, M., Sjoberg, P., Karlsson, E.N., Markides, E., 2006. Subcritical water extraction and β-glucosidase-catalyzed hydrolysis of quercetin glycosides in onion waste. Green Chem. 8, 949–959.

Wiczkowski, W., 2011. Garlic and onion: production, biochemistry, and processing. In: Sinha, N.K., Hui, Y.H., Evranuz, E.Ö., Siddiq, M., Jasim Ahmed, J. (Eds.), Handbook of Vegetables and Vegetable Processing. Blackwell Publishing Ltd., pp. 625–642.

Zhang, R., El-Mashad, H.M., 2017. Waste management in egg production. In: Roberts, J. (Ed.), Achieving Sustainable Production of Eggs Volume 2. Animal Welfare and Sustainability. Burleigh Dodds Science Publishing.

Zhao, Y., Li, W., Liu, J., Huang, K., Wu, C., Shao, H., Chen, H., Liu, X., 2017. Modification of garlic peel by nitric acid and its application as a novel adsorbent for solid-phase extraction of quinolone antibiotics. Chem. Eng. J. 326, 745–755.

CHAPTER 12
Carrots

Tyler J. Barzee*, Hamed M. El- Mashad*,†, Ruihong Zhang*, Zhongli Pan*
*Department of Biological and Agricultural Engineering, University of California - Davis, Davis, CA, United States, †Department of Agricultural Engineering, Faculty of Agriculture, Mansoura University, El-Mansoura, Egypt

Chapter Outline

1 INTRODUCTION

Carrot (*Daucus carota*) is an important food crop utilized worldwide and its production is associated with by-products such as culled carrots and carrot waste (i.e., carrot pomace). A variety of technologies aimed at adding value to the by-products or lessening the environmental impacts of current disposal strategies have been explored in the recent years. These technologies encompass physical, chemical, and biological processes and may create products in the form of extracted chemicals, biofuels, novel foods, and adsorbent materials. A recent book chapter focused on the production of carrot pomace powder, novel food products, and functional components (Sharma and Kumar, 2018). The chapter also presented some of the applications of carrot pomace in biofuel and composting operations. However, some topics were not covered by the authors including carrot cultivation/harvest and the processing technologies for functional compound extraction, stabilization, biofuel production, composting,

Integrated Processing Technologies for Food and Agricultural By-Products
https://doi.org/10.1016/B978-0-12-814138-0.00012-5

and wastewater treatment. Therefore, the objective of this chapter was to review the current trends in the carrot production throughout the world; the state of the art in carrot cultivation, harvest, storage, and processing; the production and likely amounts of carrot waste and by-products; and the application of physical, chemical, and biological technologies to the by-products. The technologies reviewed are at various stages of commercial development with some already widely commercially available and others farther off in development. The information reviewed here will help in the dissemination of research results of industrial interest to aid in the adoption of relevant technologies in commercial operations.

2 MORPHOLOGY AND COMPOSITION OF CARROT

The origins of the cultivated carrot utilized today can probably be traced back to Ancient Persia in the 10th century A.D. (Banga, 1957), where it was disseminated toward the Mediterranean from present-day Afghanistan (Banga, 1963). The most common carrot of that time was thought to have been purple in color owing to high levels of anthocyanin pigments (Banga, 1963). Since then, different colors of carrots have developed namely: yellow, white, orange, red, and purple or black. Several mechanisms of the development of different colors from the original purple carrot have been proposed (Bradeen and Simon, 2007) including: color mutants that lead from purple to yellow, yellow to white, and yellow to orange (Banga, 1963); crosses between cultivated and wild germplasms (Small, 1978); or a combination of crossing between cultivated and wild germplasm and human selection (Heywood, 1983).

The carrot plant is an herbaceous annual crop with alternate compound leaves, white umbrella-shaped collection of flowers, and a fleshy conical taproot that can extend for 5–50 cm in length (Bradeen and Simon, 2007). The main descriptive parts of the root section of a carrot can be seen in Fig. 1. The neck of the plant describes the area at the base of the petioles and the junction to the root. The collar and shoulder (crown) of the plant follow and the color, size, and slope of growth aid with distinguishing between varieties. The flesh (phloem) of the carrot extends from the outer edge toward the center and can have varying thickness and texture. The core (xylem) is the innermost part of the carrot and is described similarly to the phloem. The main root ends at the base and a thinner taproot extends further down in a "rat-tail" fashion.

Carrots are not main sources of energy in the human diet but are a significant source of nutrients among other fruits and vegetables (Alasalvar et al., 2001). Carrots are best known for their high contents of phenolics, especially carotenoids like α- and β-carotenes which impart distinctive orange colors. These are important precursors to vitamin A in human metabolism which is involved with the healthy development and function of the teeth, bones, skin, and eyes (NIH, 2018). Carotenoids function as antioxidants to deactivate free radical and singlet oxygen species and are also associated with health benefits including: inhibition of certain cancers, prevention of cardiovascular disease, decreasing risks of

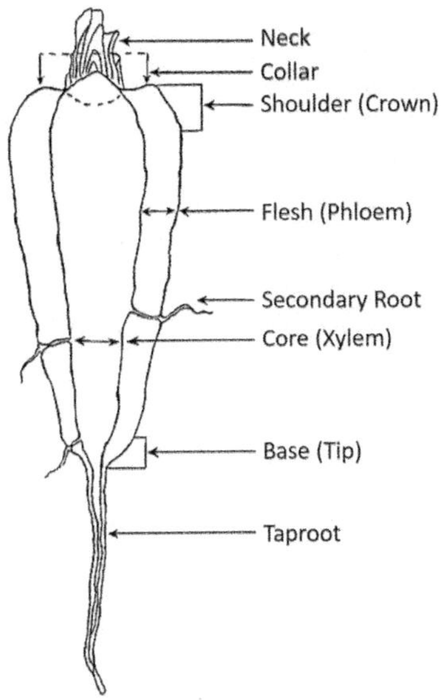

FIG. 1
Longitudinal carrot section (Babb et al., 1950).

cataract formation, prevention of muscular degeneration, and enhancing immune-system function (Sharma et al., 2012). The antioxidant activity of the carrot has been reported to be among the highest of several common vegetables at 4256 μmol vitamin C equivalents per 100 g (Nayak et al., 2015). Due to the health benefits associated with red and orange vegetables rich in carotenoid pigments, the USDA recommends adults to eat up to 350 g of these vegetables each week (USDA, 2015). Dark-colored (red or purple) carrots are rich in anthocyanin pigments, which are also important antioxidants in human nutrition and are also extracted and used as natural food colorant (Bradeen and Simon, 2007). Carrots are usually between 86% and 89% moisture, relatively low in protein and fat, and rich in carbohydrates and fiber. General biochemical contents of raw carrots are shown in Table 1.

3 CARROT CULTIVATION

Worldwide production, harvested area, and yield of carrots and turnips can be seen in Fig. 2. Total production of carrots and turnips has been increasing, having more than doubled in the past 20 years from 18 million tons in 1994 to 47 million tons in 2016 (FAO, 2018). During the same time, total harvested area and yield increased by 54% and 66%, respectively. The main carrot producing regions of the world include Asia, Europe, and the United States (Fig. 3). Asia has experienced the most notable increase in the production of any continent in the past 20 years and now accounts for more than half (53%) of total worldwide carrot and turnip production (FAO, 2018).

Table 1 Biochemical Composition of Raw Carrots

	Unit	Value	References
MC	% (w.b.)	86–89	Sharma et al. (2012), USDA (2018a)
Protein		0.7–0.93	
Fat		0.2–0.5	
Carbohydrate		6–10.6	
Crude Fiber		1.2–2.8	
Total Ash		1.1	
Ca	mg/	34–80	
Fe	100g	0.3–2.2	
P		25–53	
Na		40–69	
K		240–320	
Mg		9–12	
Cu		0.02	
Zn		0.2–0.24	
Carotenes		5.33	
Thiamin		0.04–0.066	
Riboflavin		0.02–0.058	
Niacin		0.2–0.983	
Vitamin C		4–5.9	

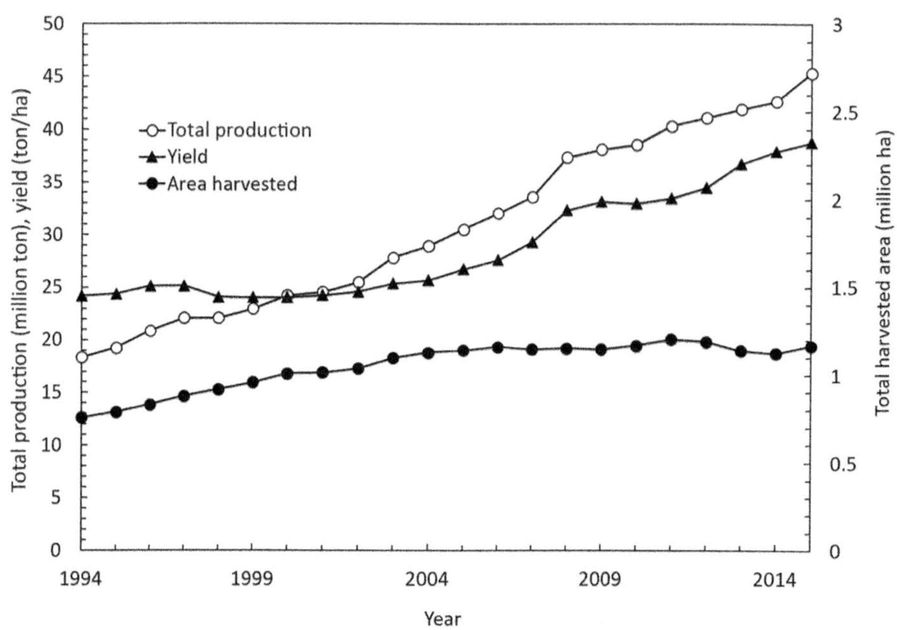

FIG. 2
Total world production, harvested areas, and yield of carrot and turnip from 1994 to 2016 (FAO, 2018).

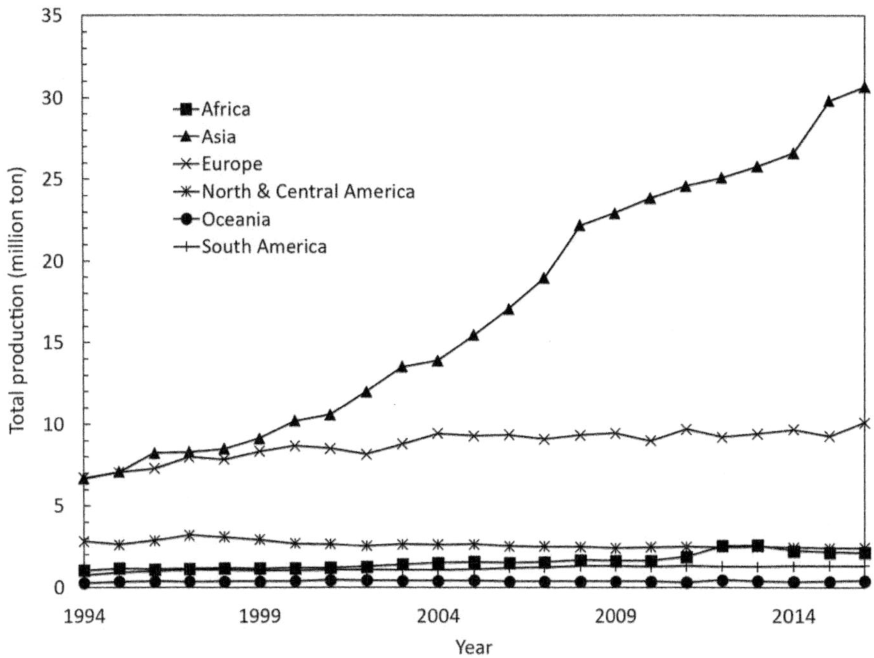

FIG. 3

Regional total production of carrot and turnip from 1994 to 2016 (FAO, 2018).

Carrot plants can be grown year-round in some climates such as California but they are optimally produced in cool temperatures (12–21°C) because warmer temperatures can lead to darkened roots and undesirable strong flavors (Nguyen and Nguyen, 2015; Nunez et al., 2008). Soil conditions such as soil type, level of compaction, and irrigation plan can affect the carrot root significantly. These factors can be reflected on the shape and length of the carrot root (Lipiec et al., 2003; Pietola and Smucker, 1998). Silty loam soil free of barriers to root growth is preferred for its optimal water-holding capacity and drainage properties (Nunez et al., 2008). Soil type and chemical composition can also affect the flavor profile of carrot roots (Nguyen and Nguyen, 2015). Carrots are generally directly seeded to fields at a rate of 0.4–0.5 million seeds per hectare and fertilized with nitrogen at a rate of 110–280 kg/ha (Nunez et al., 2008).

4 HARVESTING AND STORAGE OF CARROT

Carrots are usually harvested by hand or mechanically with a top-lifting harvester or digger elevator (Nguyen and Nguyen, 2015). Top-lifting harvesters use a blade to loosen the carrots from the soil and a series of belts to grasp the leaves (tops) and convey the whole plant to a hopper or further processing operations for dirt removal and/or top removal (Bond, 2016; Nguyen and Nguyen, 2015; Nunez et al., 2008). Digger elevator harvesters use a blade and flat shovel to loosen the earth which is then conveyed and sieved to isolate the root from the dirt (Bhattacharya et al., 2006; Singh et al., 2006). Although mechanical harvesting methods reduce manual labor requirements, they can cause higher levels of damage compared to hand harvesting (Apeland, 1974; Goncharuk and Dyachenko, 1980; Tucker, 1974). Top-lifting harvesters are the most common carrot harvesters used commercially (Bond, 2016). Following harvest, carrots are transported to a facility and stored in cold rooms if necessary awaiting washing, grading, and packing (Bond, 2016). Carrots are capable of being stored for 5–9 months under ideal conditions (approximately 0°C, 99% relative humidity) that prevent decay and dehydration (Apeland and Hoftun, 1974; Nunez et al., 2008; Sarkar and Sharma, 2011). Frequently, carrots are washed and hydrocooled in chlorine-supplemented water before storage. However, high storage life has been reported with unwashed, un-topped carrots due to the absence of harsh handling conditions that can cause surface cracks and accelerated rot (Nunez et al., 2008). Mechanical stresses such as shaking and dropping should be avoided during storage because they can increase respiration and productions of ethylene, ethanol, and bitter compounds like 6-methoxymellein (Seljåsen et al., 2001).

5 CARROT PROCESSING AND WASTE PRODUCTION

Carrots are sold in the cut and peel market with tops removed or as bunches with the tops intact (Nunez et al., 2008). Carrots destined for the bunch market are usually hand sorted in the field before washing and packaging (Nunez et al., 2008). Processing of carrots destined for the cut and peel market consists of washing, grading, and packing all performed at a processing facility, sometimes referred to as packhouses or packinghouses (Bond, 2016; Nunez et al., 2008). The carrots are first washed with water which may be supplemented with chlorine in order to reduce bacterial contamination (Nguyen and Nguyen, 2015). The washing steps may be performed by fully submersing the carrots in water, spraying the carrots with water while traveling on a conveyor or rotating drum, or a combination of these or other methods in series. The water from washing and from other operations in the processing facility are ultimately dealt with as wastewater with high nutrient or solids loading and must be treated before recycling or discharging to the environment or off-site wastewater treatment facility. Wastewater from a German carrot processing facilities was characterized by high suspended solids (435 mg/L), and high COD loading (179 mg/L) (Kern et al., 2006). An example water recycling and wastewater flow diagram can be seen in Fig. 4. At the example facility included here, suspended solids concentrations in excess of 1000 mg/L were measured between the settling ponds and a concrete-lined sump that collects material before being land applied (unpublished data).

After washing, carrots are graded and sorted for quality criteria such as: damage, shape, texture, rot, cracking, length, and width (USDA, 1965). Various grading machines have been developed to aid in the process and reduce manual labor

FIG. 4
Carrot processing wastewater pond and wastewater flow chart of a carrot processing facility.

requirements. Due to the unique shape of carrots, it is not possible to use screen grading machines that are common for other fruits and vegetables. Therefore, new grading methods were developed. Early carrot grader machine designs consisted of rollers to straighten the carrots followed by divergent conveyor belts with tapered walls (Prenveille, 1956). These machines used gravity to drop the carrots that were too long at any point along the taper onto size-specific-conveyors below. Popular carrot grading machines nowadays utilize similar principles and include methods for length and width sorting. Vibrating length graders consist of V-shaped vibrating plates that are separated from each other by a desired distance and allow carrots of desired sizes to pass through (Engineering, 2018). Lift roller graders are able to size carrots by width using height-adjustable rollers that create a desired grade size for the carrot to fall through (Engineering, 2018). Recent advances in machine vision technology have made it possible to sort carrots rapidly using optical sensors and algorithms capable of sensing color, bends, breakage, or blemishes (Bond, 2016). However, the low detection accuracies and single-index decision making of many of these machines have prompted research into methods to improve accuracy and the range of defects that can be detected (Deng et al., 2017). Carrots not meeting the desired grading criteria, carrot tops, and pomace from juicing are culled and may be used as animal feed, ensiled, or dehydrated (Wadhwa and Bakshi, 2013). In some cases, the amount of wasted material from the processing facility can be substantial. For instance, Bond (2016) reported that out-grading at Norwegian packinghouses accounted for 20% loss of the total crop with only half of that being due to decay in storage.

Following sorting and manual quality control review, carrots are either directly packed and shipped or undergo further processing. In 2015, 4.9 kg of carrots were produced per capita in the United States and 81%, 6.3%, and 12.6% were sold as fresh, canned, and frozen carrots (USDA, 2018b). There has been a steady increase in the proportion of carrots sold as fresh, while canning and especially freezing have decreased in proportion (Fig. 5). Peeling and slicing, blanching, steaming, dehydrating, and juicing are popular methods to create food products from carrots (Nguyen and Nguyen, 2015). The main source of carrot by-products

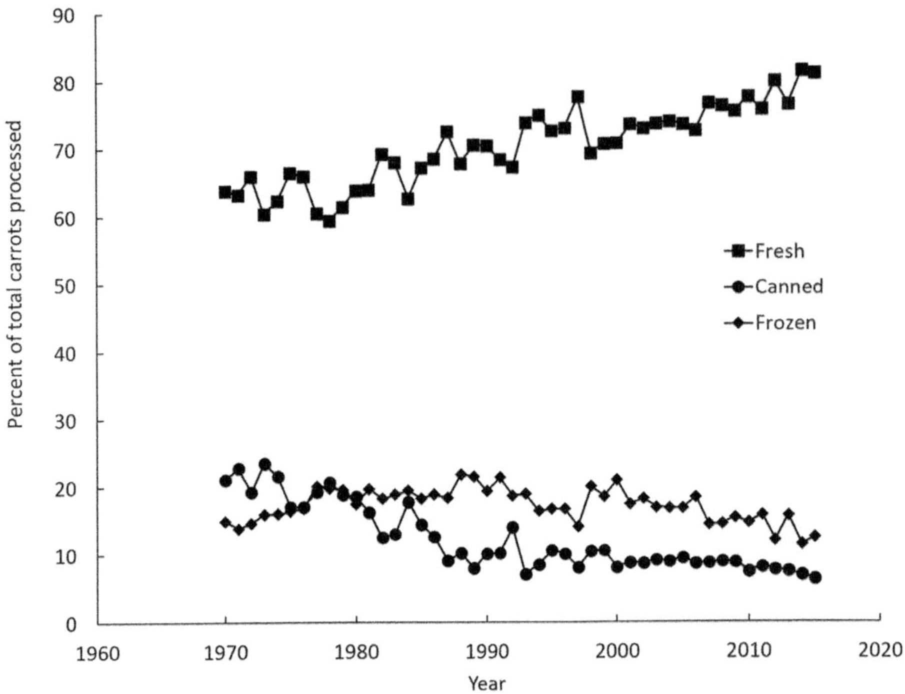

FIG. 5
Supply of fresh and processed carrots in the United States (USDA, 2018b).

is associated with juicing, which produces a large amount of carrot pomace. Carrot juice is the second most consumed vegetable juice following tomatoes (Zentek et al., 2014). Carrot juicing produces substantially more waste than tomatoes with losses up to 50% of the raw material with carrots compared to 3%–7% with tomatoes (Nguyen and Nguyen, 2015; Zentek et al., 2014). Increases in carrot juice consumption have been reported in several countries in previous years. For instance, German carrot juice production increased 69% during the period of 1995–99 with a total production of 42 million liters (42,000 tonnes) in 1999 (Schieber et al., 2001). It was reported that carrot juice consumption in Taiwan reached up to 2.3 million tonnes in 1994 (Chen and Tang, 1998) and it was estimated in 2013 that 6000 tons of carrot pomace were produced annually in Taiwan (Yu et al., 2013). The volume of carrot juice created in the United States every year is unclear but is mainly produced in California among two main firms (Lucier and Lin, 2007; Reyes-De-Corcuera et al., 2014). Carrot pomace contains the vast majority of the total carotenes present in the carrot (up to 80%) but is usually used as a fertilizer, animal feed supplement, or wasted (Schieber et al., 2001; Wadhwa and Bakshi, 2013; Yoon et al., 2005; Zentek et al., 2014).

FIG. 6
Carrot pomace conveyed from a screen separator.

Carrot pomace is created as a by-product of the juicing process. Typically, carrots to be juiced are blanched in hot water held at 80°C for 6 min to deactivate enzymes that might cause reactions that deteriorate the quality of the carrot (Nguyen and Nguyen, 2015). The carrots are then subjected to size reduction processes, often with a hammer mill and subsequent heat and/or enzyme treatment with cellulo- and pecto-lytic enzymes to increase juice yield (Nguyen and Nguyen, 2015; Reyes-De-Corcuera et al., 2014). The carrot mash is then pressed with a continuous belt press or bladder press to liberate the juice (Fig. 6) (Reyes-De-Corcuera et al., 2014). The mixture is then centrifuged to yield juice and pomace. The juice may be subject to further processing to concentrate or change flavors and inactive enzymes and microbes.

6 CHARACTERIZATION OF CARROT PROCESSING WASTE

Carrot pomace is a rich source of sugars, fiber, phenolics, and a variety of macro and micronutrients (Table 2).

7 VALUABLE PRODUCTS FROM CARROT WASTE

Research projects have been completed on a wide range of technologies that can be applied to more sustainably utilize the nutritional and energetic potential of carrot pomace, as will be discussed in the subsequent sections.

Table 2 Carrot Pomace Biochemical Characteristics

	Unit	Value	References
MC	% wet basis (w.b.)	93.7 ± 1.45	Tańska et al. (2007)
DM	% dry basis (d.b.)	9.5	Wadhwa and Bakshi (2013)
Ash		5.5 ± 0.26	Tańska et al. (2007)
		7.5	Wadhwa and Bakshi (2013)
Organic matter		92.5	
Sugars		50.8 ± 3.61	Tańska et al. (2007)
		64.3	Wadhwa and Bakshi (2013)
Reducing sugars		18.0 ± 1.20	Tańska et al. (2007)
		6.2	Wadhwa and Bakshi (2013)
Nonreducing Sugars		58.1	
Starch		6.6 ± 0.78	Tańska et al. (2007)
Total fiber		37.5 ± 2.50	
Insoluble		25.0 ± 2.97	
Neutral detergent fiber		24	Wadhwa and Bakshi (2013)
Acid detergent fiber		17.5 ± 2.50	Tańska et al. (2007)
		20	Wadhwa and Bakshi (2013)
Lignin		8.9 ± 0.50	Tańska et al. (2007)
Cellulose		8.5 ± 0.40	
		24	Wadhwa and Bakshi (2013)
Hemicellulose		12.7	
Neutral detergent solubles		76	
Total pectins		6.9 ± 0.90	Tańska et al., (2007)
Crude protein		7.2	Wadhwa and Bakshi (2013)
Albumin	% of crude	61.3	
Globulin		11.9	
Prolamin		7.9	
Glutelin		18.9	
Lipids	% d.b.	6.0 ± 0.43	Tańska et al. (2007)
Total phenolics		4.3	Wadhwa and Bakshi (2013)
Carotenoids	mg/100 g	45.6 ± 2.33	Tańska et al. (2007)
Water absorption	% w.b.	551.1 ± 0.68	
Ether extract	% d.b.	1.8	Wadhwa and Bakshi (2013)
Ca		0.42	
P		0.23	
Mg		0.17	
Na		1	
K		0.45	
S		0.22	
Fe	ppm	1816	
Cu		228.2	
Zn		710	
Mn		292.8	
Mo		2.4	
Co		3.4	
Al		297	
As		2.1	
Cr		10.4	
Ni		7.6	
Cd		0.42	
Pb		5.33	

7.1 Energy Production

7.1.1 BIOGAS

Bioenergy production from vegetable processing wastes has been explored for many years. Raynal et al. (1998) employed a two-phase liquefaction reactor to create volatile fatty acids (VFAs) from a canning wastewater containing pea and carrot residues and found that 88.5% of the particulate COD from this waste could be reduced in mesophilic conditions at a loading rate of 10g COD/L-d. The produced VFAs can be fed to a central methanogenic reactor "methanizer" for the methane production. A study on the mesophilic anaerobic digestion of carrot pomace operated in continuous mode obtained a biomethane yield of 410–420mL/g VS/d with a loading rate of 0.8–0.9g VS/L/d (Knol et al., 1978). Another research focused on downflow continuous mesophilic digestion of carrot waste and obtained average biogas yields of 205mL/g VS added when loaded at 2.3g VS/L/d (El-Shimi et al., 1992). A methane yield of 300mL CH_4/g VS could be obtained from the mesophilic digestion of whole carrots in anaerobic sequencing batch mode (Thanikal et al., 2015). A study investigating codigestion of carrots with onion waste and the use of a biological catalyst called Metaferm reported average biomethane yields of 325mL/g dry organic matter (DOM), 498mL/g DOM, and 382mL/g DOM for carrot waste, carrot waste with Metaferm, and a 50:50 mixture of carrots and onions with Metaferm, respectively (Dubrovskis and Plume, 2015). Solid carrot waste from a carrot processor in California was mesophically digested using an anaerobic-phased solids digester system (Zhang and Zhang, 1999, 2002) and achieved a biomethane yield of 330mL/g VS with a recommended hydraulic retention time of 3 days (unpublished data). In general, the biomethane yields achieved with carrot waste products tend to be relatively lower than the 440mL/g VS reported for mixed food waste (Zhang et al., 2007). Alternative anaerobic digestion strategies to produce biohydrogen reported yields of 44.9–70.7mL H_2/g VS (Mata-Alvarez et al., 2000). A study on the adaptation of microbial communities during successive anaerobic digestion trials of carrot pomace used 16S rRNA gene analysis and found that the major bacterial groups after adaptation were *Bacilli* (31%–45.3%), *Porphyromonadaceae* (12.1%–24.8%), and *Spirochaetes* (12.5%–18.5%) with the archaea community being completely dominated by an organism similar to *Mathanosarcina mazei* (Garcia et al., 2011). A summary of the anaerobic digestion results using carrot pomace is shown in Table 3.

7.1.2 BIOETHANOL

Enzymatic hydrolysis of carrot pomace and subsequent fermentation to yield bioethanol for use as a biofuel has been investigated. Lignocellulosic materials such as carrot pomace are usually pretreated to remove lignin and hemicellulose and then subject to enzymatic saccharification and then fermentation with yeast or bacteria (Sun and Cheng, 2002). Patle and Lal (2007) carried out research to produce ethanol using *Zymomonas mobilis* and *Candida tropicalis* on a variety of agricultural wastes using a separate hydrolysis and fermentation (SHF) strategy.

Table 3 Anaerobic Digestion of Carrots and Carrot Pomace

Digested material	Temperature regime	Digester operation	Organic loading rate	Biogas/ Biomethane yield	Source
Carrot pomace	Mesophilic	Continuous	0.8–0.9 g VS/L/d	410–420 mL CH_4/g VS	Knol et al. (1978)
Carrot pomace	Mesophilic	Continuous	2.3 g VS/L/d	205 mL biogas/g VS	El-Shimi et al. (1992)
Whole carrots	Mesophilic	Sequencing batch	0.5–6 g VS/L	300 mL CH_4/g VS	Thanikal et al. (2015)
Carrot pomace and metaferm	Mesophilic	Batch	N/A	498 mL CH_4/g DOM	Dubrovskis and Plume (2015)
Carrot pomace	Mesophilic	Anaerobic phased solids	N/A	330 mL CH_4/g VS	Unpublished data
Carrot pomace	Mesophilic	Continuous	N/A	151 mL CH4/g VS	Austin (2013)

The results showed that pretreatments using H_2SO_4 hydrolysis were more efficient than alkaline pretreatment with NaOH. Enzymatic hydrolysis was carried out using xylanase, pectinase, and cellulase and the sugars produced were then fermented to yield 28 g/L of ethanol (70% of theoretical yield). This value was relatively lower than the yields obtained with apple and sapota at 50 and 52 g/L, respectively. Another study reported using dilute acid pretreatment without enzymatic hydrolysis prior to fermentation with *Saccharomyces cerevisiae* or *Pichia stipites* yeasts but was only able to obtain a maximum ethanol titer of 6.91 g/L with *S. cerevisiae* (Demiray et al., 2016).

Yu et al. (2013) applied simultaneous saccharification and fermentation (SSF) to produce ethanol from carrot pomace. The SSF process does not isolate the enzymes used for saccharification from the fermenting microorganisms. Carrot pomace was not pretreated and the thermotolerant yeast *Kluyveromyces marxianus* was inoculated simultaneously with the addition of cellulase [Accelerase 1000, 15 filter paper unit (FPU)/g TS] and pectinase (52.3 FPU/g TS) enzymes. The results showed that addition of 10% (*w*/*v*) of carrot pomace after 12 h of fermentation of an initial charge of 10% (*w*/*v*) carrot pomace resulted in an ethanol concentration of 37 g/L after 42 h with an ethanol yield of 0.185 g/g (86% of theoretical yield). SSF often obtains a higher rate of reaction due to the removal of cellulose-inhibiting glucose by the microbes and simplifies industrial processes because separate reactors are not required for saccharification and fermentation. The obtained ethanol titers were higher than that (28 g/L) obtained by Patle and Lal (2007) which might be related to the different organisms used and/or the advantages of the SSF process.

A 2012 study considered two strategies for ethanol production from discarded carrots: processing to produce juice and pomace followed by enzymatic hydrolysis/fermentation and simply milling the whole carrot before enzymatic hydrolysis and fermentation (Aimaretti et al., 2012). This study considered an SHF process with enzymatic hydrolysis from seven different enzymes and fermentation by *S. cerevisiae CCUB*. It concluded that Optimase CX255L, a thermostable xylanase produced by *Trichoderma reesei*, was most effective at increasing reducing sugar concentration when loaded at 0.05% (*v*/*v*) at 70°C and pH 5.5 for 2.5 h. Ethanol titers of 40.3 g/L were obtained, which related to 77.5 L of ethanol produced per ton of discarded carrots. Although the most effective enzyme was a xylanase, high-performance anion exchange chromatography results indicated that glucose, fructose, and sucrose were the only released sugars. The authors hypothesized that the absence of xylose or arabinose hydrolysis products indicated that degradation of hemicellulose was not a main source of sugars and instead the hydrolysis effectively destabilized the cell walls of storage vacuoles that released free sugars, all of which were fermentable by *S. cerevisiae*.

7.1.3 PYROLYSIS

Pyrolysis or gasification of waste biomass products is a potential avenue for producing energy or chemical building blocks in the form of synthesis gas (Ail and Dasappa, 2016), biochar for use as a soil amendment (Lehmann et al., 2011; Yao et al., 2012) or adsorbent material (Inyang et al., 2012; Li et al., 2017), and bio-oil products (Kabir and Hameed, 2017) that can be refined and used for the development of energy or chemical products. Despite the applicability of this process for a wide range of biological waste materials, seemingly few studies have been performed on carrot waste so far. One study was found that used pyrolysis to thermally decompose carrot pomace and then created carrot pomace activated carbon by chemical activation with $ZnCl_2$ to study its effectiveness in removing Co(II) from solution (Changmai et al., 2018). This study is discussed in more depth in Section 2. Overall, the shortage of studies on this topic illuminates the need for further research in this area to better understand the possible applications of carrot pomace for pyrolysis or gasification.

7.2 Compost

Carrot pomace is nutrient rich and can be an effective addition to composting operations for stabilization and used as a fertilizer and/or soil amendment. It is widely noted in the literature that carrot pomace is usually managed as fertilizer or animal feed (Nguyen and Nguyen, 2015; Sarkar and Sharma, 2011; Sharma et al., 2012). However, academic research on the use of carrot pomace in composting operations is lacking and only a few studies were identified on the topic. Whipps and Noble (2001) reported a study that composted onion and vegetable wastes including carrot waste and investigated the compost production characteristics as well as its application in fertilizing crops and preventing or treating *Allium* white rot. Carrot waste was composted with onion shale and prepared to an initial moisture content of 80%, which related to six parts of wet carrot waste to one part of drier onion shale waste. The mixtures (700 g) were incubated at 50°C for 7 days with periodic aeration every 2 min per half hour at 250 mL/min. At the end of the experiment, minimal runoff was noted (1 mL), while a weight loss of 5.8% and low pH (3.65) was noted. The fresh compost had macronutrient contents of 0.46%, <0.01%, 0.1%, and 1.05% (w.b.) for total nitrogen, ammonia nitrogen, phosphorus, and potassium, respectively, indicating the fertilizer potential of the compost. The carrot-onion waste compost was then used in an incubation inoculated with *Allium* white rot (*Sclerotium cepivorum*) sclerotia in glasshouse pot bioassays (Coventry et al., 2005). The carrot-onion waste compost was most effective in controlling the viability of *S. cepivorum* in silt soil but less effective in sandy loam and peat soils. It was also noted that high compost incorporation rates (50%, *w*/*w*) in the soil were necessary to achieve greater than 50% sclerotia degradation. Also, possible phytotoxicity of compost at high incorporation rates can inhibit or prevent seedling germination so a longer delay (>2 months) between compost incorporation and sowing may be necessary. Austin (2013) noted the high C:N ratio (50:1) and moisture content of carrot pomace necessitated a bulking agent (wood shavings, wood chips, and straw) and ammonium chloride addition. These composting trials were carried out for 21 days and found a mass reduction of 40% (d.b.) with acceptable germination index values, indicating potential for immediate land application. However, the recommended carrot to bulking agent ratio was less than 0.5:1.

7.3 Processing and Food Production From Carrot Pomace

A number of value-added food products can also be derived from carrot pomace. Research has been carried out on several processes to disinfect and/or dry residues, extract components of interest, and incorporate pomace products into a variety of food types.

7.3.1 *PROCESSING AND DISINFECTION*

Food products that require extended shelf lives are usually subject to thermal treatment (pasteurization) using techniques such as hot water immersion for a specified period of time in order to inactivate undesired microorganisms, reduce moisture contents/concentrate, or to facilitate separation of the fruit/vegetable skin (Anaya-Esparza et al., 2017; Nayak et al., 2015; Patras et al., 2009). There are many available food processing methods using air, water, oil, and electromagnetic radiation to extend food life or impart desired characteristics to the product. Despite the importance of thermal techniques for ensuring food safety, pasteurization can affect several quality criteria of the products such as: vitamins, carotenoid, polyphenol, and organic acid contents; pH; and color (Anaya-Esparza et al., 2017). The effect of thermal treatment on the content and bioaccessibility of bioactive compounds has been a topic of interest for many years and it has been shown that different fruits, vegetables, and grains as well as various tissues within each group respond differently to thermal processing due to the location of bioactive molecules in their biomass (Barba et al., 2017; Nayak et al., 2015). Barba et al. (2017) revealed that thermal treatment damages cell walls and thus may increase the accessibility of bioactive compounds. A kinetic study found that the increase in antioxidant activity of carrot juice could be described by a first-order reaction model with an estimated Arrhenius activation energy (E_a) of 40.46±8.43 kJ/mol in the temperature range of 75–110°C (Indrawati et al., 2004). In general, antioxidant potential tends to decrease for short heating times and increase for long heating times due to the development of new antioxidant compounds during heating (Nayak et al., 2015; Nicoli et al., 1999). For short heating times, thermal degradation of natural antioxidants, early Maillard reaction product formation and the consumption of ascorbic acid and polyphenols for these reactions can describe the initial decrease in antioxidant activity (Nayak et al., 2015; Nicoli et al., 1999). However, for extended thermal processing times at adequate temperatures, antioxidant activity can increase from the development of new products with antioxidant activity, such as advanced Maillard reaction products (Kaanane et al., 1988; Nayak et al., 2015; Nicoli et al., 1997). Drying carrot pomace with hot air at an optimal temperature of 65°C has been reported to increase β-carotene concentration from 9.86 to 11.57 mg/100 g dry matter but decrease ascorbic acid concentration from 22.95 to 13.53 mg/100 g dry matter (Upadhyay et al., 2008). However, drying at temperatures above 75°C were reported to degrade β-carotene contents and sun drying was determined to greatly degrade product quality (Sarkar and Sharma, 2011). Drying the carrot pomace material is beneficial in increasing its shelf life or as a pretreatment before further processing such as grinding into powder and incorporating into food products. For continuous processes that desire to preserve the native antioxidants, nonthermal processing strategies such as ultrasonication or pulsed electric fields (discussed later) may be preferred options.

7.3.1.1 High-Pressure Processing

Many different technologies have been researched for the processing of carrot and carrot pomace. High-pressure processing (HPP) is an alternative to traditional thermal processing. HPP is characterized by the use of water to exert pressures between 100 and 800 MPa to materials in order to inactivate bacteria, spores, and viruses (Barba et al., 2017; Heinz and Buckow, 2009). The sample is placed in a flexible container and then placed in a water chamber that imparts pressure to the sample through the use of a piston (Barba et al., 2017). Adiabatic heating occurs due to the high pressure but the increase in temperature is small (for instance, 16–22°C) compared to traditional thermal processes (Barba et al., 2017; Patras et al., 2009). A study using HPP on carrot puree samples with pressures ranging from 400 to 600 MPa found that 600 MPa pressures significantly increased the antioxidant activity compared to untreated or thermally treated (70°C, 2 min) samples. An additional benefit of HPP is its ability to retain texture properties more effectively than thermal processing. In a study comparing high-pressure and high-temperature treatments, it was found that low pressure (0.1 MPa) combined with high-temperature (80–100°C) conditions led to extensive softening but high temperature (80°C) combined with high pressure (600 MPa) was able to mitigate the loss of hardness in the material (De Roeck et al., 2008). It is thought that the textural loss associated with thermal treatment is mostly due to beta-eliminative depolymerization of pectin (De Roeck et al., 2008).

7.3.1.2 Ultrasound Processing

Ultrasound technology is another method of processing carrot pomace and/or extracting valuable compounds. The basic principle of ultrasound processing is that a high-frequency ($\geq$20 kHz) electrical signal is converted to sound energy and propagates through food materials ultimately causing cavitation bubbles that grow and collapse to create shock waves that cause cell-damaging high temperatures and pressures (up to 5000°C, 500 MPa) (Anaya-Esparza et al., 2017; Barba et al., 2017; Nayak et al., 2015). When used in conjunction with elevated temperatures, the process is called "thermosonication" and often has advantages of reduced time and temperature requirements compared to traditional thermal processing (Agcam et al., 2017). This technique has been used to assist in extracting phenolic compounds from carrot pomace. Results of the response surface optimization showed that ultrasonic processing with 20 kHz at 70% amplitude, temperature of 35°C, duration of extraction of 20 min, and ethanol (solvent) concentrations of 55% resulted in an optimal total phenol extraction of 316.9 μg/g (Jabbar et al., 2015). It was reported that the ultrasound-assisted extraction technique was faster, more effective, and less energy and solvent intensive compared to traditional extraction methods. The same research group also reported enzyme inactivation as high as 90% and undetectable viable microorganisms when treating carrot juice with ultrasound at 60°C for 10 min at 20 kHz and 70% amplitude (Jabbar et al., 2015). Thermosonication was also employed to extract anthocyanins from black carrot pomace (Agcam et al., 2017). Results showed that the optimum extraction temperature and energy density were 50°C and 183.1 J/g, respectively. Carrot juice containing orange peel and pulp extracts was also subject to thermosonication in a study that found that adding orange peel to carrot juice and thermosonicating for 6.50 min at 40 kHz and 52.78°C was effective at preventing microbial growth and optimizing total phenolic content and radical-scavenging activity (Adiamo et al., 2018). A possible drawback to sonication is that it has been reported to be capable of degrading certain color compounds and anthocyanins due to oxidation reactions occurring in response to free radicals that are formed during the process or sonication-induced depolymerization or polymerization reactions, among others (Nayak et al., 2015).

7.3.1.3 Electrical/Electromagnetic Processing

Electrical or electromagnetic technologies can also be used to process carrots and carrot pomace and include processes such as: microwave heating, infrared heating, and PEF processing. Microwave energy can be used to heat materials to disinfect them or process them similarly to other thermal processing methods. Microwaves are delivered at frequencies of 0.3–300 GHz and cause dipole rotations, especially in polar molecules like water, that generate heat (Barba et al., 2017). In a study of microwave drying of carrot pomace using a central composite design, the effect of microwave power and treatment time on the moisture content, temperature, and color of carrot pomace were studied and fit to a second-degree polynomial equation (Hernández-Ortega et al., 2013). Carrot pomace moisture content and temperature predictably decreased and increased, respectively, for increasing microwave power and durations, while the color of the carrot pomace increased slightly with increased power at low drying durations and decreased substantially at increased power and drying duration values. The optimal drying conditions were reported as a microwave power of 700 W and a duration of 20 min for acceptable moisture content and color retention. In another study, microwave combined with convective drying of purple carrot pomace was performed with 300 W of microwave power with air at 40°C flowing past the samples at 3.5 m/s (Janiszewska et al., 2013). Microwave-convection drying was shown to decrease the drying time by 30% compared to convective drying and most effectively retained colored compounds such as anthocyanins upon rehydration compared to convection drying, infrared-convection, and freeze drying. On the other hand, freeze drying decreased the apparent density of the purple carrot pomace, increased the vapor adsorption rate, and increased the rate and total amount of soluble dry matter leached from the pomace, indicating structural damage and less desirable rehydration characteristics.

Microwave technology can also be used to aid in the extraction of carotenoids from carrot materials in a process called microwave-assisted extraction (MAE), where microwaves are used to superheat cell interiors causing damage and easier extractability of compounds of interest (Wang and Weller, 2006). MAE can have advantages compared to other extraction procedures such as: simplicity, short extraction times, favorable economics, and decreased solvent requirements (Saini and Keum, 2018). A life cycle assessment of MAE for β-carotene extraction from carrots showed that it was the most environmentally friendly, economical, effective, simple, and consistent extraction technique compared to conventional solvent extraction, soxhlet extraction, and ultrasound-assisted extraction techniques (Kyriakopoulou et al., 2015). However, continuous exposure of carrot materials to microwave radiation can cause thermal degradation of carotenoids and as such, intermittent application of microwave radiation has been shown to increase the extraction efficiency of carotenoids from carrot peels without causing excessive degradation of the antioxidant activity (Hiranvarachat and Devahastin, 2014). Pretreatment of carrots prior to MAE by blanching in water and/or citric acid was found to significantly increase the extractability of carotenoids due to tissue damage and softening during the blanching process (Hiranvarachat et al., 2013). A mathematical model was developed to describe the temperature and β-carotene concentration of carrot peels during continuous and intermittent MAE (Chumnanpaisont et al., 2014). In general, when using MAE, a careful balance must be struck between the magnitude of absorbed energy and power so as to facilitate rapid extraction without overheating (Chan et al., 2017). There are a number of companies that produce commercial microwave systems for use with MAE but scaling up from laboratory to commercial scale has presented complications due to variations in operations of specific systems (Chan et al., 2017).

Infrared drying is another electromagnetic technology that exposes objects to IR radiation with wavelengths of 0.75–1000 μm. A portion of this energy is absorbed by molecules in the material and converted to heat energy by changes in the molecular vibrational state (Krishnamurthy et al., 2008). Like microwave heating, infrared heating offers advantages of higher thermal efficiency and faster heating rates compared to convection heating. Doymaz (2013) studied the drying kinetics of carrot pomace with IR drying. The samples were loaded at 40 g and 10 mm thickness to an IR chamber and subjected to powers between 83 and 209 W with a halogen lamp (wavelength not reported), while mass was measured with a digital balance. The observed drying rates increased with power level and the entirety of the drying process appeared to take place in the falling-rate period, indicating that diffusion to the surface of the sample was the main rate-controlling mechanism. A summary of some drying methods can be seen in Table 4.

PEF processing has also been investigated with carrot and carrot pomace mainly to aid in the extraction of valuable components, improvement of its drying

Table 4 Advantages and Disadvantages of Four Drying Methods of Foods and Other Materials

Drying method	Advantages	Disadvantages	Reference
Hot air (convection)	Relatively simple, common, and can be scalable	Long drying times possible resulting in degradation of material quality, material shrinking and/or oxidation, energy intensive	Antal, (2015), Figiel and Michalska (2016)
Microwave	Penetration depth is high compared to IR but low compared to radio heating Can shorten drying times More efficient drying in falling rate period Improvement in quality is possible	Nonuniformity of electromagnetic field, constant motion required Rapid mass transfer can cause quality damage or "puffing" Difficulty in controlling product temperature compared to convective drying	Zhang et al. (2006)
Infrared	High thermal efficiency Fast heating rate Shorter response time Uniform Drying temperature High degree of process control Selective heating possible Reduced processing time Improved product quality compared to other drying techniques is possible	Low penetration power Prolonged exposure may cause fracturing in material Not sensitive to reflective coatings	Antal (2015), Krishnamurthy et al. (2008)
Freeze drying	Absence of air or high temperatures prevent deterioration due to oxidation or chemical modification	High energy consumption and high capital cots	Antal (2015), Vega-Mercado et al. (2001)

characteristics, and its ability to kill microbial contaminants. PEF is defined as the application of an electrical field with strengths of 0.1–80 kV/cm to a material to cause poration of biological membranes (Barba et al., 2015). In general, larger cells (~30–60 μm) require lower threshold electric field strengths than smaller cells (~1–10 μm) and there are many possible pulsing strategies that have been investigated for treating different biological materials (Barba et al., 2015). Biological membranes tend to have a transmembrane potential (u_m) of 0.5–1.5 V and depending on the strength of the applied electric field, reversible or irreversible membrane poration can occur (Barba et al., 2015; Brito et al., 2012). Carrot cells may vary in size between about 60–90 μm (Dietrich et al., 2014). Assuming a membrane width of 5 nm (Barba et al., 2015), the electric field felt on the membrane can be very large (1.8–9 MV/cm) for moderate applied electric field strengths (100–500 V/cm). Bipolar impulses, where the electric potential reverses polarity periodically has been shown to be advantageous to the efficiency of extraction of juice and intracellular components from various biological materials but are more complicated electrically, increasing their cost and possibly decreasing their economic viability with smaller-scale operations (Barba et al., 2015; Brito et al., 2012). Increasing electric field strengths, pulse widths, and number of impulses has been shown to increase extraction efficiency up to a saturation point where increasing any of these parameters no longer has much effect on extraction efficiencies but can lead to high specific electrical energy consumption and the possibility of material quality-degrading ohmic heating (Barba et al., 2015; Brito et al., 2012; Roohinejad et al., 2014). A 2014 study investigated the effect of varying electric field strength and frequency on the extractability of carotenoids from carrot purée and found that applying an electric field strength of 0.6 kV/cm at a frequency of 5 Hz (2.6 kJ/kg) was the preferred condition for maximum cell disintegration with an observed 24.5% increase in extractability of carotenoids (up to 76.34 μg/g) using a hexane solvent (Roohinejad et al., 2014). Another study used a Capmul oil-in-water microemulsion in combination with PEF treatment in order to avoid the use of environmentally hazardous organic solvents (Roohinejad et al., 2014). The study found that a β-carotene yield of 19.6 μg/g freeze dried carrot pomace was obtained at an optimal extraction condition to be 49.4 min in duration at a temperature of 52.2°C with a carrot fiber to microemulsion ratio of 1:70 (*w/w*) and microemulsion particle size of 74 nm. This condition was more efficient than pure hexane or Capmul oil alone probably due to the intracellular rehydration of carrot cells with the microemulsion (Tween 80), the small particle size of the microemulsion that was conducive to extraction of carotenoids with hydrophilic and hydrophobic properties, and the presence of a water-soluble surfactant in the microemulsion. The specific energy input of this process was substantially lower than other cell disruption technologies (mechanical, enzymatic, heating, or freezing/thawing) which have been reported as between 20 and 100 kJ/kg, depending on the technology (Barba et al., 2015). PEF treatment of carrot samples has also been reported to increase the effective water diffusion coefficient up to 16.7% and reduce convective drying times (70°C, air velocity of 2 m/s) up to 8.2% (Wiktor et al., 2015). PEF treatment also tends to shrink (decrease porosity) of carrot cells and decrease carotenoid oxidation during storage of convection-dried carrot samples (Barba et al., 2015). In general, PEF processing has several advantages such as the ability to decrease solvent and energy requirements for extractions and drying of biological materials like carrot pomace.

7.3.2 SUPERCRITICAL CARBON DIOXIDE EXTRACTION OF CAROTENOIDS

Supercritical CO_2 has been studied for at least two decades as an effective solvent for extraction of intracellular products like carotenoids with the ability to replace potentially toxic organic solvents such as hexanes that are used in traditional extractions. Creating supercritical CO_2 involves subjecting it to high temperature and pressure, above its critical point (30.85°C, 7.3 MPa) at which point it behaves similarly to nonpolar solvents (Wang and Weller, 2006). When polar substances are targeted for extraction, polar cosolvents such as methanol, ethanol, acetonitrile, acetone, water, ethyl ether, and dichloromethane may be added to the supercritical CO_2 (Wang and Weller, 2006). α- and β-Carotenes are hydrocarbon carotenoids that lack polar functional groups, making them highly soluble in nonpolar solvents (Saini and Keum, 2018). Despite its seemingly high solubility, the high-molecular weight of β-carotene (536.85) has an inhibitory effect on its solubility in supercritical CO_2 and as such, cosolvents are often added to aid in recovery of this component from carrot materials (Sun and Temelli, 2006). Vega et al. (1996) applied supercritical CO_2 at three levels of pressure (20.7, 27.6, and 34.5 MPa), temperature (40, 55, and 70°C), and ethanol cosolvent concentration (0%, 5%, and 10% wt) to extract carotenoids from carrot pomace. The highest experimental extraction of β-carotene (99.51% extraction, about 670 μg/g fresh pomace) was obtained at 55°C, 34.5 MPa, and 10% ethanol conditions. Similarly, a study using supercritical CO_2 to extract carotenoids from carrot tissue without a cosolvent found that 40°C and 50.5 MPa conditions were able to extract 108.4 and 137.4 μg/g (96.7% and 83.8%) of α- and β-carotenes, respectively (Chandra and Nair, 1997). The same study found that the addition of a chloroform solvent at 40°C, 60.6 MPa, and 5% chloroform conditions increased the yield of α- and β-carotenes only slightly up to 98.21% and 89.20%, respectively. de Andrade Lima et al. (2018) performed an optimization study on supercritical extraction of carotenoids from waste carrot peels and found optimal conditions of 59.0°C, 34.9 MPa, and 15.5% ethanol concentration for recovery of 86.1% of the total carotenoids within 30 min of processing. Sun and Temelli (2006) added canola oil continuously to the supercritical CO_2 extraction of carotenoids from carrots and found optimal conditions of 70°C, 55.1 MPa, 5% canola oil, 0.25–0.5 mm particle size, 0.8% moisture content of feed material to extract 115.4 and 119.7 μg/g fresh carrot of α- and β-carotene, respectively.

7.4 Novel Food Products From Carrot Pomace

7.4.1 FOOD PRODUCT DEVELOPMENT

Carrot pomace has been investigated for the development of new food products that increase the functional attributes of the food. Attempts at incorporating carrot pomace in bread, cake, dressing, pickles, and functional drinks are among the potential options explored before the year 2000 (Schieber et al., 2001). Since then, the investigation of food products enriched with carrot pomace has increased greatly. Some of the studies on these products have been summarized below.

7.4.1.1 Beverage

A 2003 study produced a functional beverage with hydrolyzed carrot pomace from a German processing plant which was mixed with apple juice to create a total carotene concentration of 12mg/L (Stoll et al., 2003). The results of the study showed no isomerization or degradation of α- or β-carotene contents after being stored for 20 or 24 weeks under moderate or intense illumination. In addition to the health-related functionalities of carotenes, the study also mentioned the presence of oligogalacturonic acids (OGAs) in hydrolyzed carrot pomace that may have applications as a therapeutic by the inhibition of bacterial adhesion to epithelial cells.

7.4.1.2 Carotenoid Powder

The stability of carrot pomace-derived carotenoid powder, produced by spray drying, was investigated under light and dark conditions in various temperatures (4–45°C) for storage periods up to 12 weeks (Chen and Tang, 1998). A hexane and acetone extraction and rotary evaporator (40°C) were applied prior to spray drying with a sucrose and gelatin aqueous phase. The spray drying conditions were: 15% solids content of feed, inlet air temperature of 135–145°C, outlet air temperature of 90–100°C, and a 45 min drying time. The results showed an increased degradation of α- and β-carotene and lutein molecules with increasing storage times and temperatures and in the presence of illumination. Under 25°C conditions and constant illumination (1500 lux), 28%, 55%, and 66% of trans-lutein, α-carotene, and β-carotene were degraded over 12 weeks, respectively.

7.4.1.3 Enriched Condensed Milk-Dessert

A carrot-based condensed milk product called gazrella was produced with carrot pomace using osmotic dehydration with highly concentrated sucrose syrup and powder followed by low-temperature convective drying and storage for 6 months before rehydration and cooking (Singh et al., 2006). The osmotically dehydrated-convectively dried gazrella was evaluated against gazrella prepared from convectively dried samples by a sensory panel for texture, flavor, appearance, and overall acceptability. It was found that osmotically dehydrating the carrot pomace before drying led to more acceptable final products for all sensory criteria with moderate sensory scores obtained for products even after 6 months of storage.

7.4.1.4 Yogurt

Cell wall particles (CWP) obtained from carrot pomace have been used to create a low-fat yogurt product (McCann et al., 2011). CWP were first produced using a colloid mill and high shear homogenizer to obtain particles of various sizes (34, 71, and 80 μm) and were then spray dried for particle size determination before being rehydrated and heated with milk until the desired milk solids content was obtained. The results indicated that carrot CWP accelerated pH reduction and gelation and reduced yogurt processing times by 20% (from ~280 to ~225 min). Smaller particle sizes were preferred for high gel strength properties and lower whey loss during fermentation which may be explained by a higher surface area of contact between CWP and casein particles. The yogurt produced may have added nutritional benefits including higher antioxidant and dietary fiber contents than traditional yogurt products.

7.4.1.5 Pasta

Carrot pomace was studied in the production of flour for producing ziti pasta with enhanced nutritional contents. Gull et al. (2015) evaluated the cooking characteristics of durum wheat semolina after adding different amounts (2–10 g/100 g) of dried carrot pomace. It was found that, compared to the 100% durum wheat semolina flour, increasing amounts of carrot pomace increased the amount of solids lost during cooking and led to decreased firmness and cooked weight. It was recommended that even though increasing amounts of carrot pomace addition would likely increase the nutritional value of the pasta flour, a maximum carrot pomace substitution amount of 4 g/100 g should be used to obtain acceptable pasta properties after cooking.

7.4.1.6 Cookies, Cakes, Muffins, Donuts, Bread, Rolls, and Biscuits

Several baked goods have been investigated for substitution of dough components with carrot pomace. Dried carrot pomace was used in the production of cookies and analyzed for the retention of total and β-carotenes after baking at high temperature (160°C) for 10 min (Bellur Nagarajaiah and Prakash, 2015). Carrot pomace mixtures between 4% and 12% were used for the experiment and results showed that soluble fiber, insoluble fiber, ash, and total and β-carotene contents of the cookies were substantially increased after inclusion. Cookies with 4% carrot pomace inclusion displayed 75% and 69% retention of total carotenoids and β-carotene after 60 days of storage, respectively. Sensory attributes of the cookies were not significantly affected by the inclusion of up to 8% of carrot pomace, while 12% inclusion significantly adversely affected sensory attributes. Another study on producing carrot pomace-enriched cookies considered the effect of adding carrot pomace to cookie dough between 10% and 20% (w.b.) with 260 μm (72 mesh) and 125 μm (120 mesh) dried carrot pomace particles (Ahmad et al., 2016). The results of this study showed that water absorption capacity, oil absorption capacity, solubility index, and swelling power significantly increased with incorporation of carrot pomace, especially when using the smaller, higher surface area 125 μm carrot pomace particles. However, excessive addition of carrot pomace could result in stiffer and less extensible dough due to dilution of gluten in the dough. It was also found that sensory characteristic scores increased with increasing carrot pomace inclusion with 20% carrot pomace enrichment using 125 μm particles displaying the highest sensory scores. These are contrary to the results of Bellur Nagarajaiah and Prakash (2015) who found high inclusion of carrot pomace could decrease sensory acceptability scores. Hernández-Ortega et al. (2013) dried carrot pomace by convective and microwave drying methods and produced cookies with 17.4% carrot pomace. Drying carrot pomace with microwave radiation better preserved biomolecules such as β-carotene, epicatechin, and ferulic acid and led to deeper orange-colored cookies with improved sensory scores compared to cookies with carrot pomace produced by hot-air drying. The β-carotene contents of the cookies dried by microwave radiation (2.13 μg/g DW) were 2.5 times greater than that obtained with hot-air drying (0.85 μg/g DW), which implies that carrot pomaces dried by different methods can further improve the properties of carrot pomace-enriched food products.

Investigation of gluten-free products has increased in the recent years to better accommodate the restrictive diets of patients with coeliac disease. Efforts to simultaneously counteract the lower quality of gluten-free food and increase the nutritional content of the food products can be accomplished with inclusion of vegetable by-products like carrot pomace. Majzoobi et al. (2016) and Majzoobi et al. (2017) evaluated the quality and optimal production characteristics of gluten-free batter and cakes produced with the introduction of carrot pomace. In the study of Majzoobi et al. (2016), carrot pomace powder of different particle sizes (210 and 500 μm) was substituted in the place of rice and corn flour up to 30% of the original batter recipe and then baked at 180°C for 30 min. The results showed that increasing the content and particle size of the carrot pomace in the batter increased the density, viscosity, consistency, and firmness of the batter. After baking, batter with up to 30% carrot pomace substitution decreased the cake density, hardness, and cohesiveness compared to the control cake, while the symmetry index and sensory scores increased with carrot pomace inclusion. The same research group further improved the cake batter with the inclusion of the hydrocolloids pectin and xanthan (1.5% of each) (Majzoobi et al., 2017). Similar to the previous study, the inclusion of hydrocolloids and carrot pomace increased the density of the batter, while decreasing the density of the resulting cake after baking. The authors concluded that based on the cake characteristics and sensory evaluation results, a recommended maximum carrot pomace inclusion is 20% and 30% when used in conjunction with hydrocolloids or not, respectively.

Gluten-free muffins were also produced using black carrot pomace powder of 250 μm particle size in a study that used xanthan gum as an additional thickening agent (Singh et al., 2016). The "black carrot dietary fiber concentrate" was added to muffin batter in different amounts up to 9% with and without the addition of 0.5% xanthan gum. The black carrot pomace-enriched muffins had higher water and oil absorption capacities compared to the rice flour and the incorporation of the pomace and xanthan gum increased the flour viscosity and batter viscoelasticity. Black carrot pomace decreased the L^* and b^* color values, water activity, specific volume (mL/g), and firmness of the muffins, while xanthan gum tended to increase their specific volume. Based on the physical characteristics and the sensory evaluation, the authors recommended 6% incorporation of carrot pomace with xanthan gum for the production of the most acceptable gluten-free muffins.

A recent study used response surface methodology to quantify the effects of carrot pomace powder, Persian gum, and water on the physicochemical properties of donuts (Nouri et al., 2017). The goal of the study was to optimize the levels of these factors to produce a low-fat and high-fiber donut with acceptable sensory qualities. It was found that carrot pomace and Persian gum increased the moisture retention of the donuts during frying and decreased the quantity of oil absorbed. However, carrot pomace incorporation led to decreased specific volume (mL/g) of the donuts probably from dilution of gluten and disruption of the resulting gluten matrix. This finding was in agreement with previous studies on cakes (Majzoobi et al., 2016, 2017) and muffins (Singh et al., 2016). The addition of carrot pomace also decreased the springiness and increased the firmness of the donuts. Results showed that the optimal formulation of donut contained 1.20%, 6.45%, and 48.16% of Persian gum, carrot pomace powder, and water, respectively. The sensory scores of the produced donuts had acceptable sensory properties as compared with the control.

Tańska et al. (2007) studied the effect of supplementation of wheat-based flour with dried carrot pomace (up to 10%) on bread quality characteristics. The introduction of carrot pomace increased the hardness of the wheat dough and increased the specific volume with 5% addition but decreased it for 10% carrot pomace addition. The optimal carrot pomace concentration was 5% that increased the water absorptivity of the bread by 8%, displayed a larger specific volume, and had an improved elasticity of the bread crumb. Greater additions of carrot pomace resulted in degraded rheological properties. Similarly, Kohajdová et al. (2012) reported that higher water absorptivity characteristics, lower specific volumes, and harder wheat rolls were obtained by increasing the carrot pomace powder addition. This study recommended a maximum incorporation of 3% of carrot pomace powder for rolls with acceptable properties, above which, undesirable characteristics were noticed, similarly to reported results for other food products.

To enrich the protein and dietary fiber contents of biscuits, Baljeet et al. (2014) incorporated carrot pomace powder and germinated chickpea flour into wheat flour. The incorporation of the carrot pomace powder and germinated chickpea flour increased the spread ratio (ratio of diameter to thickness of biscuits) and increased the protein, ash, and crude fiber contents. A sensory analysis showed that incorporating up to 8% of the carrot pomace and germinated chickpea flours resulted in acceptable sensory scores, while biscuits exceeding this content had significantly degraded sensory performance.

7.4.1.7 Extrudates

Extrusion is a process by which materials are continuously transferred down a barrel by means of a rotating screw to compress the material before being discharged through a die (Harper and Clark, 1979). Changes in barrel pitch angle and screw rotation speed can cause varying magnitudes of shear stress and heating to cook the material and deactivate microbes and enzymes that encourage product degradation during storage. The extrusion process has been used extensively in food manufacturing since the 1930s when it was first used industrially for pasta and breakfast cereal production (Harper and Clark, 1979). Carrot pomace was investigated for its application in the production of more nutritious and higher fiber extrudates with rice and gram pulse flour (Upadhyay et al., 2010). Carrot pomace was dried at 65°C for retention of bioactive components and then ground and sieved to 500 μm before inclusion to an extrusion machine with flour. The effect of die temperature (65–125°C), feed rate (2.5–8.5 g/s), feed moisture (10%–30% w.b.), and dried pomace powder content (1.5%–15.5%) on the extrudate moisture, expansion index, bulk density, sensory evaluation, and water absorption index were then quantified. It was found that die temperature was the most useful process variable to modify to impart desired characteristics to the extrudate. The optimum incorporation level of carrot pomace powder was recommended as 5% (w.b.). Kumar et al. (2010) evaluated the effects of moisture content (17%–21%), screw speed (270–310 rpm), and die temperature (110–130°C) on the lateral expansion, bulk density, water absorption and solubility indices, hardness, and sensory characteristics of carrot pomace-enriched rice and pulse flour extrudate products. The results of this study found significant impacts of carrot pomace proportion and moisture content on the lateral expansion qualities of the extrudate. The die temperature significantly influenced the water absorption index and along with screw speed, accounted for differences in hardness, while screw speed was the most significant variable on the sensory score. The optimum conditions derived from the results were: 16.5% carrot pomace content in rice flour, moisture content of 19.23%, screw speed of 310 rpm, and die temperature of 110°C. The extrudates prepared under the optimized conditions found in the previous study were then used to evaluate the storage characteristics of carrot pomace extrudates (Kumar et al., 2012). Over a period of 180 days of storage in low-density polyethylene (LDPE) or aluminum-lined LDPE bags, it was found that L^* and b^* color values decreased, while a^*, ΔE (overall color), moisture content, and hardness increased with storage time. Similar color changes during storage have been noted by other researchers (Dar et al., 2013, 2014a). Zero and first-order models were sufficient to predict the degradation of extrudate products during the storage in aluminum-lined LDPE bags (Kumar et al., 2012). During extrusion, color changes with increases in L^* and b^* accompanied by decreases in a^* values with increasing extrusion temperatures (Dar et al., 2013). Higher extrusion temperatures were also associated with decreases in crispiness and β-carotene and vitamin C contents, while hardness increased from increased gelatinization and microstructure density. The optimal extrudate mixtures reported by

Kumar et al. (2010) (carrot pomace content of 16.5%) were used by Dar et al. (2014b) in the investigation of deep-fried carrot pomace extrudates. Deep frying further increased L^* and b^* color values and oil absorption with increasing frying temperatures, while decreasing the a^* color value. β-Carotene content was reduced during extrusion, probably from degradation caused by high temperature, and also during storage by 49% (30.67–15.60 mg/kg). The optimal frying condition of 180°C for 15 s had the highest sensory acceptability rating. A study of extrusion products of carrot pomace and corn starch found that β-carotene contents were reduced by extrusion similarly to the reports of Dar et al. (2014a) (Kaisangsri et al., 2016). However, at low levels (5% carrot pomace content and 15% moisture content), the inclusion of carrot pomace had beneficial effects on the expansion characteristics of the extrudate product.

7.4.1.8 Commercial Food Products

There has been interest in commercially producing food products from carrot pomace across the world. A Californian company Pulp Pantry creates "Superfood Snacks" whose first ingredient is carrot pulp obtained from the juicing process (Pantry, 2018). The mixture contains other ingredients such as: sunflower seeds, coconut sugar and oil, flaxseed meal, sesame seeds, pumpkin seeds, and chickpea flour. The snack has similar applications and consistency as a granola mixture and each serving is specified to provide 12%, 70%, 6%, 4%, and 2% of the daily dietary values of dietary fiber, vitamin A, iron, calcium, and vitamin C, respectively. The Forager Project is another company that uses pulp from juiced vegetables to create crunchy vegetable chips (Project, 2018). A review of the current available products on the website indicated that carrot pomace is not currently among the vegetable pulps utilized in these chips but indicates another potential commercial interest of utilizing pomace for food products. An Irish company CyberColloids LTD released a brochure in December 2015 that indicated their research and development focus on the production of food fibers from vegetable wastes such as carrots and potatoes (LTD, 2015). The biochemical components of interest in carrot pomace were used by this company in the production of muffins and burgers with future research in meatballs, sausages, and meatloaf planned. A Swiss company called Food Solutions Team markets a functional fiber product derived from carrot pomace that can be used in a variety of recipes from hamburgers to baked goods (Team, 2018).

8 ADSORPTION AND WASTEWATER TREATMENT CAPABILITIES OF CARROT POMACE

Another possible method for utilizing carrot pomace is as a biosorbent material for removal of specific components from wastewaters. Biological materials like carrot pomace can exhibit cation exchange properties due to the presence of carboxyl, phenolic, hydroxyl, or other functional groups that are present in a variety of intracellular biopolymers (Ata et al., 2012; Guzel et al., 2008; Nasernejad et al., 2005). The removal of heavy metals from wastewaters by biological materials is of interest due to the potential negative effects of heavy metals on human health (Guzel et al., 2008). In the study of Nasernejad et al. (2005), carrot pomace was used to adsorb Cr(III), Cu(II), and Zn(II) from solution in equilibrium studies that were described by Languir and Freundlich isotherm models. The maximum removal capacities based on the Langmuir isotherm model were 45.09, 32.74, and 29.61 mg/g dry carrot residue for Cr(III), Cu(II), and Zn(II), respectively, at 25°C and optimal pH of 4.5. In a similar study, black carrot residues were used to remove Mn(II), Ni(II), Co(II), and Cu(II) from solution and found maximum removal capacities of 3.871, 5.745, 5.350, and 8.877 mg/g dry residue, respectively at 20°C and pH of 5.5 (Guzel et al., 2008). Cadmium and lead ions were removed from solution by carrot residues with maximum removal capacities of 0.421 and 0.522 mg/g of dry residue at 25°C and pH of 5 and 3, respectively (Ata et al., 2012). At a temperature of 30°C, tri- and hexavalent chromium was also removed from solution by carrot pomace with maximum removal capacities of 86.65 and 88.27 mg/g dry residue at a pH of 1 and 5, respectively (Bhatti et al., 2010). Carrot pomace was pyrolyzed and chemically activated with $ZnCl_2$ to create carrot pomace activated carbon and remove Co(II) from water (Changmai et al., 2018). Langmuir and Freundlich isotherm models were fit to equilibrium Co(II) removal data for various contact times, adsorbent doses, and pH values. Higher pH was strongly related to increased removal of Co(II) and the carrot pomace activated carbon was capable of removing 24% of the Co(II) at an adsorbent dose of 1.2 g/L with a maximum removal capacity of 56.17 mg/g dry residue. The results of all studies using carrot pomace for the removal of heavy metals from solution are summarized in Table 5.

Table 5 Removal of Heavy Metals From Solution by Carrot Residues

Sorbent material	Temperature (°C)	pH	Element removed	Maximum removal capacity, q_{max} (mg/g dry residue)	Source
Carrot Pomace	25	5	Cd(II)	0.4210	Ata et al. (2012)
Black carrot pomace	20	5.5	Co(II)	5.350	Guzel et al. (2008)
Carrot pomace activated carbon	N/A	N/A	Co(II)	56.17	Changmai et al. (2018)
Carrot pomace	25	4.5	Cr(III)	45.09	Nasernejad et al. (2005)
	30	1	Cr(III)	86.65	Bhatti et al. (2010)
	30	5	Cr(VI)	88.27	
	25	4.5	Cu(II)	32.74	Nasernejad et al. (2005)
Black carrot pomace	20	5.5	Cu(II)	8.877	Guzel et al. (2008)
	20	5.5	Mn(II)	3.871	
	20	5.5	Ni(II)	5.745	
Carrot pomace	25	3	Pb(II)	0.522	Ata et al. (2012)
	25	4.5	Zn(II)	29.61	Nasernejad et al. (2005)

References

Adiamo, O.Q., Ghafoor, K., Al-Juhaimi, F., Babiker, E.E., Mohamed Ahmed, I.A., 2018. Thermosonication process for optimal functional properties in carrot juice containing orange peel and pulp extracts. Food Chem. 245, 79–88.

Agcam, E., Akyildiz, A., Balasubramaniam, V.M., 2017. Optimization of anthocyanins extraction from black carrot pomace with thermosonication. Food Chem. 237, 461–470.

Ahmad, M., Wani, T.A., Wani, S.M., Masoodi, F.A., Gani, A., 2016. Incorporation of carrot pomace powder in wheat flour: effect on flour, dough and cookie characteristics. J. Food Sci. Technol. 53 (10), 3715–3724.

Ail, S.S., Dasappa, S., 2016. Biomass to liquid transportation fuel via Fischer Tropsch synthesis—technology review and current scenario. Renew. Sustain. Energy Rev. 58, 267–286.

Aimaretti, N.R., Ybalo, C.V., Rojas, M.L., Plou, F.J., Yori, J.C., 2012. Production of bioethanol from carrot discards. Bioresour. Technol. 123, 727–732.

Alasalvar, C., Grigor, J.M., Zhang, D., Quantick, P.T., Shahidi, F., 2001. Comparison of volatiles, phenolics, sugars, antioxidant vitamins, and sensory quality of different colored carrot varieties. J. Agric. Food Chem. 49, 1410–1416.

Anaya-Esparza, L.M., Velázquez-Estrada, R.M., Roig, A.X., García-Galindo, H.S., Sayago-Ayerdi, S.G., Montalvo-González, E., 2017. Thermosonication: an alternative processing for fruit and vegetable juices. Trends Food Sci. Technol. 61, 26–37.

Antal, T., 2015. Comparative study of three drying methods: freeze, hot air-assisted freeze and infrared-assisted freeze modes. Agron Res. 13 (4), 863–878.

Apeland, J., 1974. Storage quality of carrots after different methods of harvesting. Acta Hortic. (38), 353–358.

Apeland, J., Hoftun, H., 1974. Effects of temperature-regimes on carrots during storage. Acta Hortic. (38), 291–308.

Ata, S., Wattoo, F.H., Sidra, L.R., Sarwar Wattoo, M.H., Tirmizi, S.A., Din, I., Mohsin, I.U., 2012. Biosorptive removal of lead and cadmium ions from aqueous solution: the use of carrot residues as low cost non-conventional adsorbent. Turk. J. Biol. 37 (2), 272–279.

Austin, E., 2013. Analysis of Treatment and Disposal Methods for Vegetable Solids. Vol. Master of Applied Science, The University of Guelph, p. 50.

Babb, M.F., Kraus, J.E., Magruder, R. (Eds.), 1950. Synonymy of orange-fleshed varieties of carrots. US Department of Agriculture, Washington, DC.

Baljeet, S.Y., Ritika, B.Y., Reena, K., 2014. Effect of incorporation of carrot pomace powder and germinated chickpea flour on the quality characteristics of biscuits. Int. Food Res. J. 21 (1), 217–222.

Banga, O., 1957. Origin of the European cultivated carrot. Euphytica 6, 54–63.

Banga, O., 1963. Origin and distribution of the Western cultivated carrot. Genet. Agraria 17, 357–370.

Barba, F.J., Mariutti, L.R.B., Bragagnolo, N., Mercadante, A.Z., Barbosa-Cánovas, G.V., Orlien, V., 2017. Bioaccessibility of bioactive compounds from fruits and vegetables after thermal and nonthermal processing. Trends Food Sci. Technol. 67, 195–206.

Barba, F.J., Parniakov, O., Pereira, S.A., Wiktor, A., Grimi, N., Boussetta, N., Saraiva, J.A., Raso, J., Martin-Belloso, O., Witrowa-Rajchert, D., Lebovka, N., Vorobiev, E., 2015. Current applications and new opportunities for the use of pulsed electric fields in food science and industry. Food Res. Int. 77, 773–798.

Bellur Nagarajaiah, S., Prakash, J., 2015. Nutritional composition, acceptability, and shelf stability of carrot pomace-incorporated cookies with special reference to total and Beta-carotene retention. Cogent. Food. Agric. 1 (1), 1–10. 1039886.

Bhattacharya, A.K., Mandal, S.N., Das, S.K., 2006. Adsorption of Zn(II) from aqueous solution by using different adsorbents. Chem. Eng. J. 123 (1–2), 43–51.

Bhatti, H.N., Nasir, A.W., Hanif, M.A., 2010. Efficacy of Daucus carota L. waste biomass for the removal of chromium from aqueous solutions. Desalination 253 (1–3), 78–87.

Bond, R., 2016. Carrot Loss During Primary Production. Vol. Masters, Hedmark University of Applied Sciences.

Bradeen, J.M., Simon, P.W., 2007. Carrot. In: Kole, C. (Ed.), Genome Mapping and Molecular Breeding in Plants. Vol. 5, pp. 161–184.

Brito, P.S., Canacsinh, H., Mendes, J.P., Redondo, L.M., Pereira, M.T., 2012. Comparison Between Monopolar and Bipolar Microsecond Range Pulsed Electric Fields in Enhancement of Apple Juice Extraction. IEEE Trans. Plasma Sci. 40 (10), 2348–2354.

Chan, C.-H., Yusoff, R., Ngoh, G.C., 2017. An energy-based approach to scale up microwave-assisted extraction of plant bioactives. In: Ingredients Extraction by Physicochemical Methods in Food, Elsevier, pp. 561–597.

Chandra, A., Nair, M.G., 1997. Supercritical fluid carbon dioxide extraction of alpha- and beta-carotene from carrot (Daucus carota L.). Phytochem. Anal. 8 (5), 244–246.

Changmai, M., Banerjee, P., Nahar, K., Purkait, M.K., 2018. A novel adsorbent from carrot, tomato and polyethylene terephthalate waste as a potential adsorbent for Co (II) from aqueous solution: kinetic and equilibrium studies. J. Environ. Chem. Eng. 6 (1), 246–257.

Chen, B.H., Tang, Y.C., 1998. Processing and stability of carotenoid powder from carrot pulp waste. J. Agric. Food Chem. 46 (6), 2312–2318.

Chumnanpaisont, N., Niamnuy, C., Devahastin, S., 2014. Mathematical model for continuous and intermittent microwave-assisted extraction of bioactive compound from plant material: extraction of β-carotene from carrot peels. Chem. Eng. Sci. 116, 442–451.

Coventry, E., Noble, R., Mead, A., Whipps, J.M., 2005. Suppression of Allium white rot (Sclerotium cepivorum) in different soils using vegetable wastes. Eur. J. Plant Pathol. 111 (2), 101–112.

Dar, A.H., Kumar, N., Sharma, H.K., 2013. Physical and micro structural changes in carrot pomace-based extrudates. Ital. J. Food Sci. 25 (3), 313–321.

Dar, A.H., Sharma, H.K., Kumar, N., 2014a. Effect of extrusion temperature on the microstructure, textural and functional attributes of carrot pomace-based extrudates. J. Food Process Preserv. 38 (1), 212–222.

Dar, A.H., Sharma, H.K., Kumar, N., 2014b. Effect of frying time and temperature on the functional properties of carrot pomace, pulse powder and rice flour–based extrudates. Int. J. Food Eng. 10 (1), 139–147.

de Andrade Lima, M., Charalampopoulos, D., Chatzifragkou, A., 2018. Optimisation and modelling of supercritical CO 2 extraction process of carotenoids from carrot peels. J. Supercrit. Fluids 133, 94–102.

De Roeck, A., Sila, D., Duvetter, T., Vanloey, A., Hendrickx, M., 2008. Effect of high pressure/high temperature processing on cell wall pectic substances in relation to firmness of carrot tissue. Food Chem. 107 (3), 1225–1235.

Demiray, E., Karatay, S.E., Donmez, S., Donmez, G., 2016. The usage of carrot pomace for bioethanol production. J. Chil. Chem. Soc. 61 (2), 2996–2998.

Deng, L., Du, H., Han, Z., 2017. A carrot sorting system using machine vision technique. Appl. Eng. Agric. 33 (2), 149–156.

Dietrich, O., Hubert, A., Heiland, S., 2014. Imaging cell size and permeability in biological tissue using the diffusion-time dependence of the apparent diffusion coefficient. Phys. Med. Biol. 59 (12), 3081–3096.

Doymaz, I., 2013. Determination of infrared drying characteristics and modelling of drying behaviour of carrot pomace. Tarim Bilimleri Dergisi—J. Agric. Sci. 19 (1), 44–53.

Dubrovskis, V., Plume, I., 2015. Anaerobic digestion of vegetables processing wastes with catalyst metaferm. Agron Res. 13 (2), 294–302.

El-Shimi, S.A., El-Housseini, M., Ali, B.E., El-Shinnawi, M.M., 1992. Biogas generation from food-processing wastes. Resour. Conserv. Recycl. 6 (4), 315–327.

Engineering, T., 2018. Potato & Vegetable Sizing Equipment. https://tongengineering.com.

FAO, 2018. FAOSTAT. In: Crops. http://www.fao.org/faostat/en/#data/QC.

Figiel, A., Michalska, A., 2016. Overall quality of fruits and vegetables products affected by the drying processes with the assistance of vacuum-microwaves. Int. J. Mol. Sci. 18 (1), 1–18.

Garcia, S.L., Jangid, K., Whitman, W.B., Das, K.C., 2011. Transition of microbial communities during the adaption to anaerobic digestion of carrot waste. Bioresour. Technol. 102 (15), 7249–7256.

Goncharuk, N.S., Dyachenko, V.S., 1980. Storage performance of vegetables after mechanical harvesting and commodity processing. Acta Hortic. 116, 197–204.

Gull, A., Prasad, K., Kumar, P., 2015. Effect of millet flours and carrot pomace on cooking qualities, color and texture of developed pasta. LWT—Food Sci. Technol. 63 (1), 470–474.

Guzel, F., Yakut, H., Topal, G., 2008. Determination of kinetic and equilibrium parameters of the batch adsorption of Mn(II), Co(II), Ni(II) and Cu(II) from aqueous solution by black carrot (Daucus carota L.) residues. J. Hazard Mater. 153 (3), 1275–1287.

Harper, J.M., Clark, J.P., 1979. Food extrusion. CRC Crit. Rev. Food Sci. Nutr. 11 (2), 155–215.

Heinz, V., Buckow, R., 2009. Food preservation by high pressure. J. Verbrauch Lebensm 5 (1), 73–81.

Hernández-Ortega, M., Kissangou, G., Necoechea-Mondragón, H., Sánchez-Pardo, M.E., Ortiz-Moreno, A., 2013. Microwave dried carrot pomace as a source of fiber and carotenoids. Food Nutr. Sci. 04 (10), 1037–1046.

Heywood, V.H., 1983. Relationships and evolution in the Daucus carota complex. Isr. J. Bot. 32 (2), 51–65.

Hiranvarachat, B., Devahastin, S., 2014. Enhancement of microwave-assisted extraction via intermittent radiation: extraction of carotenoids from carrot peels. J. Food Eng. 126, 17–26.

Hiranvarachat, B., Devahastin, S., Chiewchan, N., Vijaya Raghavan, G.S., 2013. Structural modification by different pretreatment methods to enhance microwave-assisted extraction of β-carotene from carrots. J. Food Eng. 115 (2), 190–197.

Indrawati, Van Loey, A., Hendrickx, M., 2004. Pressure and temperature stability of water-soluble antioxidants in orange and carrot juice: a kinetic study. Eur. Food Res. Technol. 219(2).

Inyang, M., Gao, B., Yao, Y., Xue, Y., Zimmerman, A.R., Pullammanappallil, P., Cao, X., 2012. Removal of heavy metals from aqueous solution by biochars derived from anaerobically digested biomass. Bioresour. Technol. 110, 50–56.

Jabbar, S., Abid, M., Hu, B., Hashim, M.M., Lei, S., Wu, T., Zeng, X., 2015. Exploring the potential of thermosonication in carrot juice processing. J. Food Sci. Technol. 52 (11), 7002–7013.

Jabbar, S., Abid, M., Wu, T., Hashim, M.M., Saeeduddin, M., Hu, B., Lei, S., Zeng, X., 2015. Ultrasound-assisted extraction of bioactive compounds and antioxidants from carrot pomace: a response surface approach. J. Food Process Preserv. 39 (6), 1878–1888.

Janiszewska, E., Witrowa-Rajchert, D., Kidoń, M., Czapski, J., 2013. Effect of the applied drying method on the physical properties of purple carrot pomace. Int. Agrophys. 27(2).

Kaanane, A., Kane, D., Labuza, T.P., 1988. Time and temperature effect on stability of moroccan processed orange juice during storage. J. Food Sci. 53 (5), 1470.

Kabir, G., Hameed, B.H., 2017. Recent progress on catalytic pyrolysis of lignocellulosic biomass to high-grade bio-oil and bio-chemicals. Renew. Sustain. Energy Rev. 70, 945–967.

Kaisangsri, N., Kowalski, R.J., Wijesekara, I., Kerdchoechuen, O., Laohakunjit, N., Ganjyal, G.M., 2016. Carrot pomace enhances the expansion and nutritional quality of corn starch extrudates. LWT—Food Sci. Technol. 68, 391–399.

Kern, J., Reimann, W., Schluter, O., 2006. Treatment of recycled carrot washing water. Environ. Technol. 27 (4), 459–466.

Knol, W., van der Most, M., de Wart, J., 1978. Biogas production by anaerobic digestion of fruit and vegetable waste. A preliminary study. J. Sci. Food Agric. 29, 822–830.

Kohajdová, Z., Karovičová, J., Jurasová, M., 2012. Influence of carrot pomace powder on the rheological characteristics of wheat flour dough and on wheat rolls quality. Acta Sci. Pol. Technol. Aliment 11 (4), 381–387.

Krishnamurthy, K., Khurana, H.K., Jun, S., Irudayaraj, J., Demirci, A., 2008. Infrared heating in food processing: an overview. Compr. Rev. Food Sci. Food Saf. 7 (1), 2–13.

Kumar, N., Sarkar, B.C., Sharma, H.K., 2010. Development and characterization of extruded product of carrot pomace, rice flour and pulse powder. Afr J. Food Sci. 4 (11), 703–717.

Kumar, N., Sarkar, B.C., Sharma, H.K., Jha, S.K., 2012. Colour kinetics and storage characteristics of carrot, pulse and rice by-product based extrudates. Br. Food J. 114 (9), 1279–1296.

Kyriakopoulou, K., Papadaki, S., Krokida, M., 2015. Life cycle analysis of β-carotene extraction techniques. J. Food Eng. 167, 51–58.

Lehmann, J., Rillig, M.C., Thies, J., Masiello, C.A., Hockaday, W.C., Crowley, D., 2011. Biochar effects on soil biota—a review. Soil Biol. Biochem. 43 (9), 1812–1836.

Li, H.B., Dong, X.L., da Silva, E.B., de Oliveira, L.M., Chen, Y.S., Ma, L.N.Q., 2017. Mechanisms of metal sorption by biochars: biochar characteristics and modifications. Chemosphere 178, 466–478.

Lipiec, J., Medvedev, V.V., Birkas, M., Dumitru, E., Lyndina, T.E., Rousseva, S., Fulajtár, E., 2003. Effect of soil compaction on root growth and crop yield in Central and Eastern Europe. Int. Agrophys. 17 (2), 61–69.

LTD, Cyber Colloids, 2015. Tackling Food Loss in Ireland: New Food Fibres From Vegetable Waste. www.cybercolloids.net.

Lucier, G., Lin, B.-H., 2007. Factors Affecting Carrot Consumption in the United States. Economic Research Service/USDA.

Majzoobi, M., Poor, Z.V., Jamalian, J., Farahnaky, A., 2016. Improvement of the quality of gluten-free sponge cake using different levels and particle sizes of carrot pomace powder. Int. J. Food Sci. Technol. 51 (6), 1369–1377.

Majzoobi, M., Vosooghi Poor, Z., Mesbahi, G., Jamalian, J., Farahnaky, A., 2017. Effects of carrot pomace powder and a mixture of pectin and xanthan on the quality of gluten-free batter and cakes. J. Texture Stud. 48 (6), 616–623.

Mata-Alvarez, J., Mace, S., Llabres, P., 2000. Anaerobic digestion of organic solid wastes. An overview of research achievements and perspectives. Bioresour. Technol. 74 (1), 3–16.

McCann, T.H., Fabre, F., Day, L., 2011. Microstructure, rheology and storage stability of low-fat yoghurt structured by carrot cell wall particles. Food Res. Int. 44 (4), 884–892.

Nasernejad, B., Zadeh, T.E., Pour, B.B., Bygi, M.E., Zamani, A., 2005. Camparison for biosorption modeling of heavy metals (Cr (III), Cu (II), Zn (II)) adsorption from wastewater by carrot residues. Process Biochem. 40 (3–4), 1319–1322.

Nayak, B., Liu, R.H., Tang, J., 2015. Effect of processing on phenolic antioxidants of fruits, vegetables, and grains—a review. Crit. Rev. Food Sci. Nutr. 55 (7), 887–919.

Nguyen, H.V.H., Nguyen, L.T., 2015. Carrot processing. In: Hui, Y.H., Evranuz, E.Ö. (Eds.), Handbook of Vegetable Preservation and Processing. second ed. CRC Press.

Nicoli, M.C., Anese, M., Parpinel, M., 1999. Influence of processing on the antioxidant properties of fruit and vegetables. Trends Food Sci. Technol. 10 (3), 94–100.

Nicoli, M.C., Anese, M., Parpinel, M.T., Franceschi, S., Lerici, C.R., 1997. Loss and/or formation of antioxidants during food processing and storage. Cancer Lett. 114, 71–74.

NIH, 2018. Vitamin A. In: Medline Plus. https://medlineplus.gov/ency/article/002400.htm.

Nouri, M., Nasehi, B., Samavati, V., Mehdizadeh, S.A., 2017. Optimizing the effects of Persian gum and carrot pomace powder for development of low-fat donut with high fibre content. Bioact. Carbohydr. Diet Fibre 9, 39–45.

Nunez, J., Hartz, T., Suslow, T., McGiffen, M., Natwick, E.T., 2008. Carrot Production in California. University of California Agriculture and Natural Resources.

Pantry, 2018. Pulp. https://pulppantry.com.

Patle, S., Lal, B., 2007. Ethanol production from hydrolysed agricultural wastes using mixed culture of Zymomonas mobilis and Candida tropicalis. Biotechnol. Lett. 29 (12), 1839–1843.

Patras, A., Brunton, N., Da Pieve, S., Butler, F., Downey, G., 2009. Effect of thermal and high pressure processing on antioxidant activity and instrumental colour of tomato and carrot purées. Innov. Food Sci. Emerg. Technol. 10 (1), 16–22.

Pietola, L., Smucker, A.J.M., 1998. Fibrous carrot root responses to irrigation and compaction of sandy and organic soils. Plant Soil 200 (1), 95–105.

Prenveille, D.E. (Ed.), 1956. Carrot-Length Grader. US Patent Office, Farmco Enterprises, LTD., USA.

Project, 2018. The Forager. https://www.foragerproject.com.

Raynal, J., Delgenes, J.P., Moletta, R., 1998. Two-phase anaerobic digestion of solid wastes by a multiple liquefaction reactors process. Bioresour. Technol. 65, 97–103.

Reyes-De-Corcuera, J.I., Goodrich-Schneider, R.M., Barringer, S.A., Landeros-Urbina, M.A., 2014. Processing of fruit and vegetable beverages. In: Clark, S., Jung, S., Lamsal, B. (Eds.), Food Processing: Principles and Applications, second ed. John Wiley & Sons, Ltd.

Roohinejad, S., Everett, D.W., Oey, I., 2014. Effect of pulsed electric field processing on carotenoid extractability of carrot purée. Int. J. Food Sci. Technol. 49 (9), 2120–2127.

Roohinejad, S., Oey, I., Everett, D.W., Niven, B.E., 2014. Evaluating the effectiveness of β-carotene extraction from pulsed electric field-treated carrot pomace using oil-in-water microemulsion. Food Bioproc. Technol. 7 (11), 3336–3348.

Saini, R.K., Keum, Y.S., 2018. Carotenoid extraction methods: a review of recent developments. Food Chem. 240, 90–103.

Sarkar, B.C., Sharma, H.K., 2011. Carrots. In: Sinha, N.K. (Ed.), Handbook of Vegetables and Vegetable Processing. Blackwell Publishing Ltd, pp. 565–580.

Schieber, A., Stintzing, F.C., Carle, R., 2001. By-products of plant food processing as a source of functional compounds—recent developments. Trends Food Sci. Technol. 12 (11), 401.

Seljåsen, R., Bengtsson, G.B., Hoftun, H., Vogt, G., 2001. Sensory and chemical changes in five varieties of carrot (Daucus carota L) in response to mechanical stress at harvest and post-harvest. J. Sci. Food Agric. 81 (4), 436–447.

Sharma, K.D., Karki, S., Thakur, N.S., Attri, S., 2012. Chemical composition, functional properties and processing of carrot—a review. J. Food Sci. Technol. 49 (1), 22–32.

Sharma, H.K., Kumar, N., 2018. Utilization of carrot pomace. In: Anal, A.K. (Ed.), Food Processing By-Products and Their Utilization. John Wiley & Sons, Ltd.

Singh, J.P., Kaur, A., Singh, N., 2016. Development of eggless gluten-free rice muffins utilizing black carrot dietary fibre concentrate and xanthan gum. J. Food Sci. Technol. 53 (2), 1269–1278.

Singh, B., Panesar, P.S., Nanda, V., 2006. Utilization of carrot pomace for the preparation of a value added product. World J. Dairy Food Sci. 1 (1), 22–27.

Small, E., 1978. A numerical taxonomic analysis of the Daucus carota complex. Can. J. Bot. 56, 248–276.

Stoll, T., Schweiggert, U., Schieber, A., Carle, R., 2003. Application of hydrolyzed carrot pomace as a functional food ingredient to beverages. J. Food Agric. Environ. 1 (2), 88–92.

Sun, Y., Cheng, J.Y., 2002. Hydrolysis of lignocellulosic materials for ethanol production: a review. Bioresour. Technol. 83 (1), 1–11.

Sun, M., Temelli, F., 2006. Supercritical carbon dioxide extraction of carotenoids from carrot using canola oil as a continuous co-solvent. J. Supercrit. Fluids 37 (3), 397–408.

Tańska, M., Zadernowski, R., Konopka, I., 2007. The quality of wheat bread supplemented with dried carrot pomace. Pol. J. Nat. Sci. 22 (1), 126–136.

Team, Food Solutions, 2018. https://www.foodsolutionsteam.com/.

Thanikal, J.V., Torrijos, M., Rizwan, S.M., Yazidi, H., Senthil Kumar, R., Sousbie, P., 2015. Anaerobic co-digestion of vegetable waste and cooked oil in anaerobic sequencing batch reactor (ASBR). Int. J. Adv. Agric. Environ. Eng. 2 (1), 4–7.

Tucker, W.G., 1974. The effect of mechanical harvesting on carrot quality and storage performance. Acta Hortic. (38), 359–372.

Upadhyay, A., Sharma, H.K., Sarkar, B.C., 2008. Characterization and dehydration kinetics of carrot pomace. Agric. Eng. Int. CIGR J., 10.

Upadhyay, A., Sharma, H.K., Sarkar, B.C., 2010. Optimization of carrot pomace powder incorporation on extruded product quality by response surface methodology. J. Food Qual. 33 (3), 350–369.

USDA, 1965. United States Standards for Grades of Topped Carrots.

USDA, 2018. 11124 carrots, raw. In: National Nutrient Database for Standard Reference Legacy Release. Agricultural Research Service. https://ndb.nal.usda.gov/ndb/search/list.

USDA, 2018b. Food Availability (per capita) Data System. https://www.ers.usda.gov/data-products/food-availability-per-capita-data-system.

USDA, USDHHS and, 2015. 2015–2020 Dietary Guidelines for Americans, eighth ed. http://health.gov/dietaryguidelines/2015/guidelines/.

Vega, P.J., Balaban, M.O., Sims, C.A., Okeefe, S.F., Cornell, J.A., 1996. Supercritical Carbon Dioxide Extraction Efficiency for Carotenes From Carrots by RSM. J. Food Sci. 61(4).

Vega-Mercado, H., Gongora-Nieto, M.M., Barbosa-Canovas, G.V., 2001. Advances in dehydration of foods. J. Food Eng. 49 (4), 271–289.

Wadhwa, M., Bakshi, M.P.S., 2013. Utilization of fruit and vegetable wastes as livestock feed and as substrates for generation of other value-added products. FAO.

Wang, L., Weller, C.L., 2006. Recent advances in extraction of nutraceuticals from plants. Trends Food Sci. Technol. 17 (6), 300–312.

Whipps, J., Noble, R., 2001. Composting of Onion and Other Vegetable Wastes, With Particular Reference to Control of Allium White Rot. Consortium of Horticulture. https://horticulture.ahdb.org.uk/project/composting-onion-and-other-vegetable-wastes-particular-reference-control-allium-white-rot-4.

Wiktor, A., Nowacka, M., Dadan, M., Rybak, K., Lojkowski, W., Chudoba, T., Witrowa-Rajchert, D., 2015. The effect of pulsed electric field on drying kinetics, color, and microstructure of carrot. Dry. Technol. 34 (11), 1286–1296.

Yao, Y., Gao, B., Zhang, M., Inyang, M., Zimmerman, A.R., 2012. Effect of biochar amendment on sorption and leaching of nitrate, ammonium, and phosphate in a sandy soil. Chemosphere 89 (11), 1467–1471.

Yoon, K., Cha, M., Shin, S., Kim, K., 2005. Enzymatic production of a soluble-fibre hydrolyzate from carrot pomace and its sugar composition. Food Chem. 92 (1), 151–157.

Yu, C.-Y., Jiang, B.-H., Duan, K.-J., 2013. Production of bioethanol from carrot pomace using the thermotolerant yeast Kluyveromyces marxianus. Energies 6 (3), 1794–1801.

Zentek, J., Knorr, F., Mader, A., 2014. Reducing waste in fresh produce processing and households through use of waste as animal feed. In: Global Safety of Fresh Produce, pp. 140–152.

Zhang, R., El-Mashad, H.M., Hartman, K., Wang, F., Liu, G., Choate, C., Gamble, P., 2007. Characterization of food waste as feedstock for anaerobic digestion. Bioresour. Technol. 98 (4), 929–935.

Zhang, M., Tang, J., Mujumdar, A.S., Wang, S., 2006. Trends in microwave-related drying of fruits and vegetables. Trends Food Sci. Technol. 17 (10), 524–534.

Zhang, R.H., Zhang, Z.Q., 1999. Biogasification of rice straw with an anaerobic-phased solids digester system. Bioresour. Technol. 68 (3), 235–245.

Zhang, R., Zhang, Z., 2002. Biogasification of Solid Waste With an Anaerobc-Phased Solids Digester System. US Patent Office, The Regents of the University of California, USA.

CHAPTER 13
Sugar Beet

Steven Zicari*, Ruihong Zhang†, Stephen Kaffka‡
*California Safe Soil, LLC, McClellan, CA, United States, †Agricultural and Biological Engineering, University of California - Davis, Davis, CA, United States, ‡California Biomass Collaborative, University of California - Davis, Davis, CA, United States

Chapter Outline

1 INTRODUCTION

Sugar beets (*Beta vulgaris L.*) have been bred for over a century for increased sucrose yield and purity, and now account for almost 30% of world, and more than 50% of US, sugar production (USDA-ERS, 2018). Sugar beets are well-suited for growth in cooler climates unsuited to sugar cane and can store 15%–20% of their root mass as sucrose with average biomass yields ranging from 40 to 100 metric tons (MT) per hectare (ha) (Panella and Kaffka, 2010). Other benefits of sugar beet production are that beets are salinity tolerant and can have higher water and nitrogen use efficiencies than other industrial crops such as corn, wheat, and alfalfa (Yousefi, 2015). Beets can also be planted as one of several rotation crops to maximize land and resource use within a cropping system (Pelka et al., 2015). Beet sucrose production results in by-products such as sugar beet pulp (SBP, extracted beet fiber), and molasses or vinasse (concentrated impurities from sugar refining). These by-products have traditionally been used for animal feed in wet or dry form.

Integrated Processing Technologies for Food and Agricultural By-Products
https://doi.org/10.1016/B978-0-12-814138-0.00013-7

Several opportunities exist for integrated processing of both beet and beet-by-products for improved recovery and upcycling of these streams at industrial scale. National and regional policies greatly influence the feasibility of change with these opportunities. For example, the recent ending of market regulation using a quota system in the European Union has led to an increased supply and decreased value for beets and beet sugar in Europe, while raising interest in alternative or higher value processing options with improved economic and environmental returns. Similarly, the establishment of carbon pricing policies, such as in California and elsewhere, may prompt changes in the near future.

2 PLANT BIOLOGY AND AGRONOMY

Sugar beets (*B. vulgaris* L. ssp. *vulgaris*) are members of the family *Amaranthaceae* (formerly *Chenopodiaceae*) and order Caryophylalles which have a C_3 photosynthetic system. Cultivated beets are included in the subspecies *vulgaris* and include leafy beets (chard), garden beets (red), fodder beets (forage), and sugar beets. Leafy beets have been grown since pre-Roman times, however sugar beets were relatively recently domesticated in the late 18th century, and breeding has led to cultivated beets with a biennial reproductive schedule, flowering only after overwintered vernalization (Biancardi et al., 2010). Sugar beets grown in the northern hemisphere are typically planted in the early spring with harvesting 5–9 months later depending on soil and environmental conditions. In warmer or Mediterrenean climates, "winter beets" can be planted in the autumn and allow harvest through the following spring, summer, or fall. Where successful sugar beet industries have developed, diverse adjustments to the physiological limits to crop growth have been made, resulting in a large number of different cropping patterns worldwide (Kaffka and Grantz, 2014).

The genome of sugar beet has recently been sequenced for an industrial diploid ($2n = 18$ chromosomes) sample with estimated size of 714–758 megabases and 27,421 predicted protein coding genes (Dohm et al., 2014). Triploid and tetraploid varieties have also been developed for breeding purposes. The chloroplast genome has also been sequenced (Li et al., 2014). Sequencing should accelerate the identification of agronomically relevant traits to support molecular breeding that maximizes the plant's potential in energy biotechnology.

Sugar beets with genetically modified (GM) glyphosate resistance (Round-Up Ready; Monsanto, St. Louis, IL and KWS, Germany) were deregulated by the USDA in 2005 and by 2010 over 95% of all sugar beets grown in the United States were of this type. Planting and harvest of GM sugar beets were curtailed in 2011–12 due to legal challenges, but were deregulated again in 2012 after additional review (USDA, 2016). Adoption of GM beets for commercial use in other parts of the world has been largely delayed due to regulatory restrictions.

A 2012 census of agriculture reported 3913 sugar beet farms in the United States, 22% less than 5 years prior, although production grew 10% during this same time (McConnell, 2016). Average sugar beet yields in the United States grew over 2% per year nationally for a 15-year period, reaching a recent high of 73.5 MT/ha in 2016 (USDA-ERS, 2018). It should be noted that not only does the yield vary by location and climate, but it can also vary in quality for a single location and technology. A study of 49 hybrids of sugar and semiforage beet varieties in Poland characterized a variable concentration of sucrose (14.58% ± 1.28), root yield (68.36 ± 10.29 MT/ha) and ethanol fermentation yields (5140 ± 1220 L/ha) (Gumienna et al., 2016).

3 COMPOSITION

Sugar beet roots have a very high soluble sugar content, high pectin and hemicellulose carbohydrate contents, and relatively low lignin contents. An example composition is shown in Figs. 1 and 2 as adapted from data in Asadi (2007) and Srichuwong et al. (2010). All of these compositions vary regionally and seasonally as a function of many interacting factors including plant biology, location, agronomy, harvest, and postharvest practices (Fig. 3).

As the storage root increases in size, there is a constant translocation of sucrose from the leaves to the root where it is stored in concentric rings of vascular tissues derived from secondary cambium originated early in the roots development, and in parenchyma cells that increase in number and enlarge during growth (Kaffka and Grantz, 2014). Sugar beets have a high percentage of arabinan and galactan side chains believed to participate in binding pectins with cellulose (Zykwinska

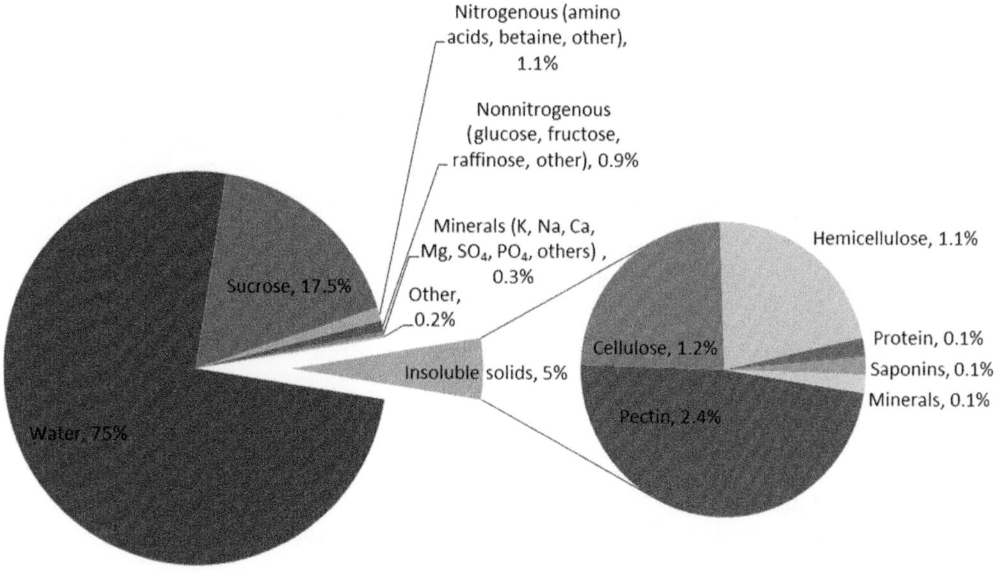

FIG. 1
Example sugar beet composition (wet basis).

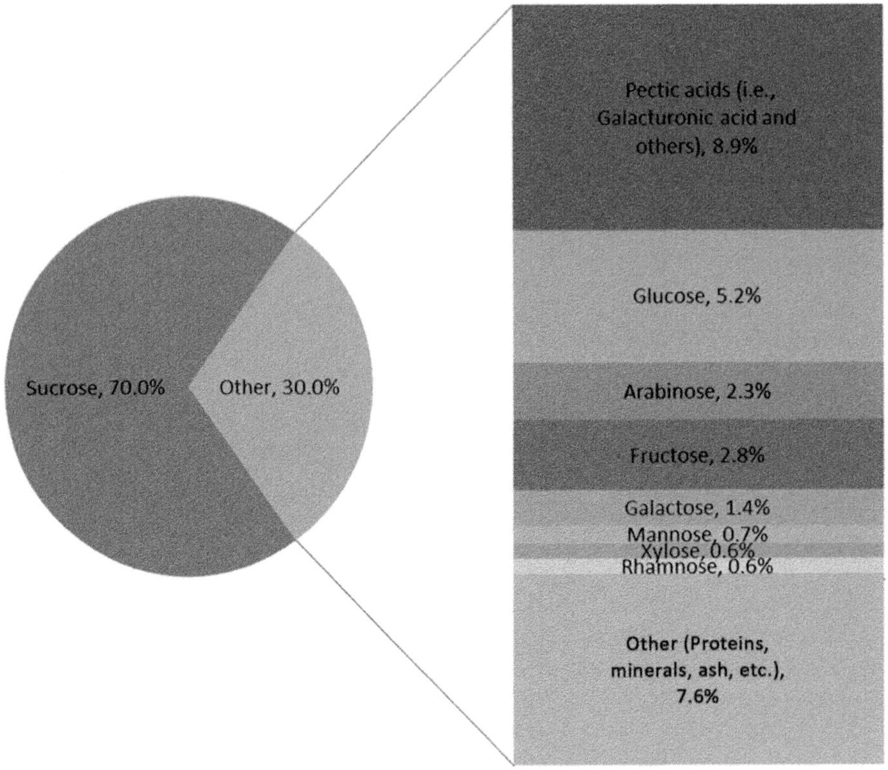

FIG. 2
Example sugar beet composition (dry basis).

FIG. 3
Examples of a sugar beet field, sugar beet, and collection bin at a sugar beet processing facility.

et al., 2006). Sugar beet pectin has been studied extensively and can be grouped into the following three main pools (Morris and Ralet, 2012):

1. high neutral sugar content and high molecular weight (Type RGI with ferulic acid bridges)

2. low in neutral sugars [Homogalacturonan (Type HG)]
3. high in neutral sugar content and low molecular weight (Type RGI)

Rhamnogalactuoronan-I (RGI), and HG regions are methyl-esterified with a degree of methylation of 70%–80%, while HG is also acetyl-esterified with a degree of acetylation around 35%. Approximately 70% of alcohol-insoluble pectin consists of galacturonic acid monomers and the remainder contains neutral sugars and about 0.5% ferulic acid.

Protein contents ranging from 2% to 10% dry basis (db) have been reported in the literature for beet roots and it is suggested that many links between pectin and protein may exist in the form of glycoproteins (i.e., extensins) (Kirby et al., 2008), which add to the difficulty in isolating sugar beet proteins due to rapid oxidation and interference with phenol and carbohydrate groups during oxidation (Parpinello et al., 2004).

A characterization of the surface of harvested sugar beet showed that it contained over 530 bacterial species with the dominant class being Proteobacteria (72.5%–77.2%), a reminder that beets support a diverse microbiota which should be considered when evaluating various downstream storage and processing options (Okazaki et al., 2014).

4 INDUSTRIAL SUGAR REFINING

Sugar beet and sugar cane are the source for nearly all industrial raw and refined crystal sugar and their processing steps are similar with only a few notable differences. Once harvested, beets can deteriorate quickly and are usually processed within a few days, or, where cold climates allow, they can be stored in large piles (clamps) under soil and straw for a period to extend the processing season. In regions with particularly cold, long winters, such as in the Northern US, uncovered piles can be processed successfully throughout the winter. Beet tops (leaves) are usually mechanically removed prior to harvest and plowed back into the soil, although beet leaves can also be considered for animal feed or as a feedstock for biofuels. Ensilng, a process whereby mainly lactic acid bacteria under anaerobic conditions decrease the pH such that further degradation does not occur, is another method that can be used to preserve large quantities of fresh beets or leaves. However, since a substantial portion of the volatile solids (11%–35%) can be consumed in this process and degradation products are problematic for sugar refining, this method is not used for beet sugar production, although may be considered for feed or energy applications (Kreuger et al., 2011).

Sugar beets are first washed to reduce unwanted soil (tare) and then sliced into thin strips (cossettes) to increase surface area upon entering a continuous countercurrent hot water extraction process known as diffusion. Prior to widespread adoption of continuous diffusion technology in the 1970s, a series of several tanks operated in batch mode (diffusion batteries) were employed to achieve the same effect (McGinnis, 1971). Beet contents are then pressed following diffusion to produce a raw juice stream, and a residual wet pulp stream with 20%–30% solids content, which is usually sold as animal feed in wet or dried form. The raw juice is then purified with multiple stages of lime and carbon dioxide addition and filtration with the clarified juice referred to as thin juice. The lime sludge produced is typically utilized as a soil amendment for pH adjustment or fertilizer in wet or dry form (May, 2010). Thin juice is further evaporated to produce a thick juice with over 60% sucrose content. Thick juice can either be stored or delivered to crystallizers for additional dewatering and centrifugation under partial vacuum and high temperatures to produce a white sugar which can be dried and packaged for sale. Typically three stages of crystallization are performed with intermediate streams, known as massecuite A, B, and C, produced with decreasing purity. These streams are separated to produce sugar crystals and molasses A, B, and C, containing the removed impurities. Molasses streams from beet facilities are typically either mixed with the pulp to supplement the animal feed or can be fermented to produce ethanol.

Unlike sugar beets, sugar cane is harvested above ground and may not contain as much soil tare as beets upon delivery to the facility. Also, the physical nature and composition of sugar cane allows it to be cut and pressed mechanically (tandem mill) to release as much as 90% of available juice. Diffusion can also be used, with various tradeoffs versus milling, although a recent survey of 455 mills in Brazil indicated that only 7% were using diffusers in 2011 (Olivero et al., 2013). As opposed to beet pulp, the pressed insoluble solids from cane mills (bagasse) have a lower moisture content (typically around 50%) and higher percentage of carbohydrates (up to 70%) and lignin (up to 30%) which makes it amenable to reliable combustion for on-site steam and electricity cogeneration (Gubicza et al., 2016). Molasses from sugar cane processing is also more palatable for human consumption and can be mixed with refined white sugar to produce brown sugar or fermented for beverage alcohol (rum) production. It is also possible for sugar cane mills to produce a less-refined light brown sugar, known as raw sugar, which can be further refined to white sugar at a separate refinery (May, 2010).

Application of membrane technologies for juice clarification and concentration is an area of active research to improve efficiency and sustainability. Also, the utilization of pulsed electric field (PEF)-assisted extraction technologies to reduce water and energy requirements during diffusion has been demonstrated at industrial scale with promising results (Vidal, 2014).

5 CURRENT BIOFUEL USE

High yields of sugars and carbohydrates in beets make them highly amenable for fermentative biofuel production. When considering processing of beets solely for biofuel production, sucrose extraction with diffusion technology may not be warranted due to its high water and energy inputs and associated costs. For biogas operations, beets can be coarsely ground or crushed prior to feeding an anaerobic digester since long residence times allow liquefaction of the material. Simple grinding and pressing of beets can yield a sizeable quantity of juice with sucrose content similar to that in the whole beet, however a substantial quantity of sugar remains in the pressed solids due to high water retention.

Calculation of ethanol productivity is dependent on many factors involving both crop and conversion metrics which vary regionally and by technology. The FAO (2008) calculated a typical yield of 5060 L ethanol/ha; however, yields potentially can be 1.5–2.0 times greater using specific location yield conditions, such as for California. High carbohydrate yields produce high theoretical ethanol yields from sucrose (first-generation ethanol), or greater when combined with additional cellulose, hemicellulose, pectic, and lignin components (second-generation ethanol). Ethanol yields per area are comparable with sugarcane and other potential second-generation biomass feedstocks, and more than double that from corn (Somerville et al., 2010). The first-generation ethanol yields of 7200 and 10,200 L/ha are estimated for the United States and California locations, respectively, with an additional 900–1400 L/ha possible when nonsucrose components are included (Fig. 4).

As nonsucrose impurities are concentrated near the top of the beet root (crown), harvesting for sugar refinining has evolved best practices for optimum defoliation, crown removal (topping), and fertilizer application

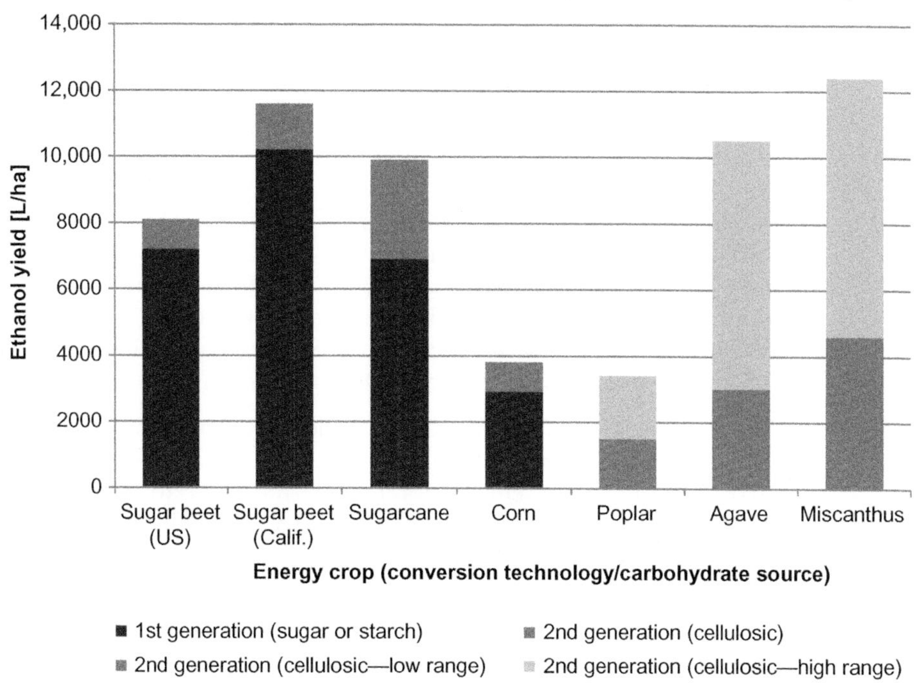

FIG. 4
Ethanol yield comparisons from selected biomass feedstocks (Zicari, 2016).

to minimize impurities. About 10%–20% of the root mass is in the crown which may be suitable for use in certain biofuel or alternative applications (Mack et al., 2007).

6 FERMENTATION OF BEET AND PROCESSING STREAMS TO ETHANOL

Extracted sugar, or whole beet and its processing by-products, is a ready fermentation substrates for a variety of valuable products. The utilization of yeasts, particularly *Saccharomyces cerevisiae*, for the production of ethanol from sugar has been exploited for thousands of years due to their high productivity, rapid growth, and robust nature. Fermentation performance can be improved through adjustment of environmental conditions, for example, minimizing glycerol by-product formation which can divert as much as 5%–8% of incoming sugars under higher osmotic stress conditions in industrial settings for sugarcane (Basso et al., 2011).

Engineered strains of *S. cerevisiae* have demonstrated the ability to ferment xylose (Jeffries, 2006) and L-arabinose sugars (Wisselink et al., 2009), among others, and the engineering of *a* D-galacturonate utilization pathway now underway is of particular interest for high pectin containing substrates such as sugar beet (Huisjes et al., 2012). Engineering of *S. cerevisiae* to prevent consumption of fructose, a more economically valuable sugar than glucose as a sweetener, has also been demonstrated such that ethanol and fructose are co-produced at yields of 59%–76% of theoretical (Atiyeh and Duvnjak, 2002).

Other organisms widely studied for industrial ethanol production include *Zymomonas mobilis and Escherichia coli. Z. mobilis* has a lower theoretical biomass production requirement and higher theoretical ethanol yield (97%), compared to *S. cerevisiae,* and generally regarded as safe (GRAS) status for feed applications (Semkiv et al., 2014). *E. coli* can ferment a wide range of hexose and pentose sugars, including galacturonic acid, and therefore has been investigated for use in conversion of pectin-rich substrates to ethanol (Edwards and Doran-Peterson, 2012). *Erwinia chrysanthemi* and *Klebsiella oxytoca* have also been tested with success using SBP (Sutton and Doran Peterson, 2001).

Several studies on fermentations with various beet sugar processing intermediates using *S. cerevisiae* have been published and are summarized in Zicari (2016). Fermentation efficiencies remain high, up to 200–250 g/L, where 12%–13% ethanol by volume is observed. To achieve high yields with 300 g/L sucrose concentrations, fed-batch operation and immobilized media reactors were implemented to achieve ethanol concentrations of 120.0–132.4 g/L (15.2%–16.8% by volume). Most batch fermentations are complete within 20–48 h, with ethanol productivity rates of 1.1–4.2 g-ethanol/L/h.

Concentrated sugar solutions need to be diluted for fermentation. Multiple researchers have investigated the effects of using recycled beet sugar refinery stillage on fermentation. Hinkova and Bubnik (2001) added 20%, 25%, and 30% stillage into thick juice dilutions for fermentations with 15%–17% sugar and found no difference in ethanol production or solid contents and predicted a commensurate volume of freshwater input could be saved.

In the 1980s, sugar beets and fodder beets were given a considerable attention by a group of researchers who published works describing submerged, diffusion, and solid-state fermentation processes. Difficulty handling the mashed beets and low ethanol concentrations during submerged fermentation were cited as reasons for the development of a solid-state method where beets were ground, fed into the hopper of a screw conveyor with steam and acid, cooled, and fermented in the conveyor (Gibbons et al., 1986). Ethanol of 8%–9% by volume was achieved with an increase in estimated yield to 87 L/MT (21 gal/ton). Several parameters were investigated for optimization during solid-state fermentation, including initial particle size (1.9 cm), initial pH (3.5), and sodium metabisulfite addition (0.25%) (Gibbons and Westby, 1986a,b 1987b). Another design, diffusion-fermentation, was researched whereby ground beets are conveyed upward in an inclined screw conveyor while water and yeast are continuously pumped countercurrent to the beets. Spent beets exit at the top end and product beer from the base reservoir. Optimum residence times were found to be 72 h for the beet solids and 264 h for the liquids (Gibbons and Westby, 1987a, 1988).

Z. mobilis has also been examined for solid-state beet fermentations, with an impressive ethanol yield of 0.48 g/g sugar, volumetric productivity of 12 g/L/h, and final ethanol concentration of 130 g/L initially reported (Amin, 1992). A more recent study of fermentation of raw sugar beet cossettes (sliced beets) in a horizontal rotating bioreactor achieved higher ethanol yields than comparable beet juice fermentations, with 54.5 g ethanol/L produced for a yield of 79.5% (Pavlecic et al., 2010).

7 ENZYME-ASSISTED PROCESSING OF BEETS FOR SUGARS AND BIOFUELS

The role of pectin in hydrolysis of ligno-cellulosic feedstocks is sometimes overlooked, but critical to consider for sugar beets, citrus wastes, and other low-lignin substrates. Sugar beet pectin can be more difficult to hydrolyze than others due to a high degree of acetylation, requiring a diversity of enzyme types for degradation. Several researchers have reported that hydrolysis of SBP with commercial fungal enzyme preparations achieves suboptimal conversions with either cellulases or pectinases and a synergistic improvement is seen using both (Martinez et al., 2009; Zicari, 2016). Depending on the enzyme mixtures utilized and test conditions, saccharification can be carried to near completion in 72 h or less and continuous hydrolysis with enzyme recycle has been demonstrated (Beldman et al., 1984; Micard et al., 1996; Spagnuolo et al., 1997; Miyaji et al., 2007). A fed-batch strategy achieved higher reducing sugar yields and demonstrated that enzyme recycling suffered primarily from inactivation due to product inhibition (Zheng et al., 2012a). Pretreatment of SBP has also been shown to drastically increase enzymatic digestibility of pulp fibers (Chamy et al., 1994).

Sugar beet leaves have also been shown to be highly degradable with enzymes and a ready source of carbohydrates and proteins. Aramrueang et al. (2017) demonstrated that 82% of theoretical monomeric sugar releases with an optimized mix of commercial cellulases and pectinases after 72h, resulting in an overall yield of 0.35g/g-TS.

Combining hydrolysis and fermentation can maximize fermentation yield while solubilizing sugars and proteins for downstream value recovery. An early study of sugar beet enzymatic hydrolysis and fermentation found a combination of pectinase and amylase effectively reduce the viscosity of the beet mash followed by heating. Pressing of raw beets recovered 79% of the initial solids and 85% of the estimated sugars and fermentative ethanol yields were 96% for beets and 89% for juices at 48h. Of the nitrogen in the juice sample, which is rich in glutamine and asparagine, 79% was retained in centrifuged supernatant and 91% of that passed through a 10-kDa MW membrane (Wu et al., 1989).

A more recent study for beet fermentations employing *S. cerevisiae* and *E coli KO11* with commercial cellulases and pectinases showed that both organisms achieved similar ethanol yields (90%) when tested at 12% solids, however *S. cerevisiae* reached a maximum in 48h compared to *E. coli KO11* in 168h. The use of combined enzyme increased free arabinose concentrations above 5g/L (86% saccharification) and GalA concentrations to over 7g/L, while reducing total suspended solids (Nahar and Pryor, 2013).

Zhang et al. (2014) examined the fermentation performance of both beet juice and juice remixed with pulp (juice-pulp-mix) and obtained after passing through a commercial juicer. Adjustment in pH (4.8 or 6.3) showed no advantage on observed liquefaction and ethanol yields of 0.48g/g were observed for juice and 0.47g/g for juice-pulp-mix with ethanol productivities up to 7.8g/L/h. Based on Arkansas field trials with energy beets yielding 115MT/ha, an ethanol productivity of 9250L/h was extrapolated.

To improve enzymatic liquefaction rate in raw beets, pretreatment can be beneficial. Srichuwong et al. (2010) examined pretreatment with sodium hydroxide (0.1N) for 1h at 25°C, followed by acid neutralization and fermentation, which resulted in faster liquefaction. Acetic acid and methanol concentrations were less than 0.8% and 0.4% g/g, respectively, and ethanol at 7%–8% g/mL was obtained from beets with 12%–13% g/mL sucrose after 24h.

Development and use of enzymes designed for a specific outcome can dramatically influence results. For example, an optimized enzyme developed for beet liquefaction (Liquebeet, Clariant Biotechnology) was tested by Zicari (2016), and at similar total soluble protein (TSP) loadings (0.1% TSP/TS) performed significantly better than other enzymes with unpretreated beets. Thermal pretreatment at 121°C for 20min, or an order of magnitude increased enzyme loading, was needed to see similar performance with the other enzymes tested.

Simultaneous saccharification and fermentation (SSF) experiments were also conducted at flask, bioreactor, and pilot scales by Zicari (2016). Bioreactor-scale experiments with ground beets at 20%–24% TS achieved ethanol yields of 0.33–0.43 g/g TS. Five metric ton (MT, wet) batches were performed in a horizontal rotary reactor and also achieved 0.43 g ethanol/g initial solids (112.2 L ethanol/MT of wet beet). Additionally, stillage digestion was performed at pilot scale and chemical oxygen demand (COD) destruction of 90 g/L achieved biogas production that was 87% of theoretically expected.

Pretreated SBP feedstock has also been used successfully as a substrate in both submerged and solid-state fermentations (Rodriguez et al., 2010; Rezic et al., 2013). Ensiling of SBP and media nutrient supplementation was shown to improve ethanol yields using *E. coli* KO11 after hydrolysis, while methods for sterilization and washing of solids were found to have a minimal effect on the overall yield. A fed-batch operation also showed better ethanol production characteristics than other modes of fermentation (Zheng et al., 2012b).

8 ANAEROBIC DIGESTION

Anaerobic digestion (AD) refers to the oxygen-deprived microbial conversion of organic matter to biogas, comprised primarily of methane (CH_4) and carbon dioxide (CO_2), and an effluent containing residual nutrients and microbiota. AD has been used for centuries for the stabilization of wastes and wastewaters and supported by relatively simple separation and energy recovery from the generated methane gas. Further treatment of effluents may be needed or nutrients are often recycled directly through land application, however researchers are continually looking to capture a greater value for these streams.

AD of sugar beets (raw or ensiled), leaves, and processing wastes from beet sugar production (pulp, tops, and tails) has been researched widely and is practiced at commercial scale. Due to a high carbon to nitrogen ratio (C:N) of around 70, raw or ensiled beets are often co-digested with other wastes, such as manures, to increase alkalinity and process stability. Co-digestion is not required as extended pilot-scale AD operation (600 days) with only beet silage fed at 3.9 g volatile solids per liter per day (g VS/L/d) was able to produce 600–700 L CH_4/kg VS (Scherer et al., 2003). Lab reactors run in parallel at 37°C, 50°C, 60°C, and 65°C showed more stability at lower temperatures, but no differences in methane yield, and are suggested that an organic loading rate (OLR) up to 12 g VS/L/d is possible.

Substrate AD potentials can be tested in the lab using biochemical methane potential (BMP) protocols, which have not been well standardized across academia (Angelidaki et al., 2009). Amon et al. (2007) report BMP values for beet silage as 430 L CH_4/kg VS and for leaves as 210 L CH_4/kg VS. Starke and Hoffmann (2011) studied AD of roots and leaves from fodder and sugar beets and found that gas potential was only correlated with dry matter content and ranged from 735 to 760 L biogas/kg TS for roots and 665 to 700 L biogas/kg TS for leaves. A practical relationship for wet basis (wb) biogas production from raw beets based on sugar content was proposed

(Eq. 1) with a suggested methane content of 53% by volume (Starke and Hoffmann, 2014).

$$\text{Biogas (L/kg wb)} = 8.88 * \text{sugar content (\%wb)} + 10.38 \quad (1)$$

Ensiling increases the percentage of alcohols and higher acids in the feedstock which increase their specific yield, however methane yields from beets are lower than many other typical forage and cereal crops. Observations of increased specific yield following ensiling should be considered in conjunction with associated mass losses for practical applicability. Kreuger et al. (2011) cautioned that standard laboratory methods for TS and VS determination can be less accurate with increasing alcohol and acid concentrations and their study found no difference between raw and ensiled maize and beet samples when controlled for this aspect.

AD of sugar beet is prone to foaming which can cause operational problems. Moeller et al. (2015) studied foam production in a silage beet digester and found that sucrose and pectins added separately displayed no additional foaming effect, but together could increase foaming. Salts of divalent ions also enhanced the foam intensity, whereas ammonium chloride and urea had a lessening effect on sugar beet-based foaming. Microbial community analyses of beet silage digesters have been performed and showed that diverse mixtures of archaeal and bacterial organisms such as Firmicutes, Proteobacteria, and Bacteroidetes adapted to various environmental conditions by optimizing their metabolism in a way that ensured efficient biogas production (Klocke et al., 2007; Langer et al., 2015).

Stillage (or vinasse) remaining after fermentation and distillation is also digestible, as has been documented for many feedstock sources (Wilkie et al., 2000). Moraes et al. (2015) reported the BMP of vinasse from a beet sugar distillery as 267 L CH_4/kg VS, but could not get steady operation of a continuous digester without the addition of manure as a co-substrate, which allowed it to reach 88% of the BMP gas yield. Addition of enzymes were found to increase biogas production from a blend of beet silage and vinasse (3,1) by as much as 28% (Zieminski and Kowalska-Wentel, 2015). Asato (2014) achieved steady operation with AD of sugar beet stillage containing 4% VS in both anaerobic sequencing batch reactor (ASBR) and anaerobic mixed biofilm reactor (AMBR) configurations at OLRs of 4 g VS/L/d. The ASBR appeared more stable and produced 320 L CH_4/kg VS, but the AMBR performed better achieving 450 L CH_4/kg VS and 84% VS reduction.

SBP has been reported to have BMPs including 295 L CH_4/kg VS (Liu et al., 2008), 391 L CH_4/kg VS (Hutnan et al., 2001), and 430–480 L CH_4/kg VS (Kryvoruchko et al., 2009). Following liquid hydrothermal pretreatment at 160°C for 25 min, 503 L CH_4/kg VS was achieved (Zieminski et al., 2014). Stoyanova et al. (2014) performed continuous AD of pressed pulp in a two-stage laboratory reactor and achieved stable operation at 7 g VS/L/D OLR with a 36 day hydraulic retention time (HRT). The same was not possible with a single-stage configuration at 50 day HRT due to foaming issues. Biogas production from SBP has also been achieved at large commercial scale (800 Mg/day) with biogas production of 305–323 L CH_4/kg VS, which could offset about 40% of the natural gas required for the thermal load at the sugar processing facility (Brooks et al., 2008).

9 ALTERNATIVE PRODUCTS

Sugar beet can be used for many different purposes other than sugar, animal feed, ethanol, and biogas production. Sugars contained in beets can be used to support fermentative conversion to hundreds of other compounds, each with unique costs and benefits. A market study identifying opportunities in the European fermentation industry predicted 5% growth of the global $127 billion market through 2020 (Deloitte, 2014). The highest growth is expected in organic acid and polymer markets, followed closely by alcohols, amino acids, vitamins, antibiotics, and industrial enzymes. Monosodium glutamate (MSG) is a common food additive that is known to utilize beet sugar and molasses as a feedstock for production (Kaffka and Grantz, 2014). Lactic acid is one of the many fermentation targets found to be suitable for production on raw and thin juices extracted from sugar beet (Koch and Venus, 2014). More so, fermentation products such as ethanol can serve as a starting point for further conversion into products that are typically only produced from fossil resources, such as ethylene (Althoff et al., 2013).

Hydrogen production from sugar beet feedstocks has also been an extensive area of research due to interest in supplying energy and transportation markets (Grabarczyk et al., 2011). Continuous hydrogen production has been achieved in photo-bioreactors (Hussy et al., 2005; Keskin and Hallenbeck, 2012; Ozkan et al., 2012) with estimated potential to supply over 300 Gg H_2/yr from SBP in the EU alone (Panagiotopoulos et al., 2010).

Many natural compounds in the beet are valuable. Pectins can be extracted from beet pulp and generally find applications as emulsifiers given a lower gelation capacity than other pectins (Ovodov, 2009). Enzymatic processing can result in functionally active prebiotic oligosaccharides which are a fast growing market (Martinez et al., 2009). Buttersack et al. (1994) examined a method of extracting enzymes (polyphenyloxidase, peroxidase, and galacturonase) directly from beetroot using mineral adsorption/desorption process. Ultrasonic-assisted extraction of phenols, anthocyanins, and antioxidants from beet molasses has also been performed (Chen et al., 2015). Betaine (trimethylglycine) is also present in beets and is considered a valuable nutritional supplement (Mikos et al., 2015).

Also, as a high-yielding plant with a wealth of historical breeding and genetic information, sugar beets offer many opportunities in the area of biotechnology.

The complete genome has been sequenced and genetic manipulation of beets has resulted in many improvements and variations with an expanding potential for use as plant-based bio-refineries (Gurel et al., 2008).

10 TECHNICAL, ECONOMIC, ENVIRONMENTAL, AND LIFE-CYCLE EVALUATIONS

Sugar beets are currently not used for dedicated biofuels production in the United States, however petitions for beet-based renewable fuel pathways have been reviewed by the EPA (2017). In 2006, the USDA estimated a cost of \$0.62/L (\$2.35/gal) to produce ethanol from sugar beets, excluding capital and transportation costs (Shapouri and Salassi, 2006). Comparable costs for ethanol from Brazilian sugar cane (\$0.21/L), wet or dry milled US corn (\$0.28/L), and US molasses (\$0.33/L), were lower, while those from US sugar cane (\$0.63/L), EU sugar beets (\$0.76), and US raw or refined sugar (\$0.92–1.04/L), were higher. Feedstock costs were 75% of the estimated cost from US sugar beets.

A recent technical and economic analysis for a facility designed to process 40,000 MT of beets per year, and producing 4 million liters per year of denatured fuel ethanol, showed a minimum ethanol selling price (MESP) of \$1.11/L would be needed to achieve a positive net present value (Zicari, 2016). Feedstock costs were 43% of production costs and enzymes a significant operating cost (\$0.12/L-ethanol). Enzyme usage reduced processing water input and facilitated production of an additional 6.5% ethanol from lignocellulosic origin. Produced biogas offset slightly over 100% of facility thermal energy demand and allowed a calculated net-energy ratio of 24.6. A total of 81% the input mass exited the facility as liquid effluent and separated solids, but only contributed 4% to revenues as modeled. Project economic viability is highly subject to the fate of these streams. Alternative process configurations were evaluated and use of an enzyme optimized for beet liquefaction offered significant potential for lowering the MESP by 10%–12%. Use of an advanced microorganism capable of producing supplemental ethanol from five-carbon sugars, such as arabinose, could produce an additional 2.5% ethanol, although utilization of a galacturonic acid consuming ethanologen would be required to enable another 5% production.

Demonstrating the effect of scale, another recent analysis for ethanol production from beets in North Dakota estimated that break-even pricing for 38- and 76-million liter per year facilities would only be \$0.45, and \$0.40 per liter, respectively, with feedstock costs (beets and molasses) comprising 70% of production costs (Maung and Gustafson, 2011). All stillage was assumed evaporated to solids and burned to recover approximately 75% of facility energy needs. Centrifugation and recovery of high-value yeast contributed 14% to revenues; feed and fertilizer another 6%; and the balance from ethanol sales.

Modeling for large facilities indicates that application of advanced technologies such as simultaneous sugar extraction, pulp fermentation, and product recovery, combined with biogas production can decrease processing energy inputs from a base case of 0.016 GJ/L of ethanol, by as much as 20 times (Šantek et al., 2010). In some cases, significant improvements to existing facilities can be made without advanced technology improvements. For existing facilities, adding ethanol production from molasses was the most attractive, and for new facilities it was favorable to produce only ethanol from thick juice, without crystal sugar production.

Ethanol production from SBP using either enzymatic hydrolysis or a combination of acid and enzymatic hydrolysis was estimated to require a MESP of $0.40–0.67/L ($1.53–$2.53/gal) for profitability (Donkoh et al., 2012). Several business cases for production of bioethylene from sugar beets were also analyzed and deemed not to be commercially viable under existing market conditions (Althoff et al., 2013).

Energy input–output accounting is often used with biofuels projects; however, care is needed when comparing results as the boundary for, and definition of, inputs and outputs and their associated values can vary greatly dependent on technology and location. When examined as a crop only, the net energy gain in a beet crop may be from 7 to 15 times that of the external energy input that was required to produce it (219–313 GJ/ha) (Koga, 2008; Reineke et al., 2013), with further improvements possible with advances in agronomic practices (Koga et al., 2009). If it is to be used for ethanol production, net energy gains of 32–37 GJ/ha have been reported for projects in the Netherlands (Langeveld et al., 2014). If it is used for dedicated biogas production, net energy gains from 160 to 279 GJ/ha have been reported for European project analyses (Gissen et al., 2014; Jacobs et al., 2014).

Water is a resource that also requires careful consideration. Beets as a crop are considered to be quite water- and nitrogen-use efficient with a robust salinity tolerance. Water use efficiencies from 0.002 to 0.0068 $Mg\,cm^{-1}\,ha^{-1}$ on a dry matter basis have been reported for beets under a diverse set of agronomic, environmental, and irrigation conditions (Kaffka and Grantz, 2014). Compared with other sugar feedstocks (maize, sugar cane, sugar beet), the water footprint for bioethanol and sweeteners varies considerably by location and utilizes different distributions of water resources such as surface and ground water ("blue"), rainwater ("green"), and water for wastewater mitigation ("gray") (Gerbens-Leenes and Hoekstra, 2012). On a global average basis, beets for bioelectricity or bioethanol production were generally preferred. Similarly, beet ethanol production was one of the lowest total water users of those evaluated at 1400 L water/L ethanol, as compared to cane and maize at 2500–2600 L/L (Gerbens-Leenes et al., 2009).

Life-cycle analyses (LCAs) are performed to more fully assess the impact of a process from beginning to end on the basis of GHG emissions, typically expressed in grams of carbon dioxide equivalents per MJ of fuel content (g CO_2e/MJ). Reviews of bioethanol production from sugar beets in Europe have estimated that GHG emissions are 28%–42% lower than gasoline (Bessou et al., 2012) and take an intermediate position when compared to other first-generation biofuel feedstocks (de Vries et al., 2010). Similarly, when compared directly with sugar production in Greece, a GHG reduction of at least 32% was estimated (Foteinis et al., 2011). GHGs are not the only basis by which to determine a projects' impact or value; possible negative impacts from land use change, higher acidification and eutrophication potentials due to reactive nitrogen losses, and freshwater consumption also need close attention.

The GHG intensity of a highly integrated enzymatic beet to ethanol pathway in California has been estimated to be 28 g CO_2e/MJ, 72% below a gasoline baseline and 19% below the lowest comparable corn ethanol pathway published at the time (Alexiades, 2014). Special accounting for the impact of enzyme use in the process was afforded, which is sometimes overlooked in other analyses.

11 SUMMARY AND FUTURE OPPORTUNITIES

Sugar beets are an established industrial crop with ample opportunities for integration with, or redirection toward, alternative feed, fuel, and chemical production technologies. Recent market changes are leading many to look more closely at sugar beets as a platform for sustainable industrial biorefinery operations and it will be interesting to follow the continued evolution of beet production and processing in the years to come.

References

Asadi, M., 2007. Beet-Sugar Handbook. Wiley-Interscience, Hoboken, NJ.

Alexiades, A.M., 2014. Applying Adaptive Agricultural Management & Industrial Ecology Principles to Produce Lower-Carbon Ethanol From California Energy Beets. MS Thesis, UC Davis.

Althoff, J., Biesheuvel, K., De Kok, A., Pelt, H., Ruitenbeek, M., Spork, G., Tange, J., Wevers, R., 2013. Economic feasibility of the sugar beet-to-ethylene value chain. ChemSusChem 6 (9), 1625–1630.

Amin, G., 1992. Conversion of sugar beet particles to ethanol by the bacterium *zymomonas mobilis* in solid-state fermentaiton. Biotechnol. Lett. 14 (6), 499–504.

Amon, T., Amon, B., Kryvoruchko, V., Machmuller, A., Hopfner-Sixt, K., Bodiroza, V., Hrbek, R., Friedel, J., Potsch, E., Wagentristl, H., Schreiner, M., Zollitsch, W., 2007. Methane production through anaerobic digestion of various energy crops grown in sustainable crop rotations. Bioresour. Technol. 98 (17), 3204–3212.

Angelidaki, I., Alves, M., Bolzonella, D., Borzacconi, L., Campos, J.L., Guwy, A.J., Kalyuzhnyi, S., Jenicek, P., van Lier, J.B., 2009. Defining the biomethane potential (bmp) of solid organic wastes and energy crops: a proposed protocol for batch assays. Water Sci. Technol. 59 (5), 927–934.

Aramrueang, N., Zicari, S.M., Zhang, R., 2017. Response surface optimization of enzymatic hydrolysis of sugar beet leaves into fermentable sugars for bioethanol production. Adv. Biosci. Biotechnol. 08 (02), 51–67.

Asato, C., 2014. Anaerobic Digestion of Sugar Beet Ethanol Stillage for Biogas Energy Production and Nutrient and Water Recovery. MS. University of California, Davis.

Atiyeh, H., Duvnjak, Z., 2002. Production of fructose and ethanol from sugar beet molasses using saccharomyces cerevisiae atcc 36858. Biotechnol. Prog. 18 (2), 234–239.

Basso, L.C., Basso, T.O., Rocha, S.N., 2011. Ethanol Production in Brazil: The Industrial Process and its Impact on Yeast Fermentation. Biofuel Production—Recent Development and Prospects. MADS Bernades, InTech.

Beldman, G., Rombouts, F.M., Voragen, A.G.J., Pilnik, W., 1984. Application of cellulase and pectinase from fungal origin for the liquefaction and saccharification of biomass. Enzyme Microb. Technol. 6 (11), 503–507.

Bessou, C., Lehuger, S., Gabrielle, B., Mary, B., 2012. Using a crop model to account for the effects of local factors on the lca of sugar beet ethanol in picardy region, France. Int. J. Life Cycle Assess. 18 (1), 24–36.

Biancardi, E., McGrath, J.M., Panella, L.W., Lewellen, R.T., Stevanato, P., 2010. Sugar beet. In: Bradshaw, J.E. (Ed.), Root and Tuber Crops, Handbook of Plant Breeding 7, Springer Science and Business Media, pp. 173–219.

Brooks, L., Parravicini, V., Svardal, K., Kroiss, H., Prendl, L., 2008. Biogas from sugar beet press pulp as substitute of fossil fuel in sugar beet factories. Water Sci. Technol. 58 (7), 1497–1504.

Buttersack, C., Nowikow, K., Schaper, A., Buchholz, K., 1994. Enzyme-production from sugar-beets. Zuckerindustrie 119 (4), 284–291.

Chamy, R., Illanes, A., Aroca, G., Nunez, L., 1994. Acid-hydrolysis of sugar-beet pulp as a pretreatment for fermentation. Bioresour. Technol. 50 (2), 149–152.

Chen, M., Zhao, Y., Yu, S., 2015. Optimisation of ultrasonic-assisted extraction of phenolic compounds, antioxidants, and anthocyanins from sugar beet molasses. Food Chem. 172, 543–550.

de Vries, S.C., van de Ven, G.W.J., van Ittersum, M.K., Giller, K.E., 2010. Resource use efficiency and environmental performance of nine major biofuel crops, processed by first-generation conversion techniques. Biomass Bioenergy 34 (5), 588–601.

Deloitte, 2014. Opportunities for the Fermentation-Based Chemical Industry: An Analysis of the Market Potential and Competitiveness of North-West Europe. Deloitte The Netherlands, Netherlands.

Dohm, J.C., Minoche, A.E., Holtgraewe, D., Capella-Gutierrez, S., Zakrzewski, F., Tafer, H., Rupp, O., Soerensen, T., Stracke, R., Reinhardt, R., Goesmann, A., Kraft, T., Schulz, B., Stadler, P.F., Schmidt, T., Gabaldon, T., Lehrach, H., Weisshaar, B., Himmelbauer, H., 2014. The genome of the recently domesticated crop plant sugar beet (*Beta vulgaris*). Nature **505** (7484), 546.

Donkoh, E., Degenstein, J., Ji, Y., 2012. Process integration and economics evaluation of sugar beet pulp conversion into ethanol. Int. J. Agric. Biol. Eng. 5 (2), 52–61.

Edwards, M.C., Doran-Peterson, J., 2012. Pectin-rich biomass as feedstock for fuel ethanol production. Appl. Microbiol. Biotechnol. 95 (3), 565–575.

EPA, 2017. Pending Petitions for Renewable Fuel Pathways. Retrieved April 11, 2017, from, https://www.epa.gov/renewable-fuel-standard-program/pending-petitions-renewable-fuel-pathways.

FAO, 2008. Biofuels: propsects, risks, and opportunities. In: The State of Food and Agriculture, 2008.

Foteinis, S., Kouloumpis, V., Tsoutsos, T., 2011. Life cycle analysis for bioethanol production from sugar beet crops in Greece. Energy Policy 39 (9), 4834–4841.

Gerbens-Leenes, W., Hoekstra, A.Y., 2012. The water footprint of sweeteners and bio-ethanol. Environ. Int. 40, 202–211.

Gerbens-Leenes, W., Hoekstra, A.Y., van der Meer, T.H., 2009. The water footprint of bioenergy. Proc. Natl. Acad. Sci. U. S. A. 106 (25), 10219–10223.

Gibbons, W.R., Westby, C.A., 1986a. Effect of pulp ph on solid-phase fermentation of fodder beets for fuel and ethanol-production. Biotechnol. Lett. 8 (9), 657–662.

Gibbons, W.R., Westby, C.A., 1986b. Solid-phase fermentation of fodder beets for ethanol production—effect of grind size. J. Ferment. Technol. 64 (2), 179–183.

Gibbons, W.R., Westby, C.A., 1987a. Effect of fodder beet cube size on ethanol production via diffusion fermentation. Biotechnol. Lett. 9 (2), 135–138.

Gibbons, W.R., Westby, C.A., 1987b. Effects of sodium meta-bisultife on diffusion fermentation of fodder beets for fuel ethanol production. Biotechnol. Bioeng. 30 (7), 909–916.

Gibbons, W.R., Westby, C.A., 1988. Preventing contamination during diffusion fermentation of fodder beet cubes by ph control. Biomass 16 (2), 119–132.

Gibbons, W.R., Westby, C.A., Dobbs, T.L., 1986. Intermediate-scale, semicontinuous solid-phase fermentaiton process for the production of fuel ethanol from sweet sorghum. Appl. Environ. Microbiol. 51 (1), 115–122.

Gissen, C., Prade, T., Kreuger, E., Nges, I.A., Rosenqvist, H., Svensson, S.E., Lantz, M., Mattsson, J.E., Borjesson, P., Bjornsson, L., 2014. Comparing energy crops for biogas production yields, energy input and costs in cultivation using digestate and mineral fertilisation. Biomass Bioenergy 64, 199–210.

Grabarczyk, R., Urbaniec, K., Koukios, E., Bakker, R., Vaccari, G., 2011. Options of sugar beet pretreatment for hydrogen fermentation. Zuckerindustrie 136 (12), 784–790.

Gubicza, K., Nieves, I.U., Sagues, W.J., Barta, Z., Shanmugam, K.T., Ingram, L.O., 2016. Techno-economic analysis of ethanol production from sugarcane bagasse using a liquefaction plus simultaneous saccharification and co-fermentation process. Bioresour. Technol. 208, 42–48.

Gumienna, M., Szwengiel, A., Szczepanska-Alvarez, A., Szambelan, K., Lasik-Kurdys, M., Czarnecki, Z., Sitarski, A., 2016. The impact of sugar beet varieties and cultivation conditions on ethanol productivity. Biomass Bioenergy 85, 228–234.

Gurel, E., Gurel, S., Lemaux, P.G., 2008. Biotechnology applications for sugar beet. Crit. Rev. Plant Sci. 27 (2), 108–140.

Hinkova, A., Bubnik, Z., 2001. Sugarbeet as a raw material for bioethanol production. Czech J. Food Sci. 19 (6), 224–234.

Huisjes, E.H., Luttik, M.A., Almering, M.J., Bisschops, M.M., Dang, D.H., Kleerebezem, M., Siezen, R., van Maris, A.J., Pronk, J.T., 2012. Toward pectin fermentation by saccharomyces cerevisiae: expression of the first two steps of a bacterial pathway for d-galacturonate metabolism. J. Biotechnol. 162 (2–3), 303–310.

Hussy, I., Hawkes, F.R., Dinsdale, R., Hawkes, D.L., 2005. Continuous fermentative hydrogen production from sucrose and sugarbeet. Int. J. Hydrogen Energy 30 (5), 471–483.

Hutnan, M., Drtil, M., Derco, J., Mrafkova, L., Hornak, M., Mico, S., 2001. Two-step pilot-scale-anaerobic treatment of sugar beet pulp. Pol. J. Environ. Stud. 10 (4), 237–243.

Jacobs, A., Brauer-Siebrecht, W., Koch, H.J., Marlander, B., Auburger, S., Bahrs, E., Pelka, N., Buchholz, M., Gotze, P., Rucknagel, J., Christen, O., 2014. The sugar beet as an energy crop in crop rotations on highly productive sites—an agronomic/economic system analysis. Sugar Ind./Zuckerindustrie 139 (2), 117–127.

Jeffries, T.W., 2006. Engineering yeasts for xylose metabolism. Curr. Opin. Biotechnol. 17 (3), 320–326.

Kaffka, S.R., Grantz, D.A., 2014. Sugar crops. In: Alfen, N.V. (Ed.), Encyclopedia of Agriculture and Food Systems. 5, Elsevier, pp. 240–260.

Keskin, T., Hallenbeck, P.C., 2012. Hydrogen production from sugar industry wastes using single-stage photofermentation. Bioresour. Technol. 112, 131–136.

Kirby, A., Macdougall, A., Morris, V., 2008. Atomic force microscopy of tomato and sugar beet pectin molecules. Carbohydr. Polym. 71 (4), 640–647.

Klocke, M., Mahnert, P., Mundt, K., Souidi, K., Linke, B., 2007. Microbial community analysis of a biogas-producing completely stirred tank reactor fed continuously with fodder beet silage as mono-substrate. Syst. Appl. Microbiol. 30 (2), 139–151.

Koch, T.J., Venus, J., 2014. Sugar beet syrups in lactic acid fermentation—part ii saving nutrients by lactic acid fermentation with sugar beet thick juice and raw juice. Sugar Ind./Zuckerindustrie 139 (11), 683–690.

Koga, N., 2008. An energy balance under a conventional crop rotation system in northern Japan: perspectives on fuel ethanol production from sugar beet. Agric. Ecosyst. Environ. 125, 101–110.

Koga, N., Takahashi, H., Okazaki, K., Kajiyama, T., Kobayashi, S., 2009. Potential agronomic options for energy-efficient sugar beet-based bioethanol production in northern Japan. Glob. Change Biol. Bioenergy 1 (3), 220–229.

Kreuger, E., Nges, I., Bjornsson, L., 2011. Ensiling of crops for biogas production: effects on methane yield and total solids determination. Biotechnol. Biofuels 4 (1), 44.

Kryvoruchko, V., Machmüller, A., Bodiroza, V., Amon, B., Amon, T., 2009. Anaerobic digestion of by-products of sugar beet and starch potato processing. Biomass Bioenergy 33 (4), 620–627.

Langer, S.G., Ahmed, S., Einfalt, D., Bengelsdorf, F.R., Kazda, M., 2015. Functionally redundant but dissimilar microbial communities within biogas reactors treating maize silage in co-fermentation with sugar beet silage. J. Microbial. Biotechnol. 8 (5), 828–836.

Langeveld, J.W.A., van de Ven, G., de Vries, S.C., van den Brink, L., de Visser, C., 2014. Ethanol from sugar beet in the Netherlands: energy production and efficiency. Int. J. Sustain. Dev. 17 (1), 78–88.

Li, H., Cao, H., Cai, Y.F., Wang, J.H., Qu, S.P., Huang, X.Q., 2014. The complete chloroplast genome sequence of sugar beet (Beta vulgaris ssp vulgaris). Mitochondrial DNA 25 (3), 209–211.

Liu, W., Pullammanappallil, P.C., Chynoweth, D.P., Teixeira, A.A., 2008. Thermophilic anaerobic digestion of sugar beet tailings. Trans. ASABE 51 (2), 615–621.

Mack, G., Hoffmann, C.M., Marlander, B., 2007. Nitrogen compounds in organs of two sugar beet genotypes (Beta vulgaris l.) during the season. Field Crop Res. 102 (3), 210–218.

Martinez, M., Gullon, B., Yanez, R., Alonso, J.L., Parajo, J.C., 2009. Direct enzymatic production of oligosaccharide mixtures from sugar beet pulp: experimental evaluation and mathematical modeling. J. Agric. Food Chem. 57 (12), 5510–5517.

Maung, T.A., Gustafson, C.R., 2011. The economic feasibility of sugar beet biofuel production in central North Dakota. Biomass Bioenergy 35 (9), 3737–3747.

May, M., 2010. Sugar Beet. Chapter 6, pp. 104–115.

McConnell, M.J., 2016. Sugar and Sweetners: Background. June 3, 2016, Retrieved August 2016, from, http://www.ers.usda.gov/topics/crops/sugar-sweeteners/background.aspx.

McGinnis, R.A., 1971. Beet-Sugar Technology. Beet Sugar Development Foundation, Fort Collins, Colo.

Micard, V., Renard, C., Thibault, J.F., 1996. Enzymatic saccharification of sugar-beet pulp. Enzyme Microb. Technol. 19 (3), 162–170.

Mikos, P., Antczak-Chrobot, A., Wojtczak, M., 2015. Free amino acids, betaine, nitrate and nitrite in the sugar beet processing—a literature review. Int. Sugar J. 117 (1403), 790–797.

Miyaji, T., Fujimura, S., Nakagawa, T., Takano, A., 2007. Practical liquefaction of potato pulp and sugar-beet pulp by commercial enzymes. J. Agric. Sci. Tokyo Univ. Agric. 52 (3), 147–150.

Moeller, L., Lehnig, M., Schenk, J., Zehnsdorf, A., 2015. Foam formation in biogas plants caused by anaerobic digestion of sugar beet. Bioresour. Technol. 178, 270–277.

Moraes, B.S., Triolo, J.M., Lecona, V.P., Zaiat, M., Sommer, S.G., 2015. Biogas production within the bioethanol production chain: use of co-substrates for anaerobic digestion of sugar beet vinasse. Bioresour. Technol. 190, 227–234.

Morris, G.A., Ralet, M.C., 2012. A copolymer analysis approach to estimate the neutral sugar distribution of sugar beet pectin using size exclusion chromatography. Carbohydr. Polym. 87 (2), 1139–1143.

Nahar, N., Pryor, S.W., 2013. Enzymatic hydrolysis and fermentation of crushed whole sugar beets. Biomass Bioenergy 59, 512–519.

Okazaki, K., Iino, T., Kuroda, Y., Taguchi, K., Takahashi, H., Ohwada, T., Tsurumaru, H., Okubo, T., Minamisawa, K., Ikeda, S., 2014. An assessment of the diversity of culturable bacteria from main root of sugar beet. Microbes Environ. 29 (2), 220–223.

Olivero, J., D'Avola, A., Faber, A., Soares, P., 2013. In: Juice Extraction Systems: Mills and Diffusers-the Brazilian Experience. XXVIII ISSCT Congress. Sao Paolo, Brazil.

Ovodov, Y.S., 2009. Current views on pectin substances. Russ. J. Bioorg. Chem. 35 (3), 269–284.

Ozkan, E., Uyar, B., Ozgur, E., Yucel, M., Eroglu, I., Gunduz, U., 2012. Photofermentative hydrogen production using dark fermentation effluent of sugar beet thick juice in outdoor conditions. Int. J. Hydrogen Energy 37 (2), 2044–2049.

Panagiotopoulos, J.A., Bakker, R.R., de Vrije, T., Urbaniec, K., Koukios, E.G., Claassen, P.A.M., 2010. Prospects of utilization of sugar beet carbohydrates for biological hydrogen production in the EU. J. Clean. Prod. 18, S9–S14.

Panella, L., Kaffka, S.R., 2010. Sugar beet (Beta vulgaris l) as a biofuel feedstock in the United States. In: Sustainability of the Sugar and Sugar-Ethanol Industries. 1058, American Chemical Society, pp. 163–175.

Parpinello, G.P., Versari, A., Riponi, C., 2004. Characterization of sugarbeet (Beta vulgaris, l.) protein. J. Sugar Beet Res. 41 (1–2), 39–46.

Pavlecic, M., Vrana, I., Vibovec, K., Santek, M.I., Horvat, P., Santek, B., 2010. Ethanol production from different intermediates of sugar beet processing. Food Technol. Biotechnol. 48 (3), 362–367.

Pelka, N., Buchholz, M., Musshoff, O., 2015. Competitiveness of energy crop rotations with and without sugar beets for biogas production considering the individual risk tolerance. Berichte Uber Landwirtschaft 93 (1), 11.

Reineke, H., Stockfisch, N., Märländer, B., 2013. Analysing the energy balances of sugar beet cultivation in commercial farms in Germany. Eur. J. Agron. 45, 27–38.

Rezic, T., Oros, D., Markovic, I., Kracher, D., Ludwig, R., Santek, B., 2013. Integrated hydrolyzation and fermentation of sugar beet pulp to bioethanol. J. Microbiol. Biotechnol. 23 (9), 1244–1252.

Rodriguez, L.A., Toro, M.E., Vazquez, F., Correa-Daneri, M.L., Gouiric, S.C., Vallejo, M.D., 2010. Bioethanol production from grape and sugar beet pomaces by solid-state fermentation. Int. J. Hydrogen Energy 35 (11), 5914–5917.

Šantek, B., Gwehenberger, G., Šantek, M.I., Narodoslawsky, M., Horvat, P., 2010. Evaluation of energy demand and the sustainability of different bioethanol production processes from sugar beet. Resour. Conserv. Recycl. 54 (11), 872–877.

Scherer, P.A., Dobler, S., Rohardt, S., Loock, R., Buttner, B., Noldeke, P., Brettschuh, A., 2003. Continuous biogas production from fodder beet silage as sole substrate. Water Sci. Technol. 48 (4), 229–233.

Semkiv, M., Dymtruk, K., Abbas, C., Sibirny, A., 2014. Increased ethanol accumulation from glucose via reduction of ATP level in a recombinant strain of saccharomyces cerevisiae overexpressing alkaline phosphatase. BMC Biotechnol. 14, 42.

Shapouri, H., Salassi, M., 2006. The Economic Feasibility of Ethanol Production From Sugar in the United States. USDA.

Somerville, C., Youngs, H., Taylor, C., Davis, S.C., Long, S.P., 2010. Feedstocks for lignocellulosic biofuels. Science 329 (5993), 790–792.

Spagnuolo, M., Crecchio, C., Pizzigallo, M.D.R., Ruggiero, P., 1997. Synergistic effects of cellulolytic and pectinolytic enzymes in degrading sugar beet pulp. Bioresour. Technol. 60 (3), 215–222.

Srichuwong, S., Arakane, M., Fujiwara, M., Zhang, Z.L., Takahashi, H., Tokuyasu, K., 2010. Alkali-aided enzymatic viscosity reduction of sugar beet mash for novel bioethanol production process. Biomass Bioenergy 34 (9), 1336–1341.

Starke, P., Hoffmann, C., 2011. Sugarbeet as a substrate for biogas production. Zuckerindustrie 136 (4), 242–250.

Starke, P., Hoffmann, C.M., 2014. Dry matter and sugar content as parameters to assess the quality of sugar beet varieties for anaerobic digestion. Sugar Ind./Zuckerindustrie 139 (4), 232–240.

Stoyanova, E., Forsthuber, B., Pohn, S., Schwarz, C., Fuchs, W., Bochmann, G., 2014. Reducing the risk of foaming and decreasing viscosity by two-stage anaerobic digestion of sugar beet pressed pulp. Biodegradation 25 (2), 277–289.

Sutton, M., Doran Peterson, J.B., 2001. Fermentation of sugarbeet pulp for ethanol production using bioengineered klebsiella oxytoca strain p2. J. Sugar Beet Res. 38 (1–2), 19–34.

USDA, 2016. About Roundup Ready Sugar Beet. From, https://www.aphis.usda.gov/aphis/ourfocus/biotechnology/brs-news-and-information/CT_Sugarbeet_about.

USDA-ERS, 2018. Sugar and Sweeteners Yearbook Tables. United States Department of Agriculture—Economic Research Service. http://www.ers.usda.gov/data-products/sugar-and-sweeteners-yearbook-tables.aspx.

Vidal, B., 2014. First report of pulsed electric field (pef) application at industrial scale in beet sugar industry. In: Sugar Industry. 1.

Wilkie, A.C., Riedesel, K.J., Owens, J.M., 2000. Stillage characterization and anaerobic treatment of ethanol stillage from conventional and cellulosic feedstocks. Biomass Bioenergy 19 (2), 63–102.

Wisselink, H.W., Toirkens, M.J., Wu, Q., Pronk, J.T., van Maris, A.J.A., 2009. Novel evolutionary engineering approach for accelerated utilization of glucose, xylose, and arabinose mixtures by engineered saccharomyces cerevisiae strains. Appl. Environ. Microbiol. 75 (4), 907–914.

Wu, Y.V., Nielsen, H.C., Bagby, M.O., 1989. Recovery of protein-rich byproducts from sugar beet stillage after alcohol distillation. J. Agric. Food Chem. 37 (4), 1174–1177.

Yousefi, M., Khoramivafa, M., Shahamat, E.Z., 2015. Water, nitrogen and energy use efficiency in major crops production systems in Iran. Adv. Plants Agric. Res 2 (3), 126–129.

Zhang, N., Steven Green, V., Ge, X., Savary, B.J., Xu, J., 2014. Ethanol fermentation of energy beets by self-flocculating and non-flocculating yeasts. Bioresour. Technol. 155, 189–197.

Zheng, Y., Cheng, Y.S., Yu, C., Zhang, R., Jenkins, B.M., VanderGheynst, J.S., 2012a. Improving the efficiency of enzyme utilization for sugar beet pulp hydrolysis. Bioprocess Biosyst. Eng. 35 (9), 1531–1539.

Zheng, Y., Yu, C., Cheng, Y.-S., Lee, C., Simmons, C.W., Dooley, T.M., Zhang, R., Jenkins, B.M., VanderGheynst, J.S., 2012b. Integrating sugar beet pulp storage, hydrolysis and fermentation for fuel ethanol production. Appl. Energy 93, 168–175.

Zicari, S.M., 2016. Enzymatic Liquefaction of Sugar Beets as a Feedstock for Biofuel Production. PhD Dissertation, University of California, Davis.

Zieminski, K., Kowalska-Wentel, M., 2015. Effect of enzymatic pretreatment on anaerobic co-digestion of sugar beet pulp silage and vinasse. Bioresour. Technol. 180, 274–280.

Zieminski, K., Romanowska, I., Kowalska-Wentel, M., Cyran, M., 2014. Effects of hydrothermal pretreatment of sugar beet pulp for methane production. Bioresour. Technol. 166, 187–193.

Zykwinska, A., Rondeau-Mouro, C., Garnier, C., Thibault, J.F., Ralet, M.C., 2006. Alkaline extractability of pectic arabinan and galactan and their mobility in sugar beet and potato cell walls. Carbohydr. Polym. 65 (4), 510–520.

SECTION 4
Olives, Tree Nuts, and Coffee

CHAPTER 14
Olive

Rebecca Milczarek*, Douglas Larson†, Yao Olive Li‡, Ivana Sedej*, Selina Wang§
*United States Department of Agriculture—Agricultural Research Service, Healthy Processed Foods Research Unit, Albany, CA, United States, †University of California - Davis, Davis, Department of Agricultural and Resource Economics, Davis, CA, United States, ‡California State Polytechnic University, Human Nutrition and Food Science Department, Pomona, CA, United States, §University of California - Davis, Olive Center, Davis, CA, United States

Chapter Outline

1 INTRODUCTION

1.1 Olive Oil Production Overview

Olive oil is obtained from the fruit of olive tree (*Olea europaea* L.), which is a small evergreen tree that averages 3–5 m in height. Olive trees thrive in moderate climates such as the warm, dry summers and mild, wet winters of the Mediterranean region. Thus, it is not surprising that Mediterranean countries (especially Spain, Italy, Greece, and Turkey) have historically dominated

Integrated Processing Technologies for Food and Agricultural By-Products
https://doi.org/10.1016/B978-0-12-814138-0.00014-9

the olive oil market (Azbar et al., 2004). However, the share of olive oil production from non-Mediterranean countries increased from 12% in 2004 to 16% in 2014, as overall production increased 5% over the same period (FAOSTAT, 2017).

1.2 Standards of Identity for Grades of Olive Oil

The Codex Alimentarius Commission and International Olive Council (2016) have set a trade standard defining different grades of olive oil and olive pomace oil and have specified methodologies to distinguish quality and purity criteria of each grade. The United States Department of Agriculture (USDA, 2010) updated its voluntary standards for grades of olive oil and olive pomace oil in 2010; the standards include both sensory evaluation and chemical analysis.

The term "virgin" universally means the oil was processed by the use of mechanical means only, with no chemical or thermal treatment (i.e., processing temperature is always <27°C). Olive pomace oil refers to oil that has been obtained from olive pomace by using solvents or other physical treatments. In the USDA standards, different grades of olive oil include extra virgin olive oil, virgin olive oil, lampante virgin olive oil, refined olive oil, and olive oil (a blend of refined olive oil and virgin olive oil); different grades of olive-pomace oil include refined olive-pomace oil, crude olive-pomace oil, and olive-pomace oil (a blend of refined olive-pomace oil and virgin olive oil). Chemical, physical, and sensory characteristics are used to determine these grades. Since extra virgin olive oil is the dominant grade, it will be the focus for the remainder of this chapter.

1.3 Processing Technologies

The modern olive oil extraction process includes (1) removing materials other than olives (such as leaves and branches) and cleaning of the olive fruit; (2) crushing the olive fruit to break the fruit's tissues and release the oil; (3) malaxing the resulting paste to allow oil droplets to coalesce; and (4) separating the oil from the rest of the paste. Within olives fruit's mesocarp cells, triglycerides are located in cytoplasmic oil bodies (Rangel et al., 1997). During crushing, the fruit's tissues are broken to allow the oils to be released from the vacuoles and to trigger enzymatic reactions affecting the volatile profile and the phenolic compounds content in the resulting oil (Inarejos-García et al., 2011). This crushing step can be achieved by stone mills, wiper blade, or hammer mills. The malaxation step allows small oil droplets to coalesce and form a continuous oil phase which can be separated by centrifugation. The most common malaxer is a horizontal channel with spiral mixing blades. Many mills now have malaxers with an atmosphere control using an inert gas such as nitrogen to reduce oxidation.

2 DESCRIPTIONS OF COMMODITY AND MAJOR BY-PRODUCTS

2.1 Commodity: Olive Oil

Olive oil is mainly a mixture of glycerides, which are esters of glycerol with fatty acids. Almost all fatty acids exist as a complex in the form of triacylglycerol, an ester derived from glycerol and 3 fatty acids. The major fatty acids present in olive oil are the monounsaturated oleic acid (55.0%–83.0%), the saturated palmitic acid (7.5%–20.0%), and the polyunsaturated linoleic acid (2.5%–20.0%) (International Olive Oil Council, 2017). In addition, the minor constituents of olive oil—hydrocarbons (e.g., squalene, carotenoids β-carotene, and lutein), phytosterols, and phenolic compounds (e.g., oleocanthal and hydroxytyrosol)—contribute to the specific properties of virgin olive oil, including its oxidative stability, unique flavor, and color. The content of the minor components of an olive oil product depends greatly on the variety of the cultivar, environmental and agronomical factors, and the processing system employed to produce the oil (Cárdeno et al., 2013).

2.2 By-Products

During olive oil processing, a huge quantity of by-products is produced. It is estimated that for every 100 kg of treated olives 35 kg of solid waste and from 55 to 200 L liquid waste are generated depending on the oil extraction process (Niaounakis and Halvadakis, 2006).

Nowadays, two processes are typically used for the extraction of olive oil. The traditional one involves a 3-phase system, where the by-products are a solid waste (cake or pomace) or *orujo* and aqueous liquid known as olive mill wastewater, olive mill process water (OMPW), vegetation water, or *alpechin*. In this 3-phase process, water is added during the malaxation step to aid the separation of olive oil. The newer and modern process is a 2-phase system in which little or no water is added. In the 2-phase system, a semisolid waste known as *alperujo* or pomace is produced, and it includes a combination of liquid and solid waste (Frankel et al., 2013; Niaounakis and Halvadakis, 2006; Obied et al., 2005). Also, in the 2-phase system, a minor aqueous by-product stream is produced after a second centrifugation step in the olive oil processing (2-phase OMPW). A schematic diagram of the two production processes is shown in Fig. 1.

The pomaces obtained from the 3- and 2-phase systems are very different in composition, especially in moisture, residual oil, and phenolic content. For the 3-phase pomace those parameters are in the following range: moisture 45%–55%, oil 3.5%–4.5%, and phenolic compounds 200–300 mg/100 g, whereas for the 2-phase pomace the ranges are: moisture 65%–75%, oil 3%–4%, and phenolic compounds 400–600 mg/100 g (Pattara et al., 2010). The 3-phase pomace has the consistency of damp soil, whereas 2-phase pomace is a thick sludge, with a doughy consistency that makes transport, storage, and handling difficult.

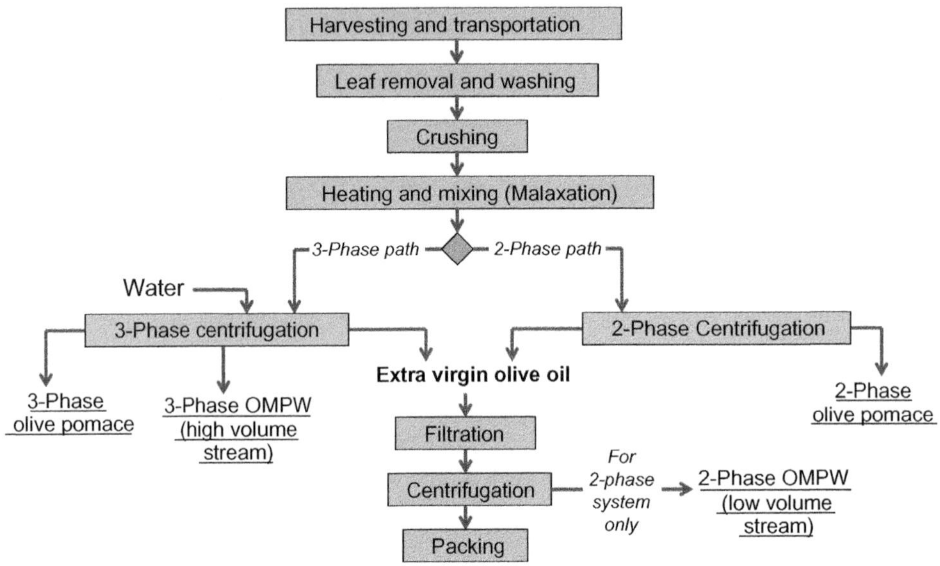

FIG. 1

Schematic diagram of the extra virgin olive oil production process. OMPW = olive mill process water. The primary product stream is in bold; by-product streams are underlined.

The OMPW from the 3-phase process (3-phase OMPW) is a liquid of violet to dark brown color, with a strong smell of fermented olives. The 3-phase OMPW is composed of vegetation water from the fruit and the water used in different stages of oil extraction, which contain olive pulp, mucilage, pectin, oil, and other suspended components in a relatively stable emulsion (Frankel et al., 2013; Paredes et al., 1999). The main 3-phase OMPW components by weight are water (83%–94%), organic compounds (4%–18%), and inorganic compounds (0.4%–2.5%) (Rahmanian et al., 2014). The physicochemical characteristics of 3-phase OMPW depend on the composition of the vegetation water (olive variety, maturity, water content, cultivation soil, harvesting time, and climate) (Niaounakis and Halvadakis, 2006), olive oil extraction process, and fruit storage time. Typical parameters for characterization of 3-phase OMPW are chemical oxygen demand (COD), biological oxygen demand (BOD), pH, conductivity, total solids, lipids, and phenolic content. The wide range of values for 3-phase OMPW parameters are reported in numerous publications and collectively presented in Table 1.

These parameters indicate a high degree of organic pollution of 3-phase OMPW if released directly to the environment without any treatment. Due to its high organic load and phytotoxic/antibacterial phenolic substances, which are resistant to biological degradation, OMPW is considered as one of the major polluting effluents produced by agrofood industries (Cassano et al., 2013; Souilem et al., 2017). OMPW is considered a "point" source of pollution, in that the generator of the pollution is readily identifiable. The principal pollutants from

Table 1 Physicochemical Characteristics of 3-Phase OMPW

Parameter	Unit	Range of Values	Reference
pH	–	2.24–5.9	Dermeche et al. (2013)
Conductivity	mS/cm	5–81	Souilem et al. (2017)
COD	g/L	16.5–190	Zbakh and El Abbassi (2012)
BOD	g/L	13.4–37.5	Souilem et al. (2017)
Dry residue	g/L	11.5–102.5	Zbakh and El Abbassi (2012)
Organic matter	g/L	16.7–81.6	Zbakh and El Abbassi (2012)
Lipids	g/L	1.64–9.8	Zbakh and El Abbassi (2012)
Phenolic compounds	g/L	0.5–24	Niaounakis and Halvadakis (2006)

OMPW are suspended solids, oil and grease, and other dissolved organics, as well as salts. The organics, in sufficient concentrations, can result in oxygen deficiencies, and salts can make groundwater and surface water unusable for drinking and other uses (Colic et al., n.d.).

In the United States, point sources of pollution are subject to water discharge regulations and are based on a combination of Federal and state oversight. Federal authority for regulation of all states, including California (where the majority of olive oil production takes place), derives principally from the Clean Water Act and Safe Drinking Water Act, among others. The goals of Federal regulation are protection of drinking water quality and availability, and maintenance of water quality levels in state surface water bodies and groundwater. Broadly, the Federal government (led by the Environmental Protection Agency or EPA) sets the national environmental goals to be achieved, and states are charged with adopting policies to see that they are achieved. The Clean Water Act made it unlawful to discharge any pollutant from point sources of pollution into navigable waters unless a permit for doing so was obtained. The EPA's National Pollutant Discharge Elimination System permit program controls allowed discharges from point sources, and industrial facilities such as olive mills must obtain permits if their discharges go directly to surface waters. The EPA helps regulated entities meet federal requirements, and holds those entities legally accountable for environmental violations, through a compliance system with periodic on-site inspections as well as off-site monitoring and analysis of compliance data submitted by regulated entities (EPA, 2017).

From the perspectives of both human health and environmental impact, the most important components of all four types of olive oil by-products are phenolic compounds, most of which are hydrophilic in nature, and thus remain in the water when portioned during olive oil processing (Sedej et al., 2016). The most abundant phenolic compounds in OMPW are hydroxytyrosol as the major component (66.5%), followed by tyrosol, oleuropein, caffeic acid, vanillic acid, verbascoside, elenolic acid, *p*-coumaric acid, catechol, and rutin (Obied et al., 2005). The major phenolic compounds identified in 2-phase pomace are tyrosol, hydroxytyrosol, *p*-coumaric acid, and caffeic acid (Niaounakis and Halvadakis, 2006), while in 3-phase pomace protocatechuic acid, hydroxybenzoic acid, sinapic acid, *p*-coumaric acid, rutin, and hesperidin are the predominant free phenolic compounds (Alu'datt et al., 2010).

3 VALUE-ADDED PROCESSING OF BY-PRODUCTS

3.1 Food/Nutritional By-Products

The phenolic compounds described above are the olive oil by-product most suitable for human food and nutrition, and these compounds possess a wide range of biological activities, including antioxidant, antiinflammatory, antibacterial, and antiviral functions (Zbakh and El Abbassi, 2012).

A recent trend in 2-phase and 3-phase OMPW management is to achieve both purification of the wastewater and concentration and extraction of phenolic compounds. For that purpose, membrane filtration technology (Paraskeva et al., 2007; Sedej et al., 2016) and liquid-liquid extraction followed by solid phase extraction (De Marco et al., 2007) have been explored. Using the latter method, from one liter of OMPW it was possible to obtain an extract containing 1 g of purified hydroxytyrosol.

Commercial hydroxytyrosol production from OMPW is described in several patented systems. Crea (2002) developed a protocol that includes acid treatment, an incubation process that converts oleuropein to hydroxytyrosol, a supercritical fluid extraction process, and freeze-drying. Another patented system to purify hydroxytyrosol (99.5%) from OMPW is using chromatographic columns filled with ion-exchange resin and absorbent nonionic resin (Fernández-Bolaños et al., 2002). Recovery of a mixture of phenolic compounds was also patented, where OMPW is defatted and concentrated prior to the extraction of phenolic compounds using ethanol in combination with an organic acid, followed by precipitation of dietary fibers (Tornberg and Galanakis, 2008). The purified hydroxytyrosol and the mixture of phenolic compounds have the potential for commercialized applications in food supplements and cosmetics, as well as food preservatives and functional ingredients. Fortification of milk beverage with phenolic compounds extracted from OMPW (Servili et al., 2011) and their potential use in the preparation of functional beverages (Zbakh and El Abbassi, 2012) are some of the possibilities for the application of OMPW phenolic compounds in food products.

In the case of the solid or semisolid by-product pomaces, findings show that phenolic compounds and other bioactive microconstituents left in these by-products after virgin olive oil extraction may be potent in the prevention of cardiovascular diseases (Boskou, 2008; Karantonis et al., 2008). Phytosterols, tocopherols, aliphatic alcohols, squalene, and triterpenic acid are some of the bioactive polar lipids extractable from olive pomace and identified with health benefits (Fernández-Bolaños et al., 2006; Visioli and Galli, 1998a). In addition, the post-extraction olive pomace is also rich in phenols and flavonoids, which present potent antioxidant and radical scavenging effects, especially with health benefits in prevention and repair of cell damages (Cardoso et al., 2005; Visioli and Galli, 1998b). Other useful components that can be extracted from the olive pomace include dietary fibers and proteins. Overall, this makes olive oil by-products a promising source of various nutraceuticals for the development of functional foods.

3.2 Nonfood By-Products

Due to their high organic matter content and the presence of phytotoxic components, both OMPW and pomace pose a major disposal burden for the industry and potentially a severe environmental hazard (Toscano and Montemurro, 2012). Therefore, traditionally these by-product streams have to be first detoxified before utilization as either animal feed (pomace) or fertilizer/irrigation water (OMPW).

As documented in the literature, there are several ways to process olive oil by-product materials to collect valuable products, such as (1) the production of fertilizers through bioremediation and/or composting, (2) the production of animal feeds by first removing antidigestive factors such as potassium, phenols, and phytates while retaining the rich contents of oil, protein, and carbohydrates, and (3) the production of biofuels or biogases by first detoxifying the feedstock through aerobic and anaerobic microbial digestions and/or other physicochemical means, such as neutralization and precipitation (Laufenberg et al., 2003). More recently, olive pomace has been explored for the production of biopolymers, such as soluble dietary fibers (oligosaccharides or prebiotics), exopolysaccharides, and polyhydroxyalkanoates (PHAs), which present great rheological properties including good solubility in water, capability of increasing viscosity of a solution even at small concentrations, pseudoplasticity, and good stabilizing effects (Coimbra et al., 2010; Fernández-Bolaños et al., 2006; Toscano and Montemurro, 2012).

4 CASE STUDIES

4.1 Case Study #1: Valorization of OMPW via Filtration

This case study examines the prospective costs and benefits of filtration systems for OMPW in California, drawing on a variety of information both from the published literature and from industry sources. First, the major operating and equipment capital costs associated with a medium-to-larger industry-scale OMPW ultrafiltration (UF) and reverse osmosis (RO) filtration system are assessed, with and without drying the retentate from RO. Then the major potential economic benefits of OMPW filtration and reuse are discussed and evaluated in the following sections.

4.1.1 OVERVIEW OF THE OMPW FILTRATION PROCESS

In this study, OMPW is first pressed through a UF system, which results in two products—a "retentate" (UF-R) that does not pass through the UF and a "permeate" that does pass through. The UF permeate is then pressed through a RO filter, which in turn produces a retentate (RO-R) and an RO permeate, which is near-pure water. Thus, UF and RO sequentially applied to the OMPW stream result in three products: UF-R, RO-R, and near-pure water. Each of these may have economic value.

In this scenario, there are three decisions that millers considering OMPW reuse must consider. The primary decision is whether to install a filtration system, which involves a significant outlay for equipment and some ongoing operation and maintenance cost but reduces the disposal cost associated with OMPW and produces pure water, which may also have a value to the miller. The second decision is whether to market wet retentates, which incurs some additional marketing cost. (This cost cannot be assessed realistically at present since there are no markets for wet retentates in the United States and little usable information on this from Europe.) The third decision is whether to install drying equipment, which incurs an additional capital outlay that can be assessed and marketing costs that at present cannot be estimated.

4.1.2 INSTALLING OMPW FILTRATION

The trade-off in the decision to install OMPW filtration is the capital and operating costs of doing so, compared with the savings from not having to haul OMPW away from the plant and the value of the pure water produced (the latter is a very minor benefit). The present analysis presumes the use of a 24,000 gal/day (equivalently, 90,850 L/day or 45.4 mt/day) OMPW filtration system, which appears to represent the needs of a medium to large olive milling operation. The useful life of this system is 20 years, and the milling season is assumed to last 3 months per year. Since amortized daily cost of filtration depends on the interest rate chosen, we consider rates of 2%, 4%, and 6%. Results are shown in Table 2. Sources for the information in this table were an estimate from the filtration company New Logic Research, Inc. (Emeryville, CA, United States), communications with California olive mill managers in 2015, and default values from the process simulation/economics software SuperPro Designer (v.8.5, Intelligen, Inc., Scotch Plains, NJ, United States), as indicated by the superscripts in the table. Thus, Table 2 presents the costs and benefits of installing filtration, on a per-operating-day and on a per-gallon-OMPW basis.

As the net benefits are positive, Table 2 suggests that installing an OMPW filtration system could pay for itself. This would depend on the course of future disposal fees as well as any differences between the engineering specifications and actual performance of the filtration system.

4.1.3 WET FILTRATION RETENTATES

The remaining product from the OMPW stream, aside from water produced, is wet retentates. There are no markets for these in California at present, but evidence from Europe suggests that it can be profitable to extract phenolics from wet retentates (Arvaniti et al., 2012). Larson (2015) explores at some length what the implications for California millers could be, drawing on mass balance and constituent analysis of the OMPW waste streams of three California millers (Sedej et al., 2016), along with the limited market information from Europe. The European information (Arvaniti et al., 2012) suggests that wet retentates can be sold in a wholesale market, and those from UF sell from the equivalent of about $110 per ton, while retentates from RO sell for about $275 per ton. Adjustments must be made for differences in the phenolics compositions of European vs California OMPW, as well as exchange rates. The prospective revenues from selling wet retentates are given in Table 3. Marketing and distribution costs cannot be assessed at present.

The daily production of phenols is derived from the data on phenolics compositions of the millers' OMPW, as determined by lab tests (Sedej et al., 2016), and scaled up to industry production levels. The differences in phenolics-equivalent prices are due to differences in phenolics concentrations (higher for 3-phase milling) and total volumes of retentates recovered from UF and RO. With a change in units for commensurability, the phenolics-adjusted price based on European market information is applied to obtain daily revenues for California millers. The prospective revenues in Table 3 are additive since wet retentate is retained from both UF and RO. The prospective revenues for marketing wet retentate are, therefore, $1438.86 per day for 2-phase mills and $3396.42 per day for 3-phase mills. While these are substantial revenues and seem promising, what would remain as profits after arrangements for delivery to buyers, once they emerge, is not known at this time.

Table 2 Costs and Benefits of Installing an OMPW Filtration System

	Interest rate					
	0.02		0.04		0.06	
	Per day	Per gal OMPW	Per day	Per gal OMPW	Per day	Per gal OMPW
Costs						
Fixed cost[a]	$354.92	$0.0148	$425	$0.0177	$502.37	$0.0209
Operating cost[a]	188.58	0.0079	188.58	0.0079	188.58	0.0079
Total cost	543.5	0.0227	613.58	0.0256	690.95	0.0288
Benefits						
Hauling cost avoided[b]	753.6	0.0314[c]	753.6	0.0314[c]	753.6	0.0314[c]
Pure water[d]	7.64	0.0003	7.64	0.0003	7.64	0.0003
Total Benefits	761.24	0.0317	761.24	0.0317	761.24	0.0317
Net Benefits	**$217.73**	**$0.009**	**$147.66**	**$0.006**	**$70.29**	**$0.003**

[a]New Logic Research, Inc. estimate.
[b]Communications with California olive mill managers.
[c]The range of cost savings is from $0.0299 for a 2-phase mill to $0.0328 for a 3-phase mill. The average of these values is used for calculations.
[d]Default value from SuperPro Designer software.

Table 3 Prospective Revenues from Wet Retentates

	2-Phase	2-Phase	3-Phase	3-Phase
Item	wet UF-R	wet RO-R	wet UF-R	wet RO-R
Daily production (g/day)	16,669,872	19,771,297	22.929,275	8.261,416
Daily production (tons/day)	18.38	21.79	25.28	9.11
Phenolics-equivalent price ($/ton)	$9.63	$57.91	$50.78	$231.91
Prospective revenue ($/day)	**$177.00**	**$1261.86**	**$1283.72**	**$2112.70**

4.1.4 DRIED FILTRATION RETENTATES

An alternative to marketing wet retentate is to dry it. Here, there is some information about the costs, since a dryer must be purchased and the operating costs have been assessed by New Logic. As with wet retentates, no markets currently exist, so the cost of getting the product to buyers, once they emerge, cannot be assessed at present. The retail values of the individual constituents of the dried retentate powder are high, especially for 3-phase powder (Sigma-Aldrich, 2015), but the costs of refining the powder are unknown, though likely high. There is, at least, a retail market for powders with one of the constituents of dried 3-phase retentate, 3-hydroxytyrosol (3-HT) (Sigma-Aldrich, 2015) [This constituent was not present in the constituent analyses performed on 2-phase retentates (Sedej et al., 2016)]. Comparing the retail value of a product with wholesale costs of production overstates the potential profit to millers since costs associated with the retail marketing and distribution would also have to be paid. Nevertheless, comparing the production costs to prospective retail revenues from the constituents of dried retentate powder provides a fairly clear picture.

Based on default values in the SuperPro Designer software, the capital cost of purchasing a retentate dryer is $160,000, and the additional daily operating cost is $1920. Using the same approach as in evaluating the daily equivalent cost of filtration equipment (20-year life, 3-month operating season, and interest rates of 2%–6%), the daily operating cost for drying retentate can be expected to be approximately $2027–$2073.

As for prospective retail values of dried retentate powder containing 3-HT, a reasonable way of estimating the possible daily production of retentate powders with these 3-HT concentrations is to determine the masses of each concentration that the daily production of pure 3-HT would support. Contacts with the pharmaceuticals industry suggest that wholesale prices for lower 3-HT concentrations would be on the order of $30/kg for 1.5% 3-HT and $465/kg for 15% 3-HT. Multiplying each of these masses by the wholesale price per unit mass generates an estimate of the prospective revenues from the dried retentates. Applying this procedure, based on the 3-phase production of 3-HT, revenues of $250–$387 per day can be expected. Comparing these prospective revenues with costs of drying retentate in the range of $2027–$2073/day, it is clear that drying retentate is not economically viable in the near term. Given the high retail values of constituents, the situation may eventually change in a more favorable direction.

4.1.5 CASE STUDY #1 CONCLUSION

The results of this analysis suggest a guarded but generally optimistic view of the eventual economic viability of OMPW filtration, though the outlook for producing dried retentate is less promising. The positives include the fact that production of wet retentates appears to be cost-effective relative to a major current disposal option for OMPW—trucking them to third-party sites. The drawbacks are that, currently, the uncertainty about the economic outlook is presently large, and there are no markets for OMPW retentate (wet or dry) in California. Thus, opportunities for profiting off the OMPW stream must await the development of these markets.

4.2 Case study #2: Valorization of Olive Pomace via Extrusion

As indicated in the literature, thermal treatments including cooking extrusion may break down the macromolecules and/or cell structures in oilseed extraction residues so that the further extraction of microconstituents could be made easier or with higher yields (Jing and Chi, 2013; Jung et al., 2009; Jung and Mahfuz, 2009; Lama-Muñoz et al., 2012). However, there is no information in the literature suggesting if the same extrusion-assisted extraction would be feasible for olive pomace. Similarly, there are some reports showing that the use of food processing waste materials such as orange pomace or carrot pomace could generate novel, functional foods when incorporated into pasta or bakery goods (Kumar et al., 2010; O'Shea et al., 2012); however, there is no study to date investigating the potential use of olive pomace in extruded food products, such as in pasta (durum wheat-based) or noodles (rice starch-based). Therefore, the objective of the present study was to explore the technical feasibility of extruding a grain flour matrix that carries a relatively high content (5%–20%) of olive pomace, followed by the examination of technological, physicochemical, and organoleptic properties of the extrudates for potential use as functional food products.

Specifically, 2-phase olive pomace, fresh-pressed, depitted, with a sludge consistency was obtained from Bondolio Olive Ranch (Winters, CA, United States), and transported in sealed containers under refrigeration to the Cal Poly Pomona Food Processing Laboratory on the same day of oil pressing. The fresh pomace was then dried using a pilot-scale forced air circulation dehydrator (Excalibur; Sacramento, CA, United States) at 165°F (74°C) overnight (approximately 24 h). The dehydrated olive pomace was then ground and sieved to obtain a fine powder with a particle size <300 μm, which was then incorporated into gluten-based or gluten-free pasta formulations using selected cereal grain flours. With certain amounts of water added, the mixture was transformed into a homogeneous dough mass suitable for extrusion.

A pilot-scale pasta extruder (La Monferrina P12, Asti, Italy) was employed to make pasta samples at typical 1-kg batch scale; various die plates were used to make different shapes, for example, shell, elbow, angel hair, fusilli, and tagliatelle. Once a steady extrusion rate was achieved with desirable pasta consistency and appearance, the samples were collected and the time was recorded. The time required to complete the extrusion run associated with the sample collected (from a typical 1-kg batch) was used to calculate the extrusion flow rate as extrusion rate (g/min) = weight of the collected samples (g)/the time required (min).

The pasta samples were then dried in the dehydrator, and the dried products were collected in zip-lock bags and stored in a dark and cool place in the lab under ambient conditions (~20°C and ~60% relative humidity) for further analyses. The physical or organoleptic characteristics, cooking quality, and sensory properties were measured, including color, appearance, residual moisture content, water activity, optimal cooking time (ready for consumption), water holding (uptake) capacity, cooking loss, and texture properties (hardness vs adhesiveness) when compared to commercial pasta products. Specifically, optimal cooking time (T) was determined following AACC 2000, method 66-50. Water uptake (holding) index was evaluated on a 50 g dry pasta sample by measuring the weight of the pasta before and after cooking (using standardized optimal cooking conditions), and then calculated as water uptake = {[(weight of cooked pasta) − (weight of dry pasta)/(weight of dry pasta)] × 100%. Cooking loss was measured based on gravimetric difference of dry matters in pasta samples before and after cooking. Specifically, the dry matter content of a pasta sample was first determined by the gravimetric method. Then the sample was cooked at T + 1 min (based on standardized optimal cooking time} and its dry matter content was determined again. The cooking loss is calculated by cooking loss = {[(dry matter of cooked sample)/(dry matter of dry sample)] − 1} × 100%. Texture characteristics of cooked pasta were determined using a Stable Micro Systems TA-XT Plus texture analyzer (Godalming, UK) equipped with the Texture Expert Software package. Hardness and adhesiveness were determined by performing a compression test using a 35 mm diameter cylindrical probe. Four replicates were measured for each pasta sample and average values were calculated. Results for the formulations with 5% (w/w) refined olive pomace are summarized in Table 4.

As indicated in Table 4, it was feasible to extrude pomace-bearing dough mixtures when using durum semolina and flours of brown rice, white rice, and whole wheat as base materials. Besides four base materials, four additional levels of olive pomace powder: 5%, 10%, 15%, and 20% (w/w, dry basis) were tested, among which 5% addition resulted in pasta samples with acceptable color, surface/texture characteristics, and cooking quality. Addition of 10% olive pomace powder did not yield desirable results; however, there is room for improvement. On contrary, 15% and 20% addition of olive pomace powder made the extrusion process difficult to proceed due to oily (slippery) consistency of the dough mass, also resulting in dark color and crumbly texture in the finished products. The pasta samples made with 5% olive pomace powder presented similar shelf stability (low a_w) and cooking performance to those of commercial pasta products, while the texture properties were slightly compromised with lower values for both hardness and adhesiveness, matching the results from the sensory test, that is, slightly softer and crumblier, with more details discussed in the following section.

Table 4 Extrusion Feasibility Study Results for the Fusilli Shape When Four Cereal Grain Flours Were Used as the Base Materials, While the Refined Olive Pomace (<300 μm) was Incorporated at 5% Level (w/w)

	Commercial pasta	Durum semolina	Whole wheat flour	Brown rice flour	White rice flour
Extrusion rate	N/A	199.5 g/min	256.6 g/min	262.3 g/min	270.0 g/min
Water activity	0.343	0.441	0.433	0.497	0.454
Moisture content	9.20%	16.50%	15.60%	10.60%	14.80%
Optimal cooking time	10–11 min	9.5–10 min	9–9.5 min	7.5–8 min	6.5–7 min
Water uptake	136%	165%	150%	159%	146%
Cooking loss	4.80%	3.60%	5.50%	8.80%	6.80%
Hardness (g force)	694 ± 18	658 ± 32	630 ± 45	589 ± 21	557 ± 32
Adhesiveness (g s)	393 ± 11	388 ± 10	402 ± 47	343 ± 12	329 ± 10

A preliminary sensory evaluation was conducted by 12 semitrained panelists (food science students from Cal Poly Pomona). The panel was asked to rank four pasta formulations made with the four base ingredients while containing olive pomace powder at the same 5% level. A 9-point-scale hedonic test was designed to assess the panel's preference over seven sensory characteristics—appearance, color, flavor, stickiness, mouthfeel, rubbery, and texture. The overall assessment was conducted to determine which of the four formulations was most preferred by the sensory panel based on the area (the greater, the better) covered in the spider's web diagram, shown in Fig. 2.

Clearly indicated by Fig. 2, the optimized whole wheat formulation containing 5% olive pomace powder made by a pilot-scale pasta extruder into a fusilli shape was most preferred by a small group of sensory panelists and is thus recommended for commercialization.

5 SUMMARY OF CURRENT OPPORTUNITIES AND FUTURE DEVELOPMENT

Traditional methods of utilizing olive oil by-products—land application or animal feed—may prove sufficient for small mills. However, as production for olive oil increases and environmental regulations become more restrictive, millers can consider the novel by-product valorization techniques which are being actively explored by researchers around the world. This chapter described the chemical properties of olive oil and its various by-products as well as how the particular oil extraction approach (2-phase vs 3-phase) affects the properties of the by-products. The two case studies explored the economic viability of valorization of OMPW via filtration and the technical feasibility and commercial potential of

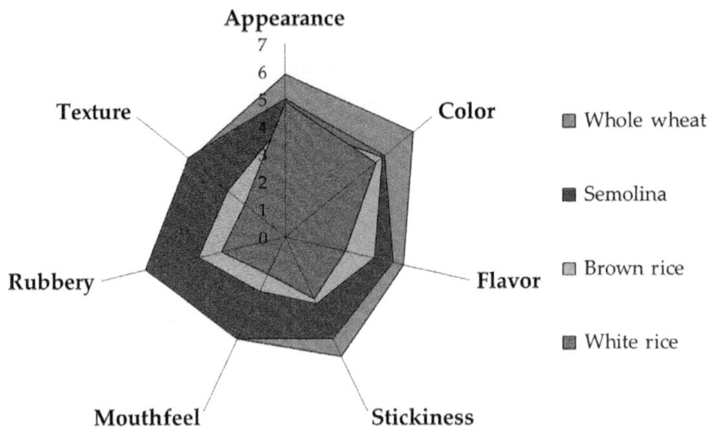

FIG. 2
Preliminary sensory evaluation on olive pomace-bearing pasta formulations using a 9-point-scale hedonic preference test against 7 characteristics from 12 panelists.

olive pomace processed by extrusion. Development of standards and markets for the valorized by-products would accelerate the full utilization of the olive fruit as a specialty cash crop in California, eventually leading to worldwide applicability.

Clearly, the treatment and management of olive oil industrial wastes is a problem with great complexity due to the chemical nature of the wastes and some economical, technological, and organizational constraints involved in the olive oil production sector. Sophisticated technical solutions proposed so far in the literature, although tested and showing some success at the laboratory scale, lack of economic viability, especially considering the majority of small-sized and geographically scattered olive mills cannot afford to adopt these methods. Therefore, as suggested by several researchers (Arvanitoyannis, 2010; Galanakis, 2012; Schieber et al., 2001), the future direction for olive oil waste management should be toward an integrated approach of simultaneous detoxification (reducing environmental impacts) and utilization for production of value-added products, subsequently converting a problem to an economic gain.

Acknowledgments

The project described in Case Study #1 was supported by the Specialty Crop Block Grant Program at the US Department of Agriculture (USDA) through Grant 12-SCBGP-CA-0040. The project described in Case Study #2 was sponsored by the following grants: California State University Agriculture Research Institute Grant Program (#14-04-212) and Southern California Institute of Food Technologists Section Education & Research Grant (2014).

Disclaimers

Mention of trade names or commercial products in this work is solely for the purpose of providing specific information and does not imply recommendation or endorsement by the US Department of Agriculture. USDA is an equal opportunity employer.

References

Alu'datt, M.H., Alli, I., Ereifej, K., Alhamad, M., Al-Tawaha, A.R., Rababah, T., 2010. Optimisation, characterisation and quantification of phenolic compounds in olive cake. Food Chem. 123, 117–122.

Arvaniti, E.C., Zaglis, D.P., Papadakis, V.G., Paraskeva, C.A., 2012. High-added value materials production from OMW: a technical and economical optimization. Int. J. Chem. Eng. https://doi.org/10.1155/2012/607219. Article ID 607219.

Arvanitoyannis, I.S., 2010. Waste Management for the Food Industries. Academic Press, Cambridge, MA.

Azbar, N., Bayram, A., Filibeli, A., Muezzinoglu, A., Sengul, F., Ozer, A., 2004. A review of waste management options in olive oil production. Crit. Rev. Environ. Sci. Technol. 34, 209–247.

Boskou, D., 2008. Olive oil, Minor Constituents and Health. CRC Press, Boca Raton (FL).

Cárdeno, A., Sánchez-Hidalgo, M., Alarcón-de-la-Lastra, C., 2013. An up-date of olive oil phenols in inflammation and cancer: molecular mechanisms and clinical implications. Curr. Med. Chem. 20, 4758–4776.

Cardoso, S.M., Guyot, S., Marnet, N., Lopes-da-Silva, J.A., Renard, C.M., Coimbra, M.A., 2005. Characterization of phenolic extracts from olive pulp and olive pomace by electrospray mass spectrometry. J. Sci. Food Agric. 85, 21–32.

Cassano, A., Conidi, C., Giorno, L., Drioli, E., 2013. Fractionation of olive mill wastewaters by membrane separation techniques. J. Hazard. Mater. 248-249, 185–193.

Coimbra, M.A., Cardoso, S.M., Lopes-da-Silva, J.A., 2010. Olive pomace, a source for valuable arabinan-rich pectic polysaccharides. Top. Curr. Chem. 294, 129–141.

Colic, M., Morse, W., & Hicks, J. Reuse of California brine olive oil processing wastewater to meet zero discharge goal. Report to California Olive Growers, n.d.

Crea, R., 2002. Method of Obtaining a Hydroxytyrosol-Rich Composition From Vegetation Water. World Intellectual Property Organization. WO/2002/0218310.

De Marco, E., Savarese, M., Paduano, A., Sacchi, R., 2007. Characterization and fractionation of phenolic compounds extracted from olive oil mill wastewaters. Food Chem. 104 (2), 858–867.

Dermeche, S., Nadour, M., Larroche, C., Moulti-Mati, F., Michaud, P., 2013. Olive mill wastes: biochemical characterizations and valorization strategies. Process Biochem. 48 (10), 1532–1552.

EPA, 2017. Drinking Water, Groundwater, Impaired Waters, Industrial Wastewater, Laws, Regulations, and Compliance sections. United States Environmental Protection Agency. Available from: http://www.epa.gov. [27 November 2017].

FAOSTAT, 2017. Crops Processed. Food and Agriculture Organization of the United Nations. Available from: http://www.fao.org/faostat/en/#data/QD. [24 November 2017].

Fernández-Bolaños, J., Heredia, A., Rodríguez, G., Rodríguez, R., Guillén, R., Jiménez, A., 2002. Method for Obtaining Purified Hydroxytyrosol From Products and By-Products Derived From the Olive Tree. World Intellectual Property Organization. WO/2002/064537.

Fernández-Bolaños, J., Rodríguez Gutiérrez, G., Rodríguez Arcos, R., Guillén Bejarano, R., Jiménez Araujo, A., 2006. Extraction of interesting organic compounds from olive oil waste. Grasas Aceites 57 (1), 95–106.

Frankel, E., Bakhouche, A., Lozano-Sanchez, J., Segura-Carretero, A., Fernandez-Gutierrez, A., 2013. Literature review on production process to obtain extra virgin olive oil enriched in bioactive compounds. Potential use of byproducts as alternative sources of polyphenols. J. Agric. Food Chem. 61 (22), 5179–5188.

Galanakis, C.M., 2012. Recovery of high added-value components from food wastes: conventional, emerging technologies and commercialized applications. Trends Food Sci. Technol. 26 (2), 68–87.

Inarejos-García, A.M., Fregapane, G., Salvador, M.D., 2011. Effect of crushing on olive paste and virgin olive oil minor components. Eur. Food Res. Technol. 232 (3), 441–451.

International Olive Council, 2016. International Trade Standard Applying to Olive Oils and Olive-Residue Oils. (COI/T.15/No.1) CODEX STAN 33-1981.

International Olive Oil Council, 2017. http://www.internationaloliveoil.org/estaticos/view/131-world-olive-oil-figures. (Accessed 12 November 2017).

Jing, Y., Chi, Y., 2013. Effects of twin-screw extrusion on soluble dietary fiber and physicochemical properties of soybean residue. Food Chem. 138 (2–3), 884–889.

Jung, S., Mahfuz, A., Maurer, D., 2009. Structure, protein interactions and in vitro protease accessibility of extruded and pressurized full-fat soybean flakes. J. Am. Oil Chem. Soc. 86, 475–483.

Jung, S., Mahfuz, A., 2009. Low temperature dry extrusion and high-pressure processing prior to enzyme-assisted aqueous extraction of full fat soybean flakes. Food Chem. 114, 947–954.

Karantonis, H.C., Tsantila, N., Stamatakis, G., Samiotaki, M., Panayotou, G., Antonopoulou, S., Demopoulos, C.A., 2008. Bioactive polar lipids in olive oil, pomace and waste byproducts. J. Food Biochem. 32, 443–459.

Kumar, N., Sarkar, B.C., Sharma, H.K., 2010. Development and characterization of extruded product of carrot pomace, rice flour and pulse powder. Afr. J. Food Sci. 4 (11), 703–717.

Lama-Muñoz, A., Rodríguez-Gutiérrez, G., Rubio-Senent, F., Fernández-Bolaños, J., 2012. Production, characterization and isolation of neutral and pectic oligosaccharides with low molecular weights from olive by-products thermally treated. Food Hydrocoll. 28 (1), 92–104.

Larson, D.M., 2015. Olive milling wastewater reuse: a look at potential costs and benefits of filtration and drying in California. In: Final Report UC Davis Olive Center. Nov. 2017.

Laufenberg, G., Kunz, B., Nystroem, M., 2003. Transformation of vegetable waste into value added products: (A) the upgrading concept; (B) practical implementations. Bioresour. Technol. 87 (2), 167–198.

Niaounakis, M., Halvadakis, C.P., 2006. Olive Processing Waste Management: Literature Review and Patent Survey, second ed. Elsevier, Oxford.

Obied, H.K., Allen, M.A., Bedgood, D.R., Prenzler, P.D., Robards, K., Stockmann, R., 2005. Bioactivity and analysis of biophenols recovered from olive mill waste. J. Agric. Food Chem. 53, 823–827.

O'Shea, N., Arendt, E.K., Gallagher, E., 2012. Dietary fiber and phytochemical characteristics of fruit and vegetable by-products and their recent applications as novel ingredients in food products. Innovative Food Sci. Emerg. Technol. 16, 1–10.

Paraskeva, C.A., Papadakis, V.G., Tsarouchi, E., Kanellopoulou, D.G., Koutsoukos, P.G., 2007. Membrane processing for olive mill wastewater fractionation. Desalination 213 (1-3), 218–229.

Paredes, C., Cegarra, J., Roig, A., Sánchez-Monedero, M.A., Bernal, M.P., 1999. Characterization of olive mill wastewater (alpechin) and its sludge for agricultural purposes. Bioresour. Technol. 67, 111–115.

Pattara, C., Cappelletti, G.M., Cichelli, A., 2010. Recovery and use of olive stones: commodity, environmental and economic assessment. Renew. Sust. Energ. Rev. 14, 1484–1489.

Rahmanian, N., Jafari, S.M., Galanakis, C.M., 2014. Recovery and removal of phenolic compounds from olive mill wastewater. J. Am. Oil Chem. Soc. 91 (1), 1–18.

Rangel, B., Platt, K.A., Thomson, W.W., 1997. Ultrastructural aspects of the cytoplasmic origin and accumulation of oil in olive fruit (Olea europaea). Physiol. Plant. 101 (1), 109–114.

Schieber, A., Stintzing, F.C., Carle, R., 2001. By-products of plant food processing as a source of functional compounds—recent developments. Trends Food Sci. Technol. 12 (11), 401–413.

Sedej, I., Milczarek, R., Wang, S.C., Sheng, R., Avena-Bustillos, R.J., Dao, L., Takeoka, G., 2016. Membrane-filtered olive mill wastewater: Quality assessment of the dried phenolic-rich fraction. J. Food Sci. 81 (4), E889–E896.

Servili, M., Rizzello, C.G., Taticchi, A., Esposto, S., Urbani, S., Mazzacane, F., Di Maio, I., Selvaggini, R., Gobbetti, M., Di Cagno, R., 2011. Functional milk beverage fortified with phenolic compounds extracted from olive vegetation water, and fermented with functional lactic acid bacteria. Int. J. Food Microbiol. 147 (1), 45–52.

Sigma-Aldrich, 2015. Web Searches for Retail Prices of Phenols Listed in Tables 8 and 9. July 2015 (https://www.sigmaaldrich.com/catalog).

Souilem, S., El-Abbassi, A., Kiai, H., Hafidi, A., Sayadi, S., Galanakis, C.M., 2017. Olive oil production sector: environmental effects and sustainability challenges. In: Galanakis, C.M. (Ed.), Olive Mill Waste: Recent Advances for Sustainable Management. Academic Press, Elsevier, pp. 1–28.

Tornberg, E., Galanakis, C.M., 2008. Olive Waste Recovery, World Intellectual Property Organization, WO/2008/082343.

Toscano, P., Montemurro, F., 2012. Olive Mill By-Products Management. Available at: http://cdn.intechopen.com/pdfs/41349/InTech-Olive_mill_by_products_management.pdf. (Accessed 26 February 2013).

USDA, 2010. United States Standards for Grades of Olive Oil and Olive Pomace Oil. Federal Register, United States Department of Agriculture, Washington, DC.

Visioli, F., Galli, C., 1998a. The effect of minor constituents of olive oil on cardiovascular disease: new findings. Nutr. Rev. 56, 142–147.

Visioli, F., Galli, C., 1998b. Olive oil phenols and their potential effects on human health. J. Agric. Food Chem. 46, 4292–4296.

Zbakh, H., El Abbassi, A., 2012. Potential use of olive mill wastewater in the preparation of functional beverages: a review. J. Funct. Foods 4 (1), 53–65.

CHAPTER 15

Almonds

Guangwei Huang, Karen Lapsley
Almond Board of California, Modesto, CA, United States

Chapter Outline

1 ALMOND PRODUCTION AND COPRODUCT PROCESSES IN CALIFORNIA

Almonds comprise the largest tree nut crop in the world, with a global production of over 1,240,000 metric tons of kernels in 2017, accounting for 31.7% of the tree nut kernel production according to the statistics by the International Nut and Dried Fruit Council (INC, 2017/18 Statistics Yearbook). Out of the 2017 production volume, 82% of almond kernels were produced by growers from California's Central Valley, where the climate of hot, dry summers and cool,

Integrated Processing Technologies for Food and Agricultural By-Products
https://doi.org/10.1016/B978-0-12-814138-0.00015-0

rainy winters provide the optimum conditions for almond production. According to the 2012 USDA Agricultural Census, there are around 6800 almond farms in California. In 2017, the California almond industry produced 1,025,000 metric tons of almond kernels from 454,000 ha of bearing orchards (ABC, 2018a).

California almond farms grow >30 sweet varieties, with >95% of the current production volume attributed to 13 major varieties: Aldrich, Butte, Butte/Padre, Carmel, Fritz, Independence, Monterey, Mission, Nonpareil, Padre, Price, Sonora, and Wood Colony. Based on shell texture, almond varieties are generally classified as hard, semihard, and soft shell. California almonds are marketed as either shelled kernels or inshell almonds.

Sweet almond nuts have three distinct parts: the inner kernel, the middle shell portion, and an outer hull, as shown in Fig. 1. Almond kernels develop within a shell that is surrounded by a hull (similar to the fleshy part of a peach). When most nuts on the trees have completely split open hulls it's the time to harvest. The harvesting is carried out by mechanically shaking the nuts off trees to the ground. The nuts remain on the orchard floor until the kernel moisture has dried to below 6%, and this typically takes 7–10 days. The nuts are then swept into long rows (windrowed) between the tree rows and collected by mechanical sweepers. The harvested almonds are loaded onto trucks and delivered to a huller-sheller facility for hulling and shelling to retrieve kernels and/or inshell nuts. There are about 160 hullers/shellers in California, and most are single-operation facilities and some are an integrated operation at handler sites.

At the huller-sheller facilities, shelled almonds or inshell almonds are recovered from the shelling and/or hulling process. At the same time, three types of

FIG. 1
Three coproducts of almonds (hull, shell, kernel).

by-products or biomass are also generated: hulls, shells, and sticks. Hulls and shells are separated from kernels by respirators into hull piles and shell piles, respectively, but there are typically some shells present in a hull pile and some hulls present in a shell pile. For semihard and hard-shell varieties, the hull piles may contain more shells. Hulling and shelling of soft-shell varieties tend to generate cleaner hull piles with fewer shells mixed in. Most huller-shellers keep soft-shell variety hulls and semihard and/or hard-shell variety (pollinizer varieties) hulls in separate piles, such as Nonpareil hull piles, California-type hull piles, and hard-shell variety hull piles, but may not separate shells by variety types. The cleaner, soft-shell variety hulls have a greater feed value, and these piles are often stored under roof protection or are covered by plastic tarps.

2 ESTIMATION OF ALMOND BIOMASS GENERATED IN CALIFORNIA

From farming and harvesting to hulling and shelling and processing, the California almond industry generates several types of biomass that can be categorized as woody biomass and fleshy biomass. The woody biomass includes old-orchard tree removal, dead trees, prunings, sticks, and shells, while the fleshy biomass includes hulls, skins, and inedibles. The old-orchard trees, dead trees, prunings, and sticks (broken branches blown or shaken off trees) are generated or accumulated at the orchard floor. Almond hulls, shells, and smaller sticks are generated and accumulated at huller-sheller facilities, while almond skins and inedible kernels are generated and accumulated at handler facilities. Fig. 2 summarizes the

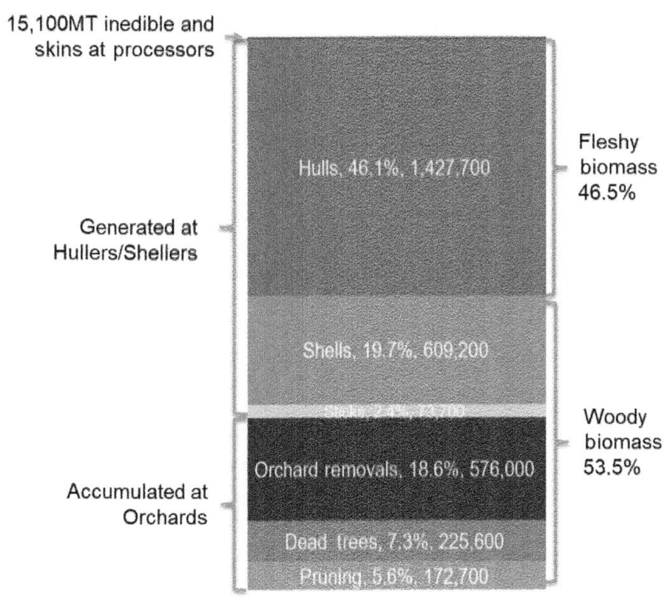

FIG. 2
2017 California almond biomass (3.1 million MT DM).

estimated volume of almond biomass generated by the California almond industry from the 2017 crop. A total of 3.1 million metric tons of biomass (dry matter) was produced by the California almond industry in 2017, with 53.5% as woody biomass and 46.5% as fleshy biomass.

2.1 Hulls and Shells

The volumes of almond hulls and shells are proportional to the almond kernels, and the ratios vary with varieties. The almond industry has been using approximate ratios of 2:1 for hulls to kernels, and 0.7:1 for shells to kernels to provide an estimation of almond hulls and shells produced. To more closely estimate the volume of almond hulls and shells, ABC contracted IEH J&L Analytical Laboratory (IEH) (Modesto, CA) to generate coproduct ratios of individual varieties from the 2017 crop. The inhull samples of 13 major varieties were collected from several growing counties and manually hulled and shelled in the laboratory to generate a set of hull, shell, and kernel samples, respectively. Each set of hull, shell, and kernel samples was weighed to obtain coproduct ratios on the basis of as-is weight. All hull, shell, and kernel samples were also tested for moisture content. The calculation of the coproduct ratio of each variety on dry matter basis was based on the fresh weight of hulls, shells, and kernels, and moisture level of each coproduct. The ratios of coproducts were found to vary greatly with variety, with hull percentages ranging from 37.5% (Mission) to 56.7% (Price), shell percentages ranging from 13.6% (Independence) to 33.5% (Mission), and kernel percentages ranging from 27.0% (Price) to 39.8% (Wood Colony). The average coproduct ratio of the 13 varieties on a dry matter basis is approximately 45:22:33 for hulls:shells:kernels. For a more accurate estimation, a weighted ratio on dry matter basis, taking into consideration the production share by each variety from the 2017 crop, could be 48:20:32 for hulls:shells:kernels. Based on the average moisture of hulls, shells, and kernels, a weighted ratio on a fresh weight basis could be 49:20:31 for hulls:shells:kernels. Fig. 3 shows the coproduct ratio of individual varieties in comparison with the averaged or weighted ratios on the dry matter and fresh weight basis. Based on the receipts of almond kernels from the 2017 crop at 1.02 million metric tons on a fresh weight basis, the volume of almond hulls and shells can be estimated as 1.46 and 0.61 million metric tons, respectively, on a dry matter basis.

2.2 Sticks

Tree shaking during harvesting may shake off dead or weakened branches (sticks) and leaves to generate biomass referred to as sticks and debris. These may be separated at orchard sites when the harvested nuts are loaded into a truck or precleaned by destoners at the huller-sheller site when the nuts are loaded to hulling-shelling line. Based on industry communication, the sticks account for about 2% of the field run weight or 8% of the kernel weight; taking into consideration a moisture content of 10%, the estimated volume of sticks from the 2017 crop is about 76,000 metric tons.

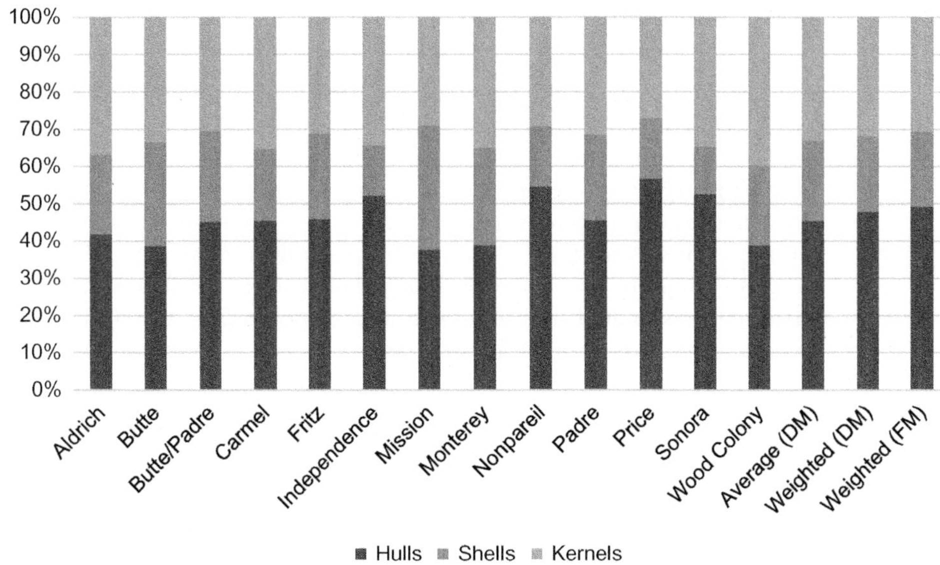

FIG. 3

Co-product ratios of almond nuts on dry matter basis.

2.3 Prunings, Dead Trees, and Old-Orchard Tree Removal

The pruning of trees after harvesting or during winter for orchard sanitation generates another group of woody biomass. As more studies have shown that pruning does not necessarily improve almond orchard productivity, less and less tree pruning is being done. Based on industry communication and estimation, 30% of bearing acreage is pruned each harvest season. Based on the bearing acreage of 454,000 ha in 2017 and a moisture level of 10%, the estimated pruning biomass on a dry matter basis is about 172,700 metric tons. Wind blow and tree disease may cause an average of 1% tree loss in an orchard each year, which is often replanted in the same or next year. Each dead tree is estimated to weigh 0.27–0.30 metric tons, with a moisture level of 10%, and the estimated biomass from the dead trees in 2017 is about 225,600 metric tons. Almond trees have 25–30 good productive years. Most orchards are replanted after 25 years and this generates lots of tree removal woody biomass widely scattered through almond planting areas. Based on the historical planting acreage of the last three decades, there will be 9,000–20,000 ha of almond orchards removed each year in the following decade. In 2017, the estimated removed acreage was about 9090 ha. As each acre of almond trees generates 32–36 metric tons of wet woody biomass with 10%–15% moisture (Industry handler communication, 2017), the estimated biomass from 2017 tree removal on dry matter basis is about 576,000 metric tons.

2.4 Skins and Inedibles

Almond kernels contain about 5% skin, and the blanching to remove skins through hot water scalding generates another group of biomass. As the demand for almond milk and beverages, made from blanched almond paste, is growing rapidly, more blanched almonds are being processed in California. Its estimated that about 159,000 metric tons of natural almonds from the 2017 crops were blanched by California processors. This generated about 7,500 metric tons of almond skins. Also, the California almond industry has a mandatory inedible disposal program to remove seriously damaged kernels from the commercial stream; the estimated amount of inedibles disposed from the 2017 crop is about 5,600 metric tons on a dry matter basis.

3 CHARACTERIZATION OF CALIFORNIA ALMOND BIOMASS

The composition of almond biomass (hulls, shells, trees, prunings, etc.) from different production regions has been published widely, and there are a limited number of papers on the composition of California almond biomass.

There are at least four publications to date on the composition of almond hulls from California orchards (Offeman et al., 2014, 2015; Holtman et al., 2015; DePeters et al., 2000). As reported, almond hulls are 12%–14% moisture, and up to 62% of the dry matter is water extractable and has approximately 32% fermentable simple sugars. The majority of nonextractable compounds from spent hulls are cellulose, hemicellulose, and lignin (25% of the total dry matter). Among the 126 samples (11–36 samples per variety collected in 2012 and 2013 from 6 varieties of 7 counties) analyzed, the fermentable sugar content showed some variations among samples, varieties, growing regions, orchards, and crop years, ranging 25%–33% (Offeman et al., 2014, 2015).

There are four publications to date regarding the composition of California shells (Aktas et al., 2015; Gong et al., 2011; Chiou et al., 2015; Klasson et al., 2010). The literature indicates that almond shells contain 81% neutral detergent fiber (NDF) (48% cellulose and 29% hemicellulose), 22% lignin, 7% moisture, and 1% ash.

3.1 Composition Analysis of California Almond Biomass by ABC

Since there is limited information on the composition of California almond biomass in literatures, this section includes biomass composition information ABC staff summarized from the analytical results by two independent analytical laboratories on hull, shell, and wood samples collected from crop years 2016 and 2017 (ABC, 2018b). Composition analysis of hull and shell samples were carried out by IEH, while the almond wood samples were analyzed by Hazen Research, Inc. (Golden, CO). Both laboratories followed official wet chemistry methods.

A total of 32 commercial hull samples (huller hulls) from 2016 season (analysis done on sorted pure hull samples) and 17 commercial hull samples (analysis done on as-is samples) and 64 hand hulls (manually separated from inhull nuts) from 2017 season were analyzed for proximate analysis, fibers [neutral detergent fibers (NDFs), acid detergent fiber sequential (ADFs), and acid detergent lignin sequential (ADLs)], minerals, and sugars. For composition analysis of shells, a total of 7 commercial shell samples (sheller shells) from 2016, and 9 sheller shell samples, and 64 hand shell samples (manually separated from inshell almonds of the 64 inhull samples for hand hulls analysis) from 2017 crop were tested for proximate analysis, fibers (NDFs, ADFs, ADLs), and minerals. In the 2017 season, the 9 sheller shell samples were also tested by IEH for elements (C, H, N, O) and heating value, while 10 almond tree wood chip samples, collected from 10 orchards of 3 counties, were tested by Hazen Research, Inc.

The compositional results of hull and shell samples (huller, sheller, and hand) from both 2016 and 2017 season are summarized in Table 1. As shown in the table, fiber (NDFs, ADFs, and ADLs) levels of both hull and shell samples were greatly influenced by the cleanliness of hulls (percentage of shells mixed) and shells (percentage of hulls mixed in). Ash or mineral contents in hull or shell samples were affected by contamination levels of dust. Dry matter or moisture levels of hull and shell samples varied with sampling times. The averaged NDFs of commercial hull samples is comparable to the literature reported values while its level in hand hull samples is lower. ADFs and ADLs in huller and hand hull samples are lower than the literature reported values, as expected, since these

Table 1 Composition of California Almond Hulls and Shells (%DM)

Analyte	Hulls			Shells		
	2016 Hullers[a] (32)	2017 Hullers (17)	2017 Handlers (64)	2016 Shellers (7)	2017 Shellers (9)	2017 Handlers (64)
Moisture	7.1±1.3	10.9±3.1	11.6±6.6	5.9±0.9	8.1±3.0	8.7±3.1
Crude protein (N X 6.25)	5.9±1.1	4.7±1.5	5.6±2.2	5.0±0.7	3.4±1.0	2.9±1.0
Fat	1.6±0.6	2.3±0.8	2.2±0.7	1.2±0.3	1.4±0.9	1.6±1.1
Ash	7.9±1.1	8.1±2.0	10.1±2.1	7.9±3.5	3.8±1.4	2.6±1.3
Potassium	2.6±0.4	2.6±0.6	3.6±0.8	1.7±0.3	1.6±0.7	0.8±0.3
Calcium	0.2±0.0	0.2±0.0	0.2±0.1	0.3±0.1	0.2±0.0	0.2±0.1
Neutral detergent fiber	22.5±4.0	28.3±10.6	24.6±3.6	65.1±5.1	68.1±14.5	85.7±4.3
Acid detergent fiber Sequential	14.7±3.0	19.7±7.8	17.0±2.5	45.3±3.3	47.2±10.2	59.0±3.1
Acid detergent lignin	3.5±1.0	4.9±2.7	3.8±1.0	13.3±1.2	14.3±3.5	19.0±2.9
Total sugars	30.2±6.9	25.8±7.1	20.7±7.3			

[a]2016 hull analysis was based on sorted, clean hulls.

Table 2 Heat Conversion Properties of Almond Tree Woods and Shells (%DM)

Analyte	Wood chips (10)	Shells (9)
Moisture	7.3±1.5	8.1±3.0
Ash	2.3±0.2	3.8±1.4
Volatile matter	79.6±1.3	72.2±11.2
Fixed carbon	18.2±1.4	20.3±0.3
MJ/kg (LHV)	17.5±0.1	16.6±0.7
MJ/kg (HHV)	18.8±0.1	18.4±0.4
Carbon	52.5±0.3	47.2±1.3
Hydrogen	5.6±0.0	8.2±2.5
Oxygen (by difference)	39.3±0.5	49.3±4.5
Nitrogen	0.2±0.1	0.3±0.2

were tested sequentially after NDF digestion in comparison with direct measurement of ADF and ADL reported in the literature (DePeters et al., 2000).

Elemental and heat value results of the almond shell and wood chip samples are summarized in Table 2. These are comparable to the literature values on shells. Aktas et al. (2015) reported on the composition of almond sticks, broken branches mixed in shell samples as they characterized thermal properties of California almond shells. Almond tree wood composition (ash, volatiles, fixed carbon, etc.) is comparable to other tree wood materials. Water-soluble alkaloids are low as the levels of Na_2O and K_2O were 0.04%, 0.03%, respectively. Sulfur was not detected. The major compounds in almond wood ash were SiO_2, CaO, and CO_2.

4 VALUE-ADDED POTENTIAL APPLICATIONS OF ALMOND FLESHY BIOMASS

4.1 High Value Feed Applications

Almond hulls generated in California are solely used as a supplement feed for dairy cows at present. There are recent papers evaluating usage of almond hulls of other origins as a dietary replacement for sheep (Rad et al., 2016; Yalchi, 2011). ABC is funding a project with Woo Kyun Kim and Fanbin Kong from the University of Georgia attempting to incorporate almond hulls into poultry feeds for broiler chicken and lay hens. Almond hulls are rich in carbohydrates, in addition to soluble sugars, they also contain a high level of cellulose, hemicellulose, pectin, etc. Kyria Bounty-Mills and Tina Yeoh at University of California, Davis, via a project funded by ABC, are trying to convert the carbohydrates from almond hulls into protein-rich yeast as functional ingredients for feed applications via hull hydrolysis, yeast selection, yeast feeding of hydrolysates. Recently, Jean VanderGheynst's team at University of California, Davis have investigated the conversion of low-quality almond hulls and shells into protein-rich black soldier fly larvae for poultry or fish feed (Palma et al., 2018). Their laboratory-scale preliminary work has demonstrated that black soldier fly (*Hermetia illucens* L.) larvae can be cultivated on almond hulls and shells (hulls) when moisture content and aeration rate are controlled. Experiments were performed to examine the role of larvae inoculation density, fed-batch processing, the impact of moisture, carbon to nitrogen ratio, and particle size on larvae growth. An optimum carbon to nitrogen ratio for larvae growth was found to be 29 using urea as a nitrogen source. Larvae growth was consistently higher on coarse hulls and shells compared to fine hulls and shells.

4.2 Sugar and Sweetener Extraction and Beyond

Holtman et al. (2015) developed a laboratory-scale process including extraction, fermentation, and anaerobic digestion to assess the bioenergy potentials of the soluble sugars from almond hulls. A countercurrent extraction procedure was developed to extract water-soluble sugars from almond hulls, similar to commercial extraction equipment used by the sugar beets industry, with 88% sugars recovery efficiency. The extracted filtrate was concentrated (131 g/L fermentable sugar) then fermented to generate ethanol (7.4%, v/v) with 86% fermentation efficiency and a potential of 31–42 gal ethanol/ton on "as-is" hulls basis. Anaerobic digestion of the stillage showed a 90% conversion efficiency and 75 mL methane/g almond hull. This research focused on ethanol conversion of fermentable sugars, but the extractable sugars could be explored for food or feed ingredient applications.

Spent hulls (solid residues after sugar extraction) may not have high feed value but it could be a source for other uses. Spent hulls should be evaluated for soil amendment, mushroom casing, anaerobic digestion to biogas, and feedstock for boiler fuel or gasification for heat/power and biochar materials. An exploratory project carried out by Zhiliang Fan's team at University of California, Davis, indicates that spent almond hulls are a good substrate for direct conversion to cellobionate (CBA) using the recombinant *N. crass* strain without any enzyme addition. Preliminary results show that a titer of 25 mM CBA and a yield >70% can be achieved in flask experiments without pH control. Avicel fermentation in bioreactors produced >51 mM CBA, which corresponded to a yield of 84% (ABC, 2016).

4.3 Phytochemicals Extraction and Food Grade Applications

The richness of phytochemicals in almond hulls and skins have been studied and reviewed by many authors (Siriwardhana and Shahidi, 2002; Wijeratne et al., 2006a, b; Takeoka et al., 2000; Takeoka and Dao, 2003; Rubilar et al., 2007; Esfahlan et al., 2010; Chen et al., 2010; Prgomet et al., 2017). Prgomet et al. (2017) provided the most recent comprehensive review on composition and biological activities of bioactive compounds from almond hulls, skins, and blanch water. The authors summarized all phenolic compounds identified from almond by-products in a table, and these compounds have demonstrated strong antioxidant, antimicrobial, prebiotic, antitumor, antiviral, and photoprotective properties. Chen et al. (2010) in their review paper recommended taking a biorefinery approach for almond hulls and skins to extract high-value chemicals, particularly antioxidants, before the residues are converted to energy products. It was suggested to process almond skins and hulls to food ingredients rich in dietary fibers and antioxidants. Based on the fact that almond hulls contain a high level of extractable phytochemicals, such as antioxidants and fibers that can be converted to functional dietary fibers, Roger Ruan's team from University of Minnesota are now working on an ABC funded project to make food and nutraceutical ingredients from almond hulls.

5 VALUE-ADDED POTENTIAL FOR ALMOND WOODY BIOMASS

5.1 Bioenergy Applications

Almond shells, tree wood, sticks, and prunings are cellulosic woody biomass materials with comparable heating values that make them good feedstocks for thermochemical conversation processes. Almond woody biomass of other origins as feedstocks for bioenergy generation from thermochemical conversion have been extensively studied and reviewed, and the optimal conditions for different thermal processes and biomass have been established (García et al., 2016; Gonzalez et al., 2002, 2005a, b, 2006a, b, 2009; Chen et al., 2010; Safari et al., 2016, 2017; Önal et al., 2017; García et al., 2017). Almond shells have been tested by Gonzalez and his colleagues (2002–2009) for bioenergy value and applications in gasification, hydrogasification, hydrolysis, and combustion system. In a most recent techno-economic review on biomass sources for thermal conversion, García et al. (2017) concluded that almond shells seem to be optimum biomass resources to use directly instead of wood pellets, charcoal, or briquettes in grate boilers. In their paper on characterization of almond processing residues from California Central Valley, Aktas et al. (2015) concluded: (1) almond hulls and shells have a relatively constant C/O ratio of 1.15 (dry matter), carry an average high heating value of 18–19 MJ/kg, and volatile matter content of 72%–76%; (2) ash content is 3.5%–22%; (3) chlorine is low (<0.05%) while nitrogen and sulfur are elevated compared to many other types of biomass; and (4) the characteristic feature of almond biomass ash is its very high potassium content, varying from 18% to 36%, suggesting both higher fouling and slagging rates in thermal systems.

There are multiple thermal processes that convert biomass into energy beyond burning, including gasification, pyrolysis, and torrefaction. Each process, operated at different temperatures, produces different outputs.

- Gasification—Converts biomass to carbon monoxide, hydrogen, and carbon dioxide at temperatures >700°C, without combustion, under a controlled amount of oxygen and/or steam.

- Pyrolysis—Converts biomass in the absence of oxygen at a temperature of 300–700°C to produce biogas, bio-oils, and biochar.
- Torrefaction—Converts biomass in the absence of oxygen at a temperature of 200–300°C to produce torrefied materials, bio-oils, biochar, etc.

Syngas or biogas or bio-oil can be feedstocks for electricity generation. The decreased prices in petroleum in recent years have slowed down development and adoption of thermochemical technologies. There have only been a few small-scale units of gasifiers installed in California for research and development purpose. With new incentives from the Bioenergy Market Adjusting Tariff (BIOMAT, n.d., Senate Bill 1122), effective from November 18, 2015, offering \$0.18 per kW with up to 20 years contract, the production of bioenergy from renewable forestry and agricultural biomass may be more feasible and make more business sense now. It is anticipated that there will be more gasification or pyrolysis systems installed in California in the coming years to process woody biomass. But any thermochemical processes must take system design into consideration to ensure that it can handle the high ash and potassium contents found in almond woody biomass.

5.2 Biochar and Activated Carbons From Almond Woody Biomass

Biochar is a coproduct from the thermochemical processing of biomass. Depending on the operating temperature ranges and different thermal conversion processes (gasification, and/or pyrolysis), different ratios of biochar, bio-oils, and syngas are generated. Biochar properties and performance are a function of feedstock material and production parameters (temperature, heating, and flow rates). Elevated temperatures increase the ratio of syngas and bio-oils, and decrease the yield of biochar. Biochar yields vary from $<10\%$ via a gasification process to $>30\%$ from slow pyrolysis.

Biochar and activated carbon derived from almond woody biomass have been studied by many groups (Klasson et al., 2010, 2014, 2015; de Yuso et al., 2014; Izquierdo et al., 2011; Nabais et al., 2011). Izquierdo et al. (2011) carried out a methodical study on chemical activation to determine optimal conditions to convert almond shells into activated carbons. It was concluded that a combination of a temperature of 550°C, an impregnation ratio of 1.10, and activation time of 112 min can prepare a high surface-activated carbon with a well-developed porosity and a high amount of oxygen surface groups. The same research group concluded in another paper (de Yuso et al., 2014) that the activating atmosphere during phosphoric acid chemical activation of almond shell has a strong influence on the characteristics and toluene adsorption capacity of the activated carbons, and activated carbons with higher surface area can be obtained by changing an inert activating atmosphere by an air atmosphere. Nabais et al. (2011) managed to produce activated carbons from almond shells by physical activation with CO_2 with good physical and chemical properties, in terms of adsorption properties, physical resistance, and surface chemistry.

The shells were carbonized at 400°C then activated at 700–800°C, and the properties of the activated carbons are comparable with the samples made at higher temperature reported by other published papers. A lower carbonization temperature of 400°C can lead to a more cost-effective process in terms of energy consumption and savings on the overall production process, leading to a decrease on the final price of the material. Klasson et al. (2010, 2014, 2015) have investigated properties of activated carbons and biochar made from almond shells of various California varieties through different pyrolysis processes at a laboratory scale. The group has evaluated properties of steam-activated carbon derived from shells of five varieties including soft and hard-shell varieties. These shells contained the same level of cellulose, but different in their lignin and hemicellulose composition. And the yield of carbon from the shells ranged from 20% to 23.5%. Regardless of the composition, the performances of the activated carbons were very similar with a capacity of 100–105 mg/g of carbons for dibromochloropropane and trichloroethylene. This finding suggests that the activated carbons can be made from almond shells with consistent quality. In one study, Klasson et al. (2014) characterized the properties of biochar from almond shells with low and high levels of ash content, and they found that biochar created from different types of almond shells (low- and high-ash) using different pyrolysis times contained significant pore volumes in the narrow micropore region. The pore distribution was more uniform for high-ash almond shell biochar compared with low-ash shell derived biochar, almost independent of pyrolysis time. Washing the biochar (mimicking rain wash) removed ash and exposed additional surface area, suggesting that adsorptive properties may change once biochar is exposed to rainwater. This indicates the importance of conditioning biochars from different resources (raw materials and thermochemical processes) prior to application for soil amendment to achieve consistent results.

5.3 Bioethanol From Almond Biomass

Syngas from gasification system or biogas from pyrolysis system can be converted to bioethanol. Aemetis Inc. (www.ametis.com) is building a bioethanol plant in Riverbank, CA, to produce bioethanol from agricultural woody biomass, including old-orchard almond tree removal, using proprietary microbe (LanzaTech) to ferment syngas generated from a gasification system (Welch, 2018). Bioethanol can also be generated from fermentation of sugars through a biochemical conversion process that involves fractionation or pretreatment, saccharification, fermentation, and ethanol recovery. Gong et al. (2011) developed a two-stage process to fractionate hemicellulose, cellulose, and lignin from almond shells involving hot water pretreatment at 195°C for 30 min followed by organic solvent pretreatment in hot water/ethanol at 195°C for 20 min. This resulted in nearly 100% glucose recovery. Kacem et al. (2016) developed a multistage process to generate bioethanol from almond shells, including acid and alkaline pretreatment, enzymatic saccharification, laccase detoxification and fermentation to ethanol using *Saccharomyces cerevisiae* yeast. Through optimization of the processing parameters, they managed to produce 5.8 g of ethanol from

100 g of almond shells. Detoxification by laccase to prevent the impact of free phenolic compounds from hydrolysates on yeasts was identified as a critical factor that had shown an increase of 3.1 g of ethanol per 100 g of shells. Messaoudi et al. (2017) also developed a process to separate hemicellulose and lignin to improve enzymatic digestibility of cellulose and to make xylose from hemicellulose and vanillin from lignin for several agricultural wastes including almond shells. The process included a toluene/ethanol extraction (2:1, v/v, 6 h) to remove extractives, a pretreatment of NaOH (1 M, 25°C, and 8 h) and H_2SO_4 (1%, 60°C, 1 h), enzymatic hydrolysis by Novozyme Cellic C-Tec2 or Cellic H-Tech 2 (45°C, 48 h), and fermentation by *S. cerevisiae* (30°C, 24 h). The published data indicate that 8.6 g of ethanol can be produced from 100 g of almond shells with this process.

Orts and McMahan (2016) carried out a thorough review on the current United States strategy on biofuels and USDA (US Department of Agriculture) ARS (Agriculture Research Services) research advancement on biorefinery development for advanced biofuels from agricultural residues. The review covered advancements in both biochemical and thermochemical conversion technologies for bioethanol production from various agricultural biomass, including improvements in microbes, enzymes, inhibitor removal and product recovery for biochemical conversion technology. The paper also quoted their colleagues' work on bioethanol production potential from almond hulls based on fermentable sugars (Holtman, et al., 2015). Based on that research finding, 49 gal (185 L) of ethanol and 75 m^3 of methane can be produced from a ton of almond hulls. To be more efficient in utilization of almond hulls, additional pretreatment and enzymatic saccharification hydrolyzing cellulose may be needed to increase bioethanol conversion.

5.4 Almond Shells and Derived Products as Plastic Fillers

Almond shells and torrefied shells have been recently studied as a filler for biodegradable plastic composites or a strengthening agent or colorant for recycled plastics (Sabbatini et al., 2017; García et al., 2015; Sabarinathan et al., 2016; Chiou et al., 2015, 2016; Zahedi et al., 2015). Sabbatini et al. (2017) studied the impact of almond shells and rice husk as fillers of poly (methyl methacrylate) (PMMA) composites. At 10% loading rate, the addition of this woody filler in PMMA-based composite did not cause significant deterioration of its mechanical properties. García et al. (2015) added up to 20% of various almond variety shells to make masterbatches based on polylactic acid (PLA). Sabarinathan et al. (2016) studied the mechanical properties of almond shell-sugarcane leaves hybrid epoxy polymer composite. Chopped sugarcane leaves (50 mm × 50 mm) and ground almond shells (average of 1-mm particle diameters) were tested in the epoxy polymer composite. Scanning electron microscopy examinations revealed consistent particle distribution and good bonding between particles and epoxy polymer. Zahedi et al. (2015) investigated the improving effects of a reinforcing agent on poor interfacial interaction between the hydrophilic lignocellulosic material (almond shell flour) and hydrophobic polypropylene matrix. Adding 3% of organically modified montmorillonite (OMMT) improved the physico-mechanical properties of the composite. At a combination of 3% OMMT and 50% almond shell flour, an optimum synergistic effect was achieved. Their findings proved that almond shell is a valuable renewable natural resource for composite production and can be utilized as a substitute for wood in composite industries.

Chiou et al. (2015, 2016) investigated the impact of torrefied almond shells on properties of polypropylene composites with loading rate up to 20%. All composites had a heat distortion temperature increase ranging from 8°C to 24°C. The increase in heat distortion temperature is attributable to the filler restricting mobility of the polymer. The composites showed lower tensile strength and elongation at break values than neat polypropylene, and this may be due to weak adhesion between torrefied shells and polypropylene. A more comprehensive study with torrefied almond shells in various low-grade plastic materials at a pilot scale is still ongoing. Scale-up trials of applications of torrefied almond shells in plastic slip sheet and plastic pallet are underway.

McCaffrey et al. (2018) recently investigated the use of torrefied almond shells in recycled polypropylene-polyethylene. The study evaluated the impact of loading rates of torrefied almond shells and particle sizes on properties of biocomposites. An increase in torrefied almond shell loading rate led to an increase in flexural and tensile module, but a decrease in strength, elongation, and toughness. Heat deflection temperature increased with a higher loading rate with small particle size.

5.5 Almond Shell and its Derived Product as Adsorbents

Almond shells and their derived products, such as activated carbons, as bio-adsorbents have been widely investigated for removals of heavy metals, pesticides, organic chemicals, dyes, etc. Recently published papers include Arfi et al. (2017) on removal of cationic and anionic dyes from aqueous solution; Banerjee et al. (2017) on removal of Cr (VI) ion from aqueous solution; Maaloul et al. (2017) on adsorption of Cu(II) ions with chemically modified almond shells [bleached almond shell (BAS) and lyophilized-bleached almond shell (L-BAS)]; Taha et al. (2017) on comparative biosorption of Hg(II) with raw and chemically activated almond shell; Thitame and Shukla (2017) on removal of lead (II) from synthetic solution and industry wastewater with almond shell activated carbon; Flores-Cano et al. (2016) on adsorption of metronidazole, dimetridazole, and diatrizoate by activated carbons from almond shells; and Saeed et al. (2016) on removal of methyl violet 2-B from aqueous solutions using untreated and magnetite-impregnated almond shells. In the comparative study by Taha et al. (2017), the maximum biosorption capacity was found to be 3.77 and 38.17 mg/g for raw and activated almond shells, respectively. The study of thermodynamic parameters showed that the biosorption of Hg(II) ions was non-spontaneous, random, and endothermic for both raw and activated almond shells. Thitame and Shukla (2017) found that the adsorption capacity for lead was 823.1 mg/g at an initial pH of 6.0. With 0.5 g/L of loading rate, a complete removal of Pb(II) ions was achieved from battery industry wastewater. Maaloul et al. 2017 focused on the adsorption of Cu(II) ions on two chemically modified almond shell-based adsorbents: BAS and L-BAS). The maximum adsorption capacity of BAS and L-BAS at 30°C and pH 6 was found to be 18.71, 28.27 mg/g, respectively. Therefore, these materials can be used as effective adsorbents for Cu(II)-containing wastewaters.

6 SUMMARY

Almond by-products or biomass from California are well utilized but their potential values have not been fully realized. The almond by-products have been widely studied for bioenergy production using thermochemical and biochemical conversion technologies. But these technologies have not been installed in California to process almond biomass. With a large amount of almond shells produced annually and accumulated at hullers/shellers facilities, California can have a steady supply for many units of thermochemical or biochemical conversion technologies across the Central Valley to operate year-round. However, for any conversion technology, a comprehensive approach is needed to fully utilize every product generated. Besides feedstock for bioenergy generation, the potential values of coproducts, such as biochar for soil amendment and bioliquids for biopesticidal applications or biochemicals are worth further investigation. Almond woody biomass as composite fillers and absorbents and torrefied shells as low-grade plastic enhancers should be further explored for scale-up or commercial applications.

ABC is funding several projects to expand hull utilization, increasing uses in dairy cow diets, expanding to other livestock, and producing protein-rich feed additives from insect larvae and yeast growth. Further research or efforts are needed for utilization of spent hulls from larvae and yeast growth and sugar extraction. Almond hulls are rich in fiber and phytochemicals. There are many papers on phytochemicals and antioxidant activities of almond hulls, but there are no serious efforts made to maximize such potentials. With a recent industry goal to reduce dust during harvest, many almond growers are thinking toward off-ground harvest. Such a concept creates an opportunity for cleaner hulls and warrants exploration of antioxidants and fiber-rich food-grade ingredients or applications.

References

Aktas, T., Thy, P., Williams, R.B., McCaffrey, Z., Khatami, R., Jenkins, B.M., 2015. Characterization of almond processing residues from the Central Valley of California for thermal conversion. Fuel Process. Technol. 140 (132), 147.

Almond Board of California (ABC), 2016. Project Report by Zhiliang Fan of UC Davis on "Preliminary Study on Conversion of Spent Hulls to Cellobionic Acid".

Almond Board of California (ABC), 2018a. Almond Almanac 2018. http://www.almonds.com/sites/default/files/Almanac%202018.pdf.

Almond Board of California (ABC), 2018b. Almond Biomass Assessment, Unpublished Data. Available from: ghuang@almondboard.com.

Arfi, R.B., Karoui, S., Mougin, K., Ghorbal, A., 2017. Adsorptive removal of cationic and anionic dyes from aqueous solution by utilizing almond shell as bioadsorbent. Euro-Mediterr. J. Environ. Integr. 2 (1), 20. https://doi.org/10.1007/s41207-017-0032-y.

Banerjee, M., Bar, N., Basu, R.K., Das, S.K., 2017. Comparative study of adsorptive removal of Cr (VI) ion from aqueous solution in fixed bed column by peanut shell and almond shell using empirical models and ANN. Environ. Sci. Pollut. Res. 24 (11), 10604–10620.

BIOMAT, n.d. California Senate Bill 1122, https://www.pge.com/en_US/for-our-business-partners/floating-pages/biomat/biomat.page.

Chen, P., Chen, Y., Deng, S., Lin, X., Huang, G., Ruan, R., 2010. Utilization of almond residues. Int. J. Agric. Biol. Eng. 3 (4), 1–18. Open access, http://www.ijabe.org.

Chiou, B.-S., Valenzuela-Medina, D., Bilbao-Sainz, C., Klamczynski, A.P., Avena-Bustillos, J., J, R., Milczarek, R.R., Du, W.X., Glenn, G.M., Orts, W.J., 2016. Torrefaction of almond shells: effects of torrefaction conditions on properties of solid and condensate products. Ind. Crop. Prod. 86, 40–48.

Chiou, B.S., Valenzuela-Medina, D., Wechsler, M., Bilbao-Sainz, C., Klamczynski, A.K., Williams, T.G., Wood, D.F., Glenn, G.M., Orts, W.J., 2015. Torrefied biomass-polypropylene composites. J. Appl. Polym. Sci. 132(10). https://doi.org/10.1002/app.41582.

DePeters, E.J., Fadel, J.G., Arana, M.J., Ohanesian, N., Etchebarne, M.A., Hamilton, C.A., Hinders, R.G., Maloney, M.D., Old, C.A., Riordan, t.J., Perez-Monti, H., Pareas, J.W., 2000. Variability in the chemical composition of seven selected by-product feedstuffs used by the California dairy industry. Prof. Anim. Sci. 16, 69–99.

de Yuso, A.M., Rubio, B., Izquierdo, M.T., 2014. Influence of activation atmosphere used in the chemical activation of almond shell on the characteristics and adsorption performance of activated carbons. Fuel Process. Technol. 119, 74–80.

Esfahlan, R.J., Jamei, R., Esfahlan, R.J., 2010. The importance of almond (*Prunus amygdalus* L.) and its by-products. Food Chem. 120 (2), 349–360.

Flores-Cano, J.V., Sánchez-Polo, M., Messoud, J., Velo-Gala, I., Ocampo-Pérez, R., Rivera-Utrilla, J., 2016. Overall adsorption rate of metronidazole, dimetridazole and diatrizoate on activated carbons prepared from coffee residues and almond shells. J. Environ. Manag. 169, 116–125.

García, A.M., García, A.I., Cabezas, M.Á.L., Reche, A.S., 2015. Study of the influence of the almond variety in the properties of injected parts with biodegradable almond shell based masterbatches. Waste Biomass Valoriz. 6 (3), 363–370.

García, A., Gandini, A., Labidi, J., Belgacem, N., Bras, J., 2016. Industrial and crop wastes: a new source for nanocellulose biorefinery. Ind. Crop. Prod. 93, 26–38.

García, R., Pizarro, C., Lavín, A.G., Bueno, J.L., 2017. Biomass sources for thermal conversion. Techno-economical overview. Fuel 195, 182–189.

Gong, D., Holtman, K.M., Franqui-Espiet, D., Orts, W.J., Zhao, R., 2011. Development of an integrated pretreatment fractionation process for fermentable sugars and lignin: application to almond (*Prunus dulcis*) shell. Biomass Bioenergy 35, 4435–4441.

Gonzalez, J.F., Gonzalez-Garcia, C.M., Ramiro, A., Ganan, J., Gonzalez, J., Sabio, E., Roman, S., Turegano, J., 2005a. Use of almond residues for domestic heating. Study of the combustion parameters in a mural boiler. Fuel Process. Technol. 86, 1351–1368.

Gonzalez, J.F., Ramiro, A., Gonzalez-Garcia, C.M., Ganan, J., Encinar, J.M., Sabio, E., Rubiales, J., 2005b. Pyrolysis of almond shells. Energy applications of fractions. Ind. Eng. Chem. Res. 44, 3003–3012.

Gonzalez, J.F., Romiro, A., Sabio, E., Encinar, J.M., Gonzalez, C.M., 2002. Hydrogasification of almond shell characteristics. Influence of operating variables and kinetic study. Ind. Eng. Chem. Res. 41 (15), 3557–3565.

Gonzalez, J.F., Ganan, J., Ramiro, A., Gonzalez-Garcia, C.M., Encinar, J.M., Sabio, E., Roman, S., 2006a. Almond residues gasification plant for generation of electric power. Preliminary study. Fuel Process. Technol. 87, 149–155.

Gonzalez, J.F., Gonzalez-Garcia, C.M., Ramiro, G.J., Ayuso, A., Turegano, J., 2006b. Use of energy crops for domestic heating with a mural boiler. Fuel Process. Technol. 87 (8), 717–726.

Gonzalez, J.F., Roman, S., Encinar, J.M., Martinez, G., 2009. Pyrolysis of various biomass residues and char utilization for the production of activated carbons. J. Anal. Appl. Pyrolysis 85, 134–141.

Holtman, K.M., Offeman, R.D., Franqui-Villanueva, D., Bayati, A.K., Orts, W.J., 2015. Countercurrent extraction of soluble sugars from almond hulls and assessment of the bioenergy potential. J. Agric. Food Chem. 63, 2490–2498.

International Treenuts and Dried Fruits Council (INC), 2017. Statistical Yearbook. https://www.nutfruit.org/industry/technical-resources?category=statistical-yearbooks.

Izquierdo, M.T., de Yuso, A.M., Rubio, B., Pino, M.R., 2011. Conversion of almond shell to activated carbons: methodical study of the chemical activation based on an experimental design and relationship with their characteristics. Biomass Bioenergy 35, 1235–1244.

Kacem, I., Koubaa, M., Maktouf, S., Chaari, F., Najar, T., Chaabouni, M., Ettis, N., Chaabouni, S.E., 2016. Multistage process for the production of bioethanol from almond shell. Bioresour. Technol. 211, 154–163.

Klasson, K.T., Ledbetter, C.A., Wartelle, L.H., Lingle, S.E., 2010. Feasibility of dibromochloropropane (DBCP) and trichloroethylene (TCE) adsorption onto activated carbons made from nut shells of different almond varieties. Ind. Crop. Prod. 31, 261–265.

Klasson, K.T., Uchimiya, M., Lima, I.M., 2015. Characterization of narrow micropores in almond shell biochars by nitrogen, carbon dioxide, and hydrogen adsorption. Ind. Crop. Prod. 67, 33–40.

Klasson, K.T., Uchimiya, M., Lima, I.M., 2014. Uncovering surface area and micropores in almond shell biochar by rainwater wash. Chemosphere 111, 129–134.

Maaloul, N., Oulego, P., Rendueles, M., Ghorbal, A., Díaz, M., 2017. Novel biosorbents from almond shells: characterization and adsorption properties modeling for cu (II) ions from aqueous solutions. J. Environ. Chem. Eng. 5 (3), 2944–2954. https://doi.org/10.1016/j.jece.2017.05.037.

McCaffrey, Z., Torres, L., Flynn, S., Cao, T., Chiou, B.S., Klamczynski, A., Glenn, G., Orts, M.J., 2018. Recycled polypropylene-polyethylene torrefied almond shell biocomposites. Ind. Crop. Prod. 25, 425–432.

Messaoudi, Y., Smichi, N., Bouachir, F., Gargouri, M., 2017. Fractionation and biotransformation of lignocelluloses-based wastes for bioethanol, xylose and vanillin production. Waste Biomass Valoriz. https://doi.org/10.1007/s12649-017-0062-3 (28 August 2017 online).

Nabais, J.M.V., Laginhas, C.E.C., Carrott, P.J.M., Carrott, R., 2011. Production of activated carbons from almond shell. Fuel Process. Technol. 92, 234–240.

Offeman, R.D., Dao, G.T., Holtman, K.M., Orts, W.J., 2015. Leaching behavior of water soluble carbohydrates from almond hulls. Ind. Crop. Prod. 65, 488–495.

Offeman, R.D., Holtman, K.M., Covello, K.M., Orts, W.J., 2014. Almond hulls as biofuels feedstock: variations in carbohydrates by variety and location in California. Ind. Crop. Prod. 4, 109–114.

Önal, E., Uzun, B.B., Pütün, A.E., 2017. The effect of pyrolysis atmosphere on bio-oil yields and structure. Int. J. Green Energy 14 (1), 1–8.

Orts, W.J., McMahan, C.M., 2016. Biorefinery developments for advanced biofuels from a sustainable array of biomass feedstocks: survey of recent biomass conversion research from agricultural research service. BioEnergy Res. 9 (2), 430–446.

Palma, L., Ceballos, S.J., Johnson, P.C., Niemeier, D., Pitesky, M., VanderGheynst, J.S., 2018. Cultivation of Black Soldier Fly Larvae on Almond Byproducts: Impacts of Aeration Moisture on Larvae Growth Composition. Published online July 2018 in Wiley Online Library.

Rad, M.I., Rouzbehan, Y., Rezaei, J., 2016. Effect of dietary replacement of alfalfa with urea-treated almond hulls on intake, growth, digestibility, microbial nitrogen, nitrogen retention, ruminal fermentation, and blood parameters in fattening lambs. J. Anim. Sci. 94 (1), 349–358.

Rubilar, M., Pinelo, M., Shene, C., Sineiro, J., Nunez, M.J., 2007. Separation and HPLC-MS identification of phenolic antioxidants from agricultural residues: almond hulls and grape pomace. J. Agric. Food Chem. 55 (25), 10101–10109.

Prgomet, I., Goncalvas, B., Domingues-Perles, R., Pascual-Seva, N., Barros, A.I.R.N.A., 2017. Valorization challenges to almond residues: phytochemical composition and functional application. Molecules 22, 1774. https://doi.org/10.3390/molecules22101774.

Sabarinathan, P., Rajkumar, K., Gnanavelbabu, A., 2016. Mechanical properties of almond shell-sugarcane leaves hybrid epoxy polymer composite. Appl. Mech. Mater. 852, 43–48.

Sabbatini, A., Lanari, S., Santulli, C., Pettinari, C., 2017. Use of almond shells and rice husk as fillers of poly (methyl methacrylate) (PMMA) composites. Materials 10 (8), 872. https://doi.org/10.3390/ma10080872.

Saeed, K., Ishaq, M., Sultan, S., Ahmad, I., 2016. Removal of methyl violet 2-B from aqueous solutions using untreated and magnetite-impregnated almond shell as adsorbents. Desalin. Water Treat. 57 (29), 13484–13493.

Safari, F., Javani, N., Yumurtaci, Z., 2017. Hydrogen production via supercritical water gasification of almond shell over algal and agricultural hydrochars as catalysts. Int. J. Hydrog. Energy. https://doi.org/10.1016/j.ijhydene.2017.05.102 (online 2 June 2017).

Safari, F., Salimi, M., Tavasoli, A., Ataei, A., 2016. Non-catalytic conversion of wheat straw, walnut shell and almond shell into hydrogen rich gas in supercritical water media. Chin. J. Chem. Eng. 24 (8), 1097–1103.

Siriwardhana, S.S.K.W., Shahidi, F., 2002. Antiradical activity of extracts of almond and its by-products. J. Am. Oil Chem. Soc. 79 (9), 903–908.

Taha, A.A., Moustafa, A.H.E., Abdel-Rahman, H.H., Abd El-Hameed, M.M.A., 2017. Comparative biosorption study of hg (II) using raw and chemically activated almond shell. Adsorpt. Sci. Technol. https://doi.org/10.1177/0263617417705473 (26 April 2017 online).

Takeoka, G., Dao, L., Teranishi, R., Wong, R., Flessa, S., Harden, L., Edwards, R., 2000. Identification of three triterpenoids in almond hulls. J. Agric. Food Chem. 48 (8), 3437–3439.

Takeoka, G.R., Dao, L.T., 2003. Antioxidant constituents of almond [*Prunus dulcis* (mill.) D.A. Webb] hulls. J. Agric. Food Chem. 51 (2), 496–501.

Thitame, P.V., Shukla, S.R., 2017. Removal of lead (II) from synthetic solution and industry wastewater using almond shell activated carbon. Environ. Prog. Sustain. Energy. https://doi.org/10.1002/ep.12616 (16 April 2017 online).

Welch, J., 2018. Aemetis Presentation to Almond Board of California. (August 2018).

Wijeratne, S.S.K., Abou-Zaid, M.M., Fereidoon, S., 2006a. Antioxidant polyphenols in almond and its coproducts. J. Agric. Food Chem. 54 (2), 312–318.

Wijeratne, S.S.K., Amarowicz, R., Shahidi, F., 2006b. Antioxidant activity of almonds and their by-products in food model systems. J. Am. Oil Chem. Soc. 83 (3), 223–230.

Yalchi, T., 2011. Determination of digestibility of almond hull in sheep. Afr. J. Biotechnol. 10 (15), 3022–3026.

Zahedi, M., Khanjanzadeh, H., Pirayesh, H., Saadatnia, M.A., 2015. Utilization of natural montmorillonite modified with dimethyl, dehydrogenated tallow quaternary ammonium salt as reinforcement in almond shell flour–polypropylene bio-nanocomposites. Compos. Part B 71, 143–151.

CHAPTER 16
Walnuts

Ragab Khir*†, Zhongli Pan*
*Department of Biological and Agricultural Engineering, University of California - Davis, Davis, CA, United States, †Department of Agricultural Engineering, Faculty of Agriculture, Suez Canal University, Ismailia, Egypt

Chapter Outline

1 INTRODUCTION

Walnuts (*Juglans regia* L.) are lauded for their health benefits as a rich source of unsaturated fats, protein, dietary fiber, phytochemicals, and micronutrients (WCRF, 1997). Cultivated throughout every continent on Earth, it is no surprise that they have various uses in the food industry. Walnuts are eaten as snacks and can be added to breakfast cereals, baked goods, salads, pastas, and soups. Thus, it is considered the number one consumer nut purchased from supermarket shelves worldwide. To extract the edible portion (kernel or meat), walnuts must undergo harvesting, hulling, drying, and shelling. During these processing operations, considerable quantities of by-products, including leaves, green hulls, shells, and

Integrated Processing Technologies for Food and Agricultural By-Products
https://doi.org/10.1016/B978-0-12-814138-0.00016-2

broken kernels are produced. Despite their potential value, most of these by-products are underutilized and discharged to the environment, resulting in serious problems. Utilization of these by-products in the production of healthy food ingredients and high value-added products can increase the profitability of the walnut processing industry and eliminate the environmental concerns.

The characteristics of walnut by-products suggest promising potentials for their use in value-added products for food and nonfood applications. Through these new applications, walnut by-products can be converted from their current low-value status to a high revenue stream of raw materials. The problem that many walnut processors face is that they are often unable to find appropriate processing methods to efficiently utilize the by-products produced from walnut processing operations. The feasibility of utilization of the walnut by-products for higher value options is also affected by several factors related to their physical and chemical characteristics, quantity produced, and technical factors.

It is important to fully understand the value of the by-products produced from walnut processing operations and determine the technical feasibility and methods to properly utilize them. This chapter discusses walnut production, economic value, health benefits, and processing operations (harvesting, dehulling, drying, storage, and shelling). Characterization of the main and by-products produced during the walnut processing operations is addressed as well. At the end of this chapter, processing technologies to efficiently utilize the by-products to produce value-added products and their food and nonfood applications are illustrated.

2 PRODUCTION AND ECONOMIC VALUE

Walnuts are the most widespread tree nut in the world. They are cultivated commercially throughout southern Europe, northern Africa, eastern Asia, the USA, and western South America (Martinez et al., 2010). Worldwide, the walnut cultivation area nearly doubled over 10 years, from 512,367 ha in 1997 to 1,186,398 ha in 2016 (FAO, 2016). The world production of whole walnuts (with shell) in 2016 was around 5,533,428 tons (FAO, 2016). China is the leading world producer with an annual production of 1,785,879 tons, followed by the United States (607,814 tons), Iran (405,281 tons), Turkey (195,000 tons), and Mexico (141,818 tons) (FAO, 2016).

Walnuts are a crop of high economic importance to the food industry. The edible part of walnut (kernel or meat) is globally recognized for its nutritional, health, and sensory characteristics. The facts related to the nutrition and health values have helped stimulate public demand of walnuts. Consequently, market development activities have expanded, leading to increased use of walnuts in the snacking, baking, and processed food sectors (USDA, 2018). In addition to their natural health benefits, the by-products produced during walnut processing have a great potential to become high-value-added products (Sürmen and Demirbas, 2003; Kar, 2011a). Unfortunately, these by-products are not fully taken advantage of.

The potential use of walnut by-products in value-added products is discussed in the different sections of this chapter.

3 HEALTH BENEFITS

Walnuts are associated with numerous health benefits. It has been proven that walnut consumption provides many health benefits since walnuts are a good source of omega-3 fatty acids, vitamin E, and other antioxidants associated with a healthy heart and a potential reduction of cancer cell growth (Iwamoto et al., 2000; Martinez et al., 2010). The fresh natural kernel is consumed as a snack item or used as an ingredient in candies, cereals, and baked goods (Martinez et al., 2010). It contains significant amounts of the omega-3 fat called alpha-linolenic acid (ALA). ALA is considered especially beneficial for heart health and helps reduce inflammation and improve the composition of blood fats (Moshfegh et al., 2007; Lino et al., 2000). Moreover, walnuts are an excellent source of several vitamins, minerals, and antioxidants that help maintain bone, nerve, and immune system function and suppress cancer growth in the breasts, prostate, colon, and kidneys (Albert et al., 2002).

A human study indicated that walnut consumption can improve brain function and help with depression and age-related decline in brain functions (Fraser et al., 1999). The walnut is also considered as a good source of flavonoids, phenolic acids, and related polyphenols. Phenolic compounds have not been clearly identified for their nutritional function; however, they may be important for human health owing to their good antioxidant, antiatherogenic, antiinflammatory, and antimutagenic properties (Anderson et al., 2001; Carvalho et al., 2010).

It has been reported that a daily intake of walnuts can significantly reduce low-density lipoprotein (LDL) cholesterol concentration and help prevent coronary heart disease (CHD) (Iwamoto et al., 2000; Zambón et al., 2000; Muñoz et al., 2001; Davis et al., 2007; Lavedrine et al., 1999). Simopoulos (2002) stated that the walnut is unique because it has a perfect 4:1 ratio of n-6 and n-3 polyunsaturated fatty acids which decreased the incidence of cardiovascular risk. The heart benefits of walnut intake include reducing inflammation and improving arterial function (Ros et al., 2004; Zhao et al., 2004). Alasalvar and Bradley (2015) mentioned that the U.S. Food and Drug Administration granted a qualified health claim that the consumption of 1.5 oz. (approximately 42.5 g)/d of most nuts, but especially walnuts, may reduce the risk of heart disease as part of a diet low in saturated fat and cholesterol. The European Food Safety Authority (EFSA) allowed an Article 13 claim that the consumption of 30 g of walnuts in the context of a balanced diet leads to the improvement of endothelium-dependent vasodilation (EFSA, 2011).

4 PROCESSING OPERATIONS

Walnut processing covers the operations from harvest to the production of edible part. These operations imply harvesting, hulling, drying, storing, and shelling.

During these operations, considerable quantities of by-products, including leaves, green hulls, broken kernels, and shells are produced.

4.1 Harvesting

The harvesting process begins when a high percentage of walnuts on the tree has split hulls and a small number of nuts have already fell to the ground. The last stage of maturation and hull split of walnuts can be hastened by applying an ethylene-producing compound, such as ethephon (Khir et al., 2012; Romas, 1998). Harvesting is then conducted by shaking the tree using mechanical shakers to remove the matured walnuts. The mechanical shakers grasp the tree trunk and vigorously shake the entire tree until the walnuts drop to the ground. The dropped nuts are then carefully swept into windrows and picked up with harvest machinery. This sweeping operation is completed quickly to reduce the amount of time that the nuts are on the ground.

Harvested nuts are transported to drying sites to be dried to a safe moisture content (MC) for storage. Before the drying process, walnuts go through hullers that remove the green hulls from the shell with wet scrubbers. The walnuts are then dried using hot air at a temperature of 43°C until the nuts reach 8% (wet basis) MC (Rumsey and Lu, 1991; Kader, 2002; Rumsey and Thompson, 1984; Khir et al., 2014a,b). Once the walnuts reach the target MC, the dried nuts are usually shipped to central facilities for further processing including shell removal. The industrial scale of centralized processing makes it more feasible to collect a large amount of the by-products.

4.2 Dehulling

Before the drying operation, the dehulling process must be conducted to remove the external green hull from walnuts using a mechanical huller. The green hull is one of the major by-products produced during the dehulling process and is mostly discarded. However, green hull is an economical source of natural compounds with antioxidant and antimicrobial properties.

An important problem associated with the dehulling is that, during the current processing practices, walnuts with and without hulls collected during the harvesting process are transported to the drying sites. Transporting walnuts with hulls from orchards to the drying site for dehulling negatively affects the processing efficiency and energy use (Atungulu and Pan, 2012). The green hulls have a high MC (85% on average) (Khir et al., 2013). Thus, the transportation of walnuts with hulls consumes a significant amount of energy and wastes labor. Moreover, the greenhouse gas emission (GGE) resulting from automobiles used for transportation causes environmental concerns. An in-field dehulling approach can mitigate issues related to the transportation of walnuts with hulls from orchards to the drying sites (Khir et al., 2013; Atungulu et al., 2013). Khir et al. (2012) reported the feasibility of using either axial dimensions or aerodynamic methods for sorting walnuts with and without hulls. They concluded that air with a terminal velocity of $10\,m\,s^{-1}$ could achieve complete separation of walnuts without hulls from those with hulls. In-field sorting, hereafter referred to as dry-hulling, would make it easy to implement in-field dehulling without soaking the walnuts in water.

4.3 Drying

After harvesting and hulling, the walnuts must thoroughly dry from their initial MC to a MC of 8% (wet basis) by normally using hot air at a temperature of about 43°C (Khir et al., 2013, 2014a; Romas, 1998). At this MC level, the walnuts are considered resistant against molding, discoloration, development of off-flavors, and rancidity (Kader, 2002; Rumsey and Thompson, 1984). Additionally, the dried walnuts can be placed in a state of equilibrium in MC for a long-term storage.

The variability in MC among individual walnuts is considered one of the biggest problems affecting drying performance and efficiency. The MC of individual walnuts varies significantly due to uneven maturation in the field. Walnuts that fall off the tree before harvest are much dryer because they have reached maturity. On the other hand, walnuts that are less mature and must be shaken off the tree typically have a high MC. About 40%–50% of harvested walnuts have high MC with adhered hulls (Khir et al., 2013, 2014a, b; Romas, 1998). Khir et al. (2013) reported that walnuts with hulls had an average MC of 32.99%, whereas walnuts without hulls had MC of 13.86%.

Despite a huge variability in the MC among individual walnuts, the current drying facilities commingle nuts with a wide range of initial MCs and dry them as a batch using 43 °C air for the entire process. Nuts that enter the dryer with a high initial MC may end up with a MC of 10% or higher, which is above the targeted safe storage MC of 8% (Kader, 2002; Rumsey and Thompson, 1984). Therefore, walnuts with higher MC must be dried longer to bring them to safe storage MC (Thompson and Grant, 1992). On the other hand, exposing the nuts with lower MC may lead to overdrying, which uses significant amounts of energy, prolongs drying time, and reduces dried product quality.

It has been reported that the sorting of freshly harvested in-shell walnuts based on their MC before drying can overcome the shortcomings of current drying practices. Moreover, Khir et al. (2013) found that walnut shells had more moisture than walnut kernels did. This means the walnuts can be dried at elevated temperatures during the first part of the drying process when the kernel temperature was significantly cooler than the drying air temperature. The elevated temperatures could be used to drive out most of the moisture in the shells of high moisture nuts without affecting the product quality (Lowe et al., 1961; Rumsey and Lu, 1991; Thompson et al., 1985; Khir et al., 2013, 2014a, b; Atungulu et al., 2013).

Additionally, it has been mentioned that there is a potential to use infrared (IR) heating to quickly remove part of the shell moisture, particularly for sorted high-moisture walnuts, to achieve partial drying and overcome the drawbacks of the current low-temperature drying method (Pan et al., 2018). The research group at the food processing laboratory in the Biological and Agricultural Engineering Department, University of California, Davis conducted comprehensive tests to develop energy-efficient drying technology using IR heating and separation of walnuts based on their MCs before drying. The group confirmed that IR heating quickly removed the walnuts' shell and surface moisture without negatively affecting the quality of dried walnuts. IR predrying of high moisture nuts for 3 and 4 min reduced a significant amount of moisture, which translated to 1 and 3 h of the conventional low-temperature drying time. During IR predrying of the high moisture nuts, the temperature of the meat at the center of the nut remained considerably below 43°C in the first 150 s, which provided time to drive out a significant amount of moisture while retaining a product quality (Atungulu et al., 2013; Khir et al., 2013, 2014a, b).

Pan et al. (2018) have proven the technical feasibility of using IR drying as an energy-efficient drying technology for walnuts. They found that significant drying time and energy in walnut drying could be saved by separating or sorting the hulled walnuts into two moisture groups before drying and predrying the walnuts using catalytic IR emitters to quickly remove the surface moisture. The sorting of walnuts and IR predrying also improved the quality of walnuts by reducing the overdrying and underdrying of walnuts.

Pan et al. (2018) also designed, built, tested, and demonstrated two IR walnut dryers. The test results from a pilot IR drying facility showed that IR heating for 150, 180, and 240 s resulted in walnut kernel temperatures of 40°C, 45°C, and 50°C, respectively. IR heating resulted in significant moisture removal from the walnut shell compared to that of the kernel. The energy consumption studies showed that sorting the walnuts into two moisture groups and drying them separately could save up to 28.80% of the energy used to dry unsorted walnuts by preventing over drying. The energy saved by IR predrying alone was found to be 11.68%, 12.81%, and 11.70% by IR heating the walnuts to kernel temperatures of 40°C, 45°C, and 50°C, respectively. Additionally, Pan et al. (2018) also tested a commercial-scale IR walnut dryer with a capacity of 10–15 ton/h to IR predry the walnuts by heating to a kernel temperature of 50°C. They found that IR predrying the walnuts reduced the hot air-drying time by 13.55%–26.50% compared to walnuts without IR predrying with the same initial MC. IR predrying resulted in energy savings of 9.96%–19.94% compared to walnuts dried without IR predrying, depending on the IR heating time and the initial MC of walnuts.

4.4 Storage

The storage of in-shell walnuts is a critical process step following the drying operation. Dried walnuts are transported to nearby packing plants and stored until needed for cracking (shelling). Proper storage helps maintain the delicious, crunchy taste, and health benefits of walnuts. If stored improperly, walnuts can quickly turn rancid due to their high oil content. MC, heat, light, and air humidity are primary factors that can cause walnut quality deterioration

(Woodroof, 1967). Enhanced stability of walnut kernels occurs when they are adjusted to the optimum moisture, relative humidity, and temperature.

Maximum flavor and color stability of kernels is at 4% moisture, 55% relative humidity, and a temperature of 4.5°C. Under the aforementioned conditions, the walnuts can be stored for up to 18 months. At temperatures below 3.3°C, walnut meats are stable over a moisture range from 1.5% to 6.0% and relative humidity range of 55% to 56% (Woodroof, 1967). For long-term storage, shelled walnuts can last up to 6 months refrigerated in an airtight container or frozen up to a year.

4.5 Shelling

Normally, dried walnuts are stored in shell form until needed to meet consumer and industrial markets demand for walnut kernels. To recover the kernels, in-shell walnuts must undergo the shelling process. Before shelling, the walnuts are initially graded and separated according to their size. In the United States, in-shell walnuts are classified as jumbo, large, medium, or baby according to USDA standards (Duke, 1989; Woodroof, 1967). The walnuts are then mechanically cracked using specialized equipment.

To remove the shell, the materials are air-separated, screened for size, and sent to electronic laser-sorting units for kernel color and shell removal. The walnut kernels are certified to meet USDA-grade standards and customer specifications (Woodroof, 1967). The product is then packed for shipment to the marketplace. Depending on whether the shelled walnut kernels are for commercial or institutional use, they are either consumer packed or placed in bulk cartons or cans. The walnut shell is the lignocellulosic material forming the thin endocarp of the walnut fruit. Shells hold a promising potential for use in additional value-added products.

5 CHARACTERIZATION OF WALNUT MAIN PRODUCTS

The main products of harvested walnuts include walnuts with hull and without hull (in-shell walnuts). The characterization of walnut properties is considered the first step to efficiently deal with several problems associated with processing cost, energy use, and product quality in the current postharvest processing operations (Khir et al., 2012, 2013, 2014a; Atungulu et al., 2013). Walnut characteristics, including moisture distribution, physical and aerodynamic properties (linear dimensions, shape, sphericity, true density, projected area, and terminal velocity), and their dependence on MC, constitute essential engineering data for developing new approaches and methods for handling, dehulling, sorting, and drying to reduce energy use, increase process efficiency, and enhance product quality (Khir et al., 2014a). The relationship between kernel weight and nut weight, or the kernel ratio, is a useful indicator of good in-shell fruit. A positive correlation between nut and kernel sizes indicates good nut filling (Malvolti et al., 1994).

High variability in the shape, size, and weight of fruit, the color and thickness of shell, and the size, weight, and external color of kernel has been reported for walnuts from different regions of the world (Khir et al., 2014b). Khir et al. (2014a) studied the relationship between MC and terminal velocity for dehulled walnuts and walnuts with hulls at harvest. They found that air with a terminal velocity of $10\,m\,s^{-1}$ could achieve complete separation of walnuts without hulls from those with hulls. Based on their findings, a pneumatic method based on terminal velocity was developed to effectively sort walnuts into different groups with desired moisture ranges. The implementation of this approach led to significant energy savings, reduced transportation and labor expenses, and improved dehulling process, drying efficiency, and end-product quality.

6 BY-PRODUCTS OF PROCESSING OPERATIONS

The by-products of walnut processing include hulls, shells, leaves, and broken kernels. These by-products are produced in large quantities during the walnut processing operations. It has been reported that the walnut shell comprises 67% of the total weight of the walnut fruit (Sürmen and Demirbas, 2003). The leading producing countries, including, China, United States, Iran, Turkey, and Mexico, generated large quantities of by-products as shown in Table 1. The numbers presented in Table 1 are calculated based on walnut production data reported by FAO (2016). The compositions of walnut by-products (shells, green hulls, leaves) are presented in Tables 2 and 3 (Kar, 2011b; Keskin et al., 2012; Uzun and Yaman, 2014). The walnut by-products have a potential to be reprocessed to produce value-added products that have food and nonfood applications but are heavily underutilized, especially on the industrial scale. This is discussed in detail in the following sections.

7 POTENTIAL APPLICATIONS OF BY-PRODUCTS

By-products produced from the walnut processing operations have been reported to have several potential applications. The leaves and green hulls have been used in both cosmetic and pharmaceutical industries (Stampar et al., 2006). Almeida et al. (2008) mentioned that the leaves have been widely used

Table 1 Estimated Quantitates of By-products Produced During Walnut Processing Operations

Country	Production (ton)	Green hull (ton)	Shell (ton)
China	1,785,879	321,458	1,196,539
United State	607,814	109,407	407,235
Iran	405,281	72,951	271,538
Turkey	195,000	35,100	130,650
Mexico	141,818	25,527	95,018
Total	3,135,792	564,443	2,100,981

Table 2 Chemical Compositions of Walnut Shell

Composition	Weight percentage (%) on dry basis
Moisture content	7.71–8.06
Hemicellulose	22.18–22.45
Cellulose	23.95–26.87
Lignin	47.68–48.11
Carbon	47.50–47.97
Hydrogen	6.35–6.35
Nitrogen	0.15–0.46
Oxygen	45.50–47.65
Ash	0.33–3.40

Table 3 Chemical Compositions of Leaves and Green Hulls of Walnuts

	Percentage (%) in extracts by (GC/MS analysis)	
Compound	Laves	Green hull
Ethylene oxide	14.74	83.67
Cyclotrisiloxane, hexamethyl	17.89	5.04
Cyclotetrasiloxane, octamethyl	10.88	2.33
Cyclopentasiloxane, decamethyl	7.56	2.62
6Aza5,7,12,14Tetrathiapentacene	1.17	0.43
Cyclohexane	4.15	0.79
Carbonic acid	3.58	–
Acetic acid, ethyl ester	3.31	–
Ethanol	9.92	–
2-Pentanone	8.71	–
Methane	3.41	–
Benzoic acid	2.08	–
Acetanone	–	1.11
Pentadecane	–	0.08
Nethyl1,3dithiosoindoline	–	0.3

in folk medicine for the treatment of skin inflammations, hyperhidrosis, and ulcers and for their antidiarrheal, anthelmintic, antiseptic, and astringent properties. Additionally, the dry walnut leaves are frequently used as infusions (Pereira et al., 2007). The walnut leaves could be used as a basic material for traditional walnut liqueur, hair shampoos, and soaps (Stampar et al., 2006). Fernández-Agullóa et al. (2013) mentioned that walnuts' green husks have the potential to be used as an economical source of antioxidant and antimicrobial agents for the food and pharmaceutical industries.

Shells have prospective uses, but they are underutilized. The majority of shells produced during walnut processing is generally combusted directly in situ for heating purposes or discarded to the surrounding environment rather than being used to make value-added products (Kar, 2011a, b).

Shells have many properties that render them as potentially suitable raw materials to produce granular activated carbons (GACs) (Soleimani and Tahereh, 2007). GACs can contribute to solving many current environmental pollution problems. GACs are increasingly used in water for removing organic chemicals and metals of environmental or economic concern. Moreover, activated carbons (ACs) have become the most effective adsorbent in treating drinking water and industrial wastewater (Yeganeh et al., 2006). Using walnut shells as a filler has notable competitive advantages such as containing lower amounts of the hygroscopic materials (cellulose and hemicellulose) and higher amounts of the hydrophobic materials (lignin and extractives) compared to wood (Hayashi et al., 2002; Soleimani and Kaghazchi, 2007). Additionally, shells possess unique abrasive qualities that make them ideal in industrial cleaning, such as for the internal parts of aircraft jet engines. Shells can also be used as polishing powders, filling materials, and filtration media to separate crude oil from water (Srinivasan and Viraraghavan, 2008). Broken walnut kernels can also be used for oil extraction. Woodroof (1967) proposed using the nut pieces that do not meet the minimum screen size specifications, known as "meal," for oil production.

8 FOOD APPLICATIONS

8.1 Edible Oil

Walnut kernels have high oil and essential fatty acid contents, making them a good source for commercial production of edible oil (Martinez et al., 2013). Martinez et al. (2010) reported that the oil content in commercial walnut varieties ranges from 620 to 740 $g\,kg^{-1}$ kernel. Walnut oil is composed mainly of triglycerides, of which monounsaturated (primarily oleic acid) and polyunsaturated fatty acids (PUFAs, linoleic and a-linolenic acids) are present in high amounts (Amaral et al., 2003; Crews et al., 2005; Martinez et al., 2006). According to Simopoulos (2002), walnut oil has a perfect 4:1 ratio of n-6: n-3 PUFAs, which was shown to decrease the incidence of cardiovascular risk (Bucher et al., 2002). Fresh walnut oil is very low in free fatty acid concentration, peroxides, and phosphatides so it may be consumed directly, without refining (Martinez et al., 2010). Although walnut oil is not described by the current Committee on Fats and Oils of the Codex Alimentarius, it is produced and commercialized in some countries such as France, Spain, Chile, and Argentina. Typically, the walnut oil is used directly (without refining) for edible purposes, mainly as a salad dressing. It is also used in the cosmetic industry as a component of dry skin creams and antiwrinkle and antiaging products.

8.2 Oil Extraction

The two major goals in walnut oil production are applying an appropriate method to recover the oil from the kernels and improving its oxidative stability. Because of the high unsaturation level of walnut oil, extreme care needs to be taken to prevent oxidative degradation reactions during the processing and storage of the oil (Martinez et al., 2013). The increasing demand for walnut oil encourages finding appropriate methods to enhance its shelf life by keeping oil oxidation at the lowest possible level. It has been reported that walnut oil can be extracted by using the solvent method, a screw press, and supercritical carbon dioxide.

Extraction by solvent was investigated by Martinez et al. (2006). They reported that oil content as high as 740 $g\,kg^{-1}$ kernel (Soxhlet extraction, n-hexane) can be achieved for some commercial walnut varieties. The solvent method has also been studied by Martinez and Maestri (2008) on a laboratory-scale using hexane, methylene chloride, or chloroform/methanol. They mentioned that, although this extraction method has a high extraction yield, it is not suitable on an industrial scale owing to the very high lipid content of the walnut kernel, which requires greater quantities of solvent to extract the oil completely.

As an alternative method to solvent extraction, walnut oil can be extracted by a screw press or a hydraulic press. Screw pressing provides a simple and reliable method of processing small batches of seed (Martinez et al., 2008). It was found that the volumes of recovered oil varied considerably between samples, apparently as a function of the physical characteristics of the kernels and operating conditions of the extraction process. Employing a pilot plant screw press, the highest oil recovery (660 $g\,kg^{-1}$ kernel) was achieved at 7.5 g/100 g kernel moisture and 50°C pressing temperature. Additionally, Martinez et al. (2008) studied the combined effects of seed MC (25, 45, and 75 $g\,kg^{-1}$) and temperature (25°C, 50°C, and 70°C) on oil recovery and quality parameters using a pilot-plant scale screw press. They found that moistening was more beneficial than heating in terms of oil recovery for the range of conditions used in the study. The highest oil recovery (893 $g\,kg^{-1}$ kernel) was obtained at 75 $g\,kg^{-1}$ MC and 50°C.

Additionally, walnut oil extraction employing supercritical carbon dioxide (SC-CO_2) has been investigated by several researchers. Oliveira et al. (2002) and Salgin and Salgin (2006) mentioned that the mass of walnut oil extracted is determined initially by the oil solubility in CO_2. Martinez et al. (2008) examined the effects of different CO_2 pressures (200 and 400 bar) and temperatures (50°C and 70°C) on oil yield and quality and the time required for oil extraction from prepressed walnut kernels. They stated that a significant effect of the operating pressure was observed: under isothermal conditions, an increase in pressure from 200 to 400 bar caused a notable increase in extraction yield and the oil recovery. Similarly, Salgin and Salgin (2006) found that the oil extraction yield increased with CO_2 pressure, obtaining the highest yield at 500 bar. An interesting application of SC-CO_2 extraction technology is the production of reduced-fat walnuts (Martinez et al., 2008).

8.3 Walnut Flour

Walnut flour has been employed in the formulation of various functional food products such as meat, dairy, and bakery products (Serrano et al., 2005). Walnut flour is mainly composed of glutelins (about 70% of the total seed proteins) together with lesser amounts of globulins (18%), albumins (7%), and prolamins

(5%) (Ayo et al., 2008). It provides appreciable amounts of proteins ($450\,g\,kg^{-1}$ on average) (Martinez et al., 2010). Sathe et al. (2009) reported that walnut flour components have positive effects on the nutritional, functional, and sensory characteristics of some developed food products. Cofrades et al. (2008) demonstrated that the incorporation of preheated defatted walnut in the formulation of meat batters improves the thermal gelling ability of myofibrillar proteins, likely because it promotes walnut-muscle protein interactions.

It is important to note that walnut proteins contain all essential AAs required for the needs of a human adult (Savage, 2001). The lysine/arginine ratio in walnut proteins is lower than those observed in other common vegetable proteins, and this fact has been identified as a positive feature in the reduction of atherosclerosis development (Venkatachalam and Sathe, 2006). Moreover, walnut glutelins have been shown to be highly digestible (Kritchevsky et al., 1982). However, protein solubility and consequently AA bioavailability are adversely influenced by phenolic compounds, especially hydrolysable and nonhydrolysable tannins such as those present in walnut kernels. When kernels are whole ground and the oil is extracted, most phenolics remain in the flour, where they may precipitate proteins through various mechanisms, such as hydrophobic and ionic interactions and hydrogen and covalent bonding (Xu and Diosady, 2000).

On the other hand, the inclusion of walnut flour in the formulation of food products could serve as a way of incorporating some specific biologically active components present in the flour. Among them, melatonin has received a special attention because of its beneficial effects on the cardiovascular system and its antioxidant and anticarcinogenic properties. The level of this compound in walnuts appears to be low ($\sim 3.5\,ng\,g^{-1}$), but significant increases in blood melatonin concentration were observed in rats fed a walnut-supplemented diet (Sauer et al., 2001). Dietary melatonin from walnut flour also showed inhibitory effects on the growth of a murine breast tumor (Reiter et al., 2005). Some studies have been carried out to evaluate storage conditions and shelf life of walnut flour (Lavedrine et al., 2000; Lopez et al., 1995). Vanhanen and Savage (2006) found that walnut flour could be preserved from oxidation for up to 26 weeks when stored below 23°C in polypropylene plastic containers with polyethylene sealing lids.

8.4 Extraction of Natural Compounds From Green Hulls

The green hulls of walnut, a major by-product generated during walnut processing, could be a valuable source of natural compounds with antioxidant and antimicrobial properties, but it is scarcely used nowadays (Fernández-Agullóa et al., 2013). Many studies have demonstrated the potential of the green hulls as an economical source of antioxidant and antimicrobial agents (Oliveira et al., 2008; Carvalho et al., 2010; Fukuda et al., 2003; Li et al., 2006; Pereira et al., 2007; Zhang et al., 2009). The beneficial effects derived from the phenolic compounds, such as their anticarcinogenic, antimutagenic, and cardioprotective activities, have been attributed to their antioxidant activity (Madhavi et al., 1996;

Balasundram et al., 2006). The phenolic compound juglone is present in all parts of the walnut and is known for its antimicrobial effect (Stampar et al., 2006). Pereira et al. (2008) also evaluated the antimicrobial activity of six different walnut varieties. Stampar et al. (2006) identified 13 phenolic compounds in walnut green hulls: chlorogenic acid, caffeic acid, ferulic acid, sinapic acid, gallic acid, ellagic acid, protocatechuic acid, syringic acid, vanillic acid, catechin, epicatechin, myricetin, and juglone. Additionally, Oliveira et al. (2008) reported that the green hulls can be an easily accessible source of compounds with health protective potential and antimicrobial activity.

The aqueous extracts of walnuts' green hulls have been studied by Oliveira et al. (2008). They studied the effect of the solvent on the properties of walnut green hull extracts. Solvents of varying polarity were used including water, methanol, ethanol, and their aqueous solutions. They found that the highest extraction yield was obtained with water. Using reducing power and DPPH assays, it was found that the highest total phenols content and antioxidant activities were obtained with 50% aqueous ethanol. In addition, aqueous extracts were able to inhibit the growth of Gram-positive bacteria, proving the antimicrobial capacity of the extracts. In another study, the highest extraction yield was achieved with water (44.11%) and high bioactive potential was shown by the samples extracted with water/ethanol (1:1) (84.46 mg GAE/g extract; EC50 = 0.95 mg/mL for reducing power and EC50 = 0.33 mg/mL for DPPH assay) (Fernández-Agullóa et al., 2013).

Additionally, the supercritical fluid extraction (SFE) technique has been applied to extract the natural compounds. The major advantages of SFE are the rapid equilibration that results in faster and more efficient extraction of analytes than liquid solvent-based extraction and the ease with which the contaminants can be separated from supercritical fluids (Ryoo, 1997). The extracts obtained by SFE technique are of outstanding quality and the yields are comparable with those from organic solvent extraction methods (Kong et al., 2009). Therefore, SFE may serve as a promising technology in extraction of natural compounds from walnuts' green hulls.

Popovici et al. (2012) and Zavoi et al. (2011) conducted a study to explore the applicability of the SFE process for effective extraction of bioactive compounds from the green hulls, examine bioactive compound compositions of walnut green husk extract using UV/Vis spectra, determine the reaction kinetics of DPPH free radical with walnut green husk extract and its scavenging activity, and establish the reducing power (EC_{50}) of the walnut green husk extracts. They found that the green hull extracts obtained by SFE showed polar properties, so chloroform was used as a solvent to analyze the antioxidant potential and content of biologically active compounds. Identification of bioactive compounds by UV/Vis spectra clearly revealed that the extracts contained phenolic acids (237 and 290 nm), flavonoids (333 nm), and carotenoids (417, 457, 484, and 537 nm). The total phenolic content (by Folin-Ciocalteu assay) was 477.59 mg/g in the green hull extracts, and the extraction yield was 5.29%. To increase the extraction efficiency and consequently reduce the extraction time of biologically active compounds and extraction yields from the green hulls, the study was proposed to increase the polarity of the carbon dioxide solvent by addition of a small amount of a liquid cosolvent (modifier).

9 NONFOOD APPLICATIONS

9.1 Reinforced Composites

The walnut by-products, especially shells, may be used for the development of natural particle-based composites. The advantages of natural particles made from walnut shells over traditional particles include low cost, acceptance with good specific strengths and modulus, economic viability, low density, reduced tool wear, enhanced energy recovery, reduced dermal and respiratory irritation, and good bio-degradability (Srivastava et al., 2013; Bolton, 1995). Consequently, the natural plant particles made from walnut shells have received increasing interest in recent years and attracted the attention of researchers due to their low density with high-specific strengths, abundance, availability, renewability, and environmental-friendly production.

Srivastava et al. (2013) found that addition of walnut shell particles improves the ultimate compressive strength of composite materials. They said that the ultimate compressive strength increased considerably due to addition of walnut shell particles. Moreover, it has been reported that the bio-based composites can potentially replace polymer-based composites and wood because of their attractive specific properties, lower cost, simple process technologies, eco-friendliness, and recyclability (Andrzej et al., 2010).

9.2 Activated Carbon

The walnut processing operations produce a huge amount of shells worldwide. However, much of these shells are underutilized. The preparation of AC from walnut shells could increase economic return and reduce pollution caused by discarding them to the surrounding environment (Alslaibi et al., 2012). AC made from walnut shells was found to be about 1.7 times larger than that of commercial ACs from Samchully and Calgon (Bae et al., 2014). Conclusively, the use of AC made from walnut shells as an adsorbent is a cost-effective alternative and reduces environmental contamination. AC has highly developed porosity, high surface area ($>1000\,m^2/g$), variable characteristics of surface chemistry, and high degree of surface reactivity, making it a very effective adsorbent for the removal of a wide variety of organic and inorganic pollutants dissolved in aqueous media or from gaseous environments (Kim et al., 2001; Bae et al., 2014). Moreover, AC is an amorphous form of carbon that is specially treated to produce a highly developed internal pore structure and a large surface area, thus producing a reasonably cheap and excellent adsorbent (Sainz-Diaz and Griffiths, 2000).

Walnut shells have been investigated as AC precursors with a highly developed internal pore structure and large surface area, making them a reasonably cheap and excellent adsorbent and a material with renewed attention. The properties of the produced AC depend on the precursor, activation method (physical, chemical), type of activating agent, and process conditions (temperature, retention time, and impregnation ratio). Bae et al. (2014) reported that the optimum conditions for producing AC from walnut shells were 1.5 h carbonization at 700°C followed by 0.5 h activation at 1000°C. They also found that the surface area and iodine number of AC made from walnut shells were 1450 and 1200 mg/g, respectively. A pore-distribution analysis revealed that most pores had a pore diameter and adsorption capacity for surfactants about 2 times larger than those of commercial AC, indicating that AC made from walnut shells can be used as an efficient adsorbent.

Because the world consumption of ACs is steadily increasing, new applications are always emerging, particularly those concerning environmental pollution remediation (Soleimani and Tahereh, 2007). This environmental concern makes AC made from walnut shells not only technically viable, but also economically feasible. Important applications of AC are related to their use in water treatment for the removal of flavor, color, odor, and other undesirable organic impurities from drinking water. AC is also used in industrial wastewater and gas treatment due to the need for environment protection and for material recovery purposes (Soleimani and Kaghazchi, 2007).

9.3 Particleboards

Walnut shells could play an important role in the manufacture of value-added wood-based panels such as particleboards. Using walnut shells as a raw material in particleboard manufacturing could help solve raw material shortage in the particleboard industry as well as diminish environmental problems regarding their burning (Zahedi et al., 2013). Furthermore, using this by-product in particleboard manufacturing could generate a second income stream for farmers (Pirayesh et al., 2012). Some studies have demonstrated the feasibility of using walnut shells to produce value-added particleboards.

Hamidreza et al. (2013) reported that the addition of walnut shells into particleboard panels significantly reduced their formaldehyde emissions and improved their water resistance. Decreasing formaldehyde emissions from particleboards containing walnut/almond shell particles is important for furniture materials used in indoor environments. The study also confirmed that the walnut shells can be an alternative raw material or filler in the manufacture of wood-based particleboards used in indoor environments due to lower thickness swelling, water absorption, and formaldehyde emission. Moreover, using these underutilized by-products could decrease the pressure on forest resources and create some job opportunities (Pirayesh et al., 2013).

10 CONCLUSIONS

Despite producing one of the healthiest foods, walnut processing operations are accompanied by the generation of considerable quantities of by-products, including leaves, hulls, broken kernels, and shells. The improper disposal of such by-products causes various environmental concerns, which negatively reflects on the sustainability of the walnut processing industry. Finding value-added applications for these by-products can not only eliminate these environmental concerns, but also create an additional revenue streams for the walnut industry. This chapter includes discussion of the production, economic value, health benefits, and processing operations of walnuts. Characterization of the walnut main and by-products and processing technologies used to produce value-added products from walnut wastes are described in this chapter as well. Although many studies have been conducted on the potential uses of the by-products produced during the walnut processing operations, most of them have not been scaled up to an industrial scale. The problem that many walnut processors face is that they are often unable to find appropriate technologies and markets to efficiently utilize the by-products. Therefore, more research is needed to fully understand the value of the by-products and advance their applications at a large scale to produce value-added products with reduced costs.

References

Alasalvar, C., Bradley, W.B., 2015. Review of nut phytochemicals, fat-soluble bioactives, antioxidant components and health effects. British J. Nutr. 113, S68–S78.

Albert, C.M., Gaziano, J.M., Willett, W.C., Manson, J.E., 2002. Nut consumption and decreased risk of sudden cardiac death in the Physicians' Health Study. Arch. Intern. Med. 162 (1382–1387), 2002.

Almeida, I.F., Fernandes, E., Lima, J.L.F.C., Costa, P.C., Bahia, M.F., 2008. Walnut (*Juglans regia*) leaf extracts are strong scavenger of pro-oxidant reactive species. Food Chem. 106, 1014–1020.

Alslaibi, T., Abustan, I., Azmeir, M., Abu Foul, A., 2012. Review: comparison of agricultural by-products activated carbon production methods using surface area response. In: Awam International Conference on Civil Engineering (AICCE'12), Park Royal Penang Resort 28th—30th August.

Amaral, J.S., Casal, S., Pereira, J.A., Seabra, R.M., Oliveira, B.P.P., 2003. Determination of sterol and fatty acid compositions, oxidative stability, and nutritional value of six walnut (*Juglans regia* L.) cultivars grown in Portugal. J. Agric. Food Chem. 51, 7698–7702.

Anderson, K.J., Teuber, S.S., Gobeille, A., Cremin, P., Waterhouse, A.L., Steinberg, F.M., 2001. Walnut polyphenolics inhibit in vitro human plasma and LDL oxidation. J. Nutr. 131, 2837–2842.

Andrzej, K., Abdullah, A.M., Volk, J., 2010. Barley husk and coconut shell reinforced polypropylene composites: the effect of fiber physical, chemical and surface properties. Compos. Sci. Technol. 70 (5), 840–846.

Atungulu, G.G., Pan, Z., 2012. In: Sacramento, C. et al., (Ed.), Development of energy efficient dehulling and drying methods for walnuts. Research Report. California Energy Commission, Energy Innovations Small Grant (EISG) Program.

Atungulu, G.G., Tech, H.E., Wang, T., Fu, R., Wang, X., Khir, R., Pan, Z., 2013. Infrared pre-drying and dry-dehulling of walnuts for improved processing efficiency and product quality. Appl. Eng. Agric. 29 (6), 961–971.

Ayo, J., Carballo, J., Solas, M.T., Jim'enez-Colmenero, F., 2008. Physicochemical and sensory properties of healthier frankfurters as affected by walnut and fat content. Food Chem. 107, 1547–1552.

Balasundram, N., Sundram, K., Sundram, S., 2006. Phenolic compounds in plants and agri-industrial by-products: antioxidant activity, ocurrence and potential uses. Food Chem. 99, 191–203.

Bae, W., Jongho, K., Jinwook, C., 2014. Production of granular activated carbon from food-processing wastes (walnut shells and jujube seeds) and its adsorptive properties. J. Air Waste Manage. Assoc. 64 (8), 879–886.

Bolton, J., 1995. The potential of plant fibres as crops for industrial use. Outlook Agric. 24, 85.

Bucher, H.C., Hengstler, P., Schindler, C., Meier, G., 2002. N-3 polyunsaturated fatty acids in coronary heart disease: a meta-analysis of randomized controlled trials. Am. J. Med. 112, 298–304.

Carvalho, M., Ferreira, P.J., Mendes, V.S., Silva, R., Pereira, J.A., Jer'onimo, C., 2010. Human cancer cell antiproliferative and antioxidant activities of *Juglans regia* L. Food Chem. Toxicol. 48, 441–447.

Cofrades, S., Serrano, A., Ayo, J., Carballo, J., Jim'enez-Colmenero, F., 2008. Characteristics of meat batters with added native and preheated defatted walnut. Food Chem. 107, 1506–1514.

Crews, C., Hough, P., Godward, J., Brereton, P., Lees, M., Guiet, S., 2005. Study of the main constituents of some authentic walnut oils. J. Agric. Food Chem. 53, 4853–4860.

Davis, L., Stonehouse, W., Loots, D.T., Mukuddem-Petersen, J., Van der Westhuizen, F., Hanekom, S.J., 2007. The effects of high walnut and cashew nut diets on the antioxidant status of subjects with metabolic syndrome. Eur. J. Nutr. 46, 155–164.

Duke, J.A., 1989. Handbook of Nuts. CRC Press, London.

EFSA Panel on Dietetic Products, Nutrition and Allergies (NDA), 2011. Scientific opinion on the substantiation of health claims related to walnuts and maintenance of normal blood LDL-cholesterol concentrations (ID 1156, 1158) and improvement of endothelium-dependent vasodilation (ID 1155, 1157) pursuant to Article 13(1) of Regulation (EC) no. 1924/2006. EFSA J. 9, 2074.

FAO, 2016. FAOSTAT Data. Food and Agriculture Organization. http://www.fao.org/faostat/en/#data/QC.

Fernández-Agullóa, A., Pereira, E., Freire, M.S., Valentão, P., Andrade, P.B., González-Álvarez, J., Pereirab, J.A., 2013. Influence of solvent on the antioxidant and antimicrobial properties of walnut (*Juglans regia* L.) green husk extracts. Ind. Crop. Prod. 42, 126–132.

Fraser, G.E., Sabaté, J., Beeson, W.L., Strahan, T.M., 1999. A possible protective effect of nut consumption on risk of coronary heart disease. The adventist health study. Arch. Intern. Med. 152, 1416–1424.

Fukuda, T., Ito, H., Yoshida, T., 2003. Antioxidative polyphenols from walnuts (*Juglans regia* L.). Phytochemistry 63, 795–801.

Hamidreza, P., Hossein, K., Ayoub, S., 2013. Effect of using walnut/almond shells on the physical, mechanical properties and formaldehyde emission of particleboard. Compos. Part B 45, 858–863.

Hayashi, J., Horikawa, T., Takeda, I., Muroyama, K., Ani, A., 2002. Preparing activated carbon from various nutshells by chemical activation with K2CO3. Carbon 40 (13), 2381–2386.

Iwamoto, M., Sato, M., Kono, M., Hirooka, Y., Sakai, K., Takeshita, A., Imaizumi, K., 2000. Walnuts lower serum cholesterol in Japanese men and women. J. Nutr. 130, 171–176.

Kader, A.A., 2002. Post-Harvest Technology of Horticultural Crops. Publication 3311, University of California, Department of Agriculture and Natural Resources, Oakland, CA.

Kar, Y., 2011a. Catalytic pyrolysis of car tire waste using expanded perlite. Waste Manag. 31, 1772–1782.

Kar, Y., 2011b. Co-pyrolysis of walnut shell and tar sand in a fixed-bed reactor. Bioresour. Technol. 102, 9800–9805.

Keskin, D., Ceyhan, N., Ugur, A., 2012. Chemical composition and in vitro antimicrobial activity of walnut (*Juglans regia*) green husks and leaves from West Anatolia. J. Pure Appl. Microbiol. 6 (2), 583–588.

Kim, J.W., Sohn, M.H., Kim, D.S., Sohn, S.M., Kwon, Y.S., 2001. Production of granular activated carbon from waste walnut shell and its adsorption characteristics for Cu^{2+} ion. J. Hazard. Mater. 85, 301–315.

Khir, R., Pan, Z., Atungulu, G.G., Thompson, J.F., 2012. Characterization of physical and aerodynamic properties of walnuts. Institute of Food Technologist (IFT) Annual Meeting and Food Expo. IFT, Chicago, IL.

Khir, R., Pan, Z., Atungulu, G.G., Thompson, J.F., Shao, D., 2013. Size and moisture distribution characteristics of walnuts and their components. Food BioTechnol. Int. J. 6 (3), 771–782.

Khir, R., Pan, Z., Atungulu, G.G., Thompson, J.F., 2014a. Characterization of physical and aerodynamic properties of walnuts. Trans. ASABE 57 (1), 53–61.

Khir, R., Pan, Z., Atungulu, G.G., Thompson, J.F., Zheng, X., 2014b. Moisture-dependent color characteristics of walnuts. Int. J. Food Prop. 17, 877–890.

Kong, Y., Fu, Y.-J., Zu, Y.-G., Liu, W., Wang, W., Hua, X., Yang, M., 2009. Ethanol modified supercritical fluid extraction and antioxidant activity of cajaninstilbene acid and pinostrobin from pigeonpea [*Cajanus cajan* (L.) Millsp.] leaves. Food Chem. 117 (1), 152–159.

Kritchevsky, D., Tepper, S.A., Czarnecki, S.K., Klurfeld, D.M., 1982. Atherogenicity of animal and vegetable proteins: influence of the lysine to arginine ratio. Atherosclerosis 41, 429–431.

Lavedrine, F., Zmirou, D., Ravel, A., Balducci, F., Alary, J., 1999. Blood cholesterol and walnut consumption: a cross-sectional survey in France. Prev. Med. 28, 333–339.

Lavedrine, F., Ravel, A., Villet, A., Ducros, V., Alary, J., 2000. Mineral composition of two walnut cultivars originating in France and California. Food Chem. 68, 347–351.

Li, L., Tsao, R., Yang, R., Liu, C.M., Zhu, H.H., Young, J.C., 2006. Polyphenolic profiles and antioxidant activities of heartnut (*Juglans ailanthifolia* var. cordiformis) and Persian walnut (*Juglans regia* L.). J. Agric. Food Chem. 54, 8033–8040.

Lino, M., Marcoe, K., Dinkins, J.M., Hiza, H., Anand, R., 2000. The role of nuts in a healthy diet. Nutr. Insight 23, 1–2.

Lopez, A., Pique, M.T., Romero, A., Aleta, N., 1995. Influence of cold-storage conditions on the quality of unshelled walnuts. Int. J. Refrig. 18, 544–549.

Lowe, E., Rockland, L.B., Yanase, K., 1961. Studies on English walnuts, Juglans regia. Food Technol. 4, 116–117.

Madhavi, D.L., Despande, S.S., Salunke, D.K., 1996. Food Antioxidants. In: Technological, Toxicological and Health Perspectives. Marcel Dekker, New York.

Martinez, M.L., Mattea, M.A., Maestri, D.M., 2006. Varietal and crop year effects on lipid composition of walnut (*Juglans regia* L.) genotypes. J. Am. Oil Chem. Soc. 83, 791–796.

Martinez, M.L., Maestri, D.M., 2008. Oil chemical variation in walnut (*Juglans regia* L) genotypes grown in Argentina. Eur. J. Lipid Sci. Technol. 110, 1183–1189.

Martinez, M.L., Mattea, M.A., Maestri, D.M., 2008. Pressing and supercritical carbon dioxide extraction of walnut oil. J. Food Eng. 88, 399–404.

Martinez, M.L., Diana, O.L., Alicia, L.L., Dami, M.M., 2010. Walnut (*Juglans regia* L.): genetic resources, chemistry, by-products. J. Sci. Food Agric. 90, 1959–1967.

Martinez, M.L., Penci, M.C., Ixtain, V., Ribotta, P.D., Maestri, D., 2013. Effect of natural and synthetic antioxidants on the oxidative stability of walnut oil under different storage conditions. LWT Food Sci. Technol. 51, 44–50.

Malvolti, M.E., Fineschi, S., Pigliucci, M., 1994. Morphological integration and genetic variability in *Juglans regia* L. J. Hered 85, 389–394.

Muñoz, S., Merlos, M., Zambón, D., Rodríguez, C., Sabaté, J., Ros, E., Laguna, J.C., 2001. Walnut enriched diet increases the association of LDL from hypercholesterolemic men with human HepG2 cells. J. Lipid Res. 42, 2069–2076.

Moshfegh, A., Ingwersen, I., Goldman, J., 2007. Nut Consumption on the U.S. and the Contribution to Nutrient Intakes. USDA, ARS, Beltsville, MD.

Oliveira, R., Rodriguez, M.F., Bernardo-Gil, M.A., 2002. Characterization and supercritical carbon dioxide extraction of walnut oil. J. Am. Oil Chem. Soc. 79, 225–230.

Oliveira, I., Sousa, A., Ferreira, I.C.F.R., Bento, A., Estevinho, L., Pereira, J.A., 2008. Total phenols antioxidant potential and antimicrobial activity of walnut (*Juglans regia* L.) green husk. Food Chem. Toxicol. 46, 2326–2331.

Reiter, R.J., Manchester, L.C., Tan, D.X., 2005. Melatonin in walnut: influence on levels of melatonin and total antioxidant capacity of blood. Nutrition 21, 920–924.

Romas, D.E., 1998. Walnut Production Manual. Publication 3373. University of California, Department of Agriculture and Natural Resources, Oakland, CA.

Ros, E.N., P'erez-Heras, A., Serra, M., Gilabert, R., Casals, E., 2004. Walnut diet improves endothelial function in hypercholesterolemic subjects: a randomized crossover trial. Circulation 109, 1609–1614.

Rumsey, T., Lu, Z., 1991. High-Temperature Walnut Drying. Research Report. Walnut Marketing Board, Sacramento, CA.

Rumsey, T.R., Thompson, J.F., 1984. Ambient air drying of English walnuts. Trans. ASAE 27 (3), 942–945.

Ryoo, K.S., 1997. Combined supercritical fluid extraction with carbon adsorption and counterflow oxidative regeneration for contaminated soil. Environ. Eng. Res. 2 (4), 251–260.

Pan, Z., Venkitasamy, C., Khir, R., El-Mashad, H., Zhang, R., McHugh, T., 2018. Demonstration and Commercial Implementation of Energy Efficient Drying for Walnuts. California Energy Commission. Publication number: CEC-500-2018-00x.

Pirayesh, H., Khazaeian, A., Tabarsa, T., 2012. The potential for using Walnut (*Juglans regia* L.) shell as a raw material for wood-based particleboard manufacturing. Compos. Part B. https://doi.org/10.1016/j.compositesb.016.

Pirayesh, H., Hossein, K., Ayoub, S., 2013. Effect of using walnut/almond shells on the physical, mechanical properties and formaldehyde emission of particleboard. Compos. Part B 45, 858–863.

Popovici, C., Gitin, L., Alexe, P., 2012. Characterization of walnut (*Juglans regia* L.) green husk extract obtained by supercritical carbon dioxide fluid extraction. J. Food Pack. 3. 9–1.

Pereira, J.A., Oliveira, I., Sousa, A., Valentão, P., Andrade, P.B., Ferreira, I.C.F.R., Ferreres, F., Bento, A., Seabra, R., Estevinho, L., 2007. Walnut (*Juglans regia* L.) leaves: phenolic compounds, antimicrobial activity and antioxidant potential of different cultivars. Food Chem. Toxicol. 45, 2287–2295.

Pereira, J.A., Oliveira, I., Sousa, A., Ferreira, I.C.F.R., Bento, A., Estevinho, L., 2008. Bioactive properties and chemical composition of six walnut (*Juglans regia* L.) cultivars. Food Chem. Toxicol. 46, 2103–2111.

Salgin, S., Salgin, U., 2006. Supercritical fluid extraction of walnut kernel oil. Eur. J. Lipid Sci. Technol. 108, 577–582.

Sauer, L.A., Dauchy, R.T., Blask, D.E., 2001. Polyunsaturated fatty acids, melatonin, and cancer prevention. Biochem. Pharmacol. 61, 1455–1466.

Sathe, S.K., Venkatachalam, M., Sharama, G.M., Kshirsagar, H.H., Teuber, S.S., Roux, K.H., 2009. Solubilization and electrophoretic characterization of select edible nut seed proteins. J. Agric. Food Chem. 57, 7846–7856.

Savage, G.P., 2001. Chemical composition of walnut (Juglans regis L) grown in New Zealand. Plant Foods Hum. Nutr. 56, 75–82.

Serrano, A., Cofrades, S., Ruiz-Capillas, C., Olmedilla-Alonso, B., Herrero-Barbudo, C., Jim'enez-Colmenero, F., 2005. Nutritional profiles of restructured beef steak with added walnut. Meat Sci. 70, 647–654.

Sainz-Diaz, C.I., Griffiths, A.J., 2000. Activated carbon from solid wastes using a pilot-scale batch flaming pyrolyser. Fuel 79 (15), 1863–1871.

Simopoulos, A.P., 2002. The importance of the ratio of omega-6/omega-3 essential fatty acids. Biomed. Pharmacother. 56, 365–379.

Soleimani, M., Tahereh, K., 2007. Agricultural waste conversion to activated carbon by chemical activation with phosphoric acid. Chem. Eng. Technol. 30 (5), 649–654.

Soleimani, M., Kaghazchi, T., 2007. Agricultural waste conversion to activated arbon by chemical activation with phosphoric acid. Chem. Eng. Technol. 30 (5), 649–654.

Srivastava, N., Singh, V.K., Bhaskar, J., 2013. Compressive behavior of walnut (Juglans L.) shell particles reinforced composite. Usak Univ. J. Mater. Sci. 1, 23–30.

Srinivasan, A., Viraraghavan, T., 2008. Removal of oil by walnut shell media. Bioresour. Technol. 99, 8217–8220.

Stampar, F., Solar, A., Hudina, M., Veberic, R., Colaric, M., 2006. Traditional walnut liqueur–cocktail of phenolics. Food Chem. 95, 627–631.

Sürmen, Y., Demirbas, A., 2003. Cofiring of biomass and lignite blends: resource facilities; technological and environmental issues. Energy Sour. 25, 175–187.

Thompson, J.F., Grant, J.A., 1992. New moisture meter could curb overdrying of walnuts. Calif. Agric. 46 (2), 31–34.

Thompson, J.F., Stone, M.L., Kranzler, G.A., 1985. Modified airflow and temperature hop drying. Trans. ASAE 28 (4), 1297–1300.

Uzun, B.B., Yaman, E., 2014. Thermogravimetric characteristics and kinetics of scrap tyre and *Juglans regia* shell co-pyrolysis. Waste Manag. Res. 32 (10), 961–970.

USDA, 2018. California Walnut Acreage Report. USDA, National Agricultural Statistics Service. https://www.nass.usda.gov/Statistics_by_State/California/Publications/Specialty_and_Other_Releases/Walnut/Acreage/2018walac.pdf.

Vanhanen, L.P., Savage, G.P., 2006. The use of peroxide value as a measure of quality for walnut flour stored at five different temperatures using three different types of packaging. Food Chem. 99, 64–69.

Venkatachalam, M., Sathe, S., 2006. Chemical composition of selected edible nut seeds. J. Agric. Food Chem. 54, 4705–4714.

WCRF, 1997. American Institute for Cancer Research and World Cancer Research Fund, Food, Nutrition and the Prevention of Cancer: A Global Perspective, BANTA Book Group, Menasha, WI. AVI Publishing Company, INC, London, England.

Woodroof, J.G., 1967. Tree Nut: Production Processing Products. vol. 2. AVI Publishing Company, INC, London, England.

Xu, L., Diosady, L.L., 2000. Interactions between canola protein and phenolic compounds in aqueous media. Food Res. Int. 33, 725–731.

Yeganeh, M.M., Kaghazchi, T., Soleimani, M., 2006. Effect of raw materials on properties of activated carbons. Chem. Eng. Technol. 29 (10), 1247–1251.

Zahedi, M., Pirayesh, H., Khanjanzadeh, H., Tabar, M.M., 2013. Organo-modified montmorillonite reinforced walnut shell/polypropylene composites. Mater. Des. 51, 803–809.

Zambón, D., Sabaté, J., Muñoz, S., Campero, B., Casals, E., Merlos, M., Laguna, J.C., Ros, E., 2000. Substituting walnuts for monounsaturated fat improves the serum lipid profile of hypercholesterolemic men and women. Ann. Int. Med. 132, 538–546.

Zavoi, S., Fetea, F., Ranga, F., Pop, R.M., Baciu, A., Socaciu, C., 2011. Comparative fingerprint and extraction yield of medicinal herb phenolics with hepatoprotective potential, as determined by UV/Vis and FT-MIR spectroscopy. Not. Bot. Hort. Agrobot. 39 (2), 82–89.

Zhang, Z., Liao, L., Moore, J., Wua, J., Wang, Z., 2009. Antioxidant phenolic compounds from walnut kernels (*Juglans regia* L.). Food Chem. 113, 160–165.

Zhao, G., Etherton, T.D., Martin, K.R., West, S.G., Gillies, P.J., Kris-Etherton, P.M., 2004. Dietary alpha-linolenic acid reduces inflammatory and lipid cardiovascular risk factors in hypercholesterolemic men and women. J. Nutr. 134, 2991–2997.

Further Reading

Bansal, R.C., Donnet, J.B., Stoeckli, F., 1988. Active Carbon. Marcel Dekker Inc., New York and Basel.

Biswal, B., Kumar, S., Singh, R.K., 2013. Production of hydrocarbon liquid by thermal pyrolysis of paper cup waste. J. Waste Manag. 2013, 1–7.

Bridgwater, A.V., 1990. Grassi, G., Gosse, G., dos Stantos, G. (Eds.), Biomass for Energy and Industry. Elsevier Applied Science, London.

Chowdhury, R., Sarkar, A., 2012. Reaction kinetics and product distribution of slow pyrolysis of Indian textile wastes. Int. J. Chem. React. Eng. 10. https://doi.org/10.1515/1542-6580.2662.

Foster, K.L., Fuerman, R.G., Economy, J., Larson, S.M., Rood, M.J., 1992. Adsorption characteristics of trace volatile organic compounds in gas streams onto activated carbon fibres. Chem. Mater. 4, 1068–1073.

Horne, P.A., Williams, P.T., 1996. Influence of temperature on the products from the flash pyrolysis of biomass. Fuel 75, 1051–1059.

Liu, A., YoonKook, P., Zhiliang, H., Baowu, W., Ramble, O.A., Prosanto, K.B., 2006. Product identification and distribution from hydrothermal conversion of walnut shells. Energy Fuel 20, 446–454.

Potter, D., Gao, F., Baggett, S., McKenna, J.R., McGranahan, G.H., 2002. Defining the sources of paradox: DNA sequence markers for North American walnut (Juglans L.) species and hybrids. Sci. Hortic. 94, 157–170.

Pereira, C.G., Meireles, M.A.A., 2010. Supercritical fluid extraction of bioactive compounds: fundamentals, applications and economic perspectives. Food Biotechnol. 3, 340–372.

Wyrzykowska-Ceradini, B., Gullett, B.K., Tabor, D., Touati, A., 2011a. Waste combustion as asource of ambient air polybrominated diphenylethers (PBDEs). Atmos. Environ. 45, 4008–4014.

Wyrzykowska-Ceradini, B., Gullett, B.K., Tabor, D., Touati, A., 2011b. PBDDs/Fs and PCDDs/Fs in the raw and clean flue gas during steady state and transient operation of a municipal waste combustor. Environ. Sci. Technol. 45, 5853–5860.

CHAPTER 17
Coffee

Adriana S. Franca, Leandro S. Oliveira
Universidade Federal de Minas Gerais, Belo Horizonte, Brazil

Chapter Outline

1 INTRODUCTION

Coffee is one of the most widely consumed pharmacologically active beverages worldwide and a rather relevant food commodity from an economic standpoint. The beverage is prepared by brewing the roasted and ground beans, with the methods of brewing being quite diverse. The end product of coffee processing, i.-e., the roasted and ground beans, is a result of the roasting of crude beans (also termed 'green beans'). Green beans are thus the commodity that is largely produced and commercialized worldwide, with the most valuable commercial species being *Coffea arabica* (or simply Arabica) and *Coffea canephora* (Robusta).

Coffee is produced solely within the commonly termed 'Coffee Belt,' the regions of the world delimited by the Tropic of Cancer (23.43695°N) and the Tropic of Capricorn (23.43695°S) (Fig. 1). According to the International Coffee Organization (ICO), in 2016, the major coffee producing countries were Brazil, with a production of 3.30 million tonnes; Vietnam, with 1.53 million tonnes; Colombia, with 0.87 million tonnes; Indonesia, with 0.69 million tonnes; Honduras,

Integrated Processing Technologies for Food and Agricultural By-Products
https://doi.org/10.1016/B978-0-12-814138-0.00017-4

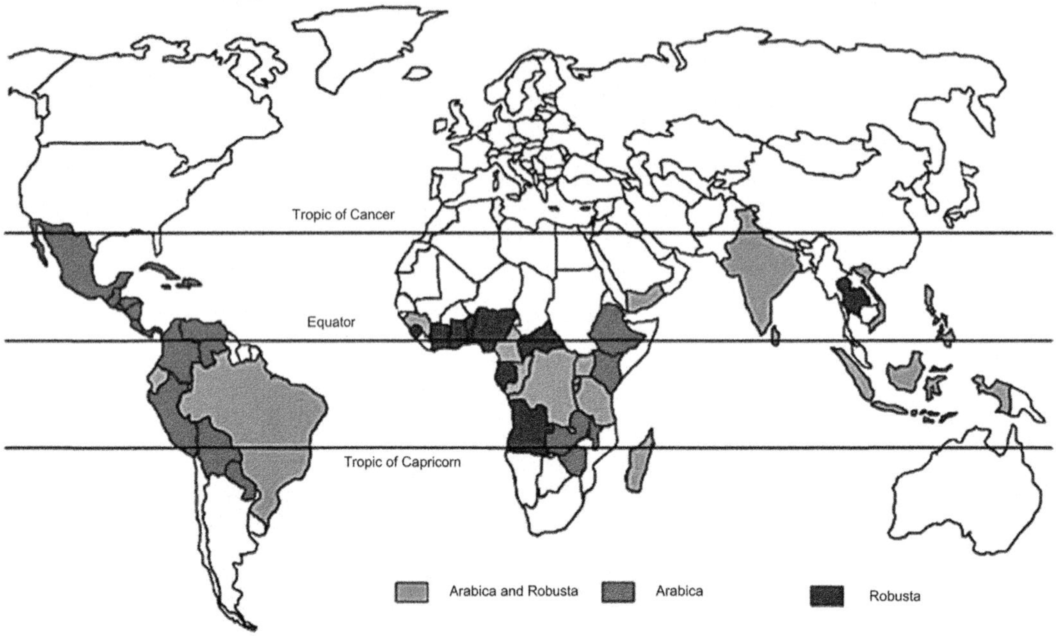

FIG. 1
Coffee producing countries (Coffee Belt).

with 0.46 million tonnes; and Ethiopia, with 0.43 million tonnes (ICO, 2018a). These same countries were also the major exporting countries, with Brazil being the major exporter (0.2 million tonnes), followed by Vietnam (0.12), Colombia (0.07), Indonesia (0.05), Peru (0.04), and Honduras (0.03) (ICO, 2018b). Coffee ranks second only to crude oil and its derivatives as the most traded commodity in the world, measured by monetary value (Obruca et al., 2015). In regard to worldwide coffee consumption, Brazil is the only producing country with a significant consumption, 1.23 million tonnes in the 2015/2016 crop years. The major individual consumer country is the United States with 1.52 million tonnes in the 2015/2016 crop years. Other major consumers are the European Union (2.56), Japan (0.47), Indonesia (0.27), and the Russia Federation (0.26) for the same 2015/2016 crop years (ICO, 2018c).

2 THE COFFEE PLANT AND FRUIT

Botanically, coffee belongs to the genus *Coffea* of the *Rubiaceae* family, with the commercially relevant species being *C. arabica* and *C. canephora*, representing, respectively, approximately two-thirds and one-third of the world coffee production. Arabica species is cultivated at high altitudes (>1000 m above sea level) in the tropics and subtropics, where the climate is cooler, whereas Robusta species is cultivated at relatively low altitudes (<800 m above sea level). Although Robusta coffees tolerate higher temperatures, heavier rainfall and presents higher yields than Arabicas, the latter are considered, by the discerning coffee cuppers and drinkers, to produce beverages of higher qualities than the former (Anzueto et al., 2005) and thus is commercialized at higher prices in international markets.

Coffee is a perennial evergreen dicotyledon plant, cultivated exclusively in tropical regions where the climate is warm (average temperatures of 18–24°C) and rainfall is abundant (ideally, 1500–2500 mm of rainfall in a 9-month period and a 3-month dry period preferably coinciding with the harvest of the fruits). The plant can reach the heights of 6 m for the Arabica species and of 10 m for the Robusta species in the wild (Farah and Santos, 2015), however, in plantations, the tree is pruned to a maximum height of 3 m to facilitate harvesting of the fruits, at which point it is usually called a shrub (Steiman, 2008).

The coffee tree has a prominent vertical trunk with primary, secondary, and tertiary lateral branches, which are termed "suckers" and "stems" at the developmental and final stages, respectively. The leaves are of a shiny dark green color with conspicuous veins and each leaf pair decussating to the pair immediately above it. The roots of the coffee plant are shallow, with its system being comprised of a main taproot with lateral and small feeder roots. The feeder roots of the Arabica species penetrate deeper into the soil than those of the Robustas, which concentrate close to the soil surface. Inflorescence takes place at the leaves axil on lateral branches, in conditions of 8–11 h of daylight, with flowers of five to six petals and of white to pink colors. Arabica flowers tend to be smaller than those of Robustas. Pollination occurs within 6 h of flowering (Anzueto et al., 2005, Murthy and Naidu, 2012a; Farah and Santos, 2015). After pollination, the flowers wither and a 10–15-mm-long fruit develops, with ripening periods of 7–9 months and of 9–11 months for the Arabica and Robusta species, respectively (Anzueto et al., 2005).

The coffee fruit, also referred to as cherry, has a green exocarp (called skin or peel) when immature, and either a red or a yellow exocarp when ripe, depending on the variety. The exocarp is comprised of a monolayer of parenchymatous compact, polygonal cells covered with a waxy substance, with the function to protect the inner content of the fruit. From the exocarp inward, the content of the cherry is comprised of an approximately 2-mm-thick gelatinous-pectic pulp (mesocarp), enclosing a pectinaceous layer of mucilage firmly adhered to a parchment, which, in turn, loosely encapsulates the beans or seeds (usually two per fruit, comprised of endosperm and embryo), the latter being covered by a thin closely-fitting sclerotic layer called silverskin (Fig. 2) (Franca and Oliveira, 2016).

The main product of commercial interest constitutes the coffee beans, representing 50%–55% of the dry matter of the fruit and varying in size, shape, and density according to the genotype (species) and the agricultural practices. Arabica beans are shaped like a half triaxial ellipsoid with a sigmoidal longitudinal slit on its flat side, whereas Robusta beans are rounder with a slightly straight central slit (Fig. 3).

FIG. 2
Transversal cut of a coffee fruit depicting its various distinguished parts.

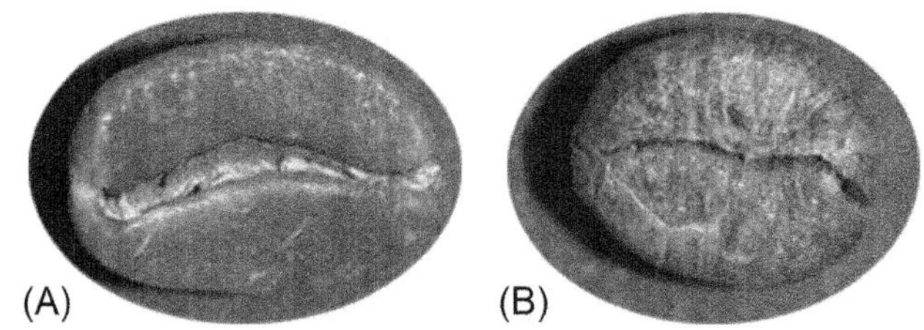

FIG. 3
Morphological traits of the (A) Arabica and (B) Robusta beans.

3 COFFEE HARVESTING

Coffee fruits of a single plant usually do not mature at the same time and, even on a single branch, it is possible to find fruits at distinct stages of maturity. Because of this heterogeneity in maturity stages, harvesting is usually carried out when the majority of fruits are ripe (Farah and Santos, 2015). There are three main types of harvesting techniques: manual selective picking; manual or mechanical strip picking; and mechanical harvesting.

Manual selective picking is the harvesting process in which only the mature and overripe coffee fruits are handpicked and the immature ones are left on the tree for future harvesting. In this type of harvesting, the laborers pick the fruits and place them into baskets and the full baskets are emptied in larger bags. At the end of the harvesting day, the daily harvest is spread out and the occasional debris are removed from the lot and the overripe fruits are separated. Subsequently, the graded fruits are weighed and payments to the pickers are apportioned by the total weight of harvested fruits. Manual stripping is the harvesting process in which the pickers place a canvas on the ground below the coffee tree to ensure the fruit does not come into contact with the ground, grab a branch laden with coffee cherries at its base, close to the trunk, and pull outward until all the fruits, irrespective of their maturity stage, are stripped from the branch and collected onto the canvas. When all the branches are completely stripped, the canvas is emptied into bags, which are taken to be weighed when full so that the pickers are paid accordingly. The mechanical stripping method is similar to the manual stripping, except that the pickers use mechanical strippers comprised of a type of grass trimmer in which the working end is a vibrating scissors-like assembly (Thurston, 2013).

In the mechanical harvesting technique, coffee cherries are stripped from the trees by a portico-shaped combine operating on top of the coffee plants row. As the machine moves along the coffee plant row, the trees are drawn into the small gap between parallel rotating vertical shafts with slightly horizontally inclined rods attached to it. In this gap, the branches are beaten and horizontally shaken with enough force to knock the fruits out of them and down into a collector, the latter comprised of retractile blades that close the space under the trees. The machine operator is positioned at a high point in the picking unit for optimal visibility and control of the harvesting process. In this technique, all the fruits are harvested from the tree, regardless of their maturity stage.

Although the technique of selective picking is the one that assures high-quality coffees and associated high revenues for the producers, for it results in lower percentages of immature fruits being harvested, it requires a large rural labor force with consequent high labor costs, potentially increasing farming costs by 40% when compared to mechanical harvesting. In countries where labor costs are high, e.g., Brazil, it is more cost-effective to harvest all the coffee in one single sweep of the plantation. In that regard, strip-picking and mechanical harvesting are the preferred techniques, regardless of the disadvantage of producing lots with fruits in various stages of maturation, consequently leading to nonuniform lower quality products and forcing coffee producers to rely on postharvest technologies to achieve higher quality products. Manual stripping can be carried out in any type of topography whereas mechanical harvesting is not suitable for terrains with reasonably steep slopes.

4 POSTHARVEST PROCESSING OF COFFEE

Subsequently to the harvesting, the coffee fruit undergoes what is termed 'primary processing,' which is basically the separation of the coffee beans from the rest of the fruit. The general steps involved in the primary processing of coffee are displayed in Fig. 4. There are three major processing types, dry, semi-wet, and wet processing. Dry processing is a relatively straightforward technique in which the recently harvested fruits (30%–65% moisture w.b.) are spread over a large concrete patio (or over suspended fine mesh nettings) to be sun-dried for a period of 15–20 days in favorable climates and up to 30 days in

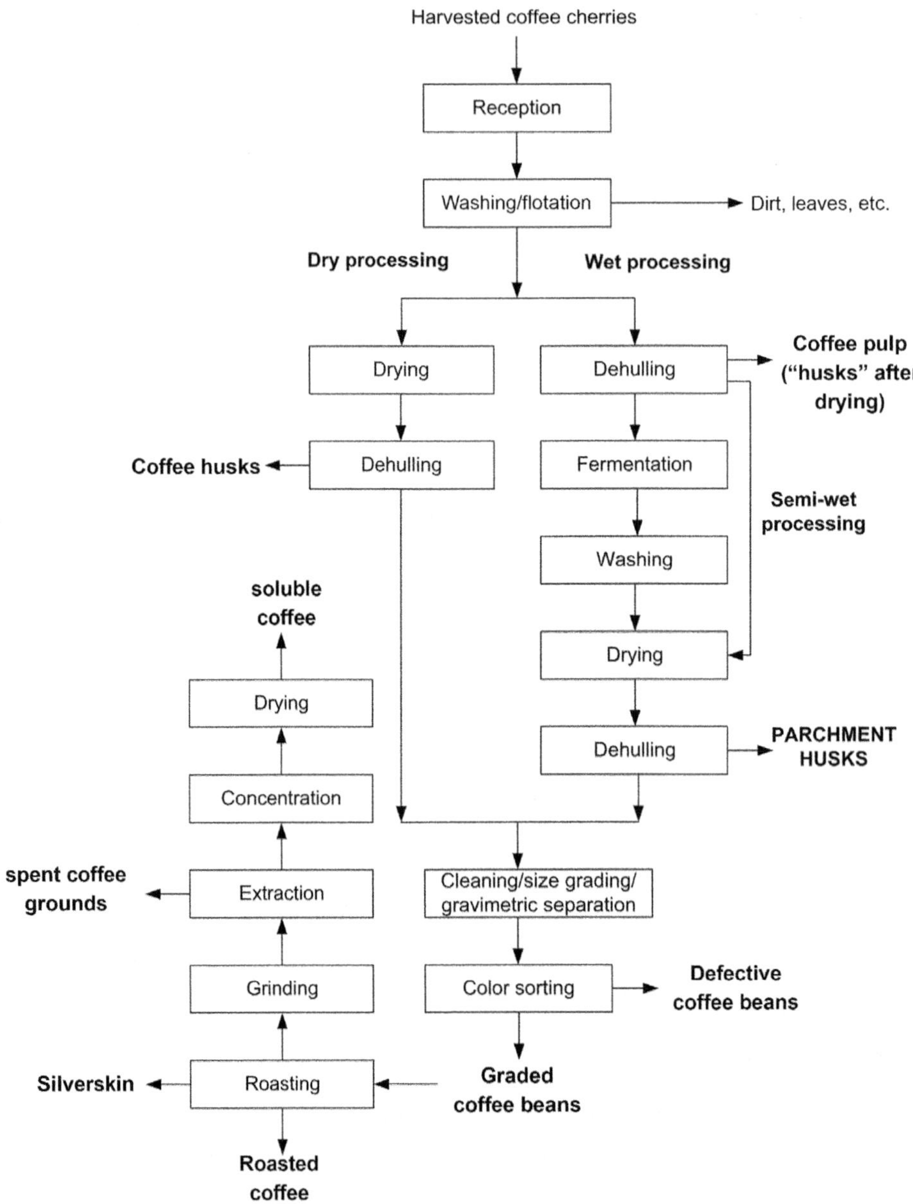

FIG. 4

Coffee postharvest processing steps. Adapted from Franca, A.S., Oliveira, L.S., 2009. Coffee processing solid wastes: current uses and future perspectives. In: Ashworth, G.S., Azevedo, P. (Eds.), Agricultural Wastes. Nova Publishers, New York, pp. 155–189.

unfavorable ones (Borém et al., 2008). The spread out cherries must be raked frequently during daylight hours to prevent mold proliferation and ensure homogeneous drying of the fruits (Franca and Oliveira, 2009). Coffee needs to be dried to 10%–11% moisture (w.b.) to comply with adequate storage conditions (i.e., to prevent spoilage) and with commercialization standards. Usually, mechanical dryers are used to remove the final percentage points of moisture. In farms where the construction of large patios is not feasible, the cherries can be dried directly in mechanical dryers, with the rotary horizontal dryer being the most appropriate for this case. After drying, the dried fruits are dehulled to separate the beans from the dry pulp and parchment. The beans are stored in large vertical rectangular bins and further bagged for the next steps of processing and commercialization.

In semi-wet processing, the exocarp and mesocarp are mechanically removed from the freshly harvested fruit (dehulling), with the parchment still remaining loosely attached to the bean. Before dehulling is performed, the inferior quality fruits are separated by flotation, a process in which the superior quality fruits sink and the inferior ones float in flowing water in a washing tank. Following flotation, the graded fruits are dehulled in a mechanical pulping machine (the pulper), in which the fruits are squeezed between a screw conveyor and a perforated stationary casing. The clearance between the conveyor and the casing is finely adjusted to allow only the mature fruits to burst into separated parchment beans and pulp that exit the system through the perforations in the casing, while the intact immature ones are pushed forward. Thus, this dehulling process also promotes the separation of immature fruits from the mature ones. The immature cherries are taken to the patios for sun-drying as whole fruits and the parchment beans and pulp are moved to a rotating sieve where they are further separated from each other. The parchment beans are then spread out in the patios for sun-drying in a period of 8–12 days, with frequent raking during daylight. The pulps are collected separately in small carts for further in-farm use, such as fertilizer or feedstock for combustion systems. A mechanical dryer is also used in this case for the removal of a few percentage points of moisture for compliance with storage and commercialization conditions. The dried parchment beans are then submitted to mechanical dehulling for the removal of the parchment from the beans and subsequently stored in vertical rectangular bins. In the wet processing of coffee, the freshly harvested fruits are dehulled exactly in the same way as in the semi-wet process, but the parchment coffee are subsequently subjected to fermentation in rectangular bins, to remove the attached residual pulp and mucilage, for a period of 12–36 h, subsequently washed and spread out in the patios for drying. From this step forward, the further processing of the dried parchment beans is identical to that of the semi-wet processing.

The processing operations carried out subsequently to drying, dehulling, and storage are denominated curing (Vincent, 1987) and include cleaning, size grading, density, and colorimetric sorting and bagging in standard 60 kg bags. Following these operations, the beans are ready for commercialization with the prices apportioned by the quality of the separated fractions. Lower prices are apportioned to smaller beans, separated in the size grading step, and even lower prices are apportioned to the defective beans (i.e., immature, sour, and black beans), which are separated from graded ones in the colorimetric sorting step. Cleaning and density sorting are carried out by gravimetric separation, in which unwanted materials, such as unhulled cherries, bits of husks, foreign seeds and misshapen, discolored or broken beans are removed from the lot. Size grading is done by vibrating sieves with distinct perforation sizes. In the last curing step, color sorter machines are employed in which the coffee beans individually pass by an electronic eye or camera system that will, according to a wavelength measurement, reject it by air blowing into a low-quality-bean lot or allow it to pass into the high-quality lot.

Following commercialization, coffee beans undergo a series of operations to generate the final products of consumption, which are the roasted and ground coffee and the soluble and instant coffees. These processing operations are usually termed 'secondary processing' (Franca and Oliveira, 2009). Roasting is a process in which the coffee beans are subjected to thermal treatment to promote a series of internal complex chemical reactions that will lead to the desired aroma and flavor of the coffee by means of significant changes in the beans chemical composition. During roasting, the beans also undergo significant physical changes due to intensive dry matter loss in the form of gases (e.g., water vapor, CO_2, and volatile organic compounds). The most commonly employed technology for roasting coffee is a batch process in which the beans are charged into rotary horizontal cylinders, where hot air is blown inside to heat the tumbling beans up to 200–220°C, for a period of time sufficient to achieve the desired degree of roast. The degree of roast is usually assessed by measurements of light reflectance from ground roasted beans or simply by visual inspection of their color by an operator (Dutra et al., 2001). Upon achieving the desired roasting degree, the lot is quenched by the spraying of water inside the roaster and discharged to be further cooled by a stream of air. Upon reaching the desired temperature (usually smaller than 40°C), the beans are sent to storage bins either by elevator cups or pneumatic transport, where they rest for a period no shorter than 8 h to equilibrate the pressure inside the beans (estimated to be ~6 atm after quenching at the end of roasting) (Clarke, 1987a), by the release of large amounts of CO_2, before they are sent to a grinder for comminution and subsequent packaging.

After roasting and grinding, coffee can be commercialized as such for domestic or commercial brewing or be further processed to produce soluble or instant coffee. In the production of both soluble and instant coffee, there is a common operation that is the extraction of solubles from the roasted and ground coffee, ideally with a concentration of soluble solids of 15%–25% (*w*/*w*) in the final extract (Clarke, 1987b). Extraction is most commonly carried out with heated liquid water (~175°C) under pressurized conditions in a series of 5–8 interconnected percolator columns (Mussatto et al., 2011). In this system, the pure feed water is firstly contacted with the most spent roasted coffee and flows from one percolator to another in a way that the most solubles-loaded water is percolated through the freshly roasted coffee bed. Subsequently to extraction, more often than not, it is necessary to concentrate the extracted solubles solution up to 40% soluble solids before submitting it to drying. The most common concentration technique in this type of industry is the multistage evaporation in short-contact-time evaporators, such as plate and centrifugal film evaporators.

Following concentration, the coffee solubles extract is submitted to drying, either by freeze drying or by spray drying, leading to the products conventionally termed instant and soluble coffee, respectively. In the freeze drying method, the concentrated extract is initially frozen, the formed frozen slabs are milled into granules, and the frozen water in the granules is subsequently removed by sublimation under high vacuum, leading to the desired final product, the instant coffee. For the production of soluble coffee, concentrated coffee extract is atomized in a spray dryer, in which the water is removed by contacting the sprayed extract with air at temperatures of the order of 200–300°C, generating a rather fine powder. Upon exiting the spray dryer, the fine powder is usually further processed in a vibrating fluidized bed to promote subsequent agglomeration of the powder, thus improving the product wettability and, consequently, its dissolution in hot water during preparation of the beverage.

5 BY-PRODUCTS OF COFFEE PROCESSING

In the coffee processing chain, including both primary and secondary processing, a series of by-products or wastes are generated with the major ones being husks, parchment husks, chaff, and spent coffee grounds (SCGs). Defective coffee beans, which impart negative qualities to the beverage and are sorted out at the last step in processing the crude beans, are dumped in internal markets in producing countries but could be considered a by-product if a proper processing operation were to be implemented for its exploitation and further valorization as precursor material for value-added products (Franca and Oliveira, 2009; Alves et al., 2017). Husks are wastes from the primary processing of coffee and are comprised of dried pulp and parchment mixed altogether in the dry processing and separately of fresh pulp and dried parchment husks in semi-wet and wet processing (Franca and Oliveira, 2009; Alves et al., 2017). Although there are conflicting data published in the literature in regard to the amount of waste generated in the primary processing, it is estimated that, for every kilogram of coffee produced, about 0.45–1 kg of by-products are generated depending on the processing practices adopted by the producers. The only waste generated in the roasting and grinding operations (secondary processing) is the chaff, collected from the roaster exhaust gas by a cyclone, and comprised mostly of charred silverskin and a small fraction of dust resultant from roasted beans friction within the roaster. It is estimated that for every 1000 kg of roasted coffee produced about 7.5 kg of chaff is generated and collected (Alves et al., 2017). In the production of both soluble and instant coffees, the generated by-products are the SCGs, estimated to be in the amount of 650 kg for every 1000 kg of green coffee originally subjected to roasting, grinding and extraction operations (Mussatto et al., 2011). SCGs are also produced in domestic and commercial brewing of coffee with the global production depending on factors such as roasting (roasting degree), grinding (particle size), and brewing (temperature and time of extraction, water chemistry, and water-to-ground-roasted-coffee ratio) practices (Navarini and Rivetti, 2010; Bhumiratana et al., 2011).

The knowledge of the physical state and the chemical composition of each specific by-product is essential to define a processing path for its adequate exploitation. To compile and present the available data for the proximate or a more detailed chemical composition of each and every by-product in the coffee production and processing chain, is a rather difficult task, because the ranges of determined values are rather wide due to significant differences in the way the by-products were produced, in the extraction of their respective analytes and in the analytical methods employed thereof. However, a common distribution of classes of compounds is found among the different types of by-products, with carbohydrates comprising the major fraction, followed by proteins, lipids, ashes, polyphenols, and caffeine. For specific compositional values of each class of compounds or of individual ones, the reader is referred to the compilations by Franca and Oliveira (2009), Mussatto et al. (2011), Murthy and Naidu (2012a), Campos-Vega et al. (2015), Alves et al. (2017), and Janissen and Huynh (2018).

6 CURRENT PROPOSALS FOR VALUE-ADDED PROCESSING OF COFFEE BY-PRODUCTS

6.1 Food Products

Coffee is one of the most widely consumed pharmacologically active beverages, and thus coffee consumption has been frequently associated to health studies. It is still a controversial subject, although more than 8000 studies were developed on this subject over the past 40 years (Bae et al., 2014; Franca and Oliveira, 2016). Apart from the controversial findings, recent studies have shown that moderate daily consumption of coffee presents inverse correlation to several diseases including Alzheimer's, Parkinson's, several types of cancer, diabetes, and depression, among others (Butt and Sultan, 2011; Bae et al., 2014; Hall et al., 2015). These health-promoting properties are attributed to coffee's biologically active compounds, including caffeine, trigonelline, niacin, and polyphenols (Hall et al., 2015). These substances are also present in the solid by-products generated during coffee processing (peel and pulp of coffee fruit, SCGs, and silverskin), and therefore recent studies have focused on the characterization of coffee by-products as sources of such bioactive compounds and on how to efficiently extract such compounds (Franca and Oliveira, 2016).

Coffee husks, CH (dry processing residue) and pulp, CP (wet processing residue), contain the hull and part of the pulp of the coffee fruits, and thus present high amounts of carbohydrates (58%–85% in CH, 45%–89% in CP) (Franca and Oliveira, 2009, 2016). Caffeine is present in coffee husks and pulps at approximately 1% concentration. Studies on characterization and extraction of bioactive compounds in such residues have focused on anthocyanins (Prata and Oliveira, 2007) and caffeine (Tello et al., 2011). The recent study by Heeger et al. (2017) evaluated the potential of employing CP for producing a refreshing beverage known as *Cascara*. CP samples were characterized and shown to present bioactive substances including chlorogenic acids, protocatechuic and gallic acids, and rutin. The produced beverage contained 226 mg/L of caffeine, 283 mg GAE/L of total extractable polyphenols (TEPs) and antioxidant capacity amounted of 8.9 mmol TE/L. According to this study, CP can be valorized by producing a functional antioxidant beverage containing caffeine and polyphenols, although it is noteworthy to point out that consumer acceptance of this potential beverage has not been yet evaluated.

SCGs are the residues resulting from water extraction of roasted and ground coffee in the preparation of coffee beverage or industrial soluble and instant coffees, thus SCG is actually roasted and ground coffee depleted of the majority of its water-soluble compounds. These by-products have been evaluated as potential sources of dietary fiber (DF), mostly insoluble (Murthy and Naidu, 2012b; Oliveira and Franca, 2016), with TDF values increasing from 50 up to 70 g/100 g after treatment with hydrogen peroxide (Vilela et al., 2016). The recovery of phenolic compounds from SCGs, as well as their respective antioxidant activities, has also been investigated (Mussatto et al., 2011; Murthy and Naidu, 2012b; Bravo et al., 2013; Zuorro, 2015; Xu et al., 2015). The study by Sampaio et al. (2013) proposes a three-step process for the production of a spirit from SCGs. The first step consists of extraction of aroma compounds by hydrothermal treatment, followed by fermentation of the obtained extract supplemented with sucrose to ethanol (step 2), and then distillation of the fermented broth (step 3). Based on the chemical composition (alcohols, esters, aldehydes, and acids), the produced spirit was considered as having organoleptic quality acceptable for human consumption. Coffee was the most representative aroma detected by olfactory analysis. Results from sensory evaluation indicated that the SCGs spirit presented features of a pleasant beverage, with smell and taste of coffee. A recent study (Martinez-Saez et al., 2017) evaluated the use of SCGs as a food ingredient in bakery products (biscuits). Characterization in terms of chemical composition, thermal stability, and food safety was also performed. Results indicated that SCGs are natural sources of antioxidant insoluble fibers (~42 g/100 g) and essential amino acids (mainly glutamic acid, threonine, aspartic acid and leucine). SCGs were also found to present low amounts of glycemic sugars, to be resistant to thermal food processing and digestion processes, and totally safe. SCGs were incorporated in biscuit formulations together with low-calorie sweeteners and oligofructose. Results confirmed that SCGs (4%, *w/w*) can be used directly as food ingredient in biscuits without affecting the conventional food preparation and the final nutritional or sensory quality of the product.

Another by-product that has been recently evaluated as a potential food ingredient is coffee silverskin (CS), the solid residue that is collected by cyclone separation from the exhaust gas of coffee beans roasting. CS presents high contents of DF (50%–60%), mostly insoluble (85%) (Narita and Inouye, 2014). Studies have confirmed the presence of phenolics and the antioxidant potential of this by-product (Bresciani et al., 2014; Costa et al., 2014). CS extract was also reported to present a high inhibitory effect against hyaluronidase, thus presenting potential for suppressing allergies and inflammation (Furusawa et al., 2011). Applications in food products include its use as source of DF in breads (Pourfarzad et al., 2013) and also in beverage formulations aiming at body fat reduction and body weight control (Martinez-Saez et al., 2014). Results obtained testing nematodes (wild type N2 *Caenorhabditis elegans*) indicated a positive dose-response effect on reducing accumulation of body fat. The most effective dose was obtained for the beverages containing 10 mg/mL of CS extracts. Body fat reduction values were 21% and 24% for Arabica and Robusta CS extracts, respectively. Sensory acceptance tests showed that 10% of the test subjects were prepared to consume the beverages as served, while 85% would drink it as long as other ingredients such as sugar or milk were added. Results confirm the potential of CS as functional ingredients in food products.

6.2 Pharmaceutical

Although most studies focus on coffee by-products as possible food ingredients, recent studies evaluated the pharmacological and cosmetic potentials of SCGs and CS (Ribeiro et al., 2013; Rodrigues et al., 2015). Ribeiro et al. (2013) developed and characterized a cream containing 10% of SCG lipid fraction and it presented improvements in sebum skin levels, with better acceptance by consumers when compared to creams containing 10% (*w*/*w*) of green coffee oil and a placebo without coffee oil. Characterization studies by Rodrigues et al. (2015) indicated that CS extracts could provide in vivo protection against free-radical damage, due to the presence of phenolics, caffeine, and high antioxidant capacity. In vitro and in vivo skin and ocular irritation assays were performed, with three different CS extracts (pure water, 50% water/50% ethanol, and pure ethanol). The in vitro results showed that extracts were not classified as irritant. The in vivo tests carried out with the most promising extract (50% water/50%) confirmed that, with respect to irritant effects, these extracts can be regarded as safe for topical application. Yang et al. (2017) prepared nanoemulsions of a mixture of SCG oil and algae oil and studied its inhibition effects on ultraviolet A-induced skin damage in mice and growth of melanoma cells B16-F10. Experiments with mice demonstrated a dose of 0.1% of the prepared nanoemulsion to be effective in mitigating transepidermal water loss, skin erythema, melanin formation, and subcutaneous blood flow. Furthermore, cytotoxicity tests showed that the emulsion was effective in inhibiting melanoma cell growth.

6.3 Polymers and Materials

Fiber reinforced polymers composites present a wide variety of applications (e.g., automotive, construction, and others), but the majority of the commercially available materials are based on petroleum-based polymers reinforced with synthetic fibers (Tarazona et al., 2017). Nonetheless, there has been a growing interest in replacing these polymers by renewable and bio-based materials, due to environmental concerns. Several automotive components are nowadays being produced using natural composites, generally based on polypropylene (PP) resin and fibers such as flax, hemp, kenaf, and sisal (Faruk et al., 2014).

SCGs consist mainly of fibers (hemicellulose, cellulose, and lignin associated compounds), and therefore recent studies have investigated the possibility of incorporating SCG particles into polymers such as polyurethane, PP, polyethylene, polyester, polyvinyl alcohol, polyhydroxyalkanoate (PHA), poly(butylene adipate-co-terephthalate) (PBAT), and maleic anhydride-grafted polylactide (Baek et al., 2013; García-García et al., 2015; Lee et al., 2015; Wu, 2015; Li et al., 2016; Moustafa et al., 2017; Wu, 2017). In general, addition of SCG provides improvements on the mechanical properties, thus being suitable for structural applications. There are also some studies on the use of coffee husks and parchment for the same purpose (Reis et al., 2015; Huang et al., 2016). The employed polymers were polyhydroxybutyrate (Reis et al., 2015) and high-density polyethylene (HDPE) with maleated polyethylene (MAPE) as a coupling agent (Huang et al., 2016). Results indicated that increasing CH and CP content improved mechanical and thermal properties of the composites, but also resulted in poor water resistance and increase in water absorption.

In order to produce a truly renewable composite, not only the fiber but also the polymer matrix should be bio-based, and all the previously discussed studies employed petroleum-based polymers. The only study in which both the polymer and reinforcement are renewable was recently developed by Tarazona et al. (2017). An epoxy matrix was produced by epoxidizing residual (frying) soybean oil and subsequently used in the preparation of composites reinforced with SCG fiber. SCGs were chemically treated with alkaline hydrogen peroxide in order to decrease hydrophilicity and increase interfacial interaction with the epoxy matrix. The results obtained indicated that the composites prepared with a molar ratio of maleic anhydride 1:1.2 and SCG mass fraction of 35% were those that presented the best values for mechanical and physical properties (tensile strength, elastic modulus, bulk density, apparent porosity, and water absorption).

Results on the addition of SCG to a polyurethane film showed that SCG provided increases in the ultraviolet protection factor and thermal comfort parameters, indicating possible applications for summer clothes and sunscreen products (Li et al., 2016). PHA is a polyester polymer that presents properties similar to those of polyethylene and PP, and thus can be used for packaging, tissue engineering, and biomedical materials. Addition of SCG to PHA not only improved the polymer mechanical properties, but also enhanced saccharide and polyphenol contents, and antioxidant properties and reduced cellular tyrosinase activities in vitro, thus providing a polymer that could be more effective for several applications, for example food packages that avoid product degradation. Another interesting effect of adding coffee residues to polymer composites is that it reduces the plastic degradation time. Polycaprolactone (PCL) is a biodegradable polyester compatible with organic fillers. Borghesi et al. (2016) employed coffee husks to prepare PCL-CH composite film. The composite film degraded faster than pure PCL film, this is being attributed to adhesion of microorganisms to the coffee husks.

Aside from using coffee wastes as ingredients or fillers for altering the properties of materials, using them as source of raw materials for the production of polymeric materials has also been recently tackled. Polymeric films were developed with the polysaccharide fraction extracted from delignified SCGs by Batista (2017). Coffee polysaccharides were comprised mostly of galactomannans (50%), arabinogalactans (25%), and cellulose (25%) and were solubilized in water by a $ZnCl_2$ solution, with $CaCl_2$ being used to improve film properties (e.g., water vapor permeability and mechanical strength) by crosslinking. The polysaccharide-rich material from delignified SCG treated with 35% H_2O_2 provided the film with the best properties, with a tensile strength of $5.971 \pm 2.107\,N/mm^2$, water vapor permeability of $1.44 \pm 0.10\,g\,mm/m^2\,h\,kPa$, and glass transition temperature of 2.26°C. SCGs were deemed a potential precursor material for the development of polysaccharide-based films.

Many studies employing CH and SCGs have focused on construction materials. Recent applications include addition of CH ashes to increase clay strength (Acchar and Dultra, 2013), SCG ashes as partial cement replacement (Lin et al., 2016) and in clay bricks to improve insulation properties (Muñoz Velasco et al., 2016), use of CS to improve acoustic insulation of buildings (Ricciardi et al., 2017) and use of SCG in cementitious geopolymers (Arulrajah et al., 2017; Kua et al., 2016, 2017).

6.4 Adsorbents

Agricultural residues in general contain high levels of cellulose, hemicellulose, lignin, and proteins and as such have been intensely investigated in the production of adsorbents (Oliveira and Franca, 2011). Thus, coffee by-products, mainly SCG and CH, have been evaluated either as biosorbents (directly used for adsorption) or as precursor materials for the production of activated carbons (AC). The major applications are toward removal of chemical pollutants, mainly heavy metals and dyes. A comprehensive review on this subject has been recently published (Anastopoulos et al., 2017), and thus only some of the most recent studies applied to wastewater treatment or other types of applications that were not mentioned in the previously cited review are addressed here.

Recent applications of coffee by-products in the production of AC aiming at wastewater treatment include: development of a polyester matrix loaded with SCG to be used as adsorbents in oil spill cleanup processes (Marques et al., 2015), development of bioelastomeric SCG composite foams for removal of heavy metals from water (Chavan et al., 2016), development of nanosized Fe_3O_4-loaded SCG with application in dye removal (Khataee et al., 2017), use of an AC prepared by acid activation of SCG for diminishing color and chemical oxygen demand (COD) of landfill leachates (Poblete et al., 2017), use of an AC prepared by basic activation of SCG for phenol and dye removal

(Laksaci et al., 2017), direct use of SCG as biosorbent for pesticide removal (Al-Zaben and Mekhamer, 2017; Mezenner et al., 2017), and use of SCG immobilized as beads for adsorption of silver ions (Jeon, 2017). Although most of the recent applications have focused on SCG, several studies have shown the potential of CH in wastewater treatment (Oliveira and Franca, 2014).

6.5 Catalysts

A heterogeneous catalyst was prepared by sulphonation of incompletely carbonized SCG with H_2SO_4 and used for esterification of caprylic acid by Ngaosuwan et al. (2016), with the highest conversion of 71% for a 4-h reaction. The catalyst was not deactivated after several cycles of use and this was attributed to its high surface area and balanced densities of strong (-SO_3H) and weak (carboxylic and phenolic) acid sites. Chung et al. (2017) produced a coffee-waste-derived carbon electrode by a single-step carbonization of SCG with $ZnCl_2$ and used it as efficient electrocatalyst for oxygen reduction reaction (ORR) and triiodide reduction reaction (IRR). The prepared catalyst presented outstanding oxygen reduction activity and durability when compared to the state-of-the-art platinum electrodes, and the half-wave potential was also comparable to the best metal-free electrocatalysts in alkaline media. Furthermore, when applied as counter electrode of dye-sensitized solar cell, its photovoltaic efficiency surpassed that of conventional platinum electrodes.

6.6 Energy and Biofuels

Coffee wastes and by-products can be either used directly as energy sources or as precursor materials for biofuels production. As energy sources, the wastes can be torrefied, pyrolyzed, or gasified to produce solid, liquid, and gaseous biofuels.

Combustion mechanisms of two types of coffee husks were studied by Saenger et al. (2001) using single particle combustion techniques as well as combustion in a pilot-scale fluidized bed facility. Devolatilization of coffee occurred rapidly and at a low temperature range of 170–200°C. Blockage of the feeder tube and nonuniform fuel flow due to pyrolysis occurred during fuel feeding in a nonwater-cooled system. Significant combustion of the volatiles in the freeboard was observed for over-bed feeding. Ash sintering and bed agglomeration were significant during combustion of the coffee husks and this was attributed to the high contents of K_2O (36%–38%) of the coffee husks that promotes lowering of the melting temperature of the ash. Coffee husk densification (briquetting) and respective combustion in bakery furnaces was studied by Suarez et al. (2003). Experimental results demonstrated that coffee husk briquettes can be used as a renewable fuel, with combustion characteristics similar to those of the firewood currently employed in Cuba for that purpose. The furnace combustion of the produced briquettes generated decreased amounts of burnt residue when compared to the currently practiced combustion of firewood in open spaces, with consequent reduction in emissions.

The performance of an improved gasifier stove that uses coffee husks as source of energy for household cooking purpose in Tanzania was evaluated by Kamoleka et al. (2015). Results showed that thermal efficiency was improved from 12.8% to 24% for the improved stove in regard to the existing ones. Both burning rate and specific fuel consumption were significantly lowered with the developed stove when compared to the existing one. Jeguirim et al. (2016) produced pellets of pure SCG and of a mixture of coffee husks and SCG to improve the low calorific value, high ash content, and low bulk density of CH, and used them as solid fuel in a modified domestic combustor. Thermogravimetric experiments demonstrated the oxidation steps of the mixture SCG/CH blend to occur earlier than that of pure SCG (with a higher oxidation rate) and this fact was attributed to higher potassium content of CH. The tests performed in a 9-kW domestic boiler showed that the addition of CH led to a reduced combustion performance and increased gas and particles emissions. Nevertheless, the prepared pellets were deemed suitable for the proposed application. Lubwama and Yiga (2018) developed briquettes from coffee and rice husks agricultural wastes, with clay or cassava starch as binders, to be used as sustainable fuel sources for domestic cooking applications in Uganda. The type of binder used in the development of the briquettes was demonstrated to significantly affect the briquette physical properties and calorific values. Higher heating values for coffee husks ranged from 21.9 to 23.0 MJ/kg and from 15.9 to 16.6 MJ/kg for rice husks briquettes with cassava starch binder. For the briquettes with clay binder, average higher heating values ranged from 13.0 to 19.5 MJ/kg and from 9.5 to 13.8 MJ/kg, respectively, for coffee husks and rice husks briquettes. The cassava starch binder was determined to impart higher drop strengths onto the briquettes than the clay binder material, with these characteristics being influenced by the physical properties of the raw biomass material as well as the high SiO_2 ash in the clay binder. Results for times for ignition and water boiling indicated the possibility of applying less amounts of briquettes for cooking in regard to the currently used sources, eventually leading to savings in domestic energy use.

The energetic characteristics and combustion behavior of pellets produced from pure SCGs or blends of it with pine sawdust were investigated by Limousy et al. (2013) in a commercial residential pellet boiler. Despite of SCG having a relatively high low-heating value, the use of pure SCG as fuel led to lower boiler efficiencies and an increase in particle and gas emissions when compared to those obtained for wood pellets. However, when a 50/50 (wt%) mixture of SCG and pine sawdust was used, combustion parameters (emissions and boiler efficiency) were very close to those for wood pellets. The bioenergy production potential of SCGs during pyrolysis at two different heating rates of 10 and 60°C/min was investigated by Li et al. (2014). Three discrete stages of pyrolysis were identified (water evaporation, pyrolysis reactions, and decomposition of residual solid) and the primary evolved volatiles were identified as CO_2, CO, and CH_4. Pyrolysis efficiency was estimated at 77%–85%, depending on the moisture content of the feedstock.

Coffee wastes have been extensively used as precursor material for the production of a diversity of gaseous and liquid biofuels, employing a variety of processes. The literature on the production of biogas by fermentative processes of coffee wastes is aplenty, with successful reports on biogas production from coffee processing wastewater published in the early 1980s (Adams and Dougan, 1987). From this point forward, only a handful of works have been published on the anaerobic digestion of coffee processing wastewater for the production of biogas (Prado and Campos, 2008; Prado et al., 2010; Selvamurugan et al., 2010; Campos et al., 2013; Lepe-Cervantes et al., 2017) with proposals of its utilization for in-farm energy supply, such as fuel for coffee dryers. Coffee pulp and mucilage and SCGs were also used as substrate, individually or mixed with other residues, for anaerobic production of biogas with focus on either methane or biohydrogen optimized production (Neves et al., 2006; Corro et al., 2013; Corro et al., 2014; Hernández et al., 2014; Battista et al., 2016; Vitěz et al., 2016; Kim et al., 2017; Luz et al., 2017). However, an economic feasibility of either on-farm or industrial use of these technologies was not encountered.

The liquid biofuels produced from coffee wastes are primarily ethanol, biodiesel, and pyrolitic bio-oil. Studies on the production of ethanol from coffee wastes are based on the fact that coffee husks, mucilage, and SCGs are rich in fermentable sugars, either readily available or as constituents of the polysaccharides fraction. For the commonly employed yeasts (e.g., *Saccharomyces cerevisiae*) to ferment the polysaccharides to ethanol, it is necessary to hydrolyze them first into the respective monomeric units (pretreatment), i.e., the monosaccharides. Coffee husks (including pulp and parchment) were used as substrate for ethanol production by Gouvea et al. (2009), Menezes et al. (2013), Bonilla-Hermosa et al. (2014), Sahu (2014), Gurram et al. (2016), and Widjaja et al. (2016), whereas mucilage was used by Yadira et al. (2014) and Woldesenbet et al. (2016), and SCGs by Mussatto et al. (2012), Kwon et al. (2013), Burniol-Figols et al. (2016), Kim et al. (2017).

Biodiesel is commonly produced by transesterification reactions of oils with short-chain alcohols and oils from coffee by-products have been considered a suitable precursor material for such. Biodiesel was produced from defective coffee beans oil (Oliveira et al., 2008) and from oils extracted from SCGs (Al-Hamamre et al., 2012; Vardon et al., 2013; Caetano et al., 2014; Jenkins et al., 2014; Rocha et al., 2014; Park et al., 2016; Efthymiopoulos et al., 2017; Go and Yeom, 2017; Jenkins et al., 2017; Liu et al., 2017; Najdanovic-Visak et al., 2017; Tuntiwiwattanapun et al., 2017; Tuntiwiwattanapun et al., 2017; Park et al., 2018). The technologies employed to carry out the transesterification reactions were in its majority using batch mechanically stirred reactors, with the exception of some using ultrasound-assisted reaction. The only application of the produced SCG-oil biodiesel in engine tests was performed by Carlucci et al. (2017), with all the aforementioned ones restricted to laboratory research only.

A bio-hydrotreated fuel was produced from spent coffee oil by Phimsen et al. (2017) using nickel sulfide, nickel phosphide and nickel carbide catalysts. Physiochemical analysis of the produced diesel fractions exhibited satisfactory properties, with density and kinematic viscosity consistent with specifications for the commercial bio-hydrogenated diesel NExBTL. The main reaction products were straight chain hydrocarbons and, thus, high cetane indexes were achieved. Yang et al. (2012) investigated the performance of a diesel engine using blends of fossil Diesel and oil produced by the pyrolysis of SCGs. Results demonstrated that when SCG pyrolysis oil was blended with fossil diesel in small proportions, diesel engines performance was superior during high-rpm operations compared to lower rpm, and pollutant emissions were also significantly improved.

6.7 Recovery and Production of Value-Added Compounds

The recovery of valuable compounds from coffee wastes have been extensively studied with the bioactive ones, such as caffeine and chlorogenic acids, as the main compounds of being targeted (Franca and Oliveira, 2016; Alves et al., 2017). Aside from recovering valuable compounds, the production of other value-added products was investigated by several researchers, with the majority being produced by bioprocesses (e.g., fermentation), such as cellulases and xylanases (Cerda et al., 2017) and PHA (Cruz et al., 2014; Obruca et al., 2014, 2015; Bhatia et al., 2017).

7 SUMMARY

Coffee wastes or by-products are generated in expressive large amounts throughout the entire coffee production and processing chain. These by-products differ among themselves in terms of physical characteristics and chemical compositions, having in common the fact that polysaccharides are the dominant class of compounds in all of them. Regardless, all by-products are rich in a diversity of chemical components and thus are suitable for a plethora of exploitation pathways. Currently, the majority of them are being used in a low-value applications such as compost and vermicompost, ingredients for animal feed or fodder for in-farm or coffee-industry-in-house combustion systems. Proposals for value-adding of these by-products have gained momentum since the beginning of the 21st century, due to worldwide spread people's awareness of environmental problems caused by these by-products which, in some places, affect how consumers select and purchase their coffee products. Furthermore, environmental laws are becoming stricter and being rigidly enforced, preventing the inadequate disposal of these by-products in water streams and landfills, as it was done since large plantations and associated processing operations were established in the beginning of the 20th century. The applications being currently proposed for profitable exploitation (valorization) of these by-products are quite diverse, including production and use of adsorbents, production of liquid and gaseous biofuels, recovery of valuable biologically active compounds, and production of composite materials and polymeric films, among others.

There are no currently implanted processes for profitable exploitation of coffee by-products, neither in farms nor in industry. Due to the fact that coffee by-products are in their majority of low density, high volume, and high moisture contents, it is not economically viable to transport them for long distances to processing facilities, thus, whatever the processing to be carried out for their proper valorization, it should be done on location or as close to the sources as possible. Also, generation of further wastes and by-products from these processing operations should be eliminated or minimized, hence, the concept of biorefinery should be considered and, if determined to be economically feasible, be implemented. To implement such a concept it is essential to have a systemic view of the entire process, starting from collection, followed by storage and transportation (for short distances to assure viability), recovery of valuable components by optimized routes (preferably employing the concept of green chemistry), removal of undesirable components (for which, when possible, a valorization pathway should be proposed), and finally the recovery of the least valuable materials, resulting in a minimum amount of residues that can be further used for low-value applications such as composting, animal feed ingredients, or feedstock for combustion systems (Mata et al., 2017).

A biorefinery concept for the valorization of SCGs was developed by Mata et al. (2017) in which bioactive and other valuable compounds are recovered by water/ethanol/supercritical CO_2 extractions in the first step and the exhausted SCG is further processed into a plethora of products, with solid and liquid biofuels (e.g., bio-oil, ethanol, biodiesel, biochar) being the main end products. With the recently developed products, such as polymeric films, this concept could be further extended. The same type of concept could be employed in the processing of crude beans and the economic feasibility of implementation of such concept would be favored by the fact that coffee producers are usually organized in cooperatives and associations in which they process their coffees in a single shared facility. At a first glance, it could be proposed to use fresh husks for the production of ethanol, after recovery of anthocyanins, extract the oil from defective coffee beans to produce biodiesel, use the oil press-cake for recovery of valuable bioactive compounds (e.g., caffeine, trigonelline, and chlorogenic acids) and use the exhausted beans either for the recovery of the polysaccharide fraction, to be processed into polymeric films or used as reinforcement for composite materials, or to produce heterogeneous catalyst for the transesterification of the extracted defective coffee oil with the produced ethanol. The biofuels therein produced could be returned to the respective farms and used in trucks, combines and any other machines used in the farm.

References

Acchar, W., Dultra, E.J.V.J., 2013. Thermal analysis and X-ray diffraction of untreated coffee's husk ash reject and its potential use in ceramics. J. Therm. Anal. Calorim. 111, 1331–1334.

Adams, M., Dougan, J., 1987. Waste products. In: Clarke, R.J., Macrae, R. (Eds.), Coffee Technology. Elsevier Applied Science Publishers Ltd, New York, pp. 257–291.

Al-Hamamre, Z., Foerster, S., Hartmann, F., Kröger, M., Kaltschmitt, M., 2012. Oil extracted from spent coffee grounds as a renewable source for fatty acid methyl ester manufacturing. Fuel 96, 70–76.

Alves, R.C., Rodrigues, F., Nunes, M.A., Vinha, A.F., Oliveira, M.B.P.P., 2017. State of the art in coffee processing by-products. In: Galanakis, C.M. (Ed.), Handbook of Coffee Processing By-Product. Academic Press, London, pp. 1–26.

Al-Zaben, M.I., Mekhamer, W.K., 2017. Removal of 4-chloro-2-methyl phenoxy acetic acid pesticide using coffee wastes from aqueous solution. Arab. J. Chem. 10, S1523–S1529.

Anastopoulos, I., Karamesouti, M., Mitropoulos, A.C., Kyzas, G.Z., 2017. A review for coffee adsorbents. J. Mol. Liq. 229, 555–565.

Anzueto, F., Baumann, T.W., Graziosi, G., Piccin, C.R., Söndahl, M.R., van der Vossen, H.A.M., 2005. The plant. In: Illy, A., Viani, R. (Eds.), Espresso Coffee: The Science of Quality, second ed. Elservier Academic Press, Amsterdam, The Netherlands. 398p.

Arulrajah, A., Kua, T.-A., Horpibulsuk, S., Mirzababaei, M., Chinkulkijniwat, A., 2017. Recycled glass as a supplementary filler material in spent coffee grounds geopolymers. Constr. Build. Mater. 151, 18–27.

Bae, J.H., Park, J.H., Im, S.S., Song, D.K., 2014. Coffee and health. Integr. Med. Res. 3, 189–191.

Baek, B.S., Park, J.W., Lee, B.H., Kim, H.J., 2013. Development and application of green composites: using coffee ground and bamboo flour. J. Polym. Environ. 21, 702–709.

Batista, M.J.P.A., 2017. Biodegradable films produced of polysaccharides extracted from delignified spent coffee grounds. Master's Dissertation, Universidade Federal de Minas Gerais, Brazil. (in Portuguese).

Battista, F., Fino, D., Mancini, G., 2016. Optimization of biogas production from coffee production waste. Bioresour. Technol. 200, 884–890.

Bhatia, S.K., Kim, J.H., Kim, M.S., Kim, J., Hong, J.W., Hong, Y.G., Kim, H.J., Jeon, J.M., Kim, S.H., Ahn, J., Lee, H., Yang, Y.H., 2017. Production of (3-hydroxybutyrate-co-3-hydroxyhexanoate) copolymer from coffee waste oil using engineered Ralstonia eutropha. Bioprocess Biosyst. Eng. https://doi.org/10.1007/s00449-017-1861-4.

Bhumiratana, N., Adhikari, K., Chambers IV, E., 2011. Evolution of sensory aroma attributes from coffee beans to brewed coffee. LWT Food Sci. Technol. 44, 2185–2192.

Bonilla-Hermosa, V.A., Duarte, W.F., Schwan, R.F., 2014. Utilization of coffee by-products obtained from semi-washed process for production of value-added compounds. Bioresour. Technol. 166, 142–150.

Borém, F.M., Reinato, C.H.R., Andrade, E.T., 2008. Coffee drying. In: Borém, F.M. (Ed.), Post-Harvest of Coffee. Editora UFLA, Lavras, Brazil, pp. 203–240 (in Portuguese).

Borghesi, D.C., Molina, M.F., Guerra, M.A., Campos, M.G.N., 2016. Biodegradation study of a novel poly-caprolactone-coffee husk composite film. Mater. Res. 19, 752–758.

Bravo, J., Monente, C., Juániz, I., Paz De Peña, M., Cid, C., 2013. Influence of extraction process on antioxidant capacity of spent coffee. Food Res. Int. 50, 610–616.

Bresciani, L., Calani, L., Bruni, R., Brighenti, F., Del Rio, D., 2014. Phenolic composition, caffeine content and antioxidant capacity of coffee silverskin. Food Res. Int. 61, 196–201.

Burniol-Figols, A., Cenian, K., Skiadas, I.V., Gavala, H.N., 2016. Integration of chlorogenic acid recovery and bioethanol production from spent coffee grounds. Biochem. Eng. J. 116, 54–64.

Butt, M.S., Sultan, M.T., 2011. Coffee and its consumption: benefits and risks. Crit. Rev. Food Sci. Nutr. 51, 363–373.

Caetano, N.S., Silva, V.F.M., Melo, A.C., Martins, A.A., Mata, T.M., 2014. Spent coffee grounds for biodiesel production and other applications. Clean Technol. Environ. Policy 16, 1423–1430.

Campos, C.M.M., Prado, M.A.C., Pereira, E.L., 2013. Anaerobic digestion of wastewater from coffee and chemical analysis of biogas produced using gas chromatography: quantification of methane, and potential energy gas exchanger. Biosci. J. 29, 570–581.

Campos-Vega, R., Loarca-Piña, G., Vergara-Castañeda, H.A., Oomah, B.D., 2015. Spent coffee grounds: a review on current research and future prospects. Trends Food Sci. Technol. 45, 24–36.

Carlucci, A.P., Ficarella, A., Tabar, F.J., Santino, A., Strafella, L., 2017. *Cynara cardunculus* and coffee grounds as promising biodiesel sources for internal combustion compression ignition engines. Energy Procedia 126, 947–954.

Cerda, A., Mejías, L., Gea, T., Sánchez, A., 2017. Cellulase and xylanase production at pilot scale by solid-state fermentation from coffee husk using specialized consortia: the consistency of the process and the microbial communities involved. Bioresour. Technol. 243, 1059–1068.

Chavan, A.A., Pinto, J., Liakos, I., Bayer, I.S., Lauciello, S., Athanassia Athanassiou, A., Fragouli, D., 2016. Spent coffee bioelastomeric composite foams for the removal of Pb2+ and Hg2+ from water. Sustain. Chem. Eng. 4, 5495–5502.

Chung, D.Y., Son, Y.J., Yoo, J.M., Kang, J.S., Ahn, C.Y., Park, S., Sung, Y.E., 2017. Coffee waste-derived hierarchical porous carbon as a highly active and durable electrocatalyst for electrochemical energy applications. Appl. Mater. Interfaces 9, 41303–41313.

Clarke, R.J., 1987a. Roasting and grinding. In: Clarke, R.J., Macrae, R. (Eds.), Coffee Technology. Elsevier Applied Science Publishers Ltd, New York, pp. 73–107.

Clarke, R.J., 1987b. Extraction. In: Clarke, R.J., Macrae, R. (Eds.), Coffee Technology. Elsevier Applied Science Publishers Ltd, New York, pp. 109–145.

Corro, G., Pal, U., Cebada, S., 2014. Enhanced biogas production from coffee pulp through deligno-cellulosic photocatalytic pretreatment. Energy Sci. Eng. 2, 177–187.

Corro, G., Paniagua, L., Pal, U., Bañuelos, F., Rosas, M., 2013. Generation of biogas from coffee-pulp and cow-dung co-digestion: infrared studies of postcombustion emissions. Energy Convers. Manag. 74, 471–481.

Costa, A.S.G., Alves, R.C., Vinha, A.F., Barreira, S.V.P., Nunes, M.A., Cunha, L.M., Oliveira, M.B.P.P., 2014. Optimization of antioxidants extraction from coffee silverskin, a roasting by-product, having in view a sustainable process. Ind. Crop. Prod. 53, 350–357.

Cruz, M.V., Paiva, A., Lisboa, P., Freitas, F., Alves, V.D., Simões, P., Barreiros, S., Reis, M.A.M., 2014. Production of polyhydroxyalkanoates from spent coffee grounds oil obtained by supercritical fluid extraction technology. Bioresour. Technol. 157, 360–363.

Dutra, E.R., Oliveira, L.S., Franca, A.S., Ferraz, V.P., Afonso, R.J.C., 2001. A preliminary study on the feasibility of using the composition of coffee roasting exhaust gas for the determination of the degree of roast. J. Food Eng. 47, 241–246.

Efthymiopoulos, I., Hellier, P., Ladommatos, N., Kay, A., Mills-Lamptey, B., 2017. Effect of solvent extraction parameters on the recovery of oil from spent coffee grounds for biofuel production. Waste Biomass Valoriz. https://doi.org/10.1007/s12649-017-0061-4.

Farah, A., Santos, T.F., 2015. The coffee plant and beans: an introduction. In: Preedy, V.R. (Ed.), Coffee in Health and Disease Prevention. Academic Press, London, pp. 5–10.

Faruk, O., Bledzki, A.K., Fink, H.P., Sain, M., 2014. Progress report on natural fiber reinforced composites. Macromol. Mater. Eng. 299, 9–26.

Franca, A.S., Oliveira, L.S., 2009. Coffee processing solid wastes: current uses and future perspectives. In: Ashworth, G.S., Azevedo, P. (Eds.), Agricultural Wastes. Nova Publishers, New York, pp. 155–189.

Franca, A.S., Oliveira, L.S., 2016. Coffee and its by-products as sources of bioactive compounds. In: Massey, J.L. (Ed.), Coffee: Production, Consumption and Health Benefits. Nova Publishers, New York, pp. 1–28.

Furusawa, M., Narita, Y., Iwai, K., Fukunaga, T., Nakagiri, O., 2011. Inhibitory effect of a hot water extract of coffee "silverskin" on hyaluronidase. Biosci. Biotechnol. Biochem. 75, 1205–1207.

García-García, D., Carbonell, A., Samper, M.D., García-Sanoguera, D., Balart, R., 2015. Green composites based on polypropylenematrix and hydrophobized spend coffee ground (SCG) powder. Compos. Part B 78, 256–265.

Go, Y.W., Yeom, S.H., 2017. Statistical analysis and optimization of biodiesel production from waste coffee grounds by a two-step process. Biotechnol. Bioprocess Eng. 22, 440–449.

Gouvea, B.M., Torres, C., Franca, A.S., Oliveira, L.S., Oliveira, E.S., 2009. Feasibility of ethanol production from coffee husks. Biotechnol. Lett. 31, 1315–1319.

Gurram, R., Al-Shannag, M., Knapp, S., Das, T., Singsaas, E., Alkasrawi, M., 2016. Technical possibilities of bioethanol production from coffee pulp: a renewable feedstock. Clean Technol. Environ. Policy 18, 269–278.

Hall, S., Desbrow, B., Anoopkumar-Dukie, S., Davey, A.K., Arora, D., McDermott, C., Schubert, M.M., Perkins, A.V., Kiefel, M.J., Grant, G.D., 2015. A review of the bioactivity of coffee, caffeine and key coffee constituents on inflammatory responses linked to depression. Food Res. Int. 76, 626–636.

Heeger, A., Kosińska-Cagnazzo, A., Cantergiani, E., Andlauer, W., 2017. Bioactives of coffee cherry pulp and its utilisation for production of Cascara beverage. Food Chem. 221, 969–975.

Hernández, M.A., Susa, M.R., Andres, Y., 2014. Use of coffee mucilage as a new substrate for hydrogen production in anaerobic co-digestion with swine manure. Bioresour. Technol. 168, 112–118.

Huang, L., Mu, B., Yi, X., Li, S., Wang, Q., 2016. Sustainable use of coffee husks for reinforcing polyethylene composites. J. Polym. Environ. 26, 48–58.

ICO, 2018a. http://www.ico.org/prices/po-production.pdf. (Accessed 3 January 2018).

ICO, 2018b. http://www.ico.org/prices/m1-exports.pdf. (Accessed 3 January 2018).

ICO, 2018c. http://www.ico.org/prices/new-consumption-table.pdf. (Accessed 3 January 2018).

Janissen, B., Huynh, T., 2018. Chemical composition and value-adding applications of coffee industry byproducts: a review. Resour. Conserv. Recycl. 128, 110–117.

Jeguirim, M., Limousy, L., Fossard, E., 2016. Characterization of coffee residues pellets and their performance in a residential combustor. Int. J. Green Energy 13, 608–615.

Jenkins, R.W., Ellis, E.H., Lewis, E.J., Paterson, M., Le, D., Ting, V.P., Chuck, C.J., 2017. Production of biodiesel from Vietnamese waste coffee beans: biofuel yield, saturation and stability are all elevated compared with conventional coffee biodiesel. Waste Biomass Valoriz. 8, 1237–1245.

Jenkins, R.W., Stageman, N.E., Fortune, C.M., Chuck, C.J., 2014. Effect of the type of bean, processing, and geographical location on the biodiesel produced from waste coffee grounds. Energy Fuels 28, 1166–1174.

Jeon, C., 2017. Adsorption and recovery of immobilized coffee ground beads for silver ions from industrial wastewater. J. Ind. Eng. Chem. 53, 261–267.

Kamoleka, M.M., Mkini, R.I., John, J.P., 2015. An efficient and environmental friendly modified natural draft coffee husks gasifier stove for coffee farmers in Tanzania. Int. J. Eng. Res. Rev. 3, 105–111.

Khataee, A., Kayan, B., Kalderis, D., Karimi, A., Akay, S., Konsolakis, M., 2017. Ultrasound-assisted removal of Acid Red 17 using nanosized Fe3O4-loaded coffee waste hydrochar. Ultrason. Sonochem. 35, 72–80.

Kim, H.M., Choi, Y.S., Lee, D.S., Kim, Y.H., Bae, H.J., 2017. Production of bio-sugar and bioethanol from coffee residue (CR) by acid-chlorite pretreatment. Bioresour. Technol. 236, 194–201.

Kua, T.-A., Arulrajah, T., Horpibulsuk, S., Du, Y.-J., Shen, S.-L., 2016. Strength assessment of spent coffee grounds-geopolymer cement utilizing slag and fly ash precursors. Constr. Build. Mater. 115, 565–575.

Kua, T.-A., Arulrajah, A., Mohammadinia, A., Horpibulsuk, S., Mirzababaei, M., 2017. 'Stiffness and deformation properties of spent coffee grounds based geopolymers. Constr. Build. Mater. 138, 79–87.

Kwon, E.E., Yi, H., Jeon, Y.J., 2013. Sequential co-production of biodiesel and bioethanol with spent coffee grounds. Bioresour. Technol. 136, 475–480.

Laksaci, H., Khelifi, A., Trari, M., Addoun, A., 2017. Synthesis and characterization of microporous activated carbon from coffee grounds using potassium hydroxides. J. Clean. Prod. 147, 254–262.

Lee, H.K., Park, Y.J., Jeong, T., Song, Y.S., 2015. Green nanocomposites filled with spent coffee grounds. J. Appl. Polym. Sci. 132, 42043.

Lepe-Cervantes, D., Leon-Becerril, E., Gomez-Romero, J., Garcia-Depraect, O., Lopez-Lopez, A., 2017. Anaerobic digestion of coffee wastewater from a fast inoculum adaptation stage replacement of complex substrate. Int. J. Chem. Mol. Eng. 11, 5.

Li, R., Shen, X., Zhou, Y., Wang, H., Jiang, S., Xu, W., 2016. Preparation and properties of polyurethane films with spent coffee grounds. Mater. Sci. Forum Online 848, 148–151.

Li, X., Strezov, V., Kan, T., 2014. Energy recovery potential analysis of spent coffee grounds pyrolysisproducts. J. Anal. Appl. Pyrolysis 110, 79–87.

Limousy, L., Jeguirim, M., Dutournié, P., Kraiem, N., Lajili, M., Said, R., 2013. Gaseous products and particulate matter emissions of biomass residential boiler fired with spent coffee grounds pellets. Fuel 107, 323–329.

Lin, L.-K., Kuo, T.-M., Hsu, Y.-S., 2016. The application and evaluation research of coffee residue ash into mortar. J. Mater. Cycles Waste Manage. 18, 541–551.

Liu, Y., Tu, Q., Knothe, G., Lu, M., 2017. Direct transesterification of spent coffee grounds for biodiesel production. Fuel 199, 157–161.

Lubwama, M., Yiga, V.A., 2018. Characteristics of briquettes developed from rice and coffee husks for domestic cooking applications in Uganda. Renew. Energy 118, 43–55.

Luz, F.C., Cordiner, S., Manni, A., Mulone, V., Rocco, V., 2017. Anaerobic digestion of coffee grounds soluble fraction at laboratory scale: evaluation of the biomethane potential. Appl. Energy 207, 166–175.

Marques, F.D., Souza, F.G., Oliveira, G.E., 2015. Oil sorbers based on renewable sources and coffee grounds. J. Appl. Polym. Sci. 133, 1097–4628.

Martinez-Saez, N., García, A.T., Domínguez Pérez, I., Rebollo-Hernanz, M., Mesías, M., Morales, F.J., Martín-Cabrejas, M.A., del Castillo, M.D., 2017. Use of spent coffee grounds as food ingredient in bakery products. Food Chem. 216, 114–122.

Martinez-Saez, N., Ullate, M., Martin-Cabrejas, M.A., Martorell, P., Genovés, S., Ramon, D., Del Castillo, M.D., 2014. A novel antioxidant beverage for body weight control based on coffee silverskin. Food Chem. 150, 227–234.

Mata, T.M., Martins, A.A., Caetano, N.S., 2017. Bio-refinery approach for spent coffee grounds valorization. Bioresour. Technol. 247, 1077–1084.

Menezes, E.G.T., Carmo, J.R., Alves, J.G.L.F., Menezes, A.G.T., Guimarães, I.C., Queiroz, F., Pimenta, C.J., 2013. Optimization of alkaline pretreatment of coffee pulp for production of bioethanol. Biotechnol. Prog. 30, 451–462.

Mezenner, N.Y., Lagha, H., Kais, H., Trari, M., 2017. Biosorption of diazinon by a pre-treated alimentary industrial waste: equilibrium and kinetic modeling. Appl. Water Sci. 7, 4067–4076.

Moustafa, H., Guizani, C., Dufresne, A., 2017. Sustainable biodegradable coffee grounds filler and its effect on the hydrophobicity, mechanical and thermal properties of biodegradable PBAT composites. J. Appl. Polym. Sci. 134, 1097–4628.

Muñoz Velasco, P., Mendívil, M.A., Morales, M.P., Muñoz, L., 2016. Eco-fired clay bricks made by adding spent coffee grounds: a sustainable way to improve buildings insulation. Mater. Struct. 49, 641–650.

Murthy, P.S., Naidu, M.M., 2012a. Sustainable management of coffee industry by-products and value addition—a review. Resour. Conserv. Recycl. 66, 45–58.

Murthy, P.S., Naidu, M.M., 2012b. Recovery of phenolic antioxidants and functional compounds from coffee industry by-products. Food Bioprocess Technol. 5, 897–903.

Mussatto, S.I., Ballesteros, L.F., Martins, S., Teixeira, J.A., 2011. Extraction of antioxidant phenolic compounds from spent coffee grounds. Sep. Purif. Technol. 83, 173–179.

Mussatto, S.I., Machado, E.M.S., Carneiro, L.M., Teixeira, J.A., 2012. Sugars metabolism and ethanol production by different yeast strains from coffee industry wastes hydrolysates. Appl. Energy 92, 763–768.

Mussatto, S.I., Machado, E.M.S., Martins, S., Teixeira, J.A., 2011. Production, composition, and application of coffee and its industrial residues. Food Bioprocess Technol. 4, 661–672.

Najdanovic-Visak, V., Leec, F.Y.L., Tavares, M.T., Armstrong, A., 2017. Kinetics of extraction and in situ transesterification of oils from spent coffee grounds. J. Environ. Chem. Eng. 5, 2611–2616.

Narita, Y., Inouye, K., 2014. Review on utilization and composition of coffee silverskin. Food Res. Int. 61, 16–22.

Navarini, L., Rivetti, D., 2010. Water quality for Espresso coffee. Food Chem. 122, 424–428.

Neves, L., Oliveira, R., Alves, M.M., 2006. Anaerobic co-digestion of coffee waste and sewage sludge. Waste Manag. 26, 176–181.

Ngaosuwan, K., Goodwin Jr., J.G., Prasertdham, P., 2016. A green sulfonated carbon-based catalyst derived from coffee residue for esterification. Renew. Energy 86, 262–269.

Obruca, S., Benesova, P., Kucera, D., Petrik, S., Marova, I., 2015. Biotechnological conversion of spent coffee grounds into polyhydroxyalkanoates and carotenoids. New Biotechnol. 32, 569–574.

Obruca, S., Benesova, P., Petrik, S., Oborna, J., Prikryl, R., Marova, I., 2014. Production of polyhydroxyalkanoates using hydrolysate of spent coffee grounds. Process Biochem. 49, 1409–1414.

Oliveira, L.S., Franca, A.S., 2011. Conventional and non-conventional thermal processing for the production of activated carbons from agro-industrial wastes. In: Kwiatkowski, J.F. (Ed.), Activated Carbon: Classifications. Properties and Applications. Nova Publishers, New York, pp. 205–238.

Oliveira, L.S., Franca, A.S., 2014. An overview of the potential uses for coffee husks. In: Preedy, V. (Ed.), Coffee in Health and Disease Prevention. Elsevier, Amsterdam, pp. 283–292.

Oliveira, L.S., Franca, A.S., 2016. Potential of spent coffee grounds as sources of dietary fiber with antioxidant activity. In: Riley, G.L. (Ed.), Food Waste: Practices, Management and Challenge. Nova Publishers, New York, pp. 51–70.

Oliveira, L.S., Franca, A.S., Camargos, R.R.S., Ferraz, V.P., 2008. Coffee oil as a potential feedstock for biodiesel production. Bioresour. Technol. 99, 3244–3250.

Park, J., Kim, B., Lee, J.W., 2016. In-situ transesterification of wet spent coffee grounds for sustainable biodiesel production. Bioresour. Technol. 221, 55–60.

Park, J., Kim, B., Son, J., Lee, J.W., 2018. Solvo-thermal in situ transesterification of wet spent coffee grounds for the production of biodiesel. Bioresour. Technol. 249, 494–500.

Phimsen, S., Kiatkittiponga, W., Yamada, H., Tagawa, T., Kiatkittipong, K., Laosiripojana, L., Assabumrungratf, S., 2017. Nickel sulfide, nickel phosphide and nickel carbide catalysts for bio-hydrotreated fuel production. Energ. Conver. Manage. 151, 324–333.

Poblete, R., Oller, I., Maldonado, M.I., Luna, Y., Cortes, E., 2017. Cost estimation of COD and color removal from landfill leachate using combined coffee-waste based activated carbon with advanced oxidation processes. J. Environ. Chem. Eng. 5, 114–121.

Pourfarzad, A., Mahdavian-Mehr, H., Sedaghat, N., 2013. Coffee silverskin as a source of dietary fiber in bread-making: optimization of chemical treatment using response surface methodology. LWT Food Sci. Technol. 50, 599–606.

Prado, M.A.C., Campos, C.M.M., 2008. Biogas production in the treatment of *Coffea arabica* L. processing wastewaters in UASB anaerobic reactor for the potential use in the coffee drying. Cienc. Agrotecnol. 32, 938–947.

Prado, M.A.C., Campos, C.M.M., Silva, J.F., 2010. A study on the variation of methane concentration in biogas produced from coffee wastewater. Cienc. Agrotecnol. 34, 475–484.

Prata, E.R.B.A., Oliveira, L.S., 2007. Fresh coffee husks as potential sources of anthocyanins. LWT Food Sci. Technol. 40, 1555–1560.

Reis, K.C., Pereira, L., Melo, I.C.N.A., Marconcini, J.M., Trugilho, P.F., Tonoli, G.H.D., 2015. Particles of coffee wastes as reinforcement in polyhydroxybutyrate (PHB) based composites. Mater. Res. 18, 546–552.

Ribeiro, H., Marto, J., Raposo, S., Agapito, M., Isaac, V., Chiari, B.G., Lisboa, P.F., Paiva, A., Barreiros, S., Pedro Simões, P., 2013. From coffee industry waste materials to skin-friendly products with improved skin fat levels. Eur. J. Lipid Sci. Technol. 115, 330–336.

Ricciardi, P., Torchia, T., Belloni, E., Lascaro, E., Buratti, C., 2017. Environmental characterisation of coffee chaff, a new recycled material for building applications. Constr. Build. Mater. 147, 185–193.

Rocha, M.V.P., Matos, L.J.B.L., Lima, L.P., Figueiredo, P.M.S., Lucena, I.L., Fernandes, F.A.N., Gonçalves, L.R.B., 2014. Ultrasound-assisted production of biodiesel and ethanol from spent coffee grounds. Bioresour. Technol. 167, 343–348.

Rodrigues, F., Pereira, C., Pimentel, F.B., Alves, R.C., Ferreira, M., Sarmento, B., Amaral, M.A., Oliveira, M.B.P.P., 2015. Are coffee silverskin extracts safe for topical use? An in vitro and in vivo approach. Ind. Crop. Prod. 63, 167–174.

Saenger, M., Hartge, E.-U., Werther, J., Ogada, T., Siagi, Z., 2001. Combustion of coffee husks. Renew. Energy 23, 103–121.

Sahu, O., 2014. Bioethanol production by coffee husk for rural area. Adv. Res. J. Biochem. Biotechnol. 1, 1–5.

Sampaio, A., Dragone, G., Vilanova, M., Oliveira, J.M., Teixeira, J.A., Mussatto, S.I., 2013. Production, chemical characterization, and sensory profile of a novel spirit elaborated from spent coffee ground. LWT—Food Sci. Technol. 54, 557–563.

Selvamurugan, M., Doraisamy, P., Maheswari, M., 2010. An integrated treatment system for coffee processing wastewater using anaerobic and aerobic process. Ecol. Eng. 36, 1686–1690.

Steiman, S., 2008. The perfect bean. growing, harvesting & processing coffee. In: Steiman, S. (Ed.), The Hawai'i Coffee Book. A Gourmet's Guide from Kona to Kaua'i'. first ed. Watermark Publishing, Hawai'i, pp. 21–39.

Suarez, J.A., Beaton, P.A., Luengo, C.A., Felfli, F.F., 2003. Coffee husk briquettes: a new renewable energy source. Energy Sources 25, 961–967.

Tarazona, E.R.T., Oliveira, L.S., Rubio, J.C., Franca, A.S., 2017. Preparation, preliminary characterization and mechanical properties of epoxy composites reinforced with spent coffee grounds. In: Proceedings of the 8th International Conference on Mechanical and Aerospace Engineering, pp. 21–24.

Tello, J., Viguera, M., Calvo, L., 2011. Extraction of caffeine from Robusta coffee (*Coffea canephora* var. Robusta) husks using supercritical carbon dioxide. J. Supercrit. Fluids 59, 53–60.

Thurston, R.W., 2013. Mechanization. In: Thurston, R.W., Morris, J., Steiman, S. (Eds.), Coffee: A Comprehensive Guide to the Bean, the Beverage, and the Industry. Rowman & Littlefield Publishers, Maryland, pp. 366–370.

Tuntiwiwattanapun, N., Monono, E., Wiesenborn, D., Tongcumpou, C., 2017. In-situ transesterification process for biodiesel production using spent coffee grounds from the instant coffee industry. Ind. Crop. Prod. 102, 23–31.

Tuntiwiwattanapun, N., Usapein, P., Tongcumpou, C., 2017. The energy usage and environmental impact assessment of spent coffee grounds biodiesel production by an in-situ transesterification process. Energy Sustain. Dev. 40, 50–58.

Vardon, D.R., Moser, B.R., Zheng, W., Witkin, K., Evangelista, R.L., Strathmann, T.J., Rajagopalan, K., Sharma, B.K., 2013. Complete utilization of spent coffee grounds to produce biodiesel, bio-oil, and biochar. Sustain. Chem. Eng. 1, 1286–1294.

Vilela, W., Leão, D.P., Franca, A.S., Oliveira, L.S., 2016. Effect of peroxide treatment on functional and technological properties of fiber-rich powders based on spent coffee grounds. ETP Int. J. Food Eng. 2 (1), 42–47.

Vincent, J.C., 1987. Green coffee processing. In: Clarke, R.J., Macrae, R. (Eds.), Coffee Technology. Elsevier Applied Science Publishers Ltd, New York, pp. 1–33.

Vitěz, T., Koutny, T., Šotnar, M., Chovanec, J., 2016. On the spent coffee grounds biogas production. Acta Univ. Agric. Silviculturae Mendelianae Brunensis 64, 1279–1282.

Widjaja, T., Altway, A., Nurkhamidah, S., Edahwati, L., Lini, F.Z., Oktafia, F., 2016. The effect of pretreatment and variety of microorganisms to the production of ethanol from coffee pulp. ARPN J. Eng. Appl. Sci. 11, 1056–1060.

Woldesenbet, A.G., Woldeyes, B., Chandravanshi, B.S., 2016. Bio-ethanol production from wet coffee processing waste in Ethiopia. Springerplus 5, 1–7.

Wu, C.S., 2015. Renewable resource-based green composites of surface treated spent coffee grounds and polylactide: characterisation and biodegradability. Polym. Degrad. Stab. 121, 51–59.

Wu, C.S., 2017. Modulation of the interface between polyester and spent coffee grounds in polysaccharide membranes: preparation, cell proliferation, antioxidant activity and tyrosinase activity. Mater. Sci. Eng. C Mater. Biol. Appl. 78, 530–538.

Xu, H., Wang, W., Liu, X., Yuan, F., Gao, Y., 2015. Antioxidative phenolics obtained from spent coffee grounds (*Coffea arabica* L.) by subcritical water extraction. Ind. Crop. Prod. 76, 946–954.

Yadira, P.S.B., Sergio, S.T., Fernando, S.E.L., Sebastian, P.J., Eapen, D., 2014. Bioethanol production from coffee mucilage. Energy Procedia 57, 950–956.

Yang, C.C., Hung, C.F., Chen, B.H., 2017. Preparation of coffee oil-algae oil-based nanoemulsions and the study of their inhibition effect on UVA-induced skin damage in mice and melanoma cell growth. Int. J. Nanomed. 12, 6559–6580.

Yang, S.Y., Wu, M.S., Wu, C.Y., Chen, K.H., Wu, T.M., Hsu, Y.H., Lin, P.H., Ku, Y.Y., 2012. The performance of a diesel engine blended with coffee bean residue pyrolysis oil. Adv. Mater. Res. 591-593, 325–332.

Zuorro, A., 2015. Optimization of polyphenol recovery from espresso coffee residues using factorial design and response surface methodology. Sep. Purif. Technol. 152, 64–69.

Further Reading

Kim, J., Kim, H., Baek, G., Lee, C., 2017. Anaerobic co-digestion of spent coffee grounds with different waste feedstocks for biogas production. Waste Manag. 60, 322–328.

Menezes, E.G.T., Carmo, J.R., Menezes, A.G.T., Alves, J.G.L.F., Pimenta, C.J., Queiroz, F., 2013. Use of different extracts of coffee pulp for the production of bioethanol. Appl. Biochem. Biotechnol. 169, 673–687.

Tan, M.Y., Kuan, H.T.N., Khan, A.A., 2017. Tensile properties of ground coffee waste reinforced polyethylene composite. Mater. Sci. Forum 880, 73–76.

Index

Note: Page numbers followed by *f* indicate figures, and *t* indicate tables.

R

S

CPI Antony Rowe
Eastbourne, UK
November 20, 2019